Power Systems

Electrical power has been the technological foundation of industrial societies for many years. Although the systems designed to provide and apply electrical energy have reached a high degree of maturity, unforeseen problems are constantly encountered, necessitating the design of more efficient and reliable systems based on novel technologies. The book series Power Systems is aimed at providing detailed, accurate and sound technical information about these new developments in electrical power engineering. It includes topics on power generation, storage and transmission as well as electrical machines. The monographs and advanced textbooks in this series address researchers, lecturers, industrial engineers and senior students in electrical engineering.

Power Systems is indexed in Scopus

Konstantin O. Papailiou · Markus Palic ·
Guntram Schultz · Herbert Lugschitz

High-Voltage Overhead Lines and Underground Cables

Springer

Konstantin O. Papailiou
Malters, Luzern, Switzerland

Markus Palic
Karlsruhe, Germany

Guntram Schultz
Karlsruhe, Germany

Herbert Lugschitz
Vienna, Austria

ISSN 1612-1287 ISSN 1860-4676 (electronic)
Power Systems
ISBN 978-3-032-07497-3 ISBN 978-3-032-07498-0 (eBook)
https://doi.org/10.1007/978-3-032-07498-0

Translation from the German language edition: "Freileitungen und Kabel in Hoch- und Höchstspannungsnetzen" by Konstantin O. Papailiou et al., © expert verlag GmbH 2022. Published by expert verlag GmbH. All Rights Reserved.

The original submitted manuscript has been translated into English. The translation was done using artificial intelligence. A subsequent revision was performed by the author(s) to further refine the work and to ensure that the translation is appropriate concerning content and scientific correctness. It may, however, read stylistically different from a conventional translation.

This Springer imprint is published by the registered company Springer Nature Switzerland AG
The registered company address is: Gewerbestrasse 11, 6330 Cham, Switzerland

Foreword

The global changes in the energy industry, driven by the rapidly growing world population with rapidly increasing per capita energy consumption versus finite resources of fossil fuels, the necessary requirements for the environmental compatibility and acceptance of systems for energy conversion, transport and distribution, as well as demands for climate neutrality and the liberalized electricity market, are not bypassing any country.

In addition, the requirements of high-tech and digitalized industrial societies in terms of supply reliability and security of energy supply are not always in line with the political stability of countries from which primary energy sources are sourced, which indicates a clear trend towards extensive independence in the provision of electrical energy from renewable energy sources. The associated spatial separation of producer and consumer locations and the volatile availability of wind and sun pose major challenges for the entire electrical supply system. This applies to its generation plants, its lines including transformer substations, right through to the equipment at the consumers. This is precisely where this book comes in. It focuses on cables and overhead lines for the transmission and distribution of electrical energy in high- and extra-high-voltage grids. The authors succeed very well in presenting the trinity of history, proven tradition and engineering experience on the one hand, and the scientific and technical fundamentals and specialist knowledge as well as current challenges for manufacturers, planners, installers, and operators on the other.

The book is aimed at learners and students as well as experienced professionals who want to further their education and prepare themselves for new tasks. It assumes that the reader has a basic knowledge of electrical engineering in the technical section, picks them up from there, and leads them to the current specialist topics and questions in an understandable way. Social, legal, administrative, economic, ecological, and acceptance issues are not left out, which clearly demonstrates the interdisciplinary nature of the subject. Many examples with concrete values are very helpful, not only to deepen the connections but also in particular to be able to estimate orders of magnitude and thus make decisions from an engineering point of view.

In terms of content, the book is dedicated to the basics of structures and networks to develop an understanding of why and to what extent electrical connections are required between the nodes of a network and which boundary conditions must be considered. This leads to the question of which type of connection to choose, overhead line or cable, and how the route should be routed for technical, economic, ecological, and legal approval reasons. The approval procedures to be observed are explained. The following chapters focus on technical explanations of the function, structure, components, and installation of overhead lines and cable systems. The reader is provided with a comprehensive overview in the necessary brevity, while references point to further reading. With the comparative chapter "Overhead lines and cables in transmission and distribution grids", the authors close an important gap in the discussion on this topic. The advantages and disadvantages of both types of electrical power connections are summarized from a neutral perspective. A final chapter provides a well-founded outlook on technical developments in overhead line and cable technology that we will have to face before long.

The structure, content, and presentation of the book make it easy to recognize that the authors, Konstantin O. Papailiou, Herbert Lugschitz, Markus Palic, and Guntram Schultz are engineers with decades of practical experience, but also solid basic knowledge, who thankfully pass on their knowledge to interested specialists, including younger ones.

Prof. ret. Dr.-Ing. Steffen Großmann

Preface

The question of the design and routing of supra-regional electrical supply lines in the high- and extra-high-voltage range has been at the center of public debate for decades. Increasing environmental awareness among the population and the obligation of grid operators to adapt the grids to the growing requirements by expanding them have led to sometimes considerable conflicts in many cases. Due to the accelerated shift in the focus of generation in the course of the energy transition, away from traditional power plants near the main load points and towards renewable electricity generation from wind and solar power, the electricity generated often has to be transported to the main load points over many hundreds of kilometers via high-performance power lines. Irrespective of this, the existing electricity grid must be strengthened and expanded at all voltage levels, for example, due to the increasing demand caused by electromobility and heat pumps. Due to the now established assessment and approval procedures, some of which involve extensive public participation, not only planners, approval authorities and public interest groups in pluralistic societies are concerned with this issue but also many citizens, who often form citizens' initiatives.

In addition to the fundamental question of the necessity of power line projects, the discussion has often focused on the urgent desire for their complete cabling. The associated technical and economic effects, particularly in the high- and extra-high-voltage grid, are multi-layered and complex. The procedural framework conditions that have been added over the past 40 years or so and their constant amendment require all those involved in the process to have a solid basic knowledge of the technical design, as well as constant learning and a high degree of flexibility.

This book provides a comprehensive yet easy-to-understand introduction to the subject and is intended to help planners and technicians as well as other parties involved in the process to familiarize themselves with the wide range of issues involved. To this end, it is divided into a detailed technical section and a process engineering section. In the technical section, the two types of equipment—cables and overhead lines—are described, their functions compared, and their interaction in the grid described. The procedural and environmental part at the beginning provides an overview of the planning and approval procedures to be carried out.

The manuscript is the result of close cooperation between the authors in the context of teaching and information events, as well as in the training and further education of young engineers. The book is suitable as teaching material for students as well as a reference work for practitioners in the field of "high and ultra-high voltage grids".

Malters, Switzerland	Konstantin O. Papailiou
Wien, Austria	Herbert Lugschitz
Karlsruhe, Germany	Markus Palic
Karlsruhe, Germany	Guntram Schultz

Acknowledgements

We would like to express our sincere thanks for the professional support of the book project with the review of chapters and passages, the provision of information, for important tips as well as for image material and its release for publication.

For the support provided by Springer-Verlag:
Merkle, Dieter, and Jantzen, Petra for their dedication and numerous forms of assistance in translating the texts.

For the technical editing and the foreword:
Großmann, Steffen, Prof. Dr.-Ing., Form. Chair of High Voltage and High Current Technology, Institute for Electrical Power Supply and High Voltage Technology, Dresden University of Technology.

For numerous suggestions, documents, notes and image releases:

Alter, Enrico, Omexom Hochspannung GmbH;
Ansorge, Samuel, Dr. sc. ETH, Tech4Speed;
Bauer, Sebastian, Prof. Dr., Christian-Albrechts-University Kiel;
Bohn, Thoralf, FNN im VDE, Berlin;
Brakelmann, Heinrich, Prof. Dr.-Ing., BCC Cables Consulting;
Cosmai, Umberto, Dr., Covis;
Durinke, Peter, Dr., Wolter Hoppenberg Rechtsanwälte Partnerschaft mbB;
Ebert, Sebastian, Südkabel GmbH;
Eduard Steiner, AG/CH, EQOS Energie Deutschland GmbH;
Knapp, Eduard, IFK GmbH;
Knauer, Martin, TenneT TSO;
Küchler, Andreas, Prof. Dr.-Ing., Hochschule Würzburg-Schweinfurt
Lamm, Matthias, Dr., Richard Bergner Elektroarmaturen GmbH & Co. KG;
LEW Verteilnetz GmbH (LVN);
Nitzke, Uwe, Cteam Consulting and Anlagenbau GmbH;
Peters, Marc, Dr.- Ing., Herrenknecht AG;
Pfisterer Switzerland AG;
Schmuck, Frank, Dr. Ing. habil., Schmuck HV Insulation Consulting;

Seifen, Guido, Omexom;
Springer Verlag;
Steevens, Stefan, Amprion GmbH;
Steiner, Eduard, AG/CH, EQOS Energie Deutschland GmbH;
van Fondern, Michael, Sweco GmbH;
VDE Kassel;
Weinlein, Andreas, Südkabel GmbH;
Wolbring, Nico, Wolbring and Feldmann PartG;
Wolff, Carsten, NKT GmbH & Co. KG

Introduction

Overhead lines and cables are both ways of transmitting electrical energy. Both technologies have their advantages and disadvantages, and both have their optimal areas of application. The immediate, regional, and sometimes supra-regional environment of a project influences the decision in favor of overhead lines or cables, as do the technical, operational, and commercial aspects. It has been shown time and again that each project must be considered individually and anew. General statements or references to other projects in which overhead lines or cables were installed for this or that reason are not useful or helpful. This is often overlooked.

There are a large number of technical publications and information brochures on overhead lines and cables from users and manufacturers, studies, and investigations from scientific institutions, specifications from national and international standardization bodies such as IEC, CENELEC, and DIN, papers, symposia, and congresses from specialist organizations, and the Internet is full of articles on the subject. The overview is not easy, quite the opposite. This is exactly where this book comes in. The authors present aspects of overhead lines and cables. The book helps readers to gain a neutral perspective on this often hotly debated topic, with a focus on Europe, but also with examples from around the world.

The authors draw on their decades of experience in the field of power transmission and electrical grids, their university teaching activities and courses, their numerous publications, contributions to discussions at congresses, symposia, and workshops.

To talk about the subject, you need a basic knowledge of general high-voltage technology and also of the special features of overhead lines and cables. The special feature of this book is that both techniques are covered in one book. It also provides up-to-date knowledge about the grid situation and the legal situation in some countries. As in many areas of our lives, today's technology has a historical background that is important for understanding its application. This aspect is also addressed in the book.

Readers can use this book to deepen their knowledge, review, consolidate, change, or calibrate their previous views—all of which can also be used to supplement existing publications and presentations. Others may find themselves immersed in a new world of our everyday technological environment. In this way, the book makes an important contribution to the public and professional debate.

Contents

Chapter 1
Basics of Electrical Energy Transmission

Summary

This chapter offers a foundational overview of key concepts in electrical engineering that are essential for understanding power transmission systems. It addresses fundamental topics such as the transformer principle, the *root mean square value* (RMS), and the various forms of electrical power-active, reactive, and apparent-which together provide the basis for analyzing power flow and efficiency.

Following this, the chapter presents the historical development of power transmission infrastructure, focusing on both overhead lines and underground cables. This evolution is outlined chronologically, beginning in the late nineteenth century and extending to contemporary advancements. Particular attention is given to the initial debate between alternating current (AC) and direct current (DC) systems, as well as to the subsequent development of standardized voltage levels that facilitated the emergence of modern supply networks and interconnected grid structures.

The chapter concludes with a discussion of key planning principles, including the (n-1) criterion, and underscores the critical role of system frequency as a fundamental parameter for ensuring the stability and security of electrical power systems.

1.1 Basics of Electrical Engineering

A sound understanding of power transmission technology requires familiarity with the fundamental principles of electrical engineering. For this reason, these basic concepts are introduced at the beginning of the book. Additional, more advanced phenomena are discussed in detail in later sections and chapters.

The presentation of elementary electrical and electromagnetic effects is designed to be accessible even to readers with little or no prior experience in electrical engineering. Throughout the explanations and illustrative calculations, standard low-voltage values of 230 V (single-phase) and 400 V (three-phase) are employed.

K. O. Papailiou et al., *High-Voltage Overhead Lines and Underground Cables*, Power Systems, https://doi.org/10.1007/978-3-032-07498-0_1

In contrast, for high-voltage and extra-high-voltage systems, rated voltages are referenced according to standardized levels: 110, 220, and 380 kV. The corresponding maximum operating voltages are 123 kV, 245 kV, and 420 kV, respectively.

Before delving deeper into the specifics of power transmission, it is essential to first clarify the distinction between direct current (DC) and alternating current (AC) transmission.

Direct current transmission systems (DC systems) are characterized by their relatively simple construction, and their operational behavior can be described entirely using Ohm's law. The total line resistance consists of the combined resistance of the forward and return conductors, each typically having the same resistance value, R. (In some early submarine cable systems, the earth was occasionally used as the return conductor; however, this configuration is not considered further here.)

For the high voltages required in electrical energy transmission, it is economically and technically advantageous to insulate both the forward and return conductors from earth potential. These are then designated as the positive and negative poles.

For example, the specification $U = \pm 100$ kV means that each of the two conductors has the *rated* or *nominal* voltage $U_r = 100$ kV against earth and the voltage of the system is thus 200 kV [1]. Together with the rated or nominal current I_N, which is determined by the conductor cross section, this results in a transferable power of $P_r = 2 \cdot U_r \cdot I_r$. The transmission losses on the line amount to $P_{loss} = 2 \cdot R \cdot I_r^2$.

In direct current (DC) systems, voltage cannot be transformed using conventional transformer technology. As a result, a single rated voltage must be maintained throughout the entire DC grid. For safety reasons, this limitation precludes the use of high voltages in areas such as commercial facilities and residential households. Conversely, the radio and entertainment industries have historically relied on direct voltage. Moreover, speed-controlled drives have traditionally been well-suited to DC applications. In contemporary contexts, direct voltage has become indispensable in several key technological domains, including photovoltaics, energy storage systems, electromobility, and the entire field of information and communication technology. However, the interruption of high DC currents presents significant challenges for switchgear systems, primarily due to the formation of electric arcs.

In *alternating current transmission systems* (AC systems), it is essential to distinguish between *root mean square value* (RMS) and *peak value*. If the same conductor cross section as for direct current transmission is assumed, the RMS value I_{RMS} of the current here is identical to the above value I_r. For the voltage, however, and regarding the insulation, the peak value with $U_{RMS} \cdot \sqrt{2}$ must be calculated; this means, the insulation must be fundamentally stronger than in the case of direct current.

A *three-phase system* is formed from three alternating voltages of equal magnitude and each 120° phase-shifted to the others. Since the instantaneous values always add up to zero, a common return conductor is not required for power transmission.

Here, the voltage between two conductors is referred to as the rated voltage and the RMS value of the current as the rated current. Thus, the transferable power value $P_r = \sqrt{3} \cdot U_r \cdot I_r$ results. Since no current flows in the return conductor during normal operation, the loss value $P_{loss} = 3 \cdot R \cdot I_r^2$ results here. To enable a comparison with direct current technology, here the topic of reactive power is excluded and only the transmission of active power is considered.

Alternating current or three phase current provides several key advantages. One of the primary benefits is its transformability, which enables power to be transmitted at high voltages while allowing it to be stepped down to lower voltages levels for commercial and residential use.

Another advantage is that two voltages are available on the consumer side, namely the phase-to-phase voltage U_r and the phase-to-neutral voltage $U_r/\sqrt{3}$. In addition, alternating or three phase currents can be switched more easily due to the natural zero crossing of their sinusoidal oscillation.

For a comparison of the different transmission methods, it should be noted that the three conductors required in the case of three phase current (under symmetrical conditions) must be replaced by two conductors with 1.5 times the cross section in the case of direct current. Under identical external conditions, this increased cross section allows the direct current to reach 1.5 times the current-carrying capacity without overloading the line. If we set for three phase current as shown above the nominal power $P_r = \sqrt{3} \cdot U_r \cdot I_r$, then for direct current with the same consumption of conductor material is true:

$$P_r = 2 \cdot U_r/\sqrt{3} \cdot 1.5 \cdot I_r = \sqrt{3} \cdot U_r \cdot I_r \tag{1.1}$$

The transmission losses are here:

$$P_{loss} = 2 \cdot 2/3 \cdot R \cdot (1.5\, I_r)^2 = 3 \cdot R \cdot I_r^2 \tag{1.2}$$

wherein is:

P_r Rated (nominal) power (W)
U_r Rated (nominal) voltage (V)
I_r Rated (nominal) current (A)
P_{loss} Loss power (W)
R Electrical resistance (Ω)

Both values are identical to those from three phase transmission. However, the comparison neglected that in the case of direct current, the thickness of the insulation of cables can be designed $\sqrt{2} \approx 1.4$ times less. In addition, direct current transmission has an advantage in terms of corridor width (right of way) for overhead lines as well as for cables due to the smaller number of conductors. For a deeper understanding of the topic, comparative calculations for overhead lines as well as for cables in both direct current and three phase systems can be found in Chap. 6.

1.1.1 Transformer Principle and RMS Value

First, to the *transformer principle.* The discovery of electrodynamics dates to the first half of the nineteenth century. Since then, its applications have been further developed and perfected. Anyone who has gone through middle grade school knows the experiment in which the teacher dipped a permanent magnet into a coil wrapped with copper wires. The voltmeter connected to both ends of the copper winding deflected, i.e., a voltage was generated. As soon as the permanent magnet came to rest, the voltmeter no longer showed any voltage. This means that a voltage is only generated when the permanent magnet is constantly moved backwards and forwards in the coil. This is the *principle of induction*, one of the possibilities of electricity generation, as used for example in generators. The characteristic sinusoidal waveform of the alternating voltage provided in generators and the associated alternating current results from the rotary motion with which a conductor loop or a winding is moved at a constant speed in a magnetic field, thus generating an alternating voltage. A special form of alternating voltage is created when three windings, each offset by 120°, generate alternating currents. In this case, the alternating currents form a system known as *three phase alternating current.* This system is used worldwide in public power supply.

For simplification, from now on we will write U instead of U_{RMS}. The following relationship between the peak value $\hat{u}$ of the alternating voltage and the RMS value U applies:

$$U^2 = \frac{\hat{u}^2}{2} \tag{1.3}$$

And therefore

$$U = \frac{\hat{u}}{\sqrt{2}} \tag{1.4}$$

Since the current follows the sinusoidal form of the voltage, the same applies

$$I = \frac{\hat{i}}{\sqrt{2}} \tag{1.5}$$

wherein is:

U RMS of the alternating voltage (V)
$\hat{u}$ Peak value of the alternating voltage (V)
I RMS of the alternating current (A)
$\hat{\imath}$ Peak value of the alternating current (A)

Thus, RMS values for current and voltage are created, which in their effect correspond to the value of the direct voltage or the direct current. This means that the sinusoidal alternating voltage and the sinusoidal alternating current perform the same amount of electrical work on a resistor over a certain period of time as an equally high direct voltage or an equally high direct current.

The RMS value is always used in the specifications of rated voltages in the public supply grid. However, the peak value of the voltage is decisive for the dimensioning of the electrical insulation.

If instead of a conductor loop, an iron core is immersed in a coil and an alternating voltage is applied, the iron core develops into a magnet that attracts iron objects like a permanent magnet. On the one hand, a voltage can be generated by moving a permanent magnet in a coil, and on the other hand, a current carrying coil in an iron core creates a magnetic field. The next step in thinking leads us to an arrangement where a coil with an iron core is connected to an alternating voltage that replaces the movement of the permanent magnet from the first consideration. If the iron core is pulled out of the coil far enough that a second coil can be pushed on, the alternating voltage in the first coil also generates an alternating voltage U with the same frequency in the second coil via the magnetic coupling. This is the basic principle of the *transformer*. According to the ratio of the number of turns N in one coil to the number of turns in the second, the respective heights of the alternating voltages in the coils result. The following applies:

$$\frac{N_1}{N_2} = \frac{U_1}{U_2} \tag{1.6}$$

Since the transferred power P results from the product of current I and voltage U, the currents on the primary and secondary side behave inversely proportional to the turns or voltages. Here applies:

$$\frac{N_1}{N_2} = \frac{U_1}{U_2} = \frac{I_2}{I_1} \tag{1.7}$$

Since transformers, like all technical devices, do not work without losses, the following applies:

$$\frac{P_2}{P_1} = \eta_T \tag{1.8}$$

wherein is:

N_1 Number of turns of the primary coil (–)
N_2 Number of turns of the secondary coil (–)
U_1 Voltage applied to the primary coil (V)
U_2 Voltage applied to the secondary coil (V)
I_1 Current through the primary coil (A)
I_2 Current through the secondary coil (A)
P_1 Power on the primary side (W)
P_2 Power on the secondary side (W)
η_T Transformer efficiency factor (–)

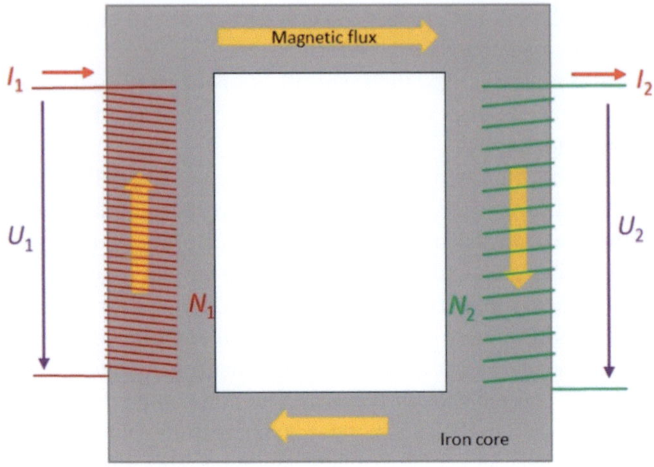

Fig. 1.1 The transformer principle

To optimize the magnetic flux within the iron core, the coils are arranged as shown in Fig. 1.1. In this arrangement, the efficiencies η_T of transformers reach 94–98%. In the three phase alternating current, the transformation happens via the same mechanism. This physical principle allowed for optimization between transmission voltages and currents according to economic considerations in single-phase or three-phase grids, as will be seen.

Example

A transformer for connection to $U_1 = 230$ V has a coil with $N_1 = 300$ turns on the primary side. On the secondary side, $U_2 = 12$ V is expected.

a. What number of turns N_2 must be chosen on the secondary side?
b. On the primary side, a current of $I_1 = 2$ A is measured. What is the power output on the secondary side if the efficiency of the transformer has a value of 97%?

To (a) The number of turns on the primary side is:

$$N_2 = N_1 \cdot \frac{U_2}{U_1}$$

$$N_2 = 300 \cdot \frac{12\text{ V}}{230\text{ V}}$$

$$N_2 = 16\,turns$$

To (b) The primary power consumption is:

$$P_1 = U_1 \cdot I_1$$

$$P_1 = 230\,\text{V} \cdot 2\,\text{A} = 460\,\text{W}$$

and the secondary power output is:

$$P_2 = P_1 \cdot \eta_T = 460\,\text{W} \cdot 0.97 = 446\,\text{W}.$$

1.1.2 Active, Reactive and Apparent Power

With the explanation of the "inductive" transformer principle, we are already on the trail of inductive reactive power. Before a current flows through a coil, the full voltage must already be present at both ends. Only then does a current begin to flow. In technical terms, this means that the current lags the voltage. With an alternating voltage, this means that the current, as soon as it has started to flow, must flow in the other direction again due to the polarity reversal. And this at a frequency of 50 Hertz (Hz) 100 times per second (in several countries with 60 Hz). This leads to a reduction in the active component of the current, as a portion of it lags behind. This constantly reversing component is called *reactive current*, as it does not contribute to the actual effective current transport but uses the conductor cross section. It is referred to as *inductive reactive current* I_L which, in conjunction with the applied alternating voltage, causes the undesired so-called *reactive power* Q. All electrical components that have a coil, such as transformers and motors, are potential causes of inductive reactive current. That's one side of the coin.

The other side acts in the opposite direction. This phenomenon can be very clearly explained using a cable operated with alternating voltage. Let's consider a so-called capacitor instead of a coil. This consists of two extended plates, which are separated by an insulating material, referred to as a *dielectric*. If a direct voltage is applied to the plates, the charge carriers distribute themselves on the two plates and a uniform electric field is created in the dielectric.

However, if an alternating voltage is applied to the two plates of a capacitor, the charge carriers must rush back and forth between the source and the two plates in time with frequency. Another special feature of this arrangement is that the charge carriers must first distribute themselves on the two plates before a voltage can arise between them. That is, the current must first flow before the voltage builds up. So here the current precedes the voltage.

High and extra-high voltage cables correspond in their designs and their essential properties to an elongated, concentric capacitor. The conductor's surface acts as one plate and the outer shell, the metallic shield, as the other (Fig. 1.2, left). Let's imagine a multitude of small capacitors between the conductor of the cable and its shield (Fig. 1.2, right). If an alternating voltage U is now applied between the conductor and the usually grounded shield, the charge carriers, as already indicated, move back and forth from one plate to the other due to the constant polarity reversal.

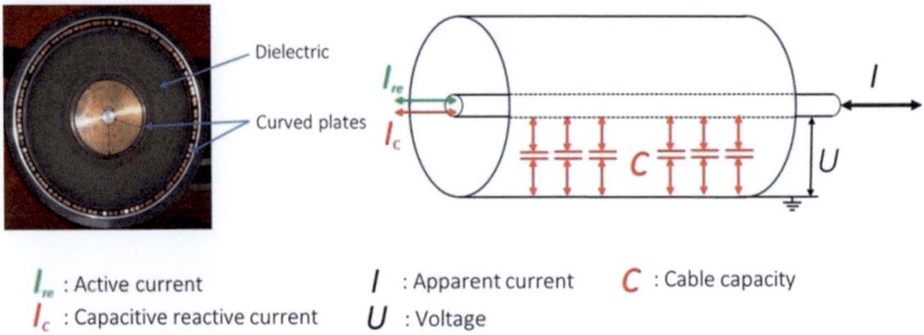

Fig. 1.2 The high-voltage cable as an elongated capacitor

In this way, they generate a current that only serves the charge exchange between the plates, burdens the cable, but does not contribute to the actual current transport. The total current through the conductor, which is referred to as apparent current I, is therefore composed of the effectively transmitted active current I_{re} and the capacitive *reactive capacitive current* I_C. However, not algebraically, but in the form of two vectors that enclose an angle of 90°. Therefore, the following applies:

$$I = \sqrt{I_{re}^2 + I_c^2} \tag{1.9}$$

wherein is:

I Apparent current (A)
I_c Capacitive reactive current (A)
I_{re} Active current (A)

With increasing length of the cable, its so-called *capacity* and thus also the capacitive reactive current increase. This phenomenon leads to length restrictions in high-voltage cables. If the entire cross section of the conductor is needed for the transport of the reactive current, no active current can be transmitted anymore. This is the reason why the integration of cables into the alternating current grid is limited at the high and highest voltage levels.

Now we have learned about two types of reactive current, which, despite their unwanted appearance, possess beneficial properties. Since the current leads the voltage in capacities and lags in inductances, they can supply or completely compensate each other. This allows a high proportion of active current and thus associated active power to be transmitted via the supply lines.

A four-quadrant system is used, in which the real components, i.e., the active components, are plotted on the abscissa and the imaginary components, i.e., the reactive components, on the ordinate. Thus, any state can be represented by geometric addition. The capacitive reactive current, which precedes the voltage, points upwards

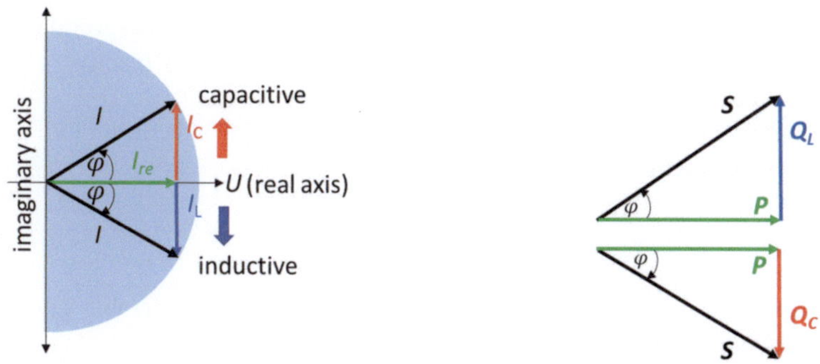

I	apparent current	A	S	apparent power	VA
I_{re}	active current	A	P	active power	W
I_C	capacitive reactive current	A	Q_C	capacitive reactive power	Var
I_L	inductive reactive current	A	Q_L	inductive reactive power	Var
$\cos\varphi$	power factor	-			

Fig. 1.3 Capacitive and inductive loads in the alternating current network (*left*: currents, *right*: powers)

and spans the angle φ between apparent and active current I_{re}. Conversely, the inductive reactive current is directed downwards. In both cases, right angled triangles are formed, which show the size ratios of the individual currents. The apparent current thus geometrically divides into the active current and the capacitive or inductive reactive current, as shown in Fig. 1.3.

AC-voltages and -currents are represented with the help of vectors, and in electrical power engineering usually represented in the Gaussian plane. Those vectors are rotating as a function of time with the *angular velocity*. The rotation direction is counterclockwise, what means positive in mathematical sense. We freeze the rotation in time by taking a "snapshot" and designate such vectors as *phasors*. The point in time for the snapshot can be chosen arbitrarily. It is usually done regarding a certain *reference phasor*, e.g. the voltage, which is assigned a phase angle of zero degrees.

For better distinction, voltage, current and apparent power as *complex quantities* are underlined in the following equation. To get the transmitted *apparent power S*, we have to form the product of $\underline{\mathrm{U}}$ and $\underline{\mathrm{I}}$. If we do so, the inductive reactive part of the apparent power gets a negative and the capacitive reactive a positive sign. But in the energy economic world the inverse is common. To be compliant with this determination, we therefore apply the following formula:

$$\underline{S} = \underline{U} \cdot \underline{I}^* \tag{1.10}$$

The asterisk symbols "conjugate complex". Chapter 6 presents several examples applying Eq. (1.10) to calculate the transmission of real and reactive power via overhead lines or underground cables.

Taking only the *magnitudes* of apparent power, voltage and current into account, it is sufficient to write

$$S = U \cdot \mathrm{I} \tag{1.11}$$

For symmetric load of the individual conductors, the transferable power for a three-phase alternating current grid is determined accordingly.

$$P = \sqrt{3} \cdot S \cdot \cos\varphi = \sqrt{3} \cdot U \cdot I \cdot \cos\varphi \tag{1.12}$$

wherein is:

S	Apparent power (VA)
U	Phase-to-phase voltage (rated voltage) (V)
I	Apparent current (A)
P	Active power (W)
$\cos\varphi$	Power factor (–)

In power transmission grids, a power factor close to one is always aimed for. In this case, a large proportion of the cross sectional area of the cables and overhead lines is available for the transport of active power.

Example

In a three-phase grid with an operating voltage of $U = 400$ V, an asynchronous motor is to be operated. The data plate indicates an active power of $P = 3.8$ kW and a power factor of $\cos\varphi = 0.8$. What is the operating current I_S?

$$I = \frac{P}{\sqrt{3} \cdot U \cdot \cos\varphi} = \frac{3800\ \mathrm{W}}{\sqrt{3} \cdot 400\ \mathrm{V} \cdot \cos\varphi} = 6.9\ \mathrm{A}$$

1.2 History of Power Transmission

As is often the case in history, significant developments begin with a dispute among experts. This was also the case with the fundamental decision about the type of voltage for long distance transmission of electrical energy. The two protagonists were Thomas Alva Edison and Nikola Tesla (Fig. 1.4) argued relentlessly about this question in the late nineteenth century.

While Edison, as an advocate of direct current (DC), vehemently advocated for this kind of transmission, Tesla proved to have greater foresight by demanding a three-phase current transmission (AC), which we use today at different voltage levels in

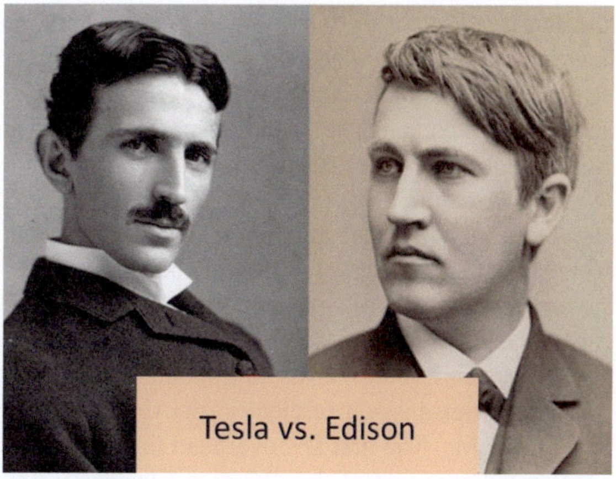

Fig. 1.4 Nikola Tesla (*left*) and Thomas Alva Edison (*right*)—two protagonists in the War of Currents [2]

public electricity supply worldwide with a frequency of 50/60 Hz. The fact that we are dealing again with direct current transmission at the highest voltage level has nothing to do with this decision at first. More on that later.

Nicola Tesla was an engineer who left his studies at the technical college in Graz, Austria, unfinished. This was followed by several jobs as a designer and inventor. Finally, he ended up in 1882 at "Continental Edison" in Paris and worked on the electric street lighting of the city. The local head of Edison's continental branch encouraged him, for better career opportunities, to move to the company's headquarters in New York. Reportedly, his recommendation letter to the boss in the USA began with the following words:

> My dear Edison: I know two great men, and you are one of them. The other is this young man.

Thus, for the 28-year-old Tesla began the work in the company of the nine years older genius, who was already famous and a successful businessman with numerous inventions in the field of the application of electricity.

The autodidact Edison began his career without any special education in telegraphy (a direct current application) and was firmly convinced that the *high voltage transmission* should also be done with DC. He recognized Tesla's genius and commissioned him, with the prospect of a substantial premium, to improve his direct current motors, so as not to have to deal further with the supposedly unsuitable alternating current technology. Despite fulfilling the task successfully, Tesla did not receive the promised premium and subsequently chose to leave the company.

Subsequently, Tesla developed the alternating current technology further together with the industrialist George Westinghouse and began to use it also for long-distance transmission of electricity. Thus, Edison and Westinghouse (who supported Tesla's alternating current research and marketed its results), clashed heavily.

In the dispute, Edison always invoked the danger posed by alternating current and repeatedly pointed to the electric chair, which was operated with alternating current and was first used in 1890 in the state of New York. Edison made several attempts to have alternating current banned by authorities because of its danger. Without success.

Since there were neither rectifiers nor inverters at that time, the applications were always dependent on the type of transport voltage when the power generation and use were geographically separated. The direct current motor was an established and also reasonably mature drive unit with numerous advantages. It could be operated with direct current generators and batteries alike. Its invention dates back to the first half of the nineteenth century. For lighting purposes, the main application of electricity at the time, the type of voltage, did not matter. The incandescent lamps lit up with direct current just as well as with alternating current. However, Tesla had just invented the three-phase current motor. Its commercial use was still to come.

For the transmission of electricity over long distances, direct current proved to be less suitable. The first DC overhead line, which Oskar von Miller (Fig. 1.5), had built on the occasion of the "Munich Electricity Exhibition" in 1881, showed this clearly. Von Miller is the later founder of the German Museum in Munich, Germany, which has become a worldwide renowned museum of technology. The direct current generator, driven by a 1.5 HP steam engine in 57 km distant small city of Miesbach, generated the necessary electrical energy with a voltage of about 2 kV, which was fed into the grid. At the end of the line, on the Munich exhibition grounds, where it drove a pump for a small waterfall, it was only about 1.5 kV. The telegraph line misused as a high-voltage line had a resistance of 3,000 ohms (Ω) and barely achieved an efficiency of around 25% [3]. The choice of a larger cross section would certainly have improved the result, but it would not have been satisfactory either. Transmitting higher powers over longer distances with direct current meant, due to the limited voltage level, large cross sections of expensive copper conductors. On top of that, the switching of high direct currents led to unwanted arcing at the switch's contacts.

Approximately a decade later, Oskar von Miller once again emerged as a pioneering figure in the field.

For the "International Electrotechnical Exhibition" 1891 (Fig. 1.6), which he organized in Frankfurt am Main, Germany, he built another overhead line. On this

Fig. 1.5 Oskar von Miller [4]

Fig. 1.6 Frankfurt international electrotechnical exhibition 1891 [6]

occasion he applied high voltage three phase current, and in a design that would prove to be standard for future high voltage overhead lines. The feed in took place in Lauffen (small city on the river Neckar) and the line length was a considerable 175 km. The trick this time was the low generator voltage of 55 V, which was transformed to 15 kV via a three-phase transformer and forwarded to Frankfurt via the overhead line. During the exhibition, the pioneers of electricity transmission presented the project as "Power Transmission Lauffen-Frankfurt". The incoming high voltage was transformed there down to 100 V and powered about 1000 light bulbs and a 74 kW three-phase synchronous motor, which drove a pump for an artificial waterfall several meters high on the exhibition grounds. And all this with a transmission efficiency of a respectable 75%. The synchronous generator driven by a water turbine at the other end of the line had a power output of 221 kW and generated an AC current with a frequency of 40 Hz [5].

This project enabled the exhibition to serve as a powerful demonstration of the capabilities of electrical energy and its wide range of applications. Most importantly, however, it conveyed a clear message: for the economical long-distance transmission of electrical power, alternating current technology was significantly more suitable than its steady-flowing direct current counterpart.

In the relevant literature, this event is often referred to as the "birth" of electrical energy transmission and supply [7]. Through the transformation based on the principle of induction from low to high and later also extra high voltages and vice versa, combined with extraordinarily high transformation efficiencies, bulk powers could be transmitted over long distances with low losses at a higher voltage level. Moreover, the current flow could be interrupted more easily since both the current and the voltage have zero crossings in their temporal course. This was the breakthrough for the transmission technology with alternating current and for the expansion of power grids at all voltage levels.

In the following years and decades, the demand for electricity rapidly increased. Light bulbs replaced oil and gas lights, and electric motors of various types took over the individual drives of machines. They rapidly displaced the steam engines together with their transmissions, which ran along the ceilings of the factory halls, susceptible to failure and accident prone. The local producers, who supplied nearby consumers, gave way to larger, more efficient generation units, which were located further away from the load centers and thus required transmission lines and later transport and distribution grids to reach them.

Since power is calculated from the product of voltage and current, transformers could generate almost arbitrarily high voltages which only required longer insulators for the lines. Such lines were significantly cheaper to manufacture than those with thick copper conductors for high currents. Thus, economical, distance-optimized and optimized heights of currents and voltages were created. The transport distances became increasingly larger and resulted in increasing transmission voltages. Soon there was the still popular rule of thumb, according to which the transmission voltage should be approximately 1 kV per kilometer distance between the power plants, transformers or switching stations.

This development made it possible to place increasingly large and more efficient generation units, which quickly grew into large power plants, where the primary energy was available. This applied and applies to hydropower and fossil resources, in the case of lignite mined in open cast mining or hard coal, which could be easily transported via maritime and inland waterways. Thus, the power plants were continuously expanded, and the generated electrical energy was transported over ever longer distances to the load centers, the large industrial plants and major cities. In this way, the transport of energy carriers could also be optimized. The raw materials for electricity production, which were heavy and required in large physical quantities, thus had short or convenient routes and electricity could be fed in and out at any grid node in almost any quantity.

1.2.1 Overhead Line Development

As early as 1912, the first 110 kV double overhead line in Europe between the Brandenburgian city Lauchhammer and the Saxon city Riesa in Germany with a length of about 50 km and a transmission power of 20 MW was put into operation. In 1929, the voltage level was raised to 220 kV. The so-called "North–South Line" stretched over a total of 600 km, from Brauweiler near Cologne to Tiengen, in South Baden, not far from the German Swiss border. In a six-year planning and construction period the Rheinisch-Westfälische Elektrizitätswerk (RWE) built the

Fig. 1.7 *Left*: setting up a concrete tower, *right*: "towers are stable", effective strength proof for the public; historical pictures, 1930s

line across several regions. It connected a total of seven transformer stations and served the energy exchange between coal power from the Rhenish coalfields and hydropower in the south of Germany and Switzerland. It is considered the first so-called interconnected line in Germany. Initially, the line was designed for operation at 380 kV but then was commissioned at 220 kV Fig. 1.7 shows two examples from the early history of tower construction.

The first 380 kV three-phase overhead line in Germany did not start its service until 1957, after the Swedes had already put a line at this voltage level with a transmission power of 1000 MW into operation five years earlier. The over 340 km long first German 380 kV overhead line was built by RWE after extensive tests on a test site in Mannheim between the substations Rommerskirchen near Cologne and Hoheneck near Stuttgart. This line formed the origin of the German highest voltage transmission grid at this voltage level. The move to the far highest transmission voltage in Europe was necessary because the lignite power plants in the Rhenish coalfields were continuously expanded, and the transmission capacities of the lines at lower voltage levels were no longer sufficient [8]. The tower configuration, referred to as *Danube tower*, which was used in this line, was later to become the standard in Germany for double overhead lines in open terrain (Fig. 1.8).

The European interconnected grid continues to operate at this voltage level and with a frequency of 50 Hz till today. It is highly unlikely that this standard will change in the foreseeable future. However, what will be added is high voltage direct current transmission (HVDC) via underground cables, which is intended to be an integral part of the existing transmission grid for the transmission of bulk power

Fig. 1.8 Tension tower along the route of Germany's first 380 kV overhead line in "Danube tower" design (*Source* Amprion)

between geographically distant points. Modern power electronics and advanced cable technology make possible what was considered technically unachievable in the early days of power transmission. In addition in short sections also DC overhead lines or even AC/DC hybrid lines are considered.

Outside of Europe, in sparsely populated countries such as Russia and Canada, three phase overhead lines with even higher transmission voltages were soon used to bridge larger distances. Thus, the Canadian company Hydro-Quebec built the first 735 kV overhead line over 500 km with a transmission power of 5300 MVA in 1965, which led from the reservoirs around Manicouagan to Montreal [9]. In 1985, the Kazakh energy supplier KEGOC put into operation the world's first three phase overhead line with 1150 kV. In the first state of expansion level, it ran from the power plant in Ekibastus over 700 km to Kokshetau. After several expansion stages, the line now measures over 1400 km. The transmission power is stated to be 5500 MVA [10]. This voltage level should remain the worldwide highest in the three phase range for the time being.

Thanks to modern semiconductor technology, high DC voltages can now also be used for the transport of very large amounts of power. As will be explained later, DC transmission, due to the lack of reactive power demand, offers an excellent opportunity to transmit energy over long distances with very low losses. However, so far this has almost exclusively involved point-to-point connections. At both ends of such a line, large scale converter stations are required which integrate these transmission lines into the existing high voltage AC grid.

In 1970, the first 500 kV direct current overhead line was built from Oregon-Celilo to California-Sylmar. The 1362 km long connection was operated with ±500 kV and could transmit a power of 3100 MW. Mercury rectifiers were used in the converter stations for the conversions between AC and DC. The next leap followed in 1979 with the ±533 kV overhead line from a hydropower plant in Cahora Bassa from Mozambique to South Africa. Over a distance of 1420 km, 1920 MW were transmitted to the industrial areas around Pretoria on two separate routes. The conversion was already done with thyristors.

In 1984, two overhead lines (785 and 805 km) from São Roque (Paraná) to São Paulo followed in Brazil, both with ±600 kV and a transmission power of 3150 MW each. As a result, several long direct current overhead lines were built worldwide, such as in 2010 in China Yunnan-Guangdong with a length of 1400 km, ±800 kV and a transmission power of 5000 MW. The highest voltage level to date for high voltage direct current transmission via overhead lines is ±1100 kV (2200 kV between the positive and negative pole!). This HVDC overhead line, built 2018 and more than 3400 km long, connects the Chinese cities of Changji and Guquan and is stated to have a transmission power of 12000 MW [11, 12].

For comparison, Fig. 1.9 shows the development of the voltage levels for AC overhead lines from the beginning.

With increasing transmission voltages and the resulting larger distances between the conductors, the resulting electric field strengths in the vicinity of overhead lines of the extra high voltage level increase sharply. They can, as will be seen later, lead to risky secondary effects, against which precautions are necessary. Regardless of the electrical phenomena, the rejection against overhead lines at the higher voltage levels increased with the growing environmental awareness in the general public, starting in the 1980s. While they had been considered necessary, if not popular, until then, the mood shifted, sometimes to the point of outright rejection. In addition to the influences on the visual impact, the alleged *electrosmog* (electric and magnetic fields in the vicinity of the line) was another reason for rejection [13]. Militant groups even carried out attacks on high voltage towers in the 1980s by sawing off the main legs. Fortunately, the spectacular actions were without personal injury, and the grid operators were able to restore immediately the stability of the damaged towers or to replace them. Figure 1.10 shows a tower that was sawn off at the main legs located opposite the resulting direction of pull, which fell in the bisecting direction and affected the adjacent support towers.

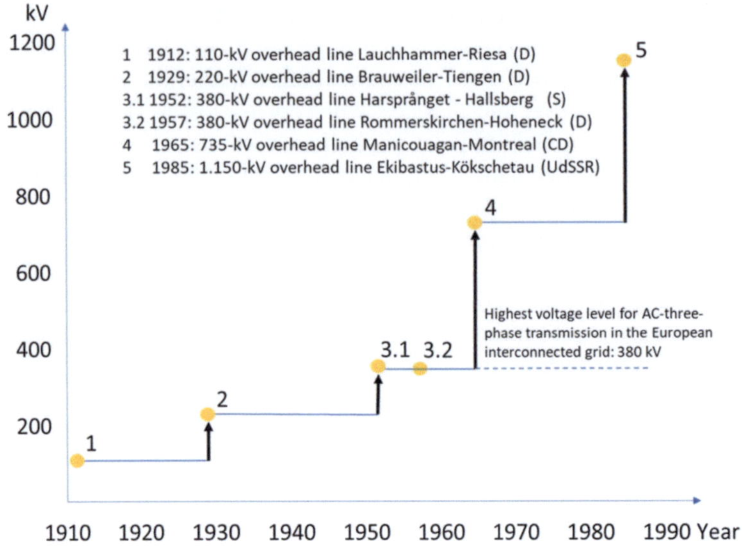

Fig. 1.9 The most important steps in the development of AC voltage levels for overhead lines worldwide

Fig. 1.10 Overturned angle tension tower, which fell after a sawing action collapsed under the strain and tore down adjacent support towers (*Source* E. Walther)

The demand for underground cables increased significantly everywhere and extended beyond newly planned transmission routes. Existing overhead lines, regardless of their voltage level, also came under growing scrutiny and were increasingly called upon to be replaced by cable systems, independent of their voltage level.

1.2.2 Underground Cable Development

The development of usable power cables probably began with an invention by the Swiss-born John Kruesi, who, as an employee of Edison, pulled three insulated conductors into a steel pipe and filled it with hot tar. The patent specification of 1883 also describes a joint in which the "cables" are connected. Until today, the design of joints resembles this basic pattern (Fig. 1.11).

The patented cable was used for power transmission from the first US power plant "Pearl Street" in New York, which supplied several hundred of Edison's filament lamps in the city, and which were operated with direct current preferred by Edison at a voltage of 110 V. The grid underwent significant expansion in the following years and, according to reports, supplied 1000 lamps after its completion [14].

Another type of insulation was already developed about 30 years earlier. In 1847, the young Werner Siemens invented an extruder press for coating underground and undersea telegraph lines with guttapercha, a rubber-like tree secretion, which was increasingly used as insulating sheathing for power cables in the 1880s.

Around the same time, Siemens & Halske developed a high voltage cable for supplying a lighting system of a post office in Berlin (Germany). The 220 V DC cable insulated with waxed jute received a band iron reinforcement and was laid directly in the ground. This type of cable laying prevailed against laying in specially prepared channels. Cast iron joints were applied for cable connections and branches, as they were common in shape and material until a few decades ago worldwide. After a brief interlude with the laying of bare copper wires on insulators in so-called "Monier channel systems" at the end of the 1880s, the power suppliers of that time returned remorsefully to earth laid cables. The broken lids of the concrete troughs continuously led to water intrusions and consequential short circuits [15].

With increasing transmission voltages, the requirements for the dielectric strength of the insulating material increased. Unlike overhead lines, which rely on air insulation over distances in the range of meters, cables require insulation over only a few centimeters. In addition, the insulation materials used until then softened at higher operating temperatures, which occurred in contrast to telegraph cables in power cables due to higher loads and the associated heating. In the following years, oil-soaked fibrous materials such as jute and hemp were increasingly used as insulating materials, which in addition to higher dielectric strength also showed better temperature performance [16]. A robust lead sheath protected the insulation from moisture ingress.

Beginning in the early twentieth century, layered paper insulation impregnated with mineral oil significantly enhanced the dielectric strength of high-voltage cables.

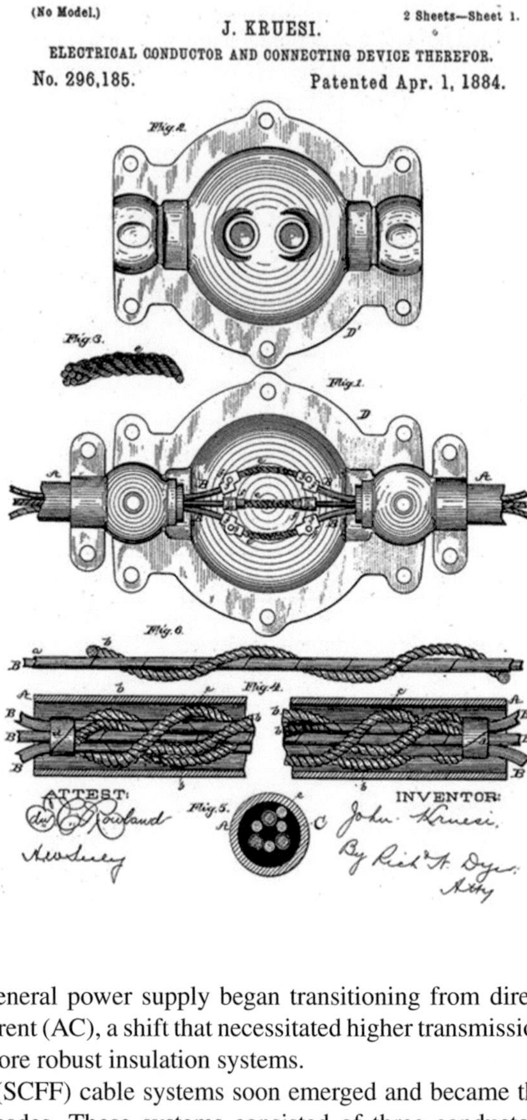

Fig. 1.11 Representation of the Kruesi tube with the three conductors in the pipe (bottom), an open (middle) and a closed joint (top) [14]

Around the same time, the general power supply began transitioning from direct current (DC) to alternating current (AC), a shift that necessitated higher transmission voltages and, consequently, more robust insulation systems.

Self-contained fluid-filled (SCFF) cable systems soon emerged and became the standard in the following decades. These systems consisted of three conductors wrapped in oil-impregnated paper, surrounded by an additional oil-soaked paper belt insulation. This assembly was enclosed within a lead sheath. For external protection, the lead sheath was supplemented by an inner protective layer, a steel tape armor for reinforcement, and an outer protective covering made of fibrous materials (Figs. 1.12 and 1.13).

In 1911, this type of cable was used for the first time on the European continent for the supply of the city of Berlin and its surrounding rural areas [16].

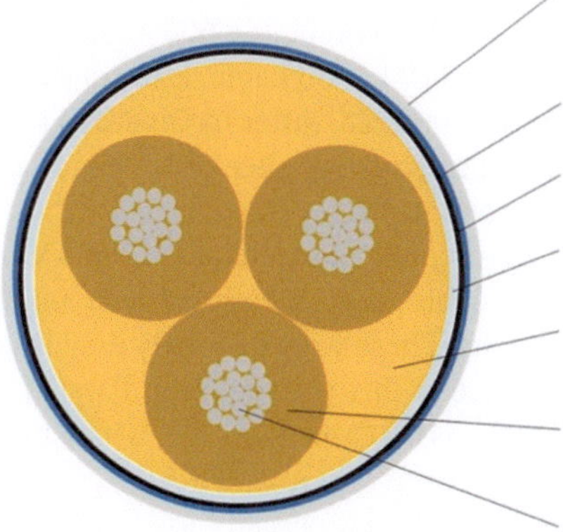

Fig. 1.12 Structure of an old three-core SCFF cable

Fig. 1.13 30 kV SCFF in 1911 [16]

In the same year, the Royal Railway Directorate in Halle (Germany) put into operation the first AC cable with 60 kV and a length of 43 km as a feeder line for the electrified test railroad track between Dessau and Bitterfeld. The next step in the development was achieved around 1920 with the "Höchstädter-paper" which consisted of perforated aluminum foil and was wound over the insulation of the individual conductors. This development was based on a suggestion by Professor Höchstädter, which he had already proposed ten years earlier. The better field distribution allowed a higher dielectric strength of the insulating material to be achieved (Chap. 5).

The next voltage level followed in 1923 with a 130 kV oil pressure cable with thin oil from the Italian cable manufacturer Pirelli, who built a 600 m long test line with three single conductors. The first German 110 kV oil cable, also consisting of three single conductors, was laid in Nuremberg in 1928. After the first 220 kV oil cables were tested in the early 1930s, the Parisian energy supplier built a large 220 kV cable grid with low pressure oil cables in the city in 1936 [9].

In the early 1950s, the rapidly increasing load demand at the 110 kV level led to the widespread use of low pressure oil cables with sheaths of lead and aluminum, both in single and three conductor versions. These were primarily used in urban areas where overhead lines could not be erected at this voltage level. Parallel to this, the Swedes built the first 380 kV cable routes as oil cables in 1952, which they operated with 425 kV a few years later.

Although the first plastic insulated cables were already used at the low voltage level in the forties, the first 110 kV polyethylene plastic (PE) insulated cables e.g. in Germany did not come onto the market until 1973. Initial problems with so-called *water trees*, or "moisture trees" in fine cracks in the plastic insulation, which caused partial discharges and subsequently cable faults, led to further developments towards longitudinally and transversely watertight cross linked polyethylene cables (XLPE). Since then, these systems have become well established at the 110 kV level and are now also widely used at higher voltage levels (Chap. 5).

In the mid-1970s, Schluchseewerk AG in Germany chose a hitherto rarely used type of line for the energy transmission with a system length of 700 m from its cavern power plant in Wehr. A gas insulated line (GIL) with an operating voltage of 420 kV transmits the current via a tube conductor supported by plastic insulators in a larger gas filled tube [17]. Due to the high costs and the use of the insulation medium SF_6—known as the strongest greenhouse gas—this technology was subsequently used for power transmission only in special cases. For example, in situations of large differences in altitude, where other cable technologies, especially oil cables, were not an option.

In 1976, the Berliner Städtische Elektrizitätswerke AG (Bewag, Germany) built the world's first over 8 km long inner city 380 kV cable connection between the substations Berlin-Reuter and -Mitte. It consisted of two paper insulated low pressure oil cable systems in single conductor design with a conductor cross section of 1200 mm^2. To dissipate the heat generated in the cables, they were laid in water flowing fiber cement pipes.

The improved technology led to the 220 kV XLPE cable in 1988. And as early as 1996, the first 380 kV XLPE cable in Germany went into operation at the Neckarwerke in Deizisau, Württemberg.

With the advent of high voltage direct current transmission, cross linked polyethylene cable technology in this segment has developed extremely rapidly over the past two decades. While such HVDC cables in 1997 allowed a power of just 3 MW to be transmitted at an operating voltage of $\pm$10 kV, in 2017 it was 3000 MW at $\pm$640 kV. The transferable power increased a thousandfold within 20 years stepwise [18].

In the next few years, cable routes from the north to the south of Germany are to be laid as HVDC cables with ±525 kV, a copper cross section of 3000 mm^2 and a transmission capacity of more than 2000 MW per system.

As part of the transition of the energy system, a fundamental further development of all grid levels is imminent in Germany and certainly also in all of Europe. The previous grid expansion was mainly oriented towards powerful connections between generation and load centers that had gradually emerged and were mostly less than 100 km apart. In the coming years and decades, a fundamental realignment is just necessary. After the abandonment of nuclear power in many countries and the foreseeable separation from coal fired power plants, more emphasis—in the following explained by the situation in Germany as example and showing at the same time similar situations in other Central European countries—will be placed on wind farms in the North- and Baltic Seas and the widely distributed wind and solar energy on the mainland, along with flexible gas power plants. The existing structure of the extra high voltage grid no longer fits with such projects. While the distribution grids need a partly massive reinforcement so that photovoltaic- and onshore wind power plants can feed into the low, medium and high voltage grid without restrictions, the transmission grid needs powerful connections between the new generation centers in the north and the energetically orphaned south of the region. The energy transmission planned a few years ago over several north south connections in Germany, which was based on a massive expansion of the extra high voltage AC grid in overhead line construction, failed due to the sometime massive opposition of the population. As soon as the planning of an extra high voltage overhead line became apparent, resistance formed on a broad front. The fragmentation and devaluation of densely populated landscapes by massive overhead lines, complained about by opponents of line construction, was strictly rejected. This led to a fundamental change in planning by the German legislator in 2016, according to which the powerful north south transmission lines are now primarily to be laid as HVDC cables. Since everything had to be planned anew and differently, the urgently needed expansion of the extra high voltage grid has been delayed by several years. It is highly likely that the grid expansion will not be able to keep pace with the expansion of offshore wind power capacities in the North- and Baltic Sea and the impending shutdown of the remaining coal fired power plants on the mainland by 2030. Figure 1.14 shows the development of direct and alternating current cables and the voltage levels achieved from the beginning.

1.2.3 Importance and Development of Standardization

In the development of transmission grids, large numbers of components from various manufacturers are employed. A range of environmental conditions, diverse technical requirements, and the differing expectations of owners and operators must all be taken into account. When transmission lines interconnect the grids of different countries or

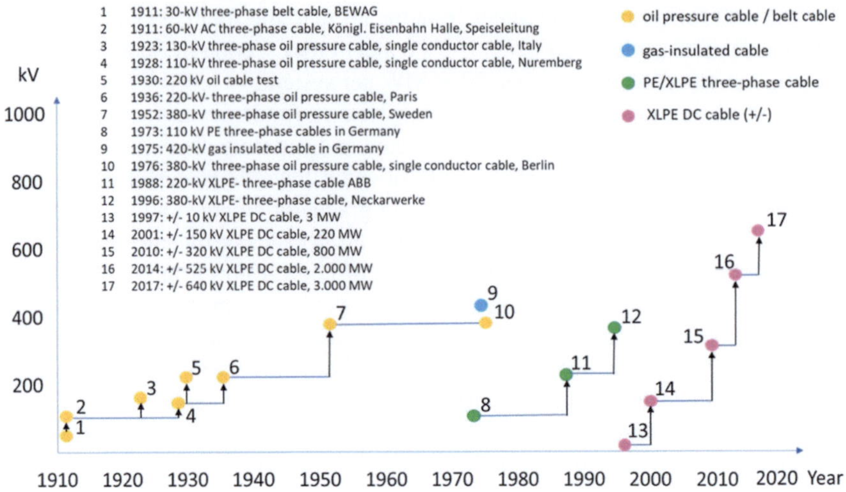

Fig. 1.14 Development of voltage levels in underground cable technology

states, additional considerations include varying operational practices and language differences among the respective grid operators.

Nevertheless, all components must fit together to ensure the safe and reliable transport of electrical energy, such as across Europe from Norway to Gibraltar. This is where standardization plays an important role. Standardization provides security for the society, for the products and for the planning technicians. It is mostly adhered to on a voluntary basis, and it reflects the current state of the art, which is constantly evolving. Compliance with established standards ensures a maximum level of functional reliability. All affected stakeholders (manufacturers, users, consumers, trade, industry, science, authorities, grid operators, utilities, etc.) are called upon to participate in the creation and ongoing development of standards.

Technicians are accustomed to relying on standards during the planning process, and they expect these standards to be readily available. At the beginning of overhead line and cable technology, however, this was not consistently the case. With the ever-increasing expansion of the grids, product standards were created (they determine minimum values for the production of towers, conductors, underground cables, joints, etc.) In parallel, standards for determining the key planning parameters of a line (e.g. wind loads on overhead lines and minimum distances to objects and terrain, laying depths and clearances for underground cables). National, and later international standardization bodies were created and supported these efforts. In Austria, for instance, overhead lines were first standardized in the so-called “Normalien für Freileitungen” in 1913 (Fig. 1.15).

Today, standardization primarily is done at the international level through organizations such as the IEC (International Electrotechnical Commission) and ISO (International Organization for Standardization), as well as at major intergovernmental levels—for example, CENELEC and CEN in Europe, and ASTM International. For

Fig. 1.15 Standards set minimum requirements. Project—specific, stricter design parameters may be required, e.g. for ice loads on conductors of overhead lines in mountainous regions

decades, these standards have served as the legal and regulatory base for line construction in many countries. The following bodies are of particular relevance for our considered use cases:

Worldwide Standardization

ISO—International Standardization Organization (e.g. steel construction, civil engineering).
IEC—International Electrotechnical Commission (electrical engineering).

European Standardization

CEN—Comité Européen de Normalisation (e.g. steel construction, civil engineering).
CENELEC—European Committee for Electrotechnical Standardization, (electrical engineering).
ETSI—European Telecommunication Standardization Institute (communications technology).

National Standardization

E.g. Germany

DIN—Deutsches Institut für Normung (e.g. steel construction, civil engineering).
VDE—Association for Electrical, Electronic and Information Technologies (science, standardization, testing, certification and application consulting).
DKE—German Commission for Electrical, Electronic and Information Technologies in DIN and VDE (electrical engineering standardization).

E.g. Austria

ON—Austrian Standards (e.g. steel construction, civil engineering).
OVE—Austrian Electrotechnical Association.

Similar standardization institutes exist in many countries. The development of a new standard typically draws upon the extensive expertise of professionals in the relevant field, and existing standards—where available—are used as a base. In addition, predecessor organizations that have previously conducted work on the subject are often involved in the process. These organizations may also be tasked with carrying out the base research or preliminary surveys required for drafting a new standard. Typical examples of such predecessor organizations include

CIGRE—Conseil International des Grands Réseaux Électriques, Paris, worldwide relevance (electrical engineering).
IEEE—Institute of Electrical and Electronics Engineers, USA, worldwide relevance (technical innovations, electrical engineering).
FNN—Forum Network Technology/Network Operation at VDE (VDE FNN).
– Germany (application rules, hints, studies and position papers, grid operation).

A notable example of cross-border coordination in standardization is the European Committee for Electrotechnical Standardization (CENELEC). Established in 1973 and headquartered in Brussels, CENELEC is responsible for developing European standards in the field of electrical engineering. These standards are published with the objective of ensuring the uniform usability of technical devices across all member states, facilitating trade within Europe, strengthening the continent's global position, and enhancing market competitiveness. Each of CENELEC's National Committees is required to adopt the European standards into their respective national standardization systems within a specified timeframe.

For the overhead line sector, CENELEC has issued the European standard EN50341 "Overhead lines over AC 1 kV-Part 1: General requirements-Common specifications". This standard applies to all 34 member countries of CENELEC (status 2024), see Fig. 1.16. The standard consists of the so-called "main part—part 1", which applies unchanged to each member country and appears in English, French, and German. However, each country has the opportunity to formulate its national peculiarities (e.g., climate, legal situation) in "National Normative Appendices—Parts 2" (NNA) as supplementary conditions to this main part. These NNAs are normatively applicable in the respective country, for all others they are informative. Figure 1.16 shows the structure of EN50341 with the NNAs.

The NNAs appear in English and in the national languages and give the individual countries the opportunity to take into account decades of successful dimensioning practice of their lines. The NNA are divided into the following groups:

A-Deviations (A-DEV) are outside the competence of the national standardization body and cannot be changed by it (e.g., laws, regulations, ordinances)
Special National Conditions (SNC) are national characteristics or procedures that cannot be changed in the long term, e.g., climate data (wind, ice, ambient temperature), grounding conditions
National Complements (NCPT) are national procedures that are neither A-DEV nor SNC.

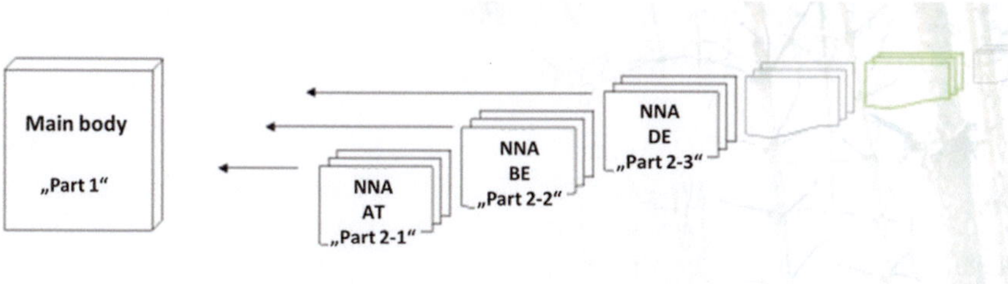

Fig. 1.16 Structure of the CENELEC European standard EN50341 "overhead AC lines over 1 kV" with "main body" part 1 and National Normative Annexes NNA

To facilitate the handling with "Part 1" and NNAs, several national standardization bodies create "consolidated versions" in which the main part of EN 50341 and the applicable NNAs are merged in one document.

The European standard EN 50341 provides a common framework for the design of overhead lines, serving as a base for both national and trans-European power transport. It also aims to eliminate trade barriers between countries. Fig. 1.17 shows the member and partner countries of CENELEC.

Fig. 1.17 Member countries of CENELEC (dark green) and partners (light green) [19]

1.3 Power Transmission via Alternative (AC) and Direct Current (DC) Lines

Power transmission via AC overhead lines enjoys all the advantages in a meshed high voltage grid that requires a nationwide, more or less homogeneous distribution of electrical generation- and load centers on land. The already mentioned shift of generation capacities to the windy coastal regions, up to offshore wind power in Europe, in the waters of the North and Baltic Sea, requires powerful line connections e.g. to the south and southwest of Germany. The very first solution, to implement these as AC overhead lines, failed due to the intense resistance of the population. To go underground with cables at the voltage level 400 kV was ruled out because, as explained in Sect. 1.2.2, they can no longer transmit active power over longer cable lengths without reactive power compensation. However, compensation is very complex and makes longer cable connections in high voltage AC grids prone to faults and uneconomical.

The situation is different with direct current cables. Since these cables have no reactive power requirement, they can be used in the highest voltage levels without compensation in any length. The limitation here is only in the electrical resistance of the conductors, which leads to losses that heat the cable and can dry out the soil around the cables in case of unfavorable dimensioning. However, this danger can be reduced by large cross sections of the conductors and thus with low electrical resistances.

Now the question arises why the forefathers of electricity supply did not opt for direct current, as Edison demanded. Direct current has two serious disadvantages compared to alternating current. On the one hand, the transformer principle does not work with direct current. Thus, an economic optimization of the grid expansion by an appropriate choice of transmission voltage and current was not possible. On the other hand, high direct currents can hardly be switched. If one wants to interrupt the current flow, the switch contacts draw an arc that is extremely difficult to extinguish. Despite much noticed developments and pilot installations in the field of direct current breakers for high currents, HVDC lines at present are still switched on the alternating current side of the converter stations.

But what has changed so much since the beginning of power transmission that what was impossible then has become possible and common today? Already at the end of the thirties of the last century, experiments were carried out on test routes with direct current transmission, which were fed from the three phase grid via mercury vapor rectifiers. In 1939, it was finally possible to use mercury vapor valves, developed by the Swiss company BBC. Referred to as "Mutators", they were able to operate as "line commutated" [20, Chap. 6]. It was possible now not only to generate high DC voltages from high AC voltages, but also to convert them back at the other end of the connection without rotating machines and thus integrate them into an existing AC grid. This allowed the advantages of DC technology for power transmission to be exploited amongst others by better utilization of the conductor's cross section.

Interrupted by the Second World War, this development was resumed in the former Soviet Union and in 1951 the first commercial DC transmission line between Kashira and Moscow (±100 kV, 30 MW, 100 km) was established. Three years later, the first DC sea cable connection between the island of Gotland and the Swedish mainland followed. The transmission took place via a single core cable with a ground return via electrodes in the seawater. Further facilities followed, and the systems with the fault prone mercury vapor valves were replaced in the 1970s by more reliable thyristor semiconductors. The first project to mention here is the "Nelson River Bipole" in Canada with almost 2000 MW over 900 km as overhead line and at a voltage of ±460 kV. After that, systems with thyristors were built worldwide on a large scale with ever higher performances and distances. Numerous such facilities are now located in Africa, Asia, and North and South America.

But the connection of grids of different frequencies or using different methods for their control was also achieved via DC back-to-back coupling. Today there are now several dozen such connections in operation worldwide [18].

Offshore wind turbines can only be connected to the mainland via sea cables. This causes problems over larger distances, which can only be economically solved with the help of DC technology. The final breakthrough in converter technology came in the nineties of the last century with bipolar transistors, so-called IGBTs. These are elements that, unlike thyristors, can be specifically switched on and off. This significantly reduces, for example, the reactive power problems and the effort required for filters. This technology is now dominant and is widely used in HVDC connections in the so-called converter stations [18, Chap. 6].

Ergo: Thanks to modern semiconductor technology, DC transmission is experiencing a renaissance and is particularly suitable for the transmission of bulk power via extra high voltage cables. These connections are almost exclusively point-to-point connections, which are predominantly switched on the AC side.

1.4 Tasks and Structures of Electrical Power Grids

In common use of language, the entirety of all power lines for the transmission and distribution of electrical energy, regardless of grid structure and voltage level, is often referred to as the "power grid". If you look at it a bit more differentiated, the cables and overhead lines connected to each other to form a whole system, which is operated with uniform voltage and frequency, established as independent power grids. In Germany and most European countries, a basic distinction is made between transmission and distribution grids (Fig. 1.21). While distribution grids serve to "distribute" the electrical energy at the voltage levels 0.4 kV up to 110 kV from the transformer substations embedded in the transmission grid to the cities, factories, craft businesses and households, the transmission grid primarily has the task of *transmitting* large amounts of energy between the feed ins from large power plants, offshore wind farms and the switching and transformer substations over larger distances. In Europe, they do this at voltage levels 150 (offshore sea cables), 220 and 380 kV.

Figure 1.18 shows the extra high voltage grid of Germany in the state of expansion at the beginning of the year 2024. It also includes the lines at the highest voltage levels that were under construction and planned at that time. Although all voltage levels above 1 kV are defined as "high voltage", and the voltage levels above 300 kV as "extra high voltage", four voltage categories have become established in common parlance among experts in the German speaking world.

- Low voltage up to 1 kV,
- Medium voltage, mainly 10 and 20 kV,
- High voltage 60 and 110 kV and
- Extra high voltage 150, 220 and 380 kV and above.

Some sources also contain differing definitions, which will not be further discussed here.

1.4.1 Interconnected Grids and Grid Associations

The extra high voltage level forms the superordinate grid level in Germany with 220 and 380 kV, which is interconnected with each other, with large power plants and via so-called coupling lines with the extra high voltage grids of the neighboring countries and thus embedded in the Central European interconnected grid (Fig. 1.18). This grid extends from the northeastern region of Poland to the southwestern tip of Portugal and is referred to as the UCTE grid (Fig. 1.19). In order to be able to support all connected participants effectively and quickly, the UCTE grid is galvanically connected at this level. This means that the frequency and the regime of the frequency in all parts of the grid are always the same. Incidentally, this grid is due to its size and performance the one with the highest frequency stability, i.e., with the smallest deviations from the 50 Hz target value, compared to the numerous worldwide existing grid associations.

The historical term UCTE stands for "Union for the Coordination of Transmission of Electricity". Meanwhile, the "European Network of Transmission System Operators for Electricity" (ENTSO-E) takes over the tasks for the coordination of the European electricity system (Fig. 1.20). Currently, this association supervises a total of 39 European Transmission System Operators (TSOs) from 35 countries. It is responsible and accountable for the safe operation of the European electricity system. The members commit themselves to maintaining supply security through establishing the technical and organizational rules for unrestricted interoperability of the subsystems regarding energy exchange and the ability to assist each other in case of disturbances. In addition to the technical cooperation of the TSOs within the association, it bundles their expertise and develops standards for expansion planning and cross border grid operation. Herein the UCTE grid area is integrated as a "Regional Group Continental Europe" (RG CE) [23].

Fig. 1.18 German extra high voltage grid, overview map, 1.1.2024 (*Source* Forum Network Technology/Network Operation at VDE (VDE FNN), Association of Electrical Engineering Electronics Information Technology e. V., [21])

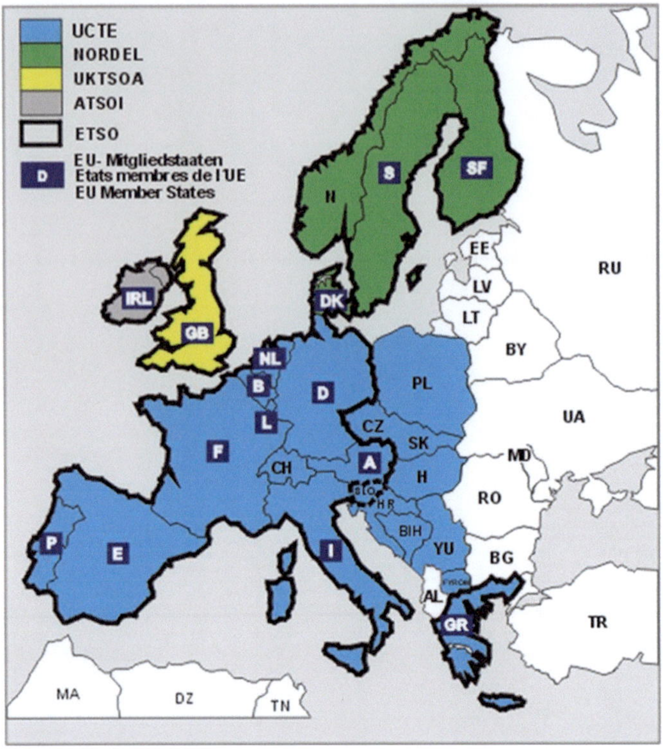

Fig. 1.19 Grid associations in Europe, UCTE member states shown in blue, status 2001 [22]

In older sources, the UC**P**TE is still mentioned. In the predecessor grid, before the liberalization of the European energy supply in 1996, the letter "P" (Production) also included the Central European electricity producers, i.e., the power plant operators. After that, the latter united in other associations due to the mandatory unbundling between the grids and the generation plants.

Back to the example of the German grid: Energy is fed into the 110 kV grids via a large number of substations from the highest voltage level. Large metropolises like Berlin, Hamburg or Munich are due to their high energy demand directly connected to the extra high voltage grid. The underlying 110 kV level takes over the supply of larger cities, rural regions, larger and medium manufacturing plants. At the same time, an independent and from the public supply independent 110 kV overhead line grid supplies the German Railways with a frequency of 16.7 Hz, which in special substations is transformed down to the voltage of 15 kV for the catenary wire of the railway grid.

Medium voltage grids with different operating modes are connected to the public 110 kV grid via countless substations. Most medium voltage grids are operated with 20 kV, partly still with 10 kV, and supply smaller factories, larger workshops and the transformer stations in the localities of the cities and municipalities. The "last mile"

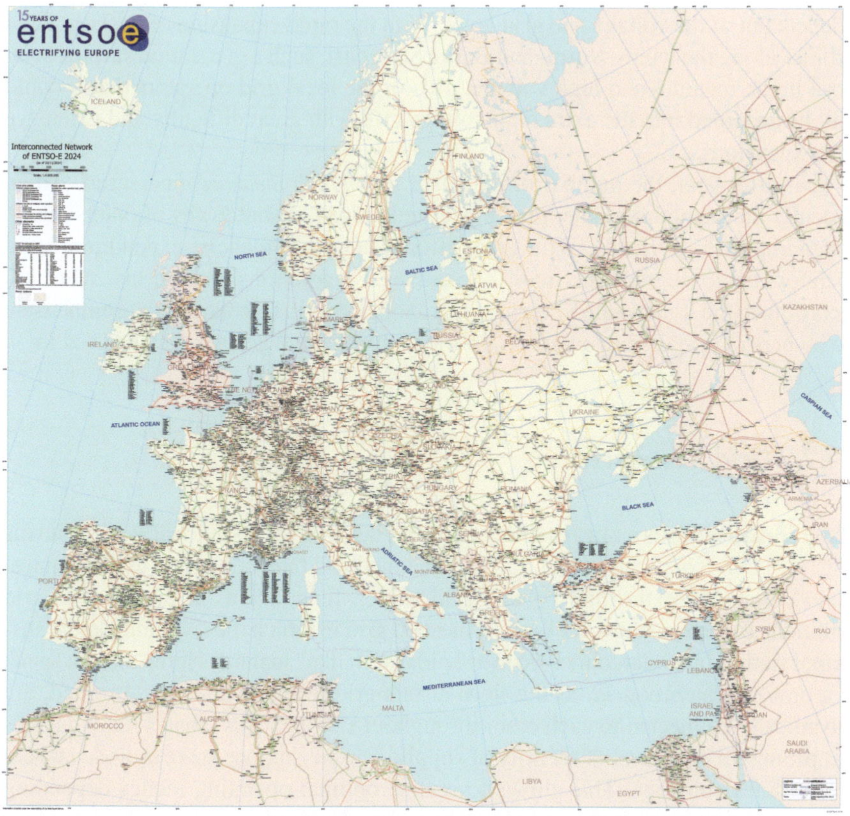

Fig. 1.20 "Interconnected grid of ENTSO-E", the yellow area is the transmission grid of ENTSO-E. 220 kV lines are green, 300–330 kV yellow, 380 kV red, 500 kV purple, 750 kV blue [24]

also originates in the transformer stations, i.e., the supply grid to the final consumers, which branches out street by street and supplies households and small businesses at the lowest voltage level of public electricity supply with 0.4 kV.

Until the 90s of the last century the energy flows always went from grids of the higher to grids with the lower voltage levels. Essentially, therefore, from the large power plants distributed over the country, feeding into the extra high voltage grid to the consumers in the underlying grid levels. Thanks to the now mostly renewable producers feeding in at all voltage levels, the energy flow directions have partly reversed. The generation surpluses in the underlying grids sometimes spill over into the overlying ones. Through the transformer principle, which equally allows step down as well as step up transformation, the energy can flow between the voltage levels vice versa according to the circumstances. Thus, the first large wind farms in the region of Eifel (Germany) already in the early 2000s generated more electrical energy within the medium voltage grid than could be taken from there, and fed back into the overlying 110 kV grid on a large scale. From now on, the energy flows were independent of the voltage and grid level. With the rapid expansion of onshore wind turbines along the windy North- and Baltic Sea coast, such a generation surplus soon arose in the medium and high voltage grids of the local grid operators that it could only be absorbed into the extra high voltage grid with great difficulty and extensive grid expansions.

Year after year, the number of so-called *redispatch measures* increased, which basically means nothing other than power reductions or shutdowns of plants due to impending grid overloads. With great efforts and high financial expenditure, grid operators try to upgrade the lines to avoid such restrictions on feed ins. After all, according to the Renewable Energy Act, grid operators in Germany must compensate the owners of curtailed or switched off plants for the outage as if they had fed in.

1.4.2 Grid Structures

The grid structures are always oriented to the tasks of the individual voltage levels (Fig. 1.21). In other words: The grids are planned and built according to the requirements. Thus, the extra high voltage grid, which functions as a transmission grid within the European network, forms a meshed grid in which the switching and transformer stations are interconnected in a "spider web like" manner. Here, one can speak of a uniform *overlay grid*, as the lines at this grid level, as already described, are galvanically connected to each other throughout Europe. This creates redundancies that promise a high level of security of supply. The substations serve as nodes where several lines converge. This allows to realize optimal operating modes depending on the load conditions in the individual branches. Furthermore, the grids of the four transmission system operators in Germany form independent *control zones*, in which the respective operators are responsible for frequency and voltage control.

1.4.3 Switching Stations and Transformer Stations

Substations at the highest voltage level, in addition to their function as switching stations within a grid level, also take the task of voltage transformation to the underlying grid level respectively grid levels. The grids of the high voltage level (usually 110 kV), the highest distribution grid level, are generally not interconnected. There are several reasons. The most important is the different operating mode of the grids from grid operator to grid operator, which cannot be arbitrarily interconnected (Fig. 1.22).

Regardless of whether it is a switching or transformer station, the arrangement of the components within such stations follows a basic pattern. The heart of a station is always at least one so-called (three phase) *busbar* in overhead line design, or as a tube busbar mounted on post insulators, as shown in Fig. 1.23. Larger stations have two or more busbars. These serve to connect the outgoing so-called *bays (or fields)*,

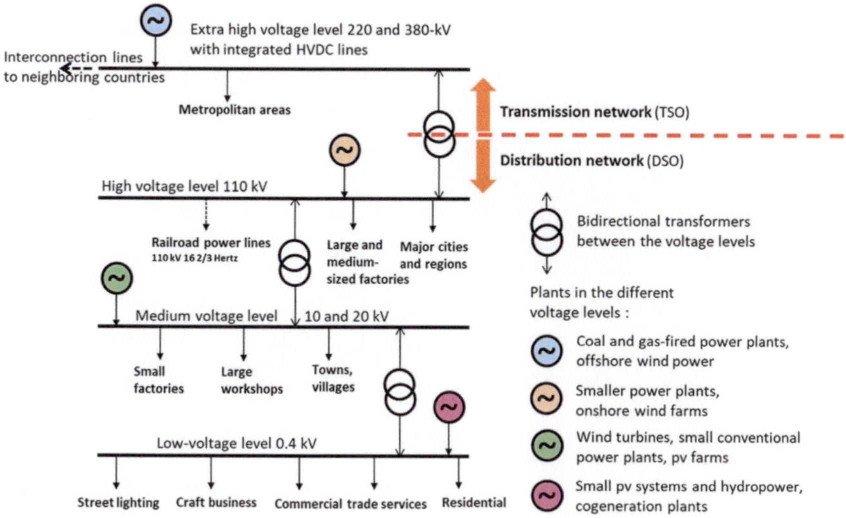

Fig. 1.21 Voltage levels in Germany

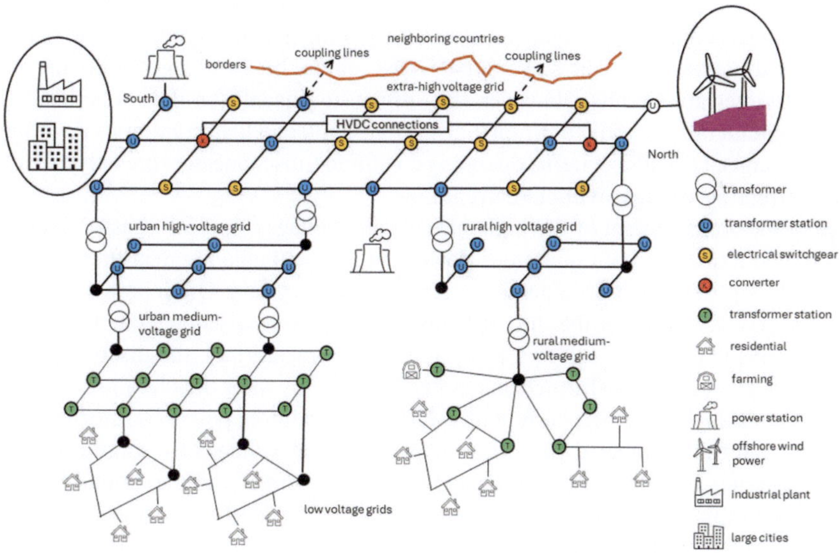

Fig. 1.22 Grid structures in Germany

with each other, depending on the switching state. The fields meet the busbar via several switching and control elements. Each bay begins, starting from the busbar, with a busbar disconnector (short: disconnector or isolator), followed by a circuit breaker and in the case of a line bay, followed by a line disconnector with integrated

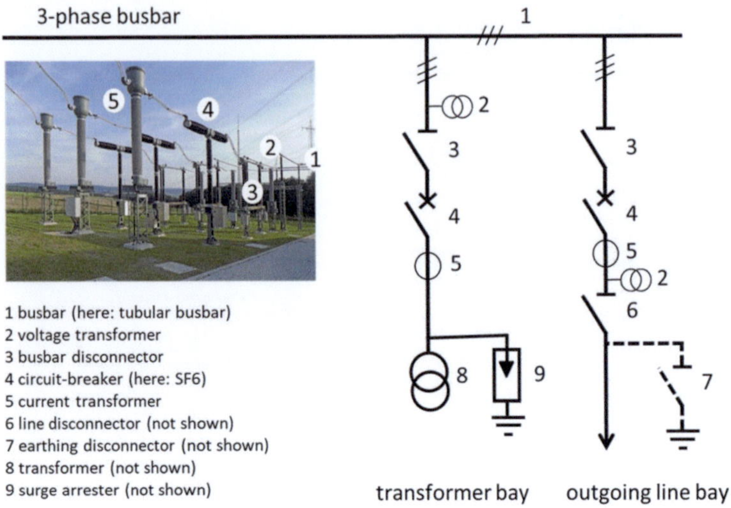

Fig. 1.23 Exemplary design of a transformer and line in switching and transformer stations (*Image* VDE Kassel (WD), [25])

grounding disconnector. The latter allows the outgoing line to be earthed in the case of line construction work. The switching devices "circuit breaker" and "disconnector" serve to interrupt circuits. However, they differ significantly in their functionality. While circuit breakers can interrupt high currents under all load conditions including short circuits, i.e. under extreme operating conditions, disconnectors are only capable of interrupting voltages without current flow.

For the measurement of the high currents and voltages, special transformers are used who transform both quantities down to a level suitable for measurement and control purposes. Their arrangement within the bay can vary. Often both are placed in the line bay between the circuit breaker and the line disconnector. In bays of substations that do not lead to a line bay but to a transformer, the disconnector is omitted. There, the transformers are connected directly after the circuit breaker, or as in the example, after the current transformer. In addition, power transformers are always carefully protected from destructive lightning voltages by surge arresters. Such arresters have the task to discharge overvoltage directly to the ground. This is done by an appropriate dimensioned spark gap which ignites before surges can cause damage to the insulation of the transformers.

1.5 Introduction to Network Planning

1.5.1 Planning Principles

The planning of infrastructure facilities regularly faces a dilemma. If they are built so that the probability of failure tends towards zero, they are unaffordable. On the other hand, if one tolerates their failure through economic expansion, the costs are lower, but the probability of failure and the associated damage are correspondingly higher. This abstract consideration applies to road infrastructures as well as to IT infrastructures or electrical grids. For the latter there is a desire concerning supply security at the highest level. However, increased expansion costs also entail increased costs for usage and maintenance. On the other hand, with low expansion costs, the costs that are caused by their failure increase. So it is, as in many other areas, an optimization task. Let's first look at the two extremes. If a failure is followed by a large damage impact, its probability of occurrence should be very low. On the other hand, if the damage impact is low, the probability of occurrence can be higher.

If we now draw a curve between the two extremes, as shown in Fig. 1.24, an area below this curve emerges in which the risk of failure is accepted. All measures that meet these conditions would be tolerable here. Measures, on the other hand, that would have a higher probability of failure and/or a greater impact of failure would be unacceptable. The "risk" of an event is the product of the size of the effect and the probability of occurrence. If we apply this principle to the reliability of power supply in electric grids, the two extremes could be described as follows.

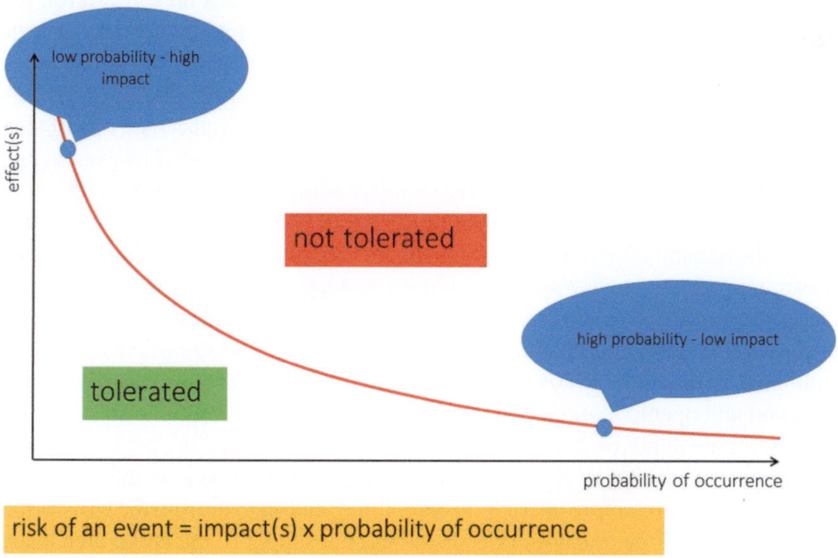

Fig. 1.24 Risk assessment in the expansion of infrastructure facilities

In the case of supplying a large metropolis with hospitals and other critical infrastructures, the case of a failure would have a high impact of damage. Consequently, grids in these areas are designed in such a way that the probability of failure is very low. If, by contrast, a single agricultural outlying farm would be concerned, the potential damage effects would be rather low. Here, therefore, a higher probability of failure is accepted, which is associated with a longer restoration time and a delayed re-entry of the supply. Chapter 6 provides a brief introduction to the methods used for calculating the *reliability* of power supply grids.

1.5.2 The (n − 1) Criterion

For the expansion planning of the grids, especially at the high and extra high voltage level, the so-called (n − 1) criterion applies in accordance with the previously explained planning logic. It is defined as follows:

> A power supply grid meets the requirements of the (n − 1) criterion if it can withstand the failure of one operating resource without unacceptable restriction of its function for any technically possible and/or operationally reasonable initial situation [26].

This planning principle therefore states that in a power grid, with actual or predicted maximum transmission and supply tasks, grid security is also guaranteed if one component, such as a transformer or a line, fails or is shut down according to plan. This means that there must not be any unacceptable supply interruption or an expansion of a disturbance, for example through overloading of remaining components. It does not matter whether the (n − 1) security is maintained within a grid level. It is possible and quite common to achieve the same level of safety by, for example, the underlying level ensuring supply security even in the event of a failure in one power feed in. In any case, the voltage must remain within the permissible limits and the remaining components must not be overloaded. This generally accepted rule of technology applies in principle to all grid levels. In distribution grids, however, supply interruptions are tolerated to a certain extent depending on the customer structure, if they can be remedied within a defined time period. On the other hand, in sensitive areas of the transmission grid, even a standard beyond the (n − 1) criterion is applied, for example, if particularly sensitive customers such as chemical or steel industry plants are supplied or when a failure would result in a large-scale disturbance or a dangerous situation [27]. This essentially preventive and deterministic rule was first formulated in the early 1980s and has since served as a planning principle in the expansion and operation of electricity grids [26].

A simple example of a medium voltage grid area planned according to the (n − 1) criterion is shown in Fig. 1.25. The network section shown on the left has no redundancy and is therefore not (n − 1)-safe. If the feed-in to or from the transformer is interrupted or if the transformer is disturbed, the supply of the entire area below can no longer be guaranteed. In the representation in the middle, there are two independent feed in points. Despite the failure of one of the grid connections to the

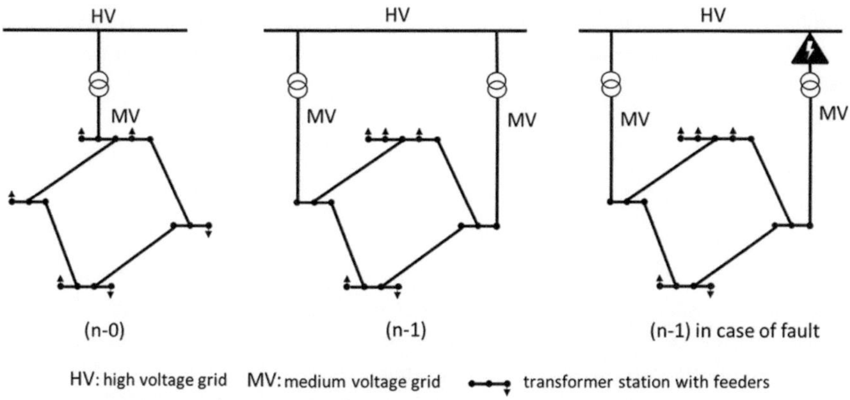

Fig. 1.25 The (n − 1) criterion in medium voltage grids

transformer (Fig. 1.25, right representation) or the breakdown of a transformer, the full supply is still guaranteed via the remaining supply path.

However, the (n − 1) criterion is not undisputed. The question often arises as to whether an overhead line with two circuits, which is standard in many countries, represents one or two components. In consideration of the (n − 1) criterion, each circuit counts as *one* component. However, if a tower is damaged, both circuits could be affected. Even when working on the towers, it could be necessary to switch off both circuits. The same applies to cable systems that run parallel to each other at a small distance. Here too, there is a risk that, for example, during construction work with an excavator, both systems could be damaged at the same time, or that both systems would have to be switched off due to work in the vicinity of the cables. Nowadays, in grid planning for the supply of facilities with high failure costs due to disturbances, probabilistic reliability calculations are increasingly being used, which e.g. in Germany can be based on the now extensive database of the FNN disturbance and availability statistics [28]. With these methods, reliability analyses related to the respective application can be carried out in a more detailed and precise manner (Chap. 6).

On the other hand, due to the rapid increase in feed ins from renewable energy sources at all voltage levels and the immense accumulation of redispatch measures, so-called *curative planning,* are being discussed, in which a temporary overload of operating resources is tolerated. In curative bottleneck management, the fact is used that most components are operated below their permissible load limit, essentially below their permissible operating temperature, during normal operation. For overhead lines, the weather conditions relevant for dimensioning, such as high summer weather conditions, occur very rarely. Under differing conditions, there is then a higher transmission capacity, which is used in the context of weather dependent overhead line operation. Therefore, in the event of a fault, they can be loaded beyond their permanent load limits for a certain (short) time until their maximum operating temperature is reached. The loads of the components after the occurrence of the (n − 1) case must be reliable and as quickly as possible reduced to their permanent limits [29].

1.5.3 Types of Planning

An electrical power supply grid is not a static structure. It is comparable to a road grid that is subject to changing and increasing demands over the years. In order to carry out the necessary high investments sensibly and in the right chronological order, particularly careful planning is required.

The past has shown that the demand for electrical energy has steadily increased in many countries in parallel with the growing economy, in Germany in the past up to seven percent annually. After consumption stagnated or only grew slightly in recent years, the energy transition is now leading to completely new requirements for both the transmission and distribution grids. In a broader sense, this also includes the topics of sector coupling and electricity storage (Chap. 7), which are expected to lead to a significant increase in demand for electrical energy.

Normally, grid planning is not about new planning, so-called *greenfield planning*, but about expansion measures, often combined with the replacement of components at the end of their life cycle. The basis for grid planning is therefore a load forecast or the planned connection of larger consumers and producer units. The exchange of electricity with neighboring countries also plays an increasing role. Some of the tasks associated with this extend into the more distant future, while others relate to rather short-term requirements. The planner must be aware that the planning process is always iterative and must adapt to changing requirements! The large area of these tasks is usually divided into the more manageable sub-areas of basic planning, expansion planning and operational planning, with the latter fluently moving into grid management.

The basic planning, which covers a period of 10–20 years, can be described with the following keywords: prediction of the development of power and energy, structure of the grid with regard to new backbone connections and connection of neighboring grids or large offshore wind parks, higher voltage levels and new technologies, such as DC technology and possibilities for storing large amounts of electricity.

The expansion planning, which covers a period of 5–10 years, can be described with the following keywords: decision for overhead lines or underground cables or partial undergrounding in alternating current or direct current. This includes the determination of routes, of cross sections, of mechanical and thermal stresses during normal operation and in case of disturbances, consideration of the above mentioned (n − 1) criterion, development of concepts for grid protection and questions of influencing neighboring facilities, neutral point treatment, grid stability, harmonics and possibilities of reactive power compensation. This requires extensive calculations using load flow and short circuit programs, in which the planner must find a solution among many possible variants that represents an optimum both in economic terms and from the points of view of environmental compatibility, future viability and supply reliability.

The operational planning refers to short periods of hours to days. As will be seen in the next section, electrical energy must be generated at the very moment when the corresponding demand exists. The provision of the necessary energy is carried out by market processes at the party responsible for the balancing. In short term operational planning, the feed in forecasts from renewable and conventional power plants, and load forecasts are analyzed to check grid security and, if necessary, adjust transmission capacities or initiate necessary countermeasures to avoid bottlenecks in the grid in good time. If grid bottlenecks are detected in advance, the grid operators early plan appropriate countermeasures (redispatch including the start-up of reserve power plants, restriction of grid capacities etc.). Operational planning also includes the switching planning for the implementation of regular maintenance measures or grid expansion measures, which must also not lead to temporary bottlenecks in the supply.

As part of the grid expansion at all grid levels, both the transmission system operators and the distribution system operators in general publish their planning principles, which are guided by the regulatory framework, on online portals. The joint grid planning principles of the transmission system operators e.g. for Germany are published under [30] and the distribution system operators as VDE application rule VDE-AR-N 4122 under [31].

1.6 Introduction to Grid Operation

In three phase systems, the grid frequency plays a central role for the stability of supply. As important keeping the stability of the voltage and provision of power (both active and reactive) are, without a stable frequency, everything would be in vain. Nevertheless, power balance and frequency within a contiguous grid area are closely linked. This will be examined in more detail below.

1.6.1 Power Balance

A big number of generators feed into the entire grid across all voltage levels (Fig. 1.26). Whether they are large power plants that feed into the highest voltage level, or photovoltaic and onshore wind power plants that feed into the medium and even low voltage grids, they collectively form a unified generation block in the broader sense.

A significantly larger number of consumers take electrical energy from the different grid levels. These consumers are highly diverse, ranging from large industrial facilities and metropolitan areas to commercial enterprises and individual households. Across this spectrum, electrical energy is taken from the grid and converted

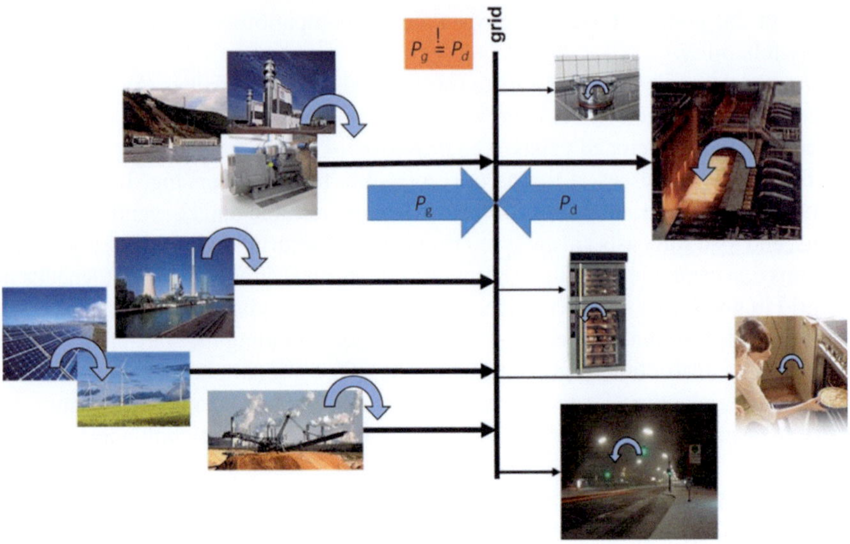

Fig. 1.26 Power balance between generation P_g (*left*) and demand P_d (*right*)

into other forms of energy—whether in steel production, heating systems, lighting, information technology applications, or numerous other uses.

The majority of electricity generation is currently provided by rotating generators with a stable speed and in some cases considerable flywheels. As will be seen later, these generators contribute significantly to grid stability. A stable supply is only guaranteed when the total power generated at any moment matches exactly the power demand of all connected consumers.

In the energy sector, the terms "energy consumption" and "energy consumers" are very often used. This terminology likely has been introduced by experts from other disciplines. It certainly does not come from physicists or engineers, because energy is always converted and never generated or consumed, as postulated in the *law of energy conservation* over 150 years ago. Nonetheless, the misleading terms have become so entrenched in language use that we will continue to use them in the following as well.

1.6.2 Importance of Frequency

The frequency is determined by the speed of the generators. With a few deviations, it is 50 Hz in large parts of the world (e.g. Europe); in other parts (e.g. the Americas, parts of Japan) it is 60 Hz. Let's stick to the target value of 50 Hz for our consideration. If the generators rotate faster, the frequency increases. If they rotate slower, it decreases. This is basically the whole secret of grid stability. Now one might wonder, how is

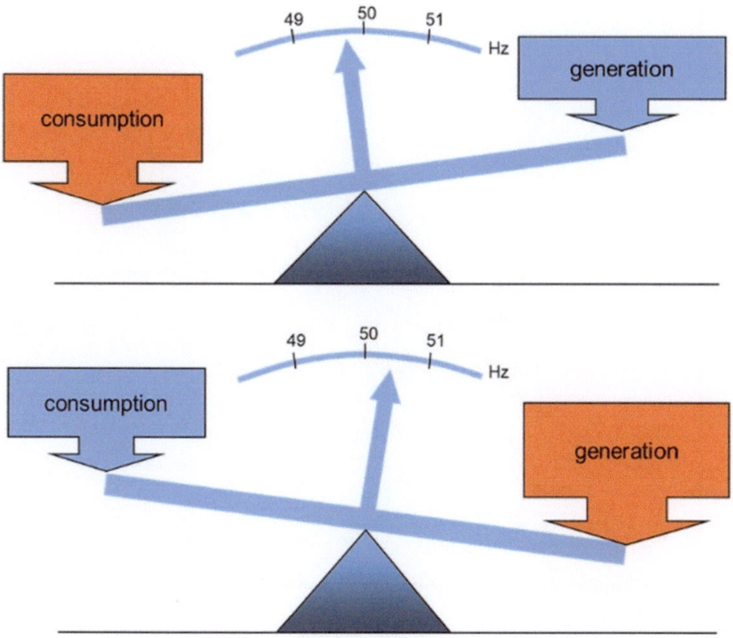

Fig. 1.27 Balance between generation and consumption

it possible to keep the frequency constant? The frequency is rarely exactly 50 Hz. Over the course of the day, however, it reaches this value quite precisely on average. The smaller the deviations from the target frequency, the higher the supply quality. But how is it possible to achieve this balance as precisely as possible? The balance between generation and consumption is that they must be equal. This can be very clearly illustrated using a scale, as shown in Fig. 1.27.

On the consumer side, energy demand fluctuates continuously—often abruptly. To maintain the balance condition, power feed-in must be constantly adjusted. When power demand increases, for example due to the connection of large consumers, it exerts a braking effect on the generator flywheels, causing the grid frequency to drop. In response, the feed-in power must be increased. Conversely, if a large consumer is disconnected, the generator flywheels accelerate, leading to a rise in frequency. In such cases, the feed-in power must be reduced accordingly. The fundamental principles of this power regulation process are illustrated in Fig. 1.28. The greater the frequency deviation from the nominal value of 50 Hz (respective 60 Hz), the greater the necessary power adjustment. In the event of an overshooting frequency, feed-in power must be reduced; if the frequency falls below the target, the feed-in power must be increased.

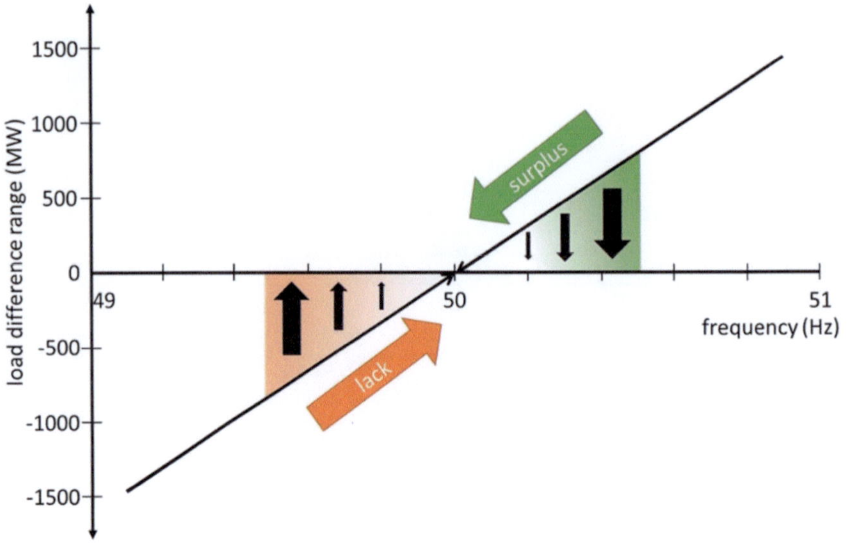

Fig. 1.28 Power–frequency characteristic in the German high voltage grid

1.6.3 Balancing Power

The power that must be constantly available for the above task is referred to as *balancing or control power*. For example, in Germany, it is procured by the four national transmission system operators responsible for frequency stability through tendering procedures. This means that operators of flexible power plants, such as storage and gas power plants, participate in the TSO's tenders, and the cheapest providers are awarded the contract. Participation in the tender requires prequalification and is linked to several technical requirements. The most important is a high as possible rate of power change P_a. Thus, the primary control power of one block of a power plant must be at least 2% of its rated power P_r, which, as formulated below, must be available within a maximum of 30 s. This means that P_a must be

$$P_a = 0.02 \cdot P_r / 30 \text{ s} \tag{1.13}$$

wherein is:

P_a Rate of power change (W/s or MW/s)
P_r Rated power (W or MW)

In the event of a power imbalance, the first line of defense against changes in system frequency is the rotational inertia- or flywheel effect of the generators. This first stabilizing effect must be immediately supplemented or replaced by primary *control*. This means that at any time, sufficient control power must be available to compensate for power deficits or surpluses within a control zone. This is also referred to as *control reserve (spinning reserve)*. To determine its level, the TSOs create a grid security planning calculation. The control reserve must be so dimensioned that the TSOs can help each other in emergency cases and within the UCTE grid. Currently, the so-called primary control power for the entire UCTE grid is 3000 MW. Hereby, the four German TSOs contribute 700 MW [7].

In Fig. 1.29(a) the different types of control power are shown in the chronological order of their use. The stabilizing effect from the *inertia* of the generator flywheels must be supplemented or replaced within 30 s by the previously described *primary control*. After 30 s the *secondary control* starts, which replaces the primary control after a maximum of 5 min. Power plants with power dynamics in the minute range are involved in providing the secondary control power. Here, mainly pump storage, gas and steam (IGCC) and fast controllable coal fired power plants are used [32].

The subsequent tertiary control, also known as minute reserve, is primarily provided by conventional power plants. Meanwhile, larger wind turbines or wind

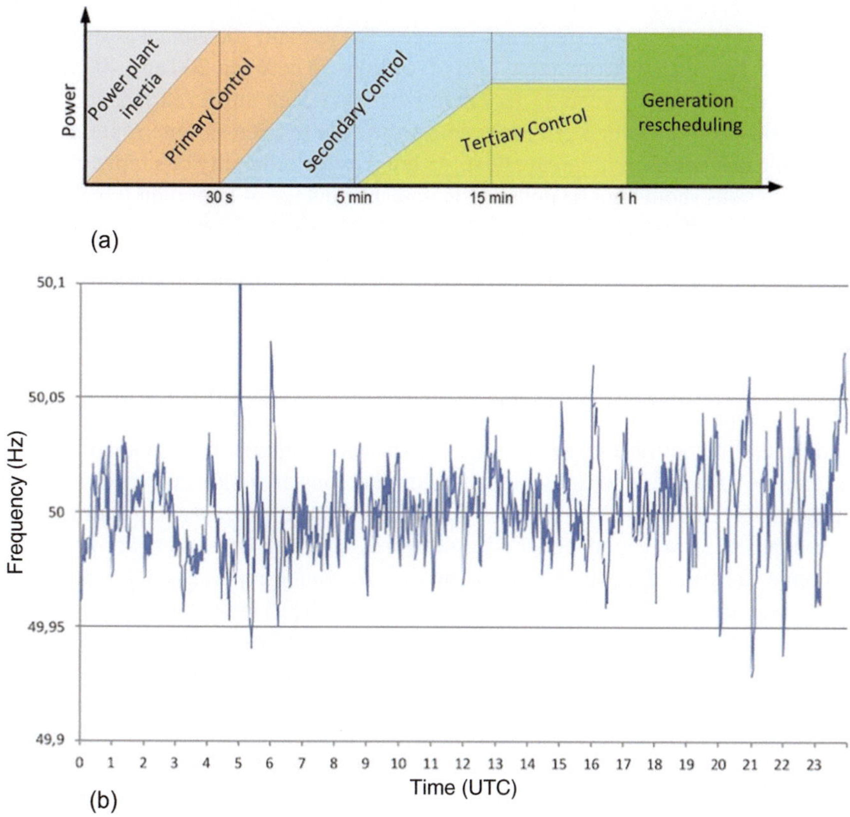

Fig. 1.29 (**a**) Various types of control power for power balancing in power grids, (**b**) actual frequency course in the UCTE grid [32]

parks are also capable of delivering this service. Tertiary control must compensate for power imbalances after a maximum of 15 min for duration of up to 1 h.

From that point onward, the party responsible for the balancing group in which the deviation occurred must restore and maintain the balance. Figure 1.29b illustrates how closely the frequency in the UCTE grid adheres to the 50 Hz target under this control regime. Frequency deviations in other European associations—and in most grids worldwide—are sometimes significantly larger.

The predictive determination of the control reserve is becoming increasingly difficult due to more and more fluctuating feed in from large wind and photovoltaic plants. This increases the volatility of the feed in and thus makes the forecasts of the supply more difficult.

Despite large fluctuations in the renewable energies to be fed into the grid, the grid operators have been very successful in recent years in maintaining the supply largely uninterrupted, except for disasters. However, it should not be concealed that with the increasing fluctuating power in recent years also an alarming increase in *redispatch* and *curtail measures* became necessary. These measures had to be carried out against grid overloads and sudden frequency changes. In Germany the annual costs of such operations have grown from a few tens of millions to between 200 and 400 million Euros over the past 15 years [33]. There have also been an increasing number of critical situations in which grid operators were able to avert an interruption in the grid at the last minute with great effort. As good as the reliability of supply has been so far, the risks should not be underestimated. The outage of important transmission lines due to a failure or bulk power plants can lead to a supply emergency situation, which in the worst case can result in a nationwide blackout.

1.6.4 Load Shedding at Underfrequency

If, despite the control reserves lasting, a disturbance causes the frequency to drop too far, the grid operators must intervene on the consumer side and carry out load adjustments. This is done via the so-called *underfrequency-dependent load shedding* (German: UFLA = Unterfrequenz-Lastabwurf), in which consumers are gradually disconnected from the grid. With this procedure, which replaces the former *automatic frequency relief* [34], the German FNN in the VDE implemented the requirements from the European "Grid Code on Emergency and Restoration (EU Regulation 2017/2196)". This sets binding requirements for automatic last measures throughout Europe. The VDE application rule VDE-AR-N 4142 implements the requirements for Germany [35]. The measure starts at 49 Hz, contains a 10-stage cascade and is decentralized, independent, and self-sufficient (Fig. 1.30). It involves all operators of public grids and users connected to it. The respective grid operators are responsible for determining the *points of shedding* that are to be disconnected from the grid and the assignment of the frequency stages. To relieve the transmission grids, loads are shed at the medium voltage level in a first step. The selection of consumers to be disconnected must be made carefully and especially without discrimination.

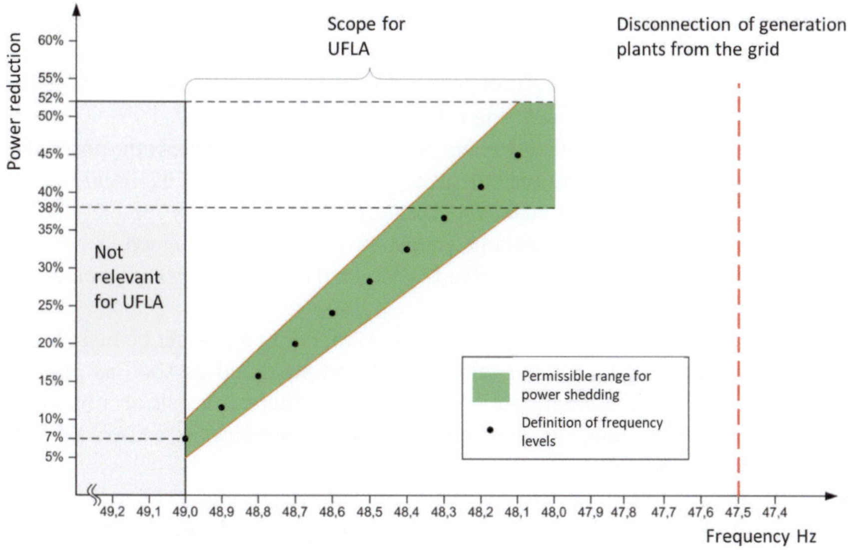

Fig. 1.30 Consumer load to be shed according to UFLA [36]

1.6.5 States of the Grid

For ongoing grid operation, it is not enough to consider the transmission facilities alone; the connected generation plants and consumers must also be taken into account.

The grid states are defined at ENTSO-E and e. g. for Germany in VDE-AR-N 4141-1 [37] as follows (Fig. 1.31):

Fig. 1.31 Grid states [37]

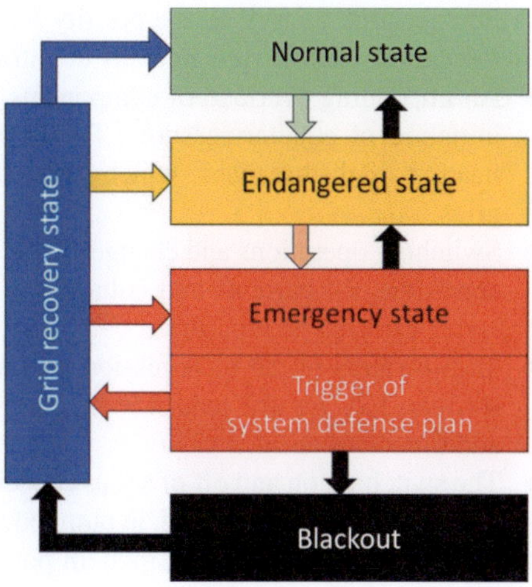

- In *normal state*, all parameters (grid frequency, voltages, currents, temperatures, pressures, levels, etc.) are within the limits set for them. The (n − 1) criterion is adhered to. The failure of any component does not lead to a violation of the above limits, and short circuits occurring are under control.
- In the *endangered state*, all customers are still supplied, but there is the possibility that the grid will now enter the emergency state in the event of the failure of one or more components. The endangered state is characterized by the fact that the above-mentioned operating parameters can no longer be maintained. A return to normal operation can only be achieved by targeted load shedding and/or extensive switching measures.
- In the *emergency state*, the supply of all consumers can no longer be maintained. Now the aim is to first restore the endangered state and as soon as possible afterwards also the normal operation. Otherwise, there is a risk of a blackout. After isolating the fault location and possibly eliminating its causes, the grid reconstruction begins.
- The fourth and last of the possible states, the *blackout*, is characterized by a total failure of the supply. Here, the power plants are just producing the power they need for internal use. In this case, the grid operation is gradually rebuilt until the supply is fully restored.

In the *normal state* of the grid, the following tasks have to be accomplished:

- In the transmission grid, system services have to be provided, which include constant information exchange with neighboring grids, maintenance of voltage and frequency, and the provision of reserve power (Sect. 1.6.3).
- Energy must be provided in a manner that ensures the economically optimal dispatch of power plants based on their generation costs.
- In the classical sense, this is achieved through the division into base load, medium load and peak load. In parallel, the very numerous transit obligations from bilateral supply contracts that have arisen in the course of liberalization, must be taken into account, as well as the priority feed in of renewable energies according to *Renewable Energy Acts* in many countries.
- An impending overload of components must be effectively prevented by timely initiation of countermeasures. The same applies to the localization and rectification of disturbances, which often require the temporary formation of sub grids.
- Switching operations and changes in the transformer ratios not only maintain the conditions for currents and voltages of individual operating resources, but grid losses can also be minimized.
- Transformers in the transmission grid are often equipped with so-called phase shifters. Hereby, the load flow in the grid can be influenced with the help of additional voltages.
- The switching on and off of series capacitors, shunt reactors and the phase shifter operation of generators (e.g. in pumped storage plants) also supports a stable grid operation. Apart from that, modern power electronics allow the use of *Thyristor*

Switched Capacitors (TSC) or *Thyristor Controlled Reactors* (TCR) with the collective term *Static Var Compensation* (SVC).

- *Flexible Alternating Current Transmission Systems* (FACTS) offer further possibilities for controlling power flow and for grid stabilization with the help of modern power electronics components.

References

1. CENELEC, Euronorm EN50341 „Freileitungen über AC 1 kV“. „Teil 1: Allgemeine Anforderungen - Gemeinsame Festlegungen, Hauptteil“, CENELEC 2012
2. Doodle Blog, Tesla gegen Edison: AC/DC – das Original (doodle.com)
3. Anon., Gleichstromfernübertragung Miesbach–München Gleichstromfernübertragung Miesbach–München, Wikipedia
4. Stiftung Werner-von-Siemens-Ring. Oskar von Miller Ringträger 1927. Stiftung Werner-von-Siemens-Ring
5. Anon., Drehstromübertragung Lauffen–Frankfurt Drehstromübertragung Lauffen–Frankfurt, Wikipedia
6. Zeag, Von Lauffen nach Frankfurt: 1891 erste Drehstromübertragung, STIMME.de
7. Heuck K et al (2005) Elektrische Energieversorgung, Vieweg, 6. Auflage
8. Anon., 380-kV-Leitung Rommerskirchen–Bürstadt–Hoheneck, Wikipedia
9. Schossig W. Chronik der Elektrotechnik, Kabel und Leitungen (vde.com)
10. Anon., Drehstromleitung Ekibastus–Kökschetau (de-academic.com)
11. Anon., Hochspannungs-Gleichstrom-Übertragung, Wikipedia
12. Schichler U (2016) Betriebsmittel für das Supergrid der Zukunft, 14. Symposium Energieinnovation Graz, 10.–12.02.2016
13. David E, Palic M (1994) Elektrosmog - Wirkungen elektrotechnischer Felder auf den Menschen. etz Bd. 115(3)
14. Anon., Johann Heinrich Kruesi. https://wikispeicher.ch/w/Kr%C3%BCsi_John
15. Biewald H, Ritter G (1992) Ein Rückblick auf die Berliner Kabeltechnik aus Sicht der Energieversorger. Elektrizitätswirtschaft 91(11/12):742–755
16. Cichowski RR (2017) Kabelhandbuch, ew Medien und Kongresse, 9. Auflage
17. Schwab AJ (2011) Elektroenergiesysteme, Erzeugung, Transport und Verteilung, 3. Auflage. Springer
18. Anon., Liste der HGÜ-Anlagen, Wikipedia
19. Anon., Europäisches Komitee für elektrotechnische Normung, Übersichtskarte, Wikipedia, Abruf Dezember 2023
20. Moglestue A. Vom Quecksilberdampf zum Hybridleistungsschalter. ABB Review 2/13:70–78
21. VDE FNN, Karte "Deutsches Höchstspannungsnetz" 2024 (vde.com)
22. Anon., Centrel-Staaten jetzt Vollmitglieder der UCTE (energie-chronik.de)
23. Anon., Union for the Co-ordination of Transmission of Electricity, Wikipedia
24. ENTSO-E, „Interconnected Network of ENTSO-E“, Netzplan 2023, ENTSO-E 2023
25. Anon., VDE Kassel, Exkursion zum neuen 400-/110-/20-kV-Umspannwerk Sandershausen der Städtischen Werke Netz + Service GmbH (vde-kassel.de)
26. Haß D et al (1981) Das (n−1)-Kriterium in der Planung von Übertragungsnetzen. Elektrizitätswirtschaft 80(25):923–926
27. Netzentwicklungsplan, n-1-Kriterium, Bundesnetzagentur
28. Backes J (1998) Bewertung der Versorgungszuverlässigkeit. Herbert Utz Verlag, München
29. Kamps K et al (2022) Wirksamkeit und Zuverlässigkeit präventiver und kurativer Maßnahmen. ew 3

30. Anon., Gemeinsame Richtlinien zur Netzplanung im Übertragungsnetz. https://www.50hertz.com/Portals/1/Dokumente/Netz/UNB-PlGrS_Juli%202022.pdf
31. Anon., Planungsgrundsätze im Hochspannungsnetz Planungsgrundsätze für Hochspannungsnetze (VDE-AR-N 4121)
32. Anon., Regelleistung (Stromnetz), Wikipedia
33. Anon., Redispatch: Prinzip, Maßnahmen, Kosten und Redispatch 2.0 Redispatch-Maßnahmen & Kosten im Strommarkt (energie-experten.org) vom 01.06.2021
34. Anon., Lösungen im Kontext der automatischen Frequenzentlastung (AFE), IDS GmbH 6/2015
35. VDE, Automatische Letztmaßnahmen, VDE-AR-N 4142
36. Bohn T (2023) Die VDE-AR-N 4140 Von der Kaskade zu den automatischen Letztmaßnahme, Infotag Kaskaden-Leitfaden 5.0, FNN, 28.03.2023
37. VDE, Technische Regeln für den Betrieb und die Planung von elektrischen Netzen, VDE-AR-N 4141-1

Chapter 2
Design of Line Corridors

Summary

The task of routing overhead lines and cables originally focused on finding cost-effective connections between geographic points for transporting electrical energy. During the early phases of grid construction around the world, these lines—regardless of type—were generally met with public enthusiasm or at least acceptance. However, as more landscapes became increasingly influenced by technical infrastructure, all of which have become essential to modern industrial society, public perception began to shift. This growing awareness led to increased opposition, particularly toward overhead lines.

This shift in public perception made it increasingly difficult to reconcile the requirements of routing and planning high and extra-high voltage lines.

As the importance, size, number, and characteristics of the overhead lines increased, acceptance decreased, particularly in populated areas or those with special landscape characteristics.

Therefore, in the planning of lines, routing principles emerged that consider the peculiarities of settlement, nature, and landscape areas, as far as technically possible. In modern line planning, in addition to technology, the perception of people and considerations for fauna and flora are increasingly included in the planning culture. The way how affected communities perceive the lines in the landscape has become a key factor in the planning process.

In this chapter, in addition to presenting tower designs and cable laying types, the relevant routing principles are explained and illustrated with examples for both overhead lines and underground cables. The chapter also provides typical dimensions for overhead lines and cable routes, along with explanations of the associated conditions and constraints.

K. O. Papailiou et al., *High-Voltage Overhead Lines and Underground Cables*, Power Systems, https://doi.org/10.1007/978-3-032-07498-0_2

2.1 Overhead Line Corridors

Planning on the drawing board is now a thing of the past. Today, advanced methods for line routing such as aerial surveys using airplanes and drones support line routing by providing plans in view and longitudinal profiles and transverse profiles of the routing area. When this data is combined with information on exclusion areas (e.g., residential buildings, protected areas), for topography, for risk areas (landslides, flood areas), with the specifications of landscape planning and the legal situation and others, and the assessment of the visual impression, a line proposal results. After further refinement by planners, this proposal serves as the basis for the approval process.

Although it is not always possible to realize a route that perfectly aligns with the desired ideal and the wishes of all participants, the modern approach shows responsible planning and construction. In any case, in modern planning culture, attention is paid to minimizing landscape consumption and encroachments during construction and later operations. As a result, routes often include numerous deviations from a straight path, requiring more angle towers and leading to significantly higher construction costs.

The development of tower design is therefore subject to a constant optimization process that must take the following factors into account:

- The power to be transmitted (voltage level, number of circuits, cross section and number of conductors)
- The terrain (slope, vegetation, clearance to settlements, possible tower locations, geology, hydrology, infrastructure facilities, etc.)
- The climate (loads on the line due to wind, ice and snow load, dimensions of materials, choice of tower material: steel, concrete, foundations, etc.)
- The bundling of lines (lines of different voltage levels together on shared lines and abandonment of no longer needed lines)
- The design standards (clearances of live conductors to each other and to the ground, loading assumption, safety, operational and maintenance specifications as well as specifications for audible noise and the electric and magnetic field)
- Operational safety (stability, reliability, load capacity, maintenance, lifespan, etc.)
- Aesthetics

When following the route of an overhead line, one will notice that towers in the straight axis differ significantly from those at angle points, even if they have the same *tower geometry*. Towers in the axis carry the conductors with suspension insulators and are only loaded by wind forces on the conductors and insulators, possible ice loads and their self-weights including conductors. They are referred to as *suspension towers*. In contrast, at towers at angle points, the conductors are tensioned from both directions. They are accordingly referred to as *angle tension towers*. In addition to the forces from weight and wind, they take the tensile forces from the conductors of both sides. These towers, due to their much more complex construction, appear more visually prominent. They weigh about two to three times as much as a suspension tower of the same category. Figure 2.1 illustrates this distinction, showing a tension tower (in the foreground) and a suspension towers (in the background) in an axis.

The components of an overhead line are shown in Fig. 2.2. Towers are predominantly built as steel lattice constructions, whose static dimensioning considers all conceivable load cases. In special designs, concrete and steel tubular poles are increasingly used.

The live conductors in conventional construction consist of aluminum/steel types as a single conductor (mainly up to 110 kV), as twin bundles (in the 220 kV level) and as triple or quad bundle (in the 380 kV level and above, Germany). This depends on the voltage level and the required current carrying capacity. Bundle conductors receive evenly distributed so-called spacers in the span, which prevent the individual sub-conductors from clashing due to wind or the effects in the event of a short circuit. Insulator strings differ mainly in their design and the building material. In Germany, Austria and Switzerland, *long rod porcelain insulators* are predominantly used. In cases of special forces or requirements, *glass cap and pin-* or *composite insulators*, the latter with a glass fiber reinforced core are also used. The ground wires, which serve to protect the conductors from lightning, are also predominantly made of aluminum/steel types. They usually have a smaller total cross section with an increased steel section. They often contain optical fibers, which are used by the

Fig. 2.1 110 kV line section with an angle tension tower in the foreground and suspension towers in the background (*Source* LVN)

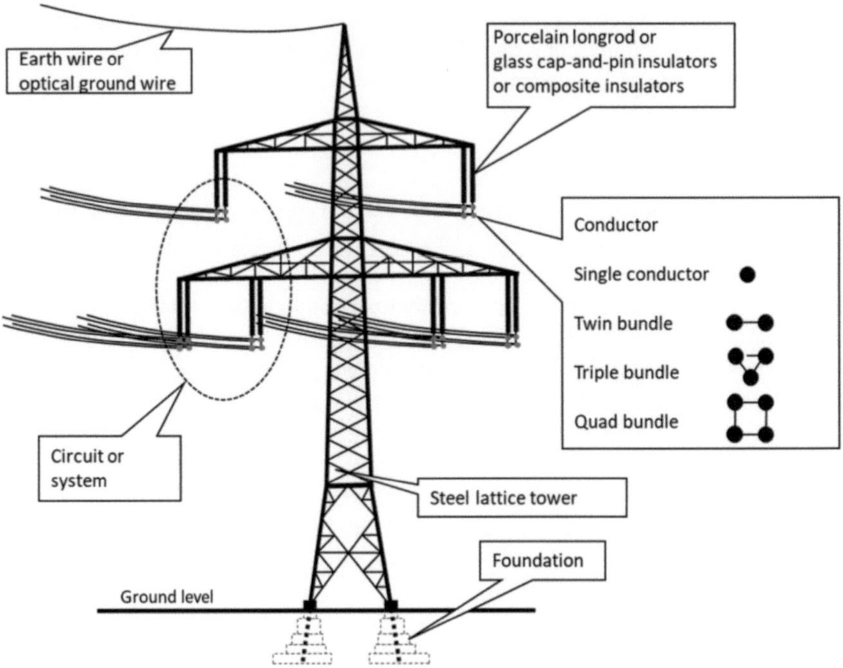

Fig. 2.2 Components and parts of a high voltage overhead line

grid operators for telecommunication and control purposes and are called *Optical Ground Wires* (OPGW).

2.1.1 Tower Design and Landscape

Overhead lines, like many other technical structures, form an integral part of the landscape and our living environment. They are perceived—whether consciously or unconsciously—as a component of modern infrastructure.

Their greatest possible integration into nature is the basis and goal of responsible overhead line construction. Unfortunately, there are no simple recipes for line construction. Each line project must be considered anew; what was suitable for one project does not necessarily fit for another [1]. As a result, a variety of tower designs have been developed across all voltage levels throughout the history of overhead line construction This was due to various types of terrain and constraints, which made certain tower designs more suitable—both visually and technically—than others. From the multitude of design variants that have been tested in the past with more or less success, three commonly used tower designs for lines with two circuits have developed, mainly in German speaking countries (Fig. 2.3).

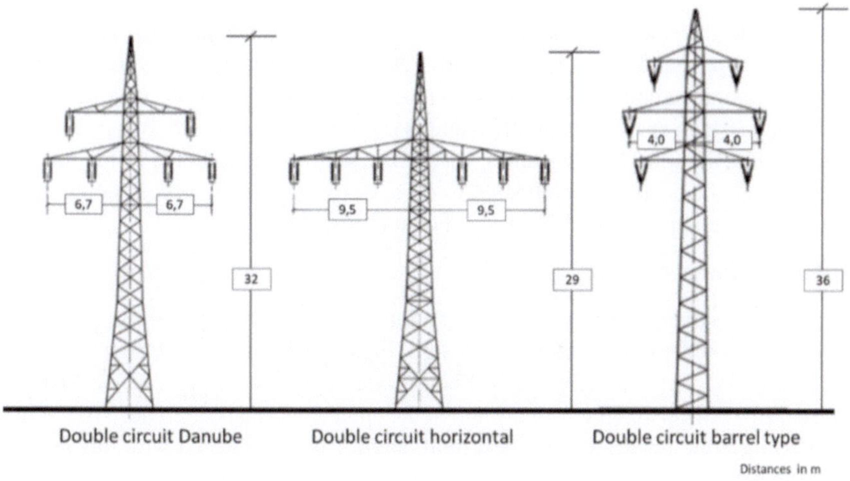

Fig. 2.3 Basic designs for double system overhead lines using the example of the 110 kV level

Other tower designs either show combinations of the basic designs as shown above or they are closely related to one of them. The choice depends in most cases on topographical conditions, as long as no higher-ranking restrictions have to be taken into account. So-called *alternative tower design* uses other ways, some of which deviate significantly from the traditional design. There are countless examples worldwide (Chaps. 4 and 7) (Fig. 2.4).

Fig. 2.4 Rarely seen: concrete lattice towers, historical picture

Danube Tower Design

The Danube tower design (Fig. 2.3 left) is particularly suitable for use in wide valleys and open landscapes with predominantly agricultural use, in which there are no restrictions in height or width. This design is often referred to as the standard tower design in German speaking countries and is considered structurally balanced. With this tower design, the lowest values for electric and magnetic fields in the vicinity of the line can be achieved compared to other designs under otherwise comparable conditions. An application example is shown in Fig. 2.5.

Horizontal Tower Design

If the line runs over a forest or in front of a forest or field grove backdrop, the choice usually falls on the horizontal tower design (Fig. 2.3 center). This design leads to a wider corridor. The advantage of this configuration is that no parts of the line rise above the backdrop into the background of the open view and is often decisive. When crossing over the forest, the smallest total height can be achieved with this design. An example in front of a field grove backdrop is shown in Fig. 2.6.

Barrel Tower Design

The barrel tower design has proven particularly effective for lines in narrow valleys or on mountain slopes (Fig. 2.3 right). Thanks to the short lengths of the lowest

Fig. 2.5 Example for a 220 kV suspension "Danube Tower design" with earth wires at the upper crossarm ends

Fig. 2.6 Horizontal design suspension towers in front of a field grove

crossarms, the height of the suspension point on steep slopes can be kept relatively low. Taking into account the deflection of the lower conductors due to wind, with this type the tower's height can be kept smaller compared to the two previously discussed designs.

Another motive for the use of this type of structure, which is gaining more and more importance, is the replacement of single lines with double circuit lines in the same route while still operating within the existing corridor width. An example of this tower design is illustrated in Fig. 2.7.

Compact Double Circuit Overhead Lines

The need to reduce the impact on nature and landscape even at the highest voltage level led to a number of concepts for compact double and multiple overhead lines in recent decades. The challenge of adapting a 380 kV line in height and span length to an existing 110 kV line, for example, led to the development of an extremely compact 380 kV structure in a horizontal design with conventional line construction components. The parallel routing of both lines resulted in the least possible landscape encroachment (Fig. 2.8). In several places buildings have now been built under both lines while maintaining safety clearances.

Another example of a compact structure with high transmission power at the highest voltage level is realized via support ropes, to which the conductor bundles are attached like garlands every approx. 20 m (Fig. 2.9). The bundle configuration of the conductors is thus supplemented by a tightly tensioned bundle of steel ropes, similar to those for cable cars. This significantly reduces the sag when using conventional conductors. At the same time, conductors with larger cross sections for high transmission capacity can be attached to the steel ropes. This leads to shorter towers and narrower corridors compared to other horizontal tower designs. The towers are designed as tubular steel poles and have significantly more massive foundations at the tension points in the line due to the increased tensile forces than conventional lattice

Fig. 2.7 Example of a 110 kV barrel tower

Fig. 2.8 Compact 380 kV overhead line with a 110 kV line run in parallel

towers. This line, erected as a pilot project by the German transmission grid operator *50 Hz* in partnership with the construction company *SPIE SAG* in the second half of the previous decade, pursues, among other things, the goal of replacing existing 220 kV lines with powerful 380 kV lines without changing the line corridor and the height of the line. It is currently still in the test phase.

Fig. 2.9 380 kV test line in horizontal arrangement and twin bundle support ropes (*Source* SPIE SAG)

Compact Multiple Circuit Overhead Lines

Compact multiple-circuit overhead lines refer to the combination of more than two circuits on a single structure, while maintaining the conventional arrangement and construction lengths of the insulators respectively insulator strings. This term also encompasses cases where compact design is achieved through the use of specially forms of insulators—for example, in V or Y shape or as a special case of the V string, or the asymmetric arrangement of the insulator strings.

Increasing lack of space for power lines and the bundling requirement for linear infrastructure facilities, such as roads, railways and other overhead lines, prompted the development of tower designs for more than the two circuits shown so far, long before the introduction of public authorization procedures.

While bundling initially meant the arrangement of several lines next to each other, the bundling compulsion increasingly led to the combination of several circuits on one tower. Here, four and sometimes six circuits were combined on a common tower and the utilization of the right of way was significantly increased. When maintenance is required, it often comes at the expense of operational safety and circuit availability. It is no longer sufficient to de-energize only the circuit being worked on; adjacent circuits—above, below, or beside it—must also be taken out of service [2].

A typical application for compact lines is the combination of circuits of different voltage levels, which can be run on a common structure for sections of the line. Another focus is on connecting lines, i.e. lines of different grid operators, which often face a shortage of available corridors due to their scale, making it necessary to combine them on shared routes.

With compact lines, especially in the highest voltage range, the principles of landscape-friendly routing can rarely be implemented satisfactorily. The natural coverage areas, topography and vegetation, are usually not sufficient to visually

Fig. 2.10 Example of a 380/110 kV quad line, here: initially equipped with one 380 kV circuit and three 110 kV circuits

screen these lines-especially in plain areas. An example of a quad line with two 380 kV and two 110 kV circuits is shown on Fig. 2.10.

2.1.2 *Safety Areas of Overhead Lines*

In the following, safety areas should be understood to mean areas that imply regulations for clearances due to the electrical, mechanical and static particularities of an overhead line, which ensure both, a failure-free and hazard-free operation of the lines. These clearances are e.g. in the European CENELEC-counties currently regulated in the European Standard EN 50341-1 [3] and in addition, e.g. in the national regulations for Germany in DIN EN 50341-2-4 [4].

When dimensioning the clearances for overhead lines, a fundamental distinction is made between internal and external clearances.

Internal Clearances

Internal clearances include all distances within the overhead line, i.e., between the conductors themselves and between conductors and earth wires or grounded parts at the towers (detailed in Chap. 4). These must be verified both at the tower itself and within the span, especially in the middle of the span. In the case of conductors with different sags within a span, it is advisable to examine the distances also between the middle of the span and the suspension points under the assumed wind loads.

Especially with long spans and in spans between different tower configurations it is possible that the smallest distance results between the suspension points and the middle of the span [5].

The minimum clearances to be maintained are specified as example from Germany in Tables 5.10–5.15 in DIN EN 50341-1 [3] and apply to the stabile states and for the deflections under influence of wind. A summarizing excerpt is shown in Table 2.1 and the following text.

External Clearances

External clearances, i.e., clearances between parts of an overhead line and objects in its immediate vicinity, must be dimensioned so that the overhead line does not pose a danger to the public and to people working on or next to the line. For the dimensioning of the external clearances, a number of cases must be considered in which different minimum distances must be maintained. To ensure that flashovers generated by overvoltages at crossings and proximities do not occur to the objects, but within the overhead line, the so-called a_{som} criterion must be considered. To meet this criterion, the distance must be 10% larger than the smallest striking distance of the insulator strings of the overhead line at the three suspension points before and at the three suspension points after the crossing point. The value determined in this way replaces D_{pp} or D_{el} in the tables. This could become decisive especially with long insulators or insulator strings, e.g. in areas with high pollution [6, 7].

In determining the clearances, the following cases must be considered according to [3, 4]. The necessary sag calculation of conductors is described in detail in Chap. 4:

- Clearances to the ground in areas remote from buildings, roads, railways and navigable waterways. It must be ensured that agricultural work with vehicles up to 4 m high below the line is possible without restriction.
- Clearances to residential and other buildings when the overhead line is above or next to the buildings or their structures. Different roof shapes and coverings are distinguished. In [4] there are also clearance specifications between overhead lines and wind turbines.
- Clearances to crossing traffic lines. This includes highways, federal, state and county roads, local connections and frequently used economic paths, trolleybus lines and cable car systems, railways with and without an overhead line system as well as waterways.

Table 2.1 *Internal clearances* according to Tables 5.6, 5.5 and 5.8 in [3]

Operating voltage	110 kV	220 kV	380 kV
Smallest distance between outer conductor D_{pp} in m	1.15	2.0	3.2
Smallest distance between outer conductor and Earth wire or earthed part D_{el} in m	1.0	1.7	2.8

- Clearances to overhead lines in proximity to clearance profiles of roads and shipping lines or to clearance profiles or parts of other overhead line systems.
- Clearances to overhead lines that cross other high voltage or telecommunication overhead lines. For the clearance verification at overhead line crossings, the following load cases must be considered. In the following example the situation is given for Germany as example [4]:

1. Load Case

Sag at the highest design temperature of the crossing overhead line (usually 80 °C). Simultaneously sag at +40 °C of the undercrossing line.

2. Load Case

Sag at −5 °C and ice load for the *crossing overhead line*. Here, the larger sag is to be considered, which results in the one hand at full ice load on the whole section and therefore also in the crossing span and on the other hand at the so-called unequal ice load. With *unequal ice load*, it is assumed that half the ice load acts in the crossing span, while the remaining spans of the section are free of ice. The tension points on the adjacent suspension strings are thus pulled into the ice-covered span and thereby increase the sag in this span. The determination of the sag under ice load in only one span is usually done using iteration algorithms via computer programs, but for rough considerations it can also be approximately determined using formula in [8]. For the under-crossing *line*, the sag at −5 °C without ice load is to be assumed.

3. Load Case

40 °C conductor temperature and vertically acting wind load on one of the lines (under-*crossing and/or the over-crossing*) and simultaneous wind load on the other line with a 40% reduced dynamic pressure (Fig. 2.11). In most cases, the smallest distance should occur in the case considered of the fully deflected under-crossing line and reduced deflection of the over-crossing line. In exceptional cases, such as large height differences of the individual suspension or tension points of both lines, other constellations could also become decisive. Due to the deflection of both lines, the crossing point is at the point of intersection of the two projections of the sag, i.e., the point of intersection of two parabolas. The determination of the intersection point and the smallest distance is usually done via computer programs [9, 10].

- Clearances *to recreational areas* at crossings of or proximity to overhead lines.

A selection of the most important minimum values for the external clearances is summarized in Table 2.2.

2.1.3 Restrictions Within an Overhead Line Corridor

The height of towers results from the sag of the conductors, a safety zone (also called safety area), which increases with increasing operating voltage, as well as the height of the objects and obstacles to be crossed, which exist as buildings or as clearance

Table 2.2 Some important *external distances* according to Tables 5.10–5.14 in [3]

		Distances in open terrain lowest conductor to the ground $5.0 + D_{el}$[a] m	Distances from roads and railway tracks lowest conductor to upper edge of road or top edge of railway edge[c] $6.0 + D_{el}$[a] m	Distances to overhead lines and clearance profiles lowest conductor to the top edge of the profile $2.0 + D_{el}$[a] m	Distances to other overhead lines lowest conductor to upper parts of the undercrossing line D_{pp}[a,b] m
Under the following conditions: – Highest conductor temperature without wind load – Extreme value of the ice load without wind load – Nominal value of the wind load	110 kV	6.0	7.0	3.0	6.0
	220 kV	6.7	7.7	3.7	6.7
	380 kV	7.8	8.8	4.8	7.8

[a] If D_{el} or D_{pp} are smaller than $1.1 \times a_{som}$, is $1.1 \times a_{som}$ to use
[b] The greater value of the two overhead lines
[c] If electrification is planned, the distance to the upper edge of the rail increases to $11.5\ \text{m} + D_{el}$ [4]

profiles of roads, railroads, and the open terrain (Fig. 2.12). The minimum clearances to be maintained are specified by the aforementioned set of regulations or standards.

An example for the determination of the heights of the fixing points of the conductors at angle tension or suspension towers in the 110, 220 and 380 kV level is shown in Table 2.3. The dimensions of the towers above the lower suspension or tensioning points depend on the chosen tower design and the voltage level relevant for the insulator length. The dimensions of the towers increase significantly with increasing the operating voltage.

In many cases, the line routing is confronted with the desire to keep overhead line towers as low as possible. This leads to shorter spans and, in relation to the length of the line, to more towers. An exaggerated handling of this well-intentioned aspect leads to an increase in construction costs and at the same time to a visual "gridlocking of the landscape". Optimization studies in the 110 kV level have shown that with spans between 280 and 300 m, both visually and cost-effectively favorable ratios can be achieved [11]. For 380 kV lines, the standard ruling spans are 380–400 m. In special topological conditions, the spans can deviate significantly from the ruling spans. For example, when crossing valleys the span length can increase to double or

Fig. 2.11 Forces due to wind and deflections of the conductors must be taken into account in the design. Here an extreme example from Slovenia during a severe storm (*Source* F. Jakl)

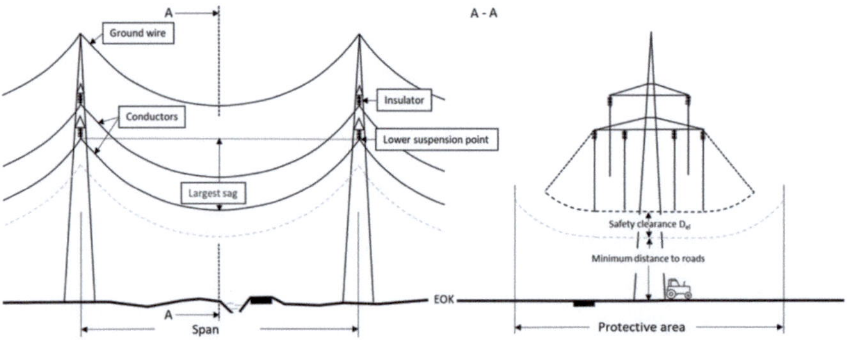

Fig. 2.12 Safety zone of high voltage overhead lines, **a** longitudinal profile, **b** cross section

more the ruling span, even if the heights of the lower tensioning points on the limiting towers remain largely unchanged compared to the standard type. Conversely, special terrain conditions, obstacles or tie off situations to existing lines may require shorter spans.

The maximum conductor temperature to be considered for the sag is usually 80 °C for conventional ACSR conductors. This maximum continuous operating temperature is specified for the common conductor types AL1/ST1A and AL3/ST1A for Germany in DIN EN50341-3-4 and Europe wide in the Euro norm EN 50182.

Table 2.3 Example: determination of the average heights of the fixing points of the lower conductors on 110, 220 and 380 kV overhead line towers

Operating voltage	110 kV	220 kV	380 kV
Sag of the conductor at the lowest point, here at 120 °C conductor temperature (depending on the span a)	App. 11 m (a = 280 m)	App. 14 m (a = 350 m)	App. 17 m (a = 380 m)
Safety distance (minimum distance D_{el}) (m)	1	1.7	2.8
Minimum distance (to roads)	6 m		
Height of the lower suspension or point above ground level (m)	18	21.7	25.8

This temperature is reached under corresponding meteorological conditions, such as low wind movement with simultaneous strong solar radiation and the maximum permissible current flowing through the conductor [12]. Furthermore, it must be considered that the conductors are subject to permanent elongation during their lifetime (creeping), which depends on the cross sectional ratio and is about 0.8‰ after 30 years for conductors with a high steel content. For those with a high aluminum content, it is around 1.2‰ of the length and results in a corresponding increase in sag. Compliance with safety distances to crossed objects over the entire lifespan of the line suggests a generous dimensioning of the tower heights. Some grid operators consider the elongation through a so-called temperature surcharge to the maximum conductor operating temperature. This means that the sag is dimensioned for a higher temperature (usually 120 °C) and thus forms a sag reserve, into which the conductor can "elongate in" during its service life. This adequately takes into account all risks associated with permanent conductor elongation.

Over the past two decades, composite conductors have come onto the market that allow higher operating temperatures and have since been used in upgrades of lines. These conductors, referred to as HTLS conductors (*High Temperature Low Sag*), allow significantly higher operating temperatures with simultaneously low sag. Their areas of application e.g. in Germany have so far been limited to transmission systems that are temporarily overloaded mainly due to fluctuating feed ins, e.g. from wind farms, and therefore would require shutdowns. For the conductor replacement (reconductoring), the existing towers usually do not need to be reinforced or raised, as the conductor cross sections and sags only change insignificantly compared to the "old" conventional conductors despite higher operating temperatures (Sect. 4.3.9). The use of HTLS conductors can thus reduce or prevent feed in restrictions. However, the increased transmission capacity leads to higher line losses, which is why this type of conductor has so far been used only rarely in new constructions. However, some new construction projects are dimensioning the towers in preparation for a future change from conventional- to HTLS conductors. Here, all expected operating parameters can be considered during planning and design, and conventional conductors can be used for the time being.

Another way to increase transmission capacity, even with conventional conductors, is the so-called line monitoring (Sect. 4.3.10). This type of increase of capacity is based on the collection of meteorological data, which, for example, at higher wind speeds or lower outside temperatures, allows the current and thus the transmission capacity to be increased without exceeding the permissible operating temperature limits of the conductors. Various monitoring methods are being installed in several lines for this mode of operation. Particularly in coastal regions with strong wind exposure, where a large number of wind turbines regularly lead to grid bottlenecks, the wind induced cooling effect of conductors correlates with the power output of the plants and can thus create increased operational margins prior to curtailment.

Further questions that arise during line routing concern both overhead lines and cables and their effects on their surroundings. For overhead lines, in addition to the impact on the landscape and the restrictions on use within the corridor (right-of-way), the question of electric and magnetic fields and possible audible noise is regularly in focus. For cables, in addition to the occurrence of magnetic fields above ground, there is the risk of soil drying out at high current loads and possible planting restrictions.

The electric and magnetic fields in the vicinity of lines have repeatedly been suspected over the past decades of promoting the development of cancer, especially in children. They are sometimes referred to as electromagnetic fields, as with high frequency fields. With the low frequency 50- or 60 Hz fields, which are exclusively involved here, the electric and magnetic field are so little coupled that we can consider the two phenomena separately with sufficient accuracy (Sect. 6.4). The voltage between the conductor and the earth potential, i.e. the earth's surface, is responsible for the electric field. The current in the conductor is responsible for the magnetic field. In numerous publications, they were also disparagingly referred to as "electro smog". This phenomenon was extensively investigated in numerous, partly epidemiological studies in the last decades. All known scientific studies have concluded that there is no significant correlation between the fields and the frequency of cancer [13]. If the limit values, which e.g. in Germany have now been set in the regulation for the Implementation of the Federal Immission Control Act (26. BImSchV), are complied with, no health impairment is to be expected. In the currently valid regulation from 2013, the limit value for static magnetic fields (DC) is 500 Micro Tesla, for magnetic alternating fields with 50 Hz 100 μT and for electric alternating fields with this frequency 5 kV/m. These limit values are considered both in the planning of overhead lines and in cables. Deviating from this, international bodies such as the ICNIRP (International Commission on Non-Ionizing Radiation Protection) have recommended limit values for the general population at 50 Hz fields for the electric field strength of 5 kV/m and for the magnetic flux density of 200 μT since 2010.

Towers and conductors are exposed to wind and can thus cause vibrations and audible noise. These can occur in strong winds or when the conductors are subjected to laminar flow.

The high electric field strengths around the conductors can also lead to ionizing discharges. This phenomenon, referred to as the *corona effect*, can be particularly noticeable as crackling or humming in rainy and foggy weather conditions. In these bad weather conditions, droplets settle on the surface of the conductors and increase the electrical edge field strength, which causes the aforementioned noises. Conductors are now available on the market whose surface appears hydrophilic due to a special pretreatment, reducing droplet formation and thus reducing the development of audible noise. Conductor arrangements in bundles can also reduce corona effects. Theoretical principles and calculations for the phenomena described will be dealt with in detail in later chapters.

2.1.4 Protection Strips

The conductors of overhead lines, influenced by wind, deflect over an area that extends beyond the lateral boundary of the conductors on largest crossarms. The vertical projection of the furthest deflected conductor plus the minimum safety clearance on both sides of the line axis is referred to as a protection strip (Fig. 2.13). Within this area, the space is only usable to a limited extent, e.g., in the form of a height restriction on plantings and structures. The use of these areas is secured for the operator of the line through a so-called easement (right-of-way, ROW), without changing the ownership of the area. The depreciation resulting from the limited usage in the protection area or from the tower area itself is compensated for the property owners by a financial one-time compensation, just like for cable lines (Sect. 3.9). There are now cases where annual payments are also being agreed upon.

2.1.5 Line Routing and Use of Land

Line planners have always strived to adapt the tower designs to the surrounding landscape type and to consider the respective land use in order to minimize conflicts of usage. In addition to spatial planning criteria, considerations related to nature and landscape protection play a particularly important role.

Agricultural Areas

For agriculturally used areas, regardless of the type of use, the aim is to place towers near agricultural roads. The reduction of obstruction in the cultivation of fields (to facilitate the allowance) is just as important as avoiding unnecessary damage to the field during line construction. Otherwise, the tower setting is carried out with largely optimal use of the topography. The conductor sag and the heights of the lower suspension points on the towers are chosen in a way that professional cultivation of the fields is possible without impairment. At most, there are restrictions for the growth of field shrubs and high growing cultures.

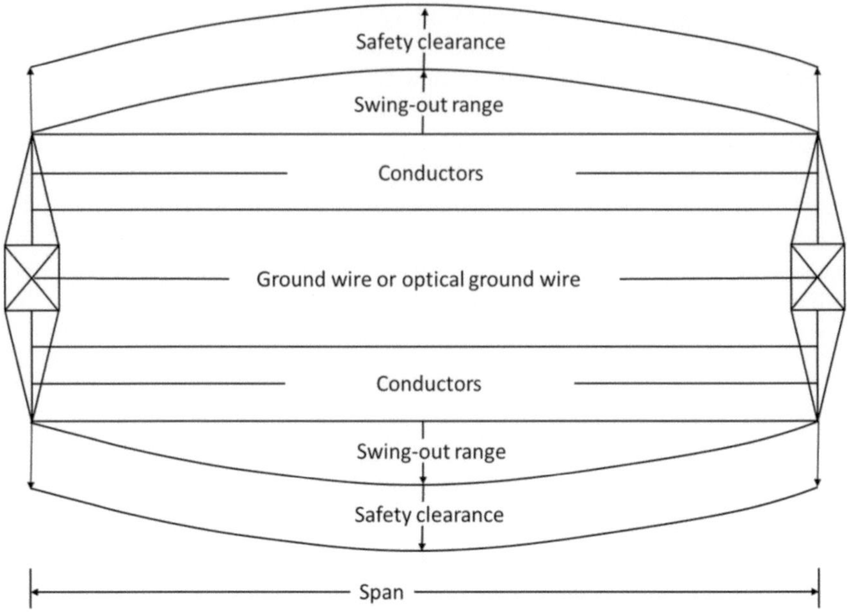

Fig. 2.13 Schematic sketch of the protection strip of a span in top view

Forest Areas

If a line is to pass through a forested area, the question initially arises as to whether closed forest areas can be avoided, even if this involves additional route lengths, or whether they should be crossed or traversed. If bypassing proves to be complex or impossible for other reasons, the question arises: *Forest Traversing* or *Clearing*?

Traversing of a forest and taking into account the final heights of the growth, allows unrestricted forestry operations except for the areas of the tower locations. Clearings are of limited use for forestry. Although clearings can show a high biodiversity with appropriate management, forestry authorities usually demand traversing to avoid restrictions of use. On the other hand, traverses are often negatively assessed from the perspective of landscape protection. Forest-traversing of multiple circuit high and highest voltage lines usually become dominant elements of the landscape with a massive, long distance effect. To a neutral observer, this may evoke a sense of irresponsible defilement upon the natural landscape. The practice of line routing in the past has shown that there is no patent remedy for such cases. Here, carefully balanced particular decisions have to be made [14].

Forest Traversing

In rugged forest stands, a forest traversing can be a good solution. In this way, open spaces can be used for the route and the felling of trees can be minimized. For the cuts between the open spaces, narrow places should be used if possible. Assuming that narrow places can be found in a stretched line without major detours, the decision is usually in favor of a clearing in view of the moderate felling areas. In such cases, it represents the most favorable solution in terms of line routing technology. In stands that are ecologically valuable biotopes or where there is an increased risk of wind breakage, the approval of the forestry authorities for a clearing is usually not obtainable, even for smaller areas.

When determining the dimensions of forest clearings, a distinction is made between two cases. For new clearings, the required width is determined by considering the fall of a final grown tree reaching the line axis, plus an additional safety distance (Fig. 2.14, right part of the picture). With increasing final growth height and increasing operating voltage, the clearing thus becomes larger. In newly created clearings, there is a risk of collapse for the newly created edge trees in the forest aisles, which previously belonged to the forest and do not have pronounced eaves. It must be ensured here that the falling tree does not reach the protection area of the not deviated conductor. If, on the other hand, the line runs along a grown forest edge or in an existing clearing, which has eaves on the side to the line, due to the comparatively high stability of such consolidated edge trees, a collapse is significantly less likely.

The forest areas used depend on the voltage level of the line and the resulting size of the towers and their crossarm length, from the type of conductor and the span length between two towers and thus the conductor sag and the conductor deflection due to wind which has been assumed in the project and from the intended use of the line. The fundamental relationship among these considerations is shown in the sketch in Fig. 2.15. It comes from a study in Great Britain from the 1960s for the three voltage levels 132, 275 and 400 kV.

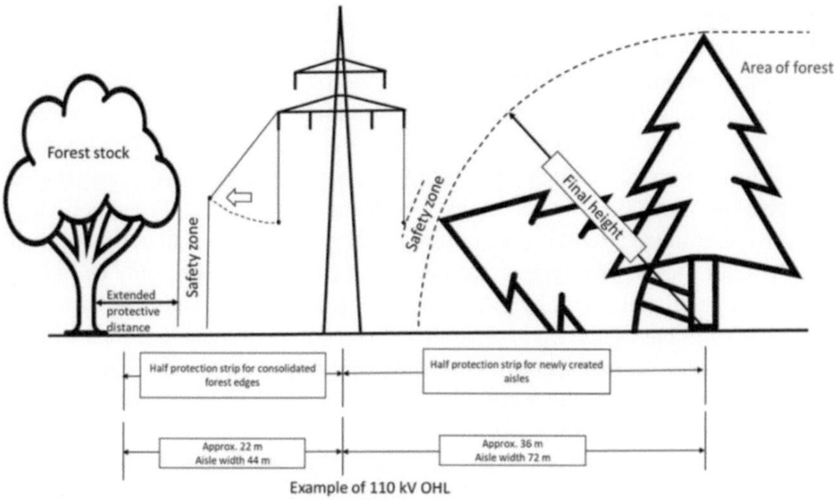

Fig. 2.14 Dimensions of forest clearings, *left* at consolidated forest edges, *right* at unconsolidated forest edges with consideration of treefall

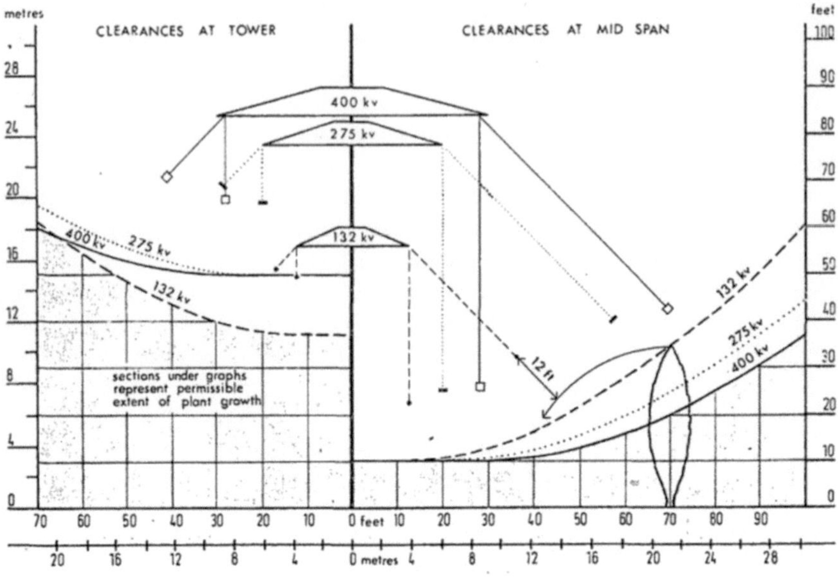

Fig. 2.15 Forest aisle by overhead lines of the three voltage levels 132, 275 and 400 kV. Basic representation from 1969 [15]

As a result, wind breakage does not necessarily have to be considered. This applies to the proximity to solitary trees as well as for the after-treatment of clearings of existing lines, where the eaves have grown back. The dimensioning of the protection strip can in this case be simplified by adding an extended protection distance added to the deflected conductor and to the electrical safety distance, which prevents the long-term growth of trees into the protection area (Fig. 2.14, left part of the picture). The extended protection distance in the specific example is 5.5 m and—according to studies by a southwestern German grid operator—corresponds to the 95% frequency value of the crown radii of all there occurring woody plants. For particularly impressive trees, the protection distance must be extended as needed. Protection strips, which are determined after this method, are partly much narrower than when considering the fall of trees. This type of determination can be, however, only be justified from a line operational point of view. In areas of crossings with roads and railways or in the vicinity of buildings, where there is indirect danger to people from a windfall of trees on lines is to be expected, complete safety must be placed. This can only be achieved in clearings which consider the fall of trees. Borderline cases require careful professional assessments in any case.

Forest clearings for overhead lines often also have the function of firebreaks in countries where forest fires are common and pose a great danger. Such aisles can prevent sparks and the spread of fire and can be several hundred meters wide. From Canada clearing of 1000 m widths are known, the growth is deliberately kept small to keep the fire load as low as possible.

Forest Crossing

In many countries clearings in closed forest stands can rarely be enforced because of disproportionately strong forestry use restriction, although they deserve certainly attention from an ecological point of view. In widely closed forest stands, the importance of long distance effects decreases and makes forest crossings appear bearable from this point of view. It should be noted, "however, that clear-cuts and young stands are also included in the forestry rotation period." These are temporally extended periods in which the excessively appearing line can seem as a massive foreign object.

In forest crossings the area needed for the construction of a line is limited to the areas for the assembling of the towers on ground. It can consist of squares with 25–30 m side length depending on the voltage level. The cleared square can subsequently become overgrown up to the immediate area around the tower, which—for maintenance work—is to be kept free of growth. During the tower spotting their positions near forest roads is always desired to avoid unnecessary cutting and to facilitate access to the tower locations.

Depending on the type of operational monitoring of the overhead line, continuous site clearing can also be avoided. That means that after the completion of the erection of the line, the entire aisle is available for unrestricted forestry use—except for the remaining free areas around the towers. Figure 2.16 shows the conditions in a forest crossing over with a 110 kV double line in horizontal configuration. Using this tower design forest crossings can achieve the lowest possible total height of the towers.

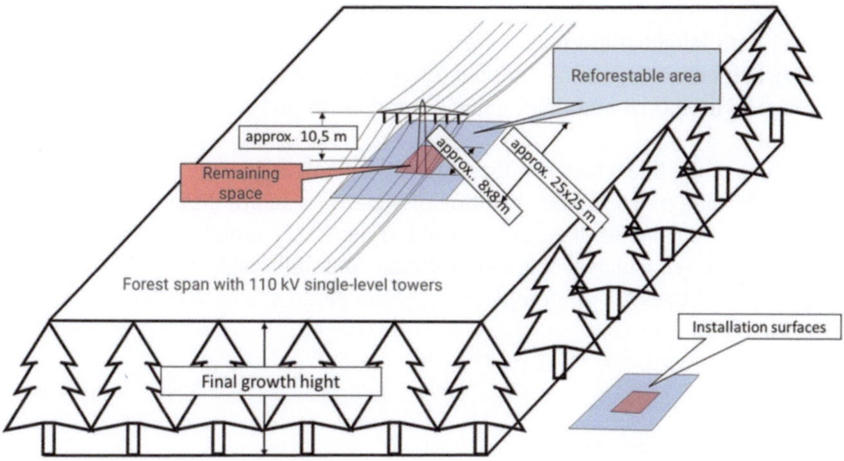

Fig. 2.16 Area requirement for forest crossings by a 110 kV overhead line

Fig. 2.17 Line routing in horizontal arrangement in the approach area of an airfield with daytime marking (*left* 220 kV, *right* 110 kV)

2.1.6 Areas with Height Restrictions

In areas with height restrictions—such as zones near airfields and airports where building heights are regulated—special tower types are typically used. These towers are designed in a horizontal single-plane configuration and are reduced in both height and span.

Sections of the line which are directly in the takeoff or landing area also receive a daytime marking (red and white painting of the towers and/or flight warning spheres on the highest conductor or earth wire) and, if necessary, a night marking (red light on the tower tops), which enable both safe line and flight operations. In such areas, multiple circuit lines are often separated and run parallel to each other. An example is shown in Fig. 2.17.

2.1.7 Residential Areas

In most countries, avoiding residential areas when routing overhead lines—especially those of higher voltage levels—is standard practice.

This is not because the electric or magnetic fields of an overhead line pose a health risk—below specified limits, they do not, as explained in detail earlier and in Chap. 6—but because land use restrictions within the corridor (right-of-way) can have potential for conflicts between the grid operator and the landowner, which should be avoided whenever possible.

This topic regularly triggers disputes between territorial bodies and authorities or interest groups representing the conservation of nature and landscape. While municipalities always demand a distant routing of lines, nature and landscape conservationists want lines to be avoided in untouched landscapes. Thus, compromises often lead to a line near settlements, which has no special ecological value. Existing or intended zoning maps that are sufficiently specified in a land-use plan must not be hindered or restricted. A relevant example from Germany illustrates that, under the 26th Federal Immission Control Ordinance (26.Bundes-Immissionsschutzverordnung), newly constructed 50 Hz overhead lines with a nominal voltage of 220 kV or above are prohibited from crossing over residential buildings.

For precautionary reasons, new high-voltage lines must be planned and constructed in such a way that buildings—or parts of buildings intended for permanent human occupancy—are always located outside the designated protection strip. However, the subsequent construction of buildings beneath the line is permitted, provided that the required safety clearances are maintained. An example of this is illustrated in Fig. 2.18.

In large cities, where the demand for electricity requires high and highest voltage connections within urban areas, but overhead lines cannot be realized, underground

Fig. 2.18 Inhabited area under a 110 kV double system overhead line

cables are used up to the highest voltage levels, with the associated limitations on cable-length.

2.1.8 Integrating Overhead Lines into the Landscape

As mentioned at the beginning, route planners always strive to design routes that allow the lines to blend in optimally with their surroundings. In hilly and mountainous landscapes, it is often possible to lay overhead lines in such a way that they disappear into the background and are barely visible. A successful example of the integration of a 110 kV overhead line in a hilly landscape is shown in Fig. 2.19.

2.2 Routes for Underground Cables

While high voltage overhead lines can be massive encroachments in the landscape, cables are usually no longer visible after their installation. However, depending on the type of installation, cables at these voltage levels may be subject to building prohibitions and restrictions on planting vegetation with deep roots. Invisibility is probably the reason why almost every overhead line project, regardless of the length of the line,

Fig. 2.19 Successful integration of a 110 kV overhead line into the landscape of a Northern Italian Wine Region

is accompanied by calls for underground cabling. In the past two decades, several citizens' initiatives have been formed in many countries, which have vehemently advocated the cabling of planned overhead lines.

This situation led, among other consequences, to a significant delay in the urgently needed expansion of the transmission grid using conventional overhead line technology. As a result—using Germany as an example—the legislator mandated a preference for underground cabling in the mid-2010s for the construction of high-capacity north–south DC transmission corridors.

2.2.1 AC Cable Routes

When planning high and highest voltage AC three phase cable lines, grid operators often use two separately laid cable systems at these voltage levels to substitute one overhead line system for reasons of high transmission performance and availability. This variant is often selected for intermediate cabling to limit the thermal effects in the ground despite high transmission power. This means that each electrical system of an overhead line is divided into two cable systems in this case. It is schematically shown in Fig. 2.20 for a double system overhead line.

Depending on the company's operating philosophy or the importance of the cable connection, other design forms can be chosen. Thus, the continuation of a double system overhead line can also be achieved by three cable systems. In this variant, one reserve system can take over the complete power transmission of one cable system in case of disturbances or maintenance work (Fig. 2.21).

In a stripped-down variant, a single cable phase serves as a reserve. In this way, each individual cable phase requiring maintenance or repair work can be temporarily replaced by the reserve connection until the work on the line is completed (Fig. 2.22). Sufficient clearances between the phases and systems are the precondition.

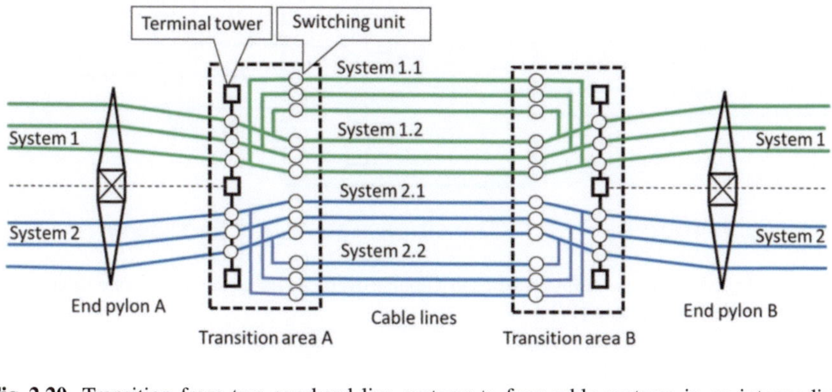

Fig. 2.20 Transition from two overhead line systems to four cable systems in an intermediate cabling [16]

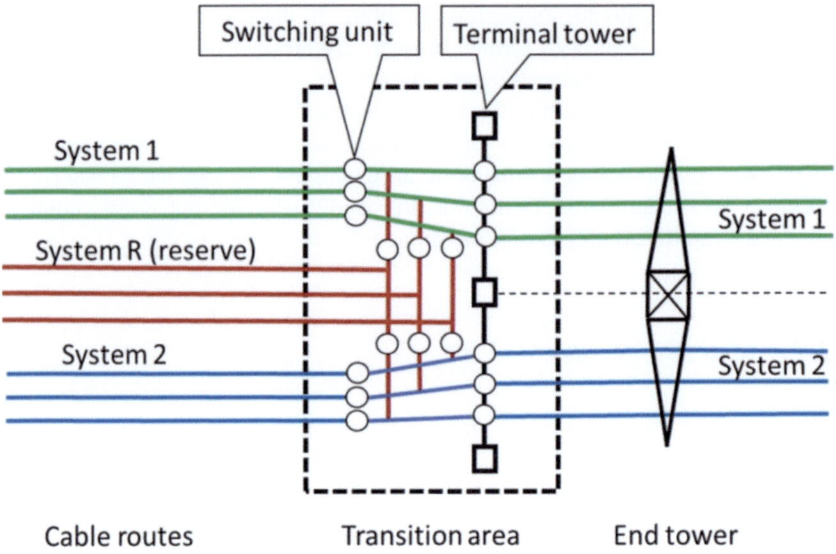

Fig. 2.21 Transfer from three cable systems to a double overhead line system, with one cable system being used as a reserve [17]

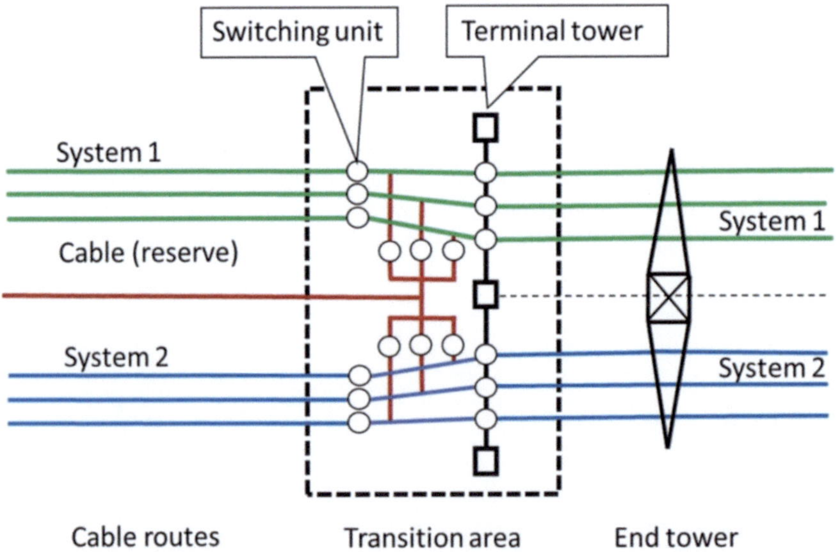

Fig. 2.22 Transfer from two cable systems to a double overhead line system, with one single cable phase being used as a reserve [17]

The routing of cable systems is primarily influenced by the laying method and the laying arrangement. The first key decision is whether to install the cables in an open cable trench or to apply methods without digging, using drilling or pressing methods. Once this is determined, the next consideration is whether the cables should be laid directly into the soil or be pulled into protective plastic pipes. Other options, which so far have mainly been used in urban areas or at frequent intersections with other infrastructure facilities, are so-called ducts or accessible concreted cable ducts or tunnels, which can be realized using open or semi-open methods. In agriculturally used areas, the preferred method is regularly one of the in-soil embedded laying variants (Fig. 2.23a, b). Both variants have advantages and disadvantages. The advantages of directly laying in soil are good heat transfer from the cable sheath to the surrounding bedding material and good fixation of the cables. This reliably prevents movements due to operational mutual force effects. However, a disadvantage is that the trench must be reopened at specific points or along the entire line in the event of cable faults or their replacement. In addition, soil settlements or shearing due to tectonic disturbances directly affect the cable. These can cause unwanted compression effects that impair the function of the cable or even destroy the cable.

One advantage of laying cables in plastic pipes is that the trenches can be backfilled immediately after the pipes are laid. The laying of the pipes and the pulling in and jointing of the cables can be carried out separately. During the laying process, there is also the option to lay reserve pipes, allowing additional cables to be pulled in at a later stage. The most important advantage of this type of laying, however is a possible later replacement of cables without the need to reopen the cable trench. The disadvantage is the air-filled gap between the cable sheath and the protective pipe, which inhibits the heat dissipation to the bedding material. This reduces the load carrying capacity of the cables. Filling the protective pipes with water can improve heat dissipation, but it is technically very complex and requires intensive maintenance (Chap. 5).

Alternatively, or in combination with the water filling, modern monitoring systems can counteract this disadvantage by providing information about both the temperatures and the spots along the cable system where disturbances or faults occur. This

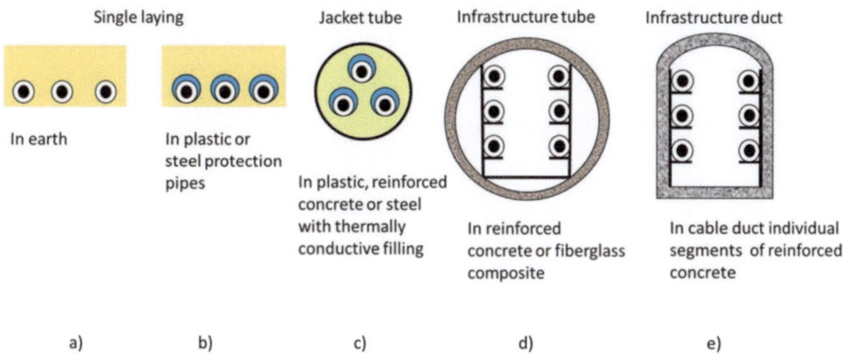

Fig. 2.23 Possible laying arrangements for three phase high and extra high voltage cables [17]

approach allows for the detection of localized overheating in the cables, helping to prevent potential damage. Monitoring can also determine the load condition of the cable and thus possible reserves for power transmission.

2.2.2 DC Cable Routes

In the future, the use of direct current cables at high and extra-high voltage levels is expected to increase. Due to the legally mandated preference for cables of this voltage type and level, e.g. in Germany, the focus will initially be on point-to-point cable connections between AC/DC converter stations, which establish the integration into the high voltage three phase grid (Sect. 6.8).

In principle, both the laying arrangements and the types of laying are comparable to those for three phase cable systems (Fig. 2.24). Only the number and cross sections of the individual conductors may differ. For many established HVDC cable connections concepts, installations in open rural areas—with electrical systems placed in separate trenches and protective conduits—either already exist or are currently being planned. At intersections with significant infrastructure facilities, such as railway lines and highways, alternative laying methods may also be required. This means that the same basic conditions apply to HVDC cables as to AC cable systems (Chap. 5).

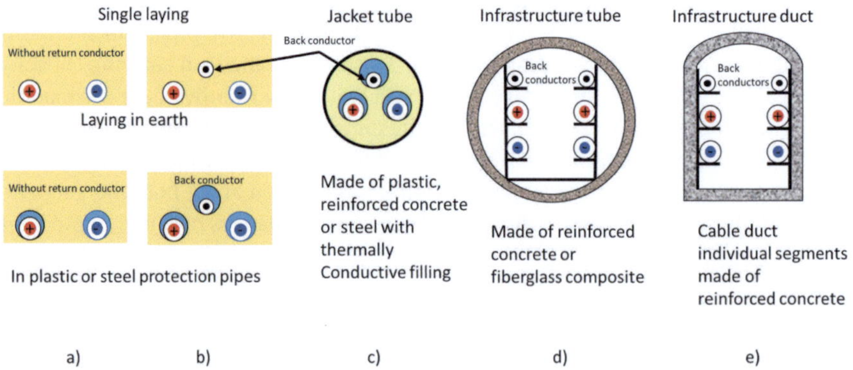

Fig. 2.24 Possible laying arrangements for HVDC cables with and without metallic return conductor [17]

2.3 Cable Laying Techniques

Cable laying techniques refer to the various methods used to lay cables. These range from simple laying in soil in open trench to trenchless, closed laying at varying depth. The techniques differ in terms of soil encroachment as well as in the efforts, costs, and speed of the laying.

2.3.1 Open Construction

When the cables are laid in an open trench, with or without protective pipes or ducts, it is assumed that the trench will subsequently be backfilled with thermally stabilized material, which retains a high thermal conductivity even at higher temperatures. This type of installation can be referred to as the standard variant. For the backfill, materials such as lean concrete or the so-called liquid soil are used. Liquid soil is created by mixing the excavated material with additives and water. The material is first brought into a plastic to flowable state and then back-solidified again—either with the properties of the starting material or with deliberately changed properties for better heat dissipation. Bedding materials still have the minimum thermal conductivity required by relevant standards even when dry. In special cases, for example in foreseeable thermal hot spots, improved materials (e.g. Powercrete) can be used, which have a very high thermal conductivity also in a completely dry state [17].

Figure 2.25 Exemplarily shows the profile and the protection strip during the installation of two AC systems 110 kV and one reserve cable in an open trench.

In Fig. 2.26, the trench profile for an intermediate cabling of two AC systems 380 kV is shown according to the structure from Fig. 2.20.

Figure 2.27 shows a comparable profile for the installation of three AC systems, as Fig. 2.20 basically represents.

The dimensions of the trench profile and the protection strip for the installation of two systems of ±525 kV HVDC cables without return conductor are shown in Fig. 2.28.

The cable installation in an open trench requires a slow construction progress and involves extensive restoration of the removed surface areas. Subsequently, it needs a regeneration time for the cultivation with the same yields as before. However, this method offers the significant advantage to select the bedding material in type, shape and size to optimize heat dissipation into the surrounding soil. A practical example of the installation of two adjacent systems is shown in Fig. 2.29.

Temporary building roads, which can consist of wood, aluminum or mats of steel—depending on the load bearing capacity of the subsoil—are being removed after commissioning of the cable system (Fig. 2.29, left edge of the picture).

The backfill material in the cable trench consists primarily of porous material, whose cavities are partially filled with water depending on the material properties, the distance to the groundwater level, the respective season and the possibilities of

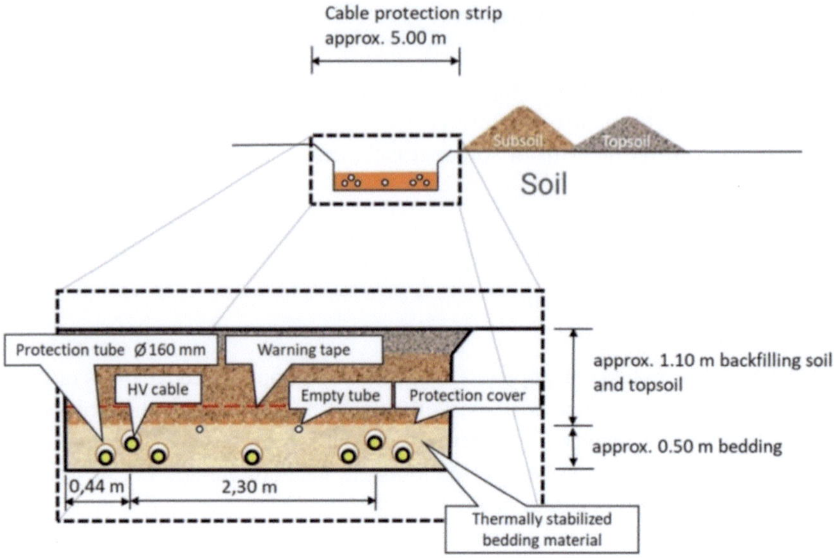

Fig. 2.25 Exemplary trench profile and protection strip for a 110 kV AC cable installation in pipes with two systems and a reserve cable shown during installation and in final operational state

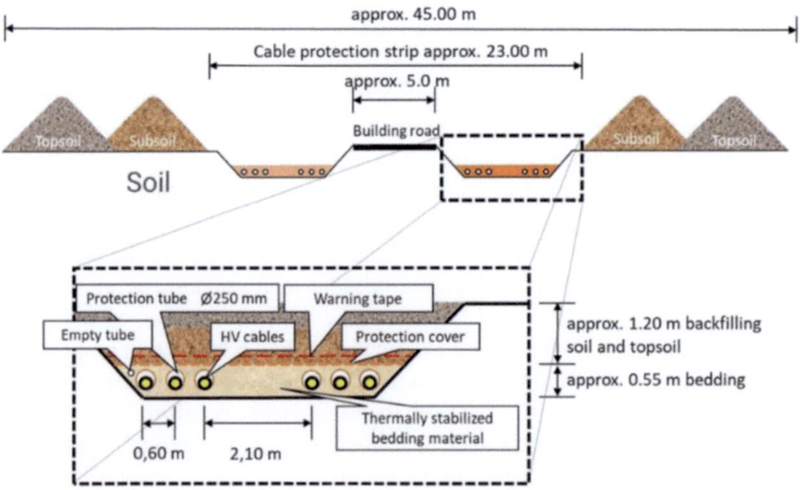

Fig. 2.26 Exemplary trench profile and protection strip for a 380 kV AC cable installation in pipes with four systems shown during installation and in final operational state (cable protection strip approx. 23 m) [16]

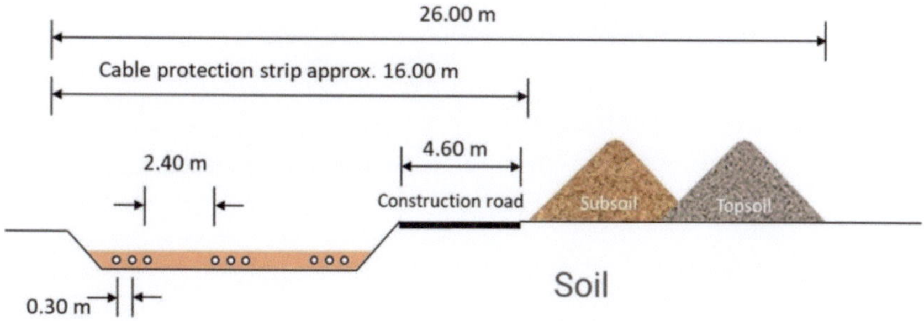

Fig. 2.27 Exemplary trench profile and protection strip for a 380 kV AC cable installation with three systems shown during installation and in final operational state (cable protection strip approx. 16 m) [17]

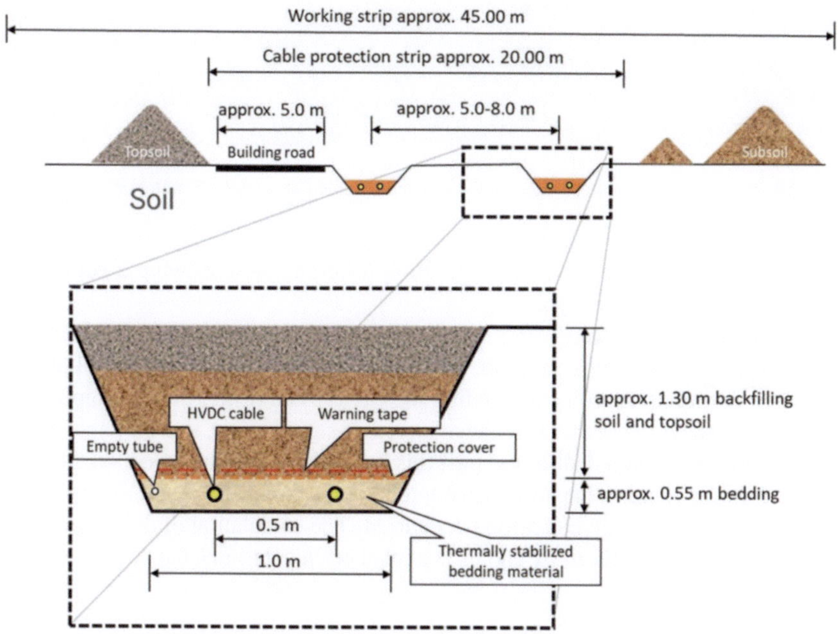

Fig. 2.28 Trench profile and protection strip for an HVDC cable ±525 kV shown during installation "working strip" and in final operational state (cable protection strip approx. 20 m) [18]

infiltration. The higher the water content, the higher is also the thermal conductivity and thus the heat dissipation from the cable. The heat emission of the cables at high cable temperatures can cause a process in the soil, which can be compared to a heat pipe. The water in the immediate vicinity of the cable evaporates due to the cable's heat, migrates to cooler surrounding areas where it condenses, and then

Fig. 2.29 Route for two cable systems 380 kV when installed in an open trench [19]

returns toward the cable driven by saturation differences—following a continuous evaporation–condensation cycle.

The enthalpy of the evaporation extracts heat from the cable environment. If the cable temperature becomes too high, insufficient moisture returns to the cable area, leading to drying of the soil surrounding the cable. As a result, the heat pipe effect breaks down, the heat transfer resistance increases significantly, and the cable can no longer adequately be cooled. In the past this has often led to the so-called "thermal death" of cables. Due to the absence of temperature monitoring the issue was only detected after the damage had already occurred. The calculation of the thermal and hydraulic processes induced by cables is explained in detail in Chap. 5.

2.3.2 Ploughing In

An alternative to the installation in an open trench is ploughing in protective pipes and the subsequent pulling in of cables. Modern cable ploughs can install three plastic pipes for cables up to the 380 kV level plus several pipes for telecommunication and diagnostic technology simultaneously in one single operation (Fig. 2.30). The arrangement of the pipes or ducts can be in triangle or side by side. However, this type of installation is limited to areas free of crossings as possible, where the soil conditions allow ploughing in. It can be used in soils up to soil class 6 (light rock) and

Fig. 2.30 Cable plough during the installation of empty ducts for cables and telecommunication pipes [20]

allows laying depths of over 2 m. With this trenchless, so-called semi-open method, the slot created from the ploughing is immediately refilled. Generally, ploughing in allows a limited amount of bedding material for the pipes' surrounding area. Compared to open trench installations, this method is faster, more cost-effective, and causes minimal disturbance to the soil surface. Often, only such trenchless methods are permissible in conservation areas for water, landscape and nature. Despite all its advantages, ploughing requires a very careful investigation of possible infrastructure facilities in the ground located within the route, such as other power cables, sewer pipes, drainages, pipelines and telecommunication cables. Special attention should be paid to military supply lines, whose locations are often omitted from publicly accessible documentation and plans for sabotage protection reasons.

2.3.3 Installation in Casing Pipes and Infrastructure Pipes

The method known from the installation of district heating and natural gas pipes can also be used for the laying of high and extra high voltage cables. In this case, the casing pipe contains three plastic protective pipes for accommodating a three phase system, which are evenly distributed in the enclosing duct by spacers. The cavities are then backfilled with thermally conductive material (Figs. 2.23 and 2.24c). Unlike gas pipelines, the casing pipe in this case can be made not only of steel, but also of high-strength plastic or reinforced concrete segments. Afterwards, the cables can be pulled into the protective pipes in sections, similar to the previously described laying methods. A particular advantage of this method is that casing pipes with diameters up to 600 mm can be installed not only by conventional laying in an open trench, but also in semi-open and closed drilling methods [17].

Round infrastructure ducts, primarily made of reinforced concrete or glass fiber composite elements, can have nominal widths of 2000 mm or more (see Figs. 2.23 and 2.24d) and are preferably installed using semi-open or closed laying methods. After installation, the cable systems are mounted on cable trays or at intervals of several meters with brackets on the duct's walls.

Semi-open Construction Method

A semi-open construction method (also known as semi-open pipe jacking) is a drilling pressing technique in which the excavated material is transported to the surface through a slot positioned above or behind the drill head. This process can be carried out using a support vehicle as in the Pipe Express® method [21], which travels above the drill head, which mills a 300–400 mm wide slot and conveys the excavated material to the surface. The slot is immediately backfilled behind the vehicle. This minimizes the disturbance of the soil structure to a narrow strip (Fig. 2.31). The speed in this method is 500–1000 m per day. Lengths up to 2000 m can be achieved; this method is primarily used in open terrain.

The conventional method of semi-open laying as it is mainly used for installations of urban sewers, consists of an approximately 1000–1500 mm wide, usually shored slot on the surface (Fig. 2.32). The removal of the soil is done by a drill head with an opening on the front. A digger or a gripper removes the excavation when the cutting drill head with the opening stops upwards. Even with this traditional method, the encroachment at the surface can be significantly minimized compared to the

Fig. 2.31 Laying of a casing duct in semi-open construction method with the Pipe Express® method [21]

Fig. 2.32 Conventional pipe laying in a semi-open construction method. The picture shows the launch pit and the milled slot [22]

completely open trench construction method. The speeds for new installations are given in [22] as 35–95 m per day, the maximum lengths between the stations are a about 200 m. Construction pits with pit lining are required at both ends, at the launch pit, and the target pit.

In both variants, water retention can be omitted—unlike in the open construction method, where they may be required in areas with high groundwater levels. These methods are limited in the laying depth by the slot to the earth's surface, typically between 2 and 4 m. The advance is achieved by pressing the pipe's interconnected elements, together with the previously docked ones towards the target pit.

Closed Construction Methods

In urban areas where surface access routes are unavailable, only fully closed laying methods come into consideration. The same applies to particularly sensitive protected areas or crossings of railways, roads and rivers. For the laying of high and extra high voltage cables in pipes and tubes, the so-called wash drilling method can be used, which is preferably used in pipeline construction, or the worldwide used guided pipe jacking, particularly as for sewer construction.

Horizontal Wash Drilling Method

This laying technique, known as the HDD method (Horizontal Directional Drilling), is particularly suitable for crossings under wide infrastructure facilities as well as rivers and water bodies. It is very often used in combination with the open construction method for crossings. This allows the laying of pipes of steel or plastic, or even casing pipes with a diameter of approx. 600 mm to be laid trenchless up to 1700 m [21]. The method is particularly well-suited for the installation of individual protective or casing pipes. The method is implemented using mobile drilling rigs such as inclined so-called drilling frames with 3D location and control options in connection with a hydromechanical drilling fluid technique, in which the drilled holes can be widened as required. Drilling is done with thin, sharp, rock dissolving water jets or drilling suspension jets, which emerge from nozzles at the drill head's tip and cause a hydromechanical crushing. A part of the dissolved material is transported out through the return flow along the drill rod, while another part results in a partial relocation of the loose rock in the surrounding area of the drilled hole [23]. Subsequently, the borehole is widened in multiple steps and a protective pipe is inserted, into which a cable or bundle of cables can then be pulled in (Fig. 2.33). The speed of the drilling process is given in [24] as approximately 50–120 m per day, depending on the soil conditions.

Guided Pipe Jacking

This method, based on tunnel drilling technology and also called micro tunneling for small dimensions, offers the greatest flexibility in route selection. Guided pipe jacking encompasses a range of methods—from the two stage E-PowerPipe® process,

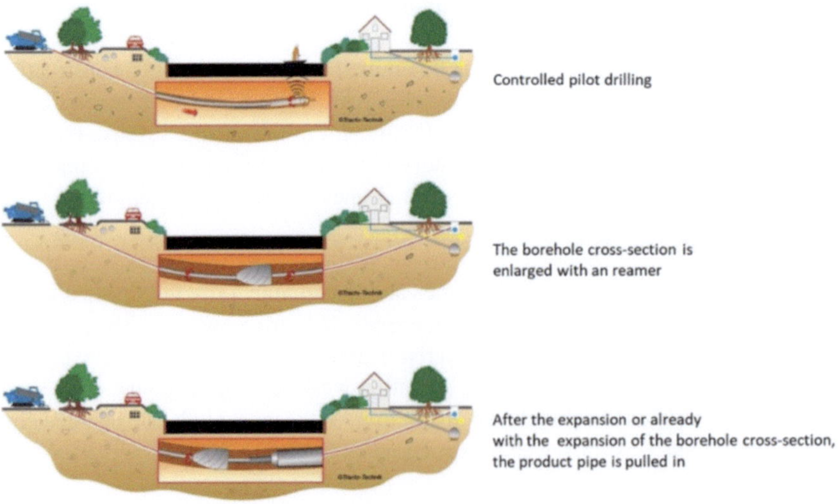

Fig. 2.33 The three steps in the horizontal directional drilling process HDD [23]

like the HDD method, for near surface laying with diameters of 500 mm, to large pipe jacking with diameters up to 4000 mm at virtually unrestricted laying depths under urban and extensively protected areas.

The two stage E-PowerPipe® laying technique with small diameters for individual, even non-pressure-resistant protective pipes starts with a launch pit with an installed jacking frame. Here, the two-stage laying technique of the HDD method is combined with the drilling technology of micro tunneling. This means that a fluid-assisted micro tunneling machine takes over the jacking between the launch and target pit. The remotely controlled jacking unit is equipped with a jet pump and a hydraulic unit for the realization of particularly long bores in addition to the cutting wheel. In the target pit, the tunnel boring machine is removed, the protective pipe is docked and pulled in the opposite direction, in which the steel jacking pipe segments, as shown in Fig. 2.34, are withdrawn and individually dismantled. The tensile steel jacking pipes keep the borehole mechanically supported during the pulling process. The annular gap between the protective pipe and the ground can be filled with filler material for mechanical and thermal stabilization during the operating phase if required. The method can be used in varying soil conditions, from silty soft soils to medium-hard rock, with jacking lengths of over 1000 m. The jacking speed can reach up to 100 m per day in light rock and up to 300 m per day in non-cohesive soils [25].

In conventional pipe jacking, precast pipe walls made of reinforced concrete or fiberglass composite are driven through the ground from a launch shaft using a

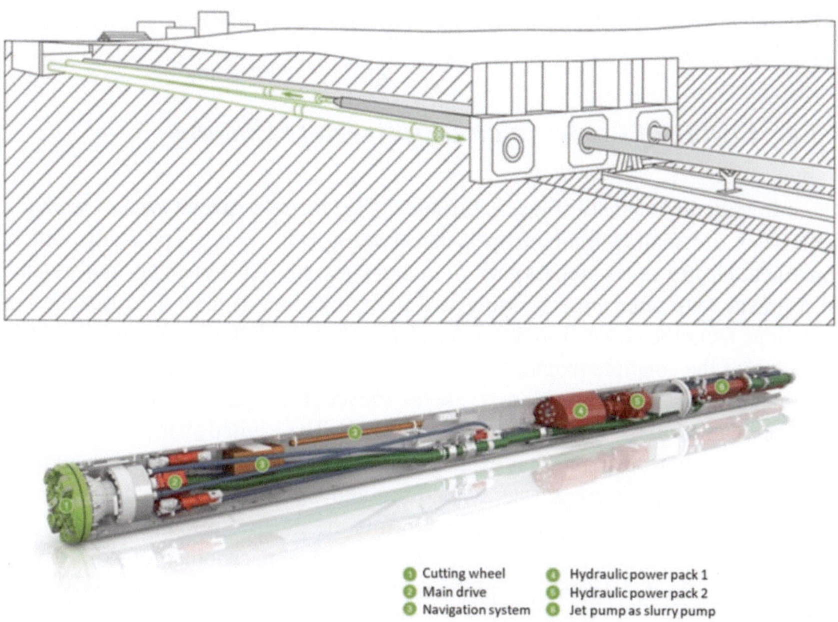

Fig. 2.34 Two stage pipe jacking according to the E-Power Pipe® method by Herrenknecht [25]

Fig. 2.35 Guided pipe jacking with shield machine, launch pit, and press station [21]

press station, like the semi-open method, towards the target shaft (Fig. 2.35). The drilling material worked out with the shield machine is crushed, made pumpable, and conveyed in a pipe system through the driven pipe string to the launch pit and from there to the surface. The controllable shield machine can accurately follow slightly curved lines in all directions. The jacking speed depends very much on the type of subsoil material and averages over 100 m per day.

This method enables the bundling of infrastructure facilities by laying casing pipes and infrastructure pipes, for example, parallel to highways and railway lines, without disturbing operations or endangering operational safety.

2.3.4 Laying of Cables in an Infrastructure Channel

The infrastructure channel (Figs. 2.23e and 2.24e) forms a special type of walkable duct. They are mainly built in open construction (Fig. 2.36). The shafts can be constructed using either precast parts or in-situ concrete. The trench normally stands on a gravel reinforced base bedding and must maintain at least 1 m distance on both sides from the channel to allow for formwork installation, or for unobstructed operation of the concrete machine.

When using concrete machines, these are moved in sections and achieve a laying speed of up to 15 m per day in both directions [26]. The vaulted shape eliminates the need for steel reinforcement, thus extending the durability of the cover. The addition of fly ash in the concrete accelerates curing and achieves very high mechanical strength [17]. The infrastructure channel offers the greatest number of options in terms of possible cable capacity, accessibility, and replaceability of components compared to all other laying possibilities. However, the large excavation volume and the large amount of excess excavation material that must be disposed of—if the channel is not laid above ground—are disadvantages (Fig. 2.37).

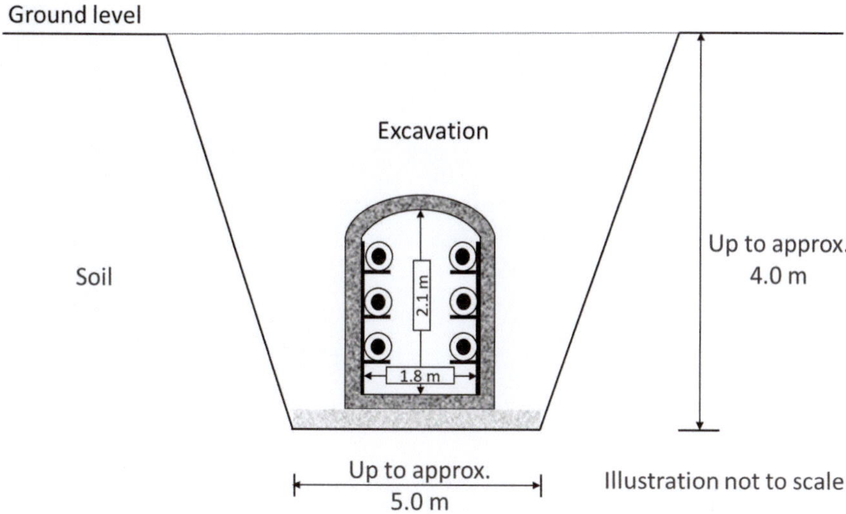

Fig. 2.36 Infrastructure channel laid in open trench construction for two 380 kV three phase cable systems (after [17])

Fig. 2.37 Continuous production of an infrastructure channel (*left*: by concrete machines on site, *right:* built in precast construction) [27]

2.3.5 *Joints*

For transport reasons the delivery lengths of high and extra high voltage cables are usually limited to 1000 m (in exceptional cases 1500 m as special transport). There are two reasons for this limitation. First, the transport weight on public roads is usually restricted to 50 tons. Second, the cable drums must have diameters that allow for safe and permissible transport on public roadways. This implies that the individual partial lengths must be connected with cable joints. Depending on the type of installation, different possibilities exist for the arrangement of joints. In walkable infrastructure pipes or channels, the joints can be integrated into the straight cable line and attached to the walls just like the cables (Fig. 2.38). Here they are freely accessible and can be

Fig. 2.38 Arrangement of 380 kV three phase cable joints in an infrastructure channel [28]

inspected at any time if necessary. For all other types of installations, joint pits are necessary, in which the joints are either laid or directly buried in the ground. The two possibilities are used differently. Currently, for many projects, the direct laying of the joints together with the cables in the bedding material is preferred. Their position, like that of the cables, is carefully measured and documented to enable quick and accurate localization in the event of a defect.

For 380 kV cables, joints reach diameters of 600 mm and lengths of over 2,000 mm. Direct current joints are comparable in size to those used for alternating current cables at the same voltage level. To avoid exceeding the width of the trench, the joint assembly is mostly staggered (Chap. 5).

Since joint assembly requires the highest standards of care and cleanliness, this work to connect individual cable sections is preferably performed in an enclosure or in a dust-free, air-conditioned container, which is relocated to each jointing point and removed after the entire line is completed.

The backfilling of the joint pits is then carried out in the same manner as for the cable trenches with bedding material, warning tapes and protective concrete slabs. Figure 2.39 shows a joint pit for four 380 kV AC systems (12 single conductors) after the joint assembly and before backfilling (Fig. 2.39, left picture) as well as a clean room container for the fabrication of HVDC cable joints (Fig. 2.39, right picture). Here the staggered arrangement of the joints is clearly visible and also the precautions for the water retention in the cable trench (bottom right).

2.4 Transition Compounds from Cable to Overhead Line

In partial cabling, i.e. the transitions from cables to overhead lines and vice versa, transition facilities called "transition compounds" are necessary (for both kinds, for underground cables from power plants or substations to overhead lines, and for

Fig. 2.39 Joint pit for four 380 kV alternating current systems, *left* [29] and an air-conditioned container above the joint pit for two HVDC cables, *right* [30]

intermediate cable sections in an overhead line). In the high and extra-high voltage range they can take on landscape-defining dimensions depending on the number of systems to be transferred. This mainly affects high and extra high voltage alternating current systems. Partial cabling is also a solution for HVDC connections, due to the smaller number of conductors their dimensions are generally smaller compared to those for AC lines.

The transition compounds contain all technical components to enable the transition. Two compounds are required for each intermediate underground cable section, Fig. 2.40 shows the principle. Depending on project-specific requirements and local conditions, these facilities typically require a footprint of approximately 50 × 70 to 130 × 150 m [31].

For longer cable lengths, reactive power compensation is required for efficient active power transmission, as already described in Chap. 1. For 380 kV AC cable sections of several 10 km system length, so-called compensation coils (also called chokes) are installed in the compounds and thus ensure the full transmission power of the cables. An additional switchgear field must be provided for the chokes. The overall facility requires more space in this case than a simple transition compound for short cable sections. For very long cable lines additional separate facilities for compensation may become necessary.

During the operation of transition compounds, the legal limits for electric and magnetic fields are already significantly undercut at the compound's fence. Since power lines and compensation coils can cause noise, the grid operator must

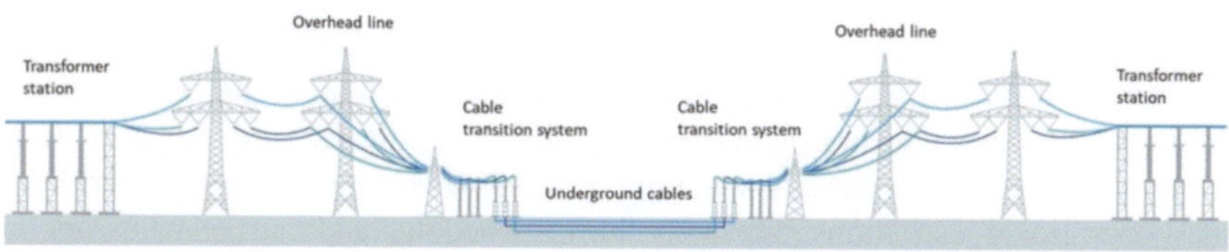

Fig. 2.40 Principal sketch of the intermediate undergrounding of a 380 kV line [31]

Fig. 2.41 Transition compound from a 380 kV AC double system overhead line to four cable systems (facility without compensation devices) [32]

also ensure compliance with the values given in the relevant standards and other publications (e.g. in Germany: Technical Instruction for Noise Protection).

Figure 2.41 shows an example of a transition compound from four 380 kV AC cable systems to a double system overhead line. The technical setup of partial cabling inevitably entails operational risks. Thus, the overhead line section remains exposed to atmospheric influences which can cause overvoltages, earth faults and even short circuits.

The overhead line sections can be controlled by the so-called *automatic reclosure.* High lightning impulse voltages, which encroach onto the cable systems, can, however, lead to the destruction of the cable's insulation if these impulses are not reliably eliminated by an efficient overvoltage protection system. This must therefore be placed as close as possible to the entry of the overhead line into the compound.

2.5 Utilization of the Line Corridor

The transmission capacity related to the width of the corridor of electrical lines serves as a measure of "line utilization". It differs between overhead line and cable due to different cross sections, distances, materials, and operating methods. Moreover, due to their routing above-ground, overhead lines are only partially comparable to underground cables in terms of visual impact. In the case of multiple circuit overhead lines, operational restrictions may apply during scheduled and fault related work on individual circuits. This is always the case when other circuits have to be switched off for safety reasons during work on a circuit and the (n − 1) criterion may not be met. In the case of underground cables, however, it can be assumed in certain cases that work can be carried out on a system without restrictions, even if the systems laid parallel at a sufficient distance are still in operation. Possible influences must be considered here. Anyhow, like for overhead lines also for cables, each work on lines has to be checked on a case-by-case basis.

Figure 2.42 exemplary shows the maximum possible line utilizations for the common overhead line tower designs and equipment as well as for the common types of cable laying arrangements in voltage levels 110, 380 kV AC and ±525 kV DC. In this example of undergrounding of overhead lines, the most common conventional conductor types (in Germany) 264-Al1/34-ST1A and 550-Al1/71-ST1A (equivalent to ACSR) are considered. The choice of bundle arrangement (single conductor, double or quadruple bundles) takes into account the different transmission capacities and the electric field strength. Thus, a 110 kV double circuit overhead line in conventional design with single conductors can transmit a total power of around 260 MVA. A cable connection at the same voltage level with approximately equivalent transmission reliability with two cable systems and a reserve phase transmits the same power at a multiple of the line utilization. Similar ratios are also evident in the other cable corridors. The line utilization is always higher for cables than for overhead lines, although this advantage noticeably decreases with increasing nominal voltages. This is mainly due to the fact that the cable insulation, compared to the insulation of overhead lines, is only a few cm and the distance between the phases and the systems can therefore be significantly smaller. In the example, the distances are determined by the best possible heat dissipation from the cables to the surrounding soil.

Overhead lines

	Tower form	Cross section	Conductor Design	P_{max} MVA	Route width m	**MVA/m**
110 kV AC		264-Al1/34-ST1A	●	260	44	**6**
			●–●	520	45	**12**
			●	520	50	**10**
			●–●	1.040	52	**20**
380 kV AC		264-Al1/34-ST1A		3.580	68	**53**
		550-Al1/71-ST1A		5.370	68	**79**
		264-Al1/34-ST1A		7.160	72	**99**
		550-Al1/71-ST1A		10.740	72	**150**

Underground cable

	Installation type	Cross section mm²	P_{max}	Route width m	**MVA or MW/m**
110 kV AC	Two systems +spare cable	630 Cu	260 MVA	5	**52**
380 kV AC	Four Systems	2.500 Cu	4.000 MVA	23	**174**
+/-525 kV DC	Two Systems	2.500 Cu	4.700 MW	20	**235**

Fig. 2.42 Maximum line utilization of overhead lines and underground cables in the voltage levels 110 kV AC, 380 kV AC and ±525 kV DC [2, 33–35]

2.6 Crossings, Proximities and Parallel Routes

Crossings, proximities and parallel routes with or to other infrastructure facilities play a significant role in the routing, design and selection of a line. Crossings with other cables or overhead lines, railway routes, federal highways and waterways are particularly noteworthy. For these facilities, special coordination is required for both technical and legal reasons to enable an agreement specifically tailored to the individual crossing.

Main components are regulations for the installation and subsequent operating phase as well as the dismantling of facilities (follow-up obligations and costs). During the negotiations, a number of technical regulations, guidelines, laws and rules for the affected crossed objects must be observed. The results of the negotiations are recorded in crossing contracts and technical documents respectively in permits.

When creating crossings, in addition to the technical safety measures (e.g. scaffolds or temporary relocations of electrical circuits and facilities), the influences of crossed facilities on each other must be reduced to a minimum. In the case of overhead line crossings such influences may include alternating current corrosion, touch voltages (both long- and short-term influences) as well as thermal influences. The latter mainly refer to cable crossings. In such cases, a model-based calculation must demonstrate that no overheating occurs in the crossing area under conditions of maximum transmitted power (Chap. 5). The consequence of mutual thermal influence can be, as already described, a reduced transmission capacity or damage to the crossing facilities.

For overhead lines, in addition to possible influences on third party facilities, distances to crossing and proximities to objects must be maintained. The minimum clearances applicable here are specified e.g. EU-wide in the standard DIN EN 50341-1 [3] and its National Normative Annexes (NNA) [4] and serve as a basis for determining clearances.

The following minimum load conditions in EN50341-1 must be taken into account for the calculation of electrical clearances, which can be specified differently in the NNAs:

- *Highest conductor temperature*: Highest constant operating temperature of conductors.
- *Wind load cases*: In addition to the load case without wind, the nominal value of the 10-min wind load with a return period of 3 years and the extreme value of the 10-min wind load with a return period of 50 years are used as a basis. A possible decrease in operating temperature due to lower ambient temperatures and/or wind action may be taken into account in the design of a line and in line operation.
- *Ice loads*: Extreme value of the ice load with a return period of 50 years.
- *Simultaneous wind and ice load*: The assumptions and calculation methods for this load case are referred to in the National Normative Annexes. The aim is to optimize the assumptions for wind speeds and ice loads to such an extent that an economic design of transmission and distribution grids with maximum stability

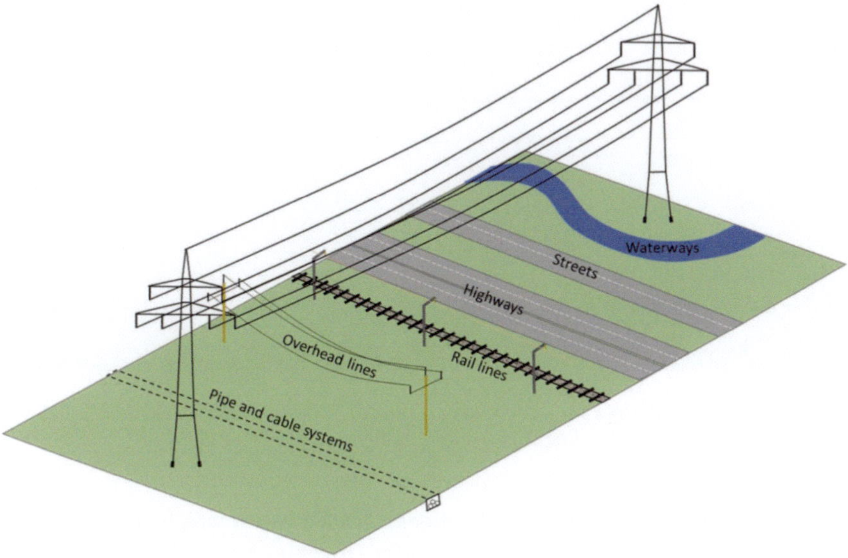

Fig. 2.43 Overhead line crossings with linear infrastructure facilities

is made possible. In the case of expected extraordinarily extreme weather events, the safety of persons is paramount.

In Fig. 2.43, possible overhead line crossings with various objects are schematically represented.

Major infrastructure operators such as railway companies or federal highway authorities often issue separate guidelines for such crossings, which allow standardized processes of crossing procedures. For crossings of objects that must remain operational undisturbed during the construction of an overhead line, such as other overhead lines, railways and motorways, extensive protective scaffolding is sometimes used. These structures are temporarily erected at locations agreed upon in advance with the landowners. In general, they require a significant amount of space and are dismantled after completion of the work. Figure 2.44 illustrates an example of a protective scaffold using a safety net to cover both a motorway and a railway line during stringing of the conductors.

2.7 Lines with Temporary Overhead Line Structures

Wherever there are existing or potential bottlenecks in the transmission grid, there is a need for expansion or upgrading. This is especially the case for connections that are heavily loaded and form an indispensable part of the transmission or distribution grid. How can they be rebuilt or upgraded if—due to their importance—they cannot

Fig. 2.44 Protective scaffold for the crossing of a high voltage overhead line with a highway (*left*) and an electrified railway line (*right*) during stringing of conductors [36]

be taken out of service for extended periods? In such cases, which are expected to become more common in the future, temporary structures are used, which take over the transmission function during the reconstruction phase. Another key application for temporary structures is the repair of faults and weather-related disasters, where lines are damaged. For example, lines damaged by falling trees or fail in whole sections due to overloads from wind, snow and ice can be replaced provisionally with such systems to rapidly restore supply.

Two fundamentally different construction methods are used for temporary structures: anchored towers and portals on the one hand, and structures with weight (or gravity) foundations on the other (Fig. 2.45).

At anchored towers and portals, the footing points are placed on wooden or metal plates for a better and evenly distribution of the soil pressure. Some of those constructions have movable connections at the foundations. Stability is achieved through lateral anchoring with steel ropes, which are typically attached to ground anchors or to wooden or metal beams embedded in the ground. The key advantage of this construction method is the use of lightweight, easily transportable individual components—typically lattice segments made of steel or aluminum. The disadvantage, however, is the large temporary use of land required for the anchoring and the high susceptibility to sabotage.

The temporary structures with ballast foundations usually also consist of individual components made of lattice steel or are solid steel poles. The footing points of the shafts are rigidly connected to a crossed foundation structure, the extensions of which are weighted with performed concrete blocks to provide the necessary stability. The advantage of this type of construction is the small space requirement, which is confined to the cross-shaped foundation area. By placing protective foils and piling up grit beneath the ends of the foundation cross-designed as footing points, soil disturbance can be minimized, provided the subsoil has sufficient load-bearing capacity. The disadvantage is the need to transport the heavy ballast weights to the

Fig. 2.45 Temporary structures with anchored towers (**a**) and with ballast foundations (**b**) [36, 37]

tower locations, which typically requires the construction of temporary access roads. Figure 2.46 shows examples for temporary structures of both variants.

2.8 Re-routing of Lines

Re-routing refers to the subsequent surveying and documentation of all line parameters of an overhead line or a cable. This is recommended after noticeable changes in the line area, identified during line inspections, and which are not reflected in the

Fig. 2.46 Examples of temporary overhead lines with anchored tension portals (*left*) [38] and tension tower with ballast foundations (*right*) [39, 40]

original as-built-drawings or plans. In addition to vegetation, it is essential to document changes in land use—particularly any underbuilding or overbuilding within the corridor—that may have gradually occurred over time.

At the same time, any changes in the operating conditions of the lines over time must also be considered. Many overhead lines were planned and built based on construction regulations of that time with e.g. conductor temperatures of a maximum of 40 °C. Although at that time the designers in general were very generous and generally opted for increased clearances, e.g. by using the next larger constructively predetermined element of the tower body.

However, it must be checked in each individual case whether the line can be operated with a higher conductor temperature. The individual heights of the bodies' elements were—in case of steel lattice structures—usually at least 2 m and were connected in a cascade like manner. The permissible operating temperature for ACSR conductors was already 60 °C in the standards for conductors at that time. This often served the grid operators as a benchmark for line operation. The associated increases in conductor sag have often led to reinforcements of towers and the insertion of intermediate elements in the tower bodies to increase their heights over the past decades. With such adaptations the operators can maintain the necessary clearances to the ground and to other crossed objects at these conductor temperatures and possibly even higher ones.

Re-routing will also be necessary prior to re-conductoring with high temperature conductors (HTLS conductors, see Chap. 4). Meanwhile, re-routing can be carried out easily and efficiently by flying over the lines using laser scanning.

Re-routing is necessary for cable lines for similar reasons if necessary. Subsequent developments on and in the vicinity of the line may have occurred, as well as crossings or parallel routings of other infrastructure facilities (including other cables, water pipes, sewer pipes, district heating and cooling pipes, roads and paths, etc.). Ground sinkage and changes in groundwater levels can also affect the operation of a cable system and must be carefully recorded during re-routing.

2.9 Biodiversity in Line Corridors

The environmental discussion includes not only visual impression, audible noise and electric, magnetic and electromagnetic fields, but also fauna and flora in the area of power lines. In this context, changes in biodiversity play a prominent role. The areas in the corridors used by both transmission techniques can continue to be used for agriculture, but with restrictions, as previously described. For overhead lines, vegetation is limited in height to maintain the required clearances to the conductors. In the case of underground cables, no deep-rooted plants are allowed in order to protect the cable from damages.

If the corridor is not used for agriculture, valuable new habitats for various species of life can be created in the route aligning with the principles of a modern ecologic planning culture. Routes of overhead lines are usually wider than routes of cables and thus offer good opportunities for creating new habitats. Many line operators are open to this approach and engage experts with planning, creation and care of the biotopes. Through this concept of "extended ecology", desolate or alienated areas can be rehabilitated by ecological corridor management.

For example, in Belgium and France over 20 projects with a total of several hundred hectares were implemented by the LIFE-ELIA initiative [41]. In the routes of overhead lines, moors were (re)established, pastures, natural fruit areas, biotopes, hiking routes were created.

In several areas, it has even been possible to increase the number of particularly protected species of plants through new settlements, as demonstrated by an example from Upper Austria [42]. The construction of a 220 kV overhead line as a replacement for a 110 kV line runs 1200 m in the designated landscape protection area of a river floodplain. This project offered the opportunity to restore the existing partly denatured areas of this floodplain in a natural way. After the restoration, a total of 680 species of plants were recorded, among them 17 endangered ones from the *Red List—Data Inventories* that had newly established themselves in the area (Fig. 2.47).

The success of the project and the resulting *habitat overhead line corridor* can be largely attributed to the intensive involvement of the responsible authorities, whose suggestions and requirements were actively incorporated. Through such a procedure, a route not only has its function for the power line, but also offers a valued habitat for many—also new—plants and animals. It is also used as recreation area for the local population. In many cases, this form of corridor management is required as part of the approval process, serving as ecological compensation for land use impacts and being formally stipulated in official permits (Chap. 3).

During the operation of the mentioned line, maintenance follows the *Sustainable Line Management* with the following principles and thus takes into account the preservation of the desired biodiversity:

- The safety of line operation has top priority.
- Maintenance work takes into account the needs of the environment and the landowners in the best possible way.

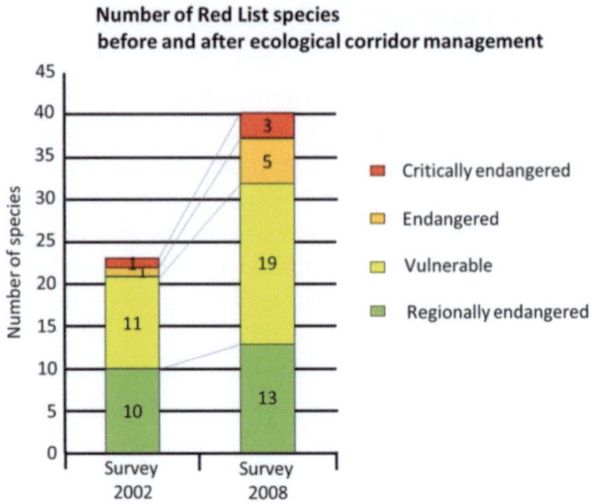

Fig. 2.47 Improved biodiversity in a line corridor: 17 new species of plants from the "Red List—Data Inventories" have become native in a few years [42]

- Maintenance work takes into account the natural and cultural background of the region, as well as the ecological situation.
- There is a permanent dialogue with the landowners and representatives of the various interest groups.
- Measures are taken as far as possible by the local farmers and foresters.

The measures illustrated through the example of overhead lines can also be used for underground cables and can be implemented in the relevant areas equally to maintain or improve biodiversity. As part of vegetation management, which line operators regularly carry out, aspects of biodiversity can be integrated to support ecological goals.

CIGRE has published a noteworthy technical brochure (TB) on sustainable corridor management for overhead lines and underground cables, which contains data from 41 survey responses from 32 countries, descriptions of best practice examples and experiences from Working Group Members [43]. These are of great benefit to planners and operators of transmission lines. The TB is also of benefit to urban planners, landowners and their organizations, as well as other interest groups that carry out activities in the corridors of the electricity infrastructure.

References

1. Gehmacher, Piereder-Hagen. Integration von Freileitungen in die Natur, NANU Forschung im Verbund Schriftenreihe Band 34, Wien
2. Boos K-V, Müller K, Palic M, Schenk G (1986) Betriebserfahrungen mit Mehrfach-Hochspannungsfreileitungen in kompakter Bauweise. Elektrizitätswirtschaft 85(21):814–819
3. DIN EN 50341-1 Freileitungen über AC 1 kV - Teil 1: Allgemeine Anforderungen - Gemeinsame Festlegungen, EN 50341-1 (2012)

4. DIN EN 50341-2-4 (VDE 0210-2-4), Freileitungen über AC 1 kV - Teil 2–4: Nationale Normative Festlegungen (NNA) für Deutschland, EN 50341-2-4 (2019)
5. Palic M (1982) Programmierte Abstandsermittlung zwischen Freileitungsseilen. Elektrizitätswirtschaft 81(9):289–293
6. Dyk U, Hirsch F (2002) Freileitungsnorm EN 50341 Bemessung und Nachweis der Elektrischen Abstände für Freileitungen, Freileitungsnorm in neuer Gestalt EN 50341 (VDE 0210). ETG-Fachtagung 15.–16. Mai 2002, Würzburg
7. van Fondern M (2014) Abstände, Vortag anlässlich des Seminars "Grundlagen des Hoch- und Höchstspannungsnetzbaus". Hochschule Karlsruhe, 26.09.2014
8. Göhring A (1971) Diskussionsbeitrag auf der VDEW-Freileitungstagung 1970. Elektrizitätswirtschaft 70:5–6
9. Kießling F, Nefzger P, Kaintzyk U (2001) Freileitungen - Planung, Berechnung, Ausführung, 5. Auflage. Springer
10. Palic M (1989) Abstandsermittlung bei Freileitungskreuzungen. Elektrizitätswirtschaft 88(23):1601–1604
11. Geißler J (1987) Optimierung von 110 kV Gestängen. Diplomarbeit, Fachhochschule Karlsruhe
12. Kirn H, Palic M (1990) Messung und Berechnung der Freileitungsseiltemperatur bei veränderlicher Strombelastung. etz 111(1)
13. Peschke A, Siebert M (2016) 380 kV Leitung Stade – Sottrum – Wechold – Landesbergen (Ersatzneubau). TenneT TSO, Sweco
14. Fleckenstein K, Rhiem W, Palic M (1991) Prüfungsrahmen zur Findung von Freileitungstrassen in Waldbereichen. Elektrizitätswirtschaft 90(21/22):1135–1139
15. Goultry GA (2014) The transmission of electricity in England and Wales; land use, and amenity aspects, 1969. Durham University
16. Gehlen C (2015) Netzbau Zwischenverkabelung in Raesfeld (Pilotprojekt), Vortrag beim Forum Netzbau und Netzbetrieb, Frankfurt, 07.–08. Mai 2015
17. Brakelmann H, Jarass LJ (2019) Erdkabel für den Netzausbau, BoD Norderstedt
18. Anon., Unterlagen zur Antragskonferenz Südlink, TransnetBW (2019)
19. Wolbring N. Was kommt da auf uns zu? DLG-Mitteilungen 9/2015
20. Anon., Hochspannungs Blog NetzOÖ, Netz Oberösterreich GmbH, Linz
21. Anon., Pipe Express sowie Sonder- und Tunnelbauverfahren, Herrenknecht AG, Schwanau (2015)
22. Stein D, Stein R. Erneuerung in halboffener Bauweise Instandhaltung von Kanalisationen, Prof. Dr.-Ing. Stein und Partner GmbH, Bochum
23. Bayer H-J et al (2016) HDD-Praxishandbuch. Vulkan-Verlag, Essen
24. BFR-Abwasser, Baufachliche Richtlinien Abwasser - A-6.4.3.4 Horizontal-Spül-Bohr-Verfahren Bundesbauministerium
25. Peters M (2021) E-Power Pipe® als Standardverfahren für eine wachsende Anzahl von Projekten in der Erdkabelverlegung. Herrenknecht AG, Schwanau
26. Brakelmann H et al (2011) Power-Tubes – innovatives Kabel- und Installationssystem. ew 110(25–26):74–81
27. Brakelmann H (2015) HuH-Kabel - Ein Vergleich unterschiedlicher Kabeltechnologien, Vortrag beim Forum Netzbau und Netzbetrieb, Frankfurt, 07.–08. Mai 2015
28. Henningsen CG (2015) 380-kV Diagonale Berlin - ein Erfahrungsbericht Vortrag beim Forum Netzbau und Netzbetrieb, Frankfurt, 07.–08. Mai 2015
29. Anon., Drehstrom-Erdkabel - Chancen und Herausforderungen bei 380 kV Erdkabeln im Drehstromnetz Broschüre, TenneT TSO GmbH, Bayreuth (2016)
30. Keussen U (2017) Netzausbau in Bayern und Deutschland. Tennet TSO, Bayreuth
31. Anon., Erdkabel im 380 kV Drehstromnetz, Broschüre TenneT TSO GmbH, Bayreuth (2020)
32. Anon., Dörpen West – Niederrhein, Mitteilungen der Amprion, Dortmund
33. Palic M et al (1992) Kabel und Freileitungen in Überregionalen Versorgungsnetzen. expert Verlag, Ehingen
34. Palic M (2021) Grundlagen des Hoch- und Höchstspannungsnetzbaus - Einführung in die HuH-Netztechnik, Vortrag anlässlich des Seminars "Ausgewählte Themen des Hoch- und Höchstspannungsnetzbaus". Hochschule Karlsruhe, 9. und 10. März 2021

35. Anon., SüdOstLink Höchstspannungsleitung Wolmirstedt-Isar Gleichstrom, Technische Vorhabensbeschreibung Abschnitt D
36. Knauer M. Kreuzungen mit Infrastrukturbetreibern. TenneT TSO GmbH, Bayreuth
37. Anon., Notgestänge und Provisorien, Broschüre equos Energie, Bieberach
38. Eggert S et al (2019) 110 kV Leitung Schuby-Schuby/West, Erläuterungsbericht. Schleswig-Holstein Netz AG, Quickborn
39. Anon., Für den Notfall gut gerüstet - Avacon testet neues kV Notgestänge, Pressemitteilung avacon 07.02.2018
40. Anon., CP-Gestänge, Broschüre cteam, Ummendorf
41. Anon., Creating green corridors under overhead lines. http://www.lifeelia.eu/en/Projects-sites
42. Hagen A, Lenglacher F, Aberle S, Minichberger H, Lugschitz H (2010) Overhead line in a protected floodplain woodland "extended ecology" as basis for OHL planning and maintenance. CIGRE 2010, B2-107-2010
43. CIGRE TB 966 (2025) Sustainable corridor management. Paris

Chapter 3
Expansion Needs and Approval Procedures

Summary

The need for the expansion of power grids will increase enormously worldwide in the coming years and decades. This growing need is globally driven by the decarbonization of power generation and the simultaneous substitution of fossil energy sources by electricity applications, among other things in the areas of industry, mobility, building heating, and hydrogen production. It affects transmission and distribution grids equally. Cross-border electricity exchange plays a special role in this context (Chap. 7).

This chapter examines the forecasts of the expansion needs for power grids in the major economic nations and describes the approval practice in the European Union and presents it exemplary in some countries. The approval procedures in Germany are described in greater detail due to their complexity and extensive nature. Many of these procedural elements can also be found in the approval procedures of other countries, sometimes in a modified form.

In most countries of the world, environmental impact assessments play a central role in line projects, especially at the high and highest voltage levels, alongside considerations of energy-related and economic necessity. Their purpose is to ensure that interventions in nature and landscape are kept as low as possible during the necessary grid expansions.

3.1 Global Expansion Needs

Relevant forecasts for the expansion of power grids in the major economic nations indicate a considerable need. For example, the USA predicts that their transmission grids will need to be expanded by 60% of the stock by 2030 based on the year 2023. The State Grid Corporation of China has announced investments in power grids amounting to 1020 trillion CNY (132 billion EUR) for the years 2022 and

K. O. Papailiou et al., *High-Voltage Overhead Lines and Underground Cables*, Power Systems, https://doi.org/10.1007/978-3-032-07498-0_3

2023. According to estimates by the International Energy Agency, the global power grid expansion requirement by 2040 amounts to more than 80 million km, which corresponds to the total length of the global power grid in 2023. At least 3000 gigawatts (GW) of renewable energy projects, of which 1500 GW are in an advanced stage, are also in the queue for grid connection [1].

3.2 Expansion Needs and Approval Procedures in the European Union

The European Union has one of the densest and most resilient power grids in the world, extending over more than 11 million kilometers throughout the EU-internal market. Based on the EU Regulation on Trans European Energy Infrastructure (TEN-E), the EU has selected more than 100 projects of common interest (PCI) in the electricity sector and facilitated their approval and construction, among other things by providing funds for them. This has helped to develop a physical electricity infrastructure suitable for a real EU-internal market [2, 3].

Development of the European Grid Association

The Europe-wide grid expansion is the central instrument for achieving the goals of the Energy Union. To this end, ENTSO-E, as an EU wide association of transmission system operators, prepares the European grid development plan every two years based on Regulation (EU) No. 347/2013 [4], which shows how the transmission grid should be expanded in the next 10–20 years to meet future requirements. The main focus of the evaluation is on all transmission or storage projects in Europe.

This development plan, known as the *10-year network development plan* (TYNDP), is based on the regulation and takes into account the guidelines specified therein for the timely development and interoperability of priority trans-European energy infrastructure corridors and areas. These are listed in its Annex I. However, the TYNDP only has a recommendatory character, into which, for example, the projects planned in Germany according to the Federal Requirements Plan Act are embedded. Figure 3.1 schematically shows the relationship between European and German grid development.

Now that the EU markets are fully coupled, the completion of the infrastructure grid is the next step to maximize the benefits from the power grid. Despite these advances, the European power grids face considerable challenges. They must also satisfy the increasing demand associated with electromobility, heating, cooling, electrification of industry, and the boost of low-CO_2 hydrogen production. Electricity consumption is expected to increase by around 60% across the EU by 2030. The grids must also be able to integrate large amounts of volatile renewable energy. The capacities for generating wind and solar energy must increase from 400 GW in 2022 to at least 1000 GW by 2030, coupled with a comprehensive expansion of renewable offshore energy to 317 GW, with a powerful and reliable connection to the mainland

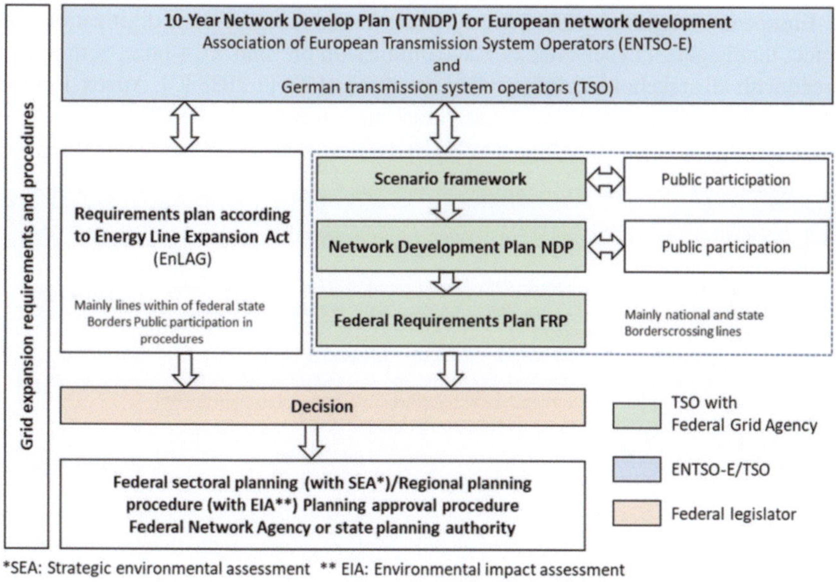

Fig. 3.1 Determination of the expansion needs in the European transmission grids

[5]. Thus, the grids must be adapted to a more decentralized, more digitized, and more flexible electricity system. According to the Ten-Year Network Development Plan (TYNDP) of the European Grid of Transmission System Operators (ENTSO-E), the cross-border transmission infrastructure would have to be doubled in the next seven years, with an additional 23 GW of capacity to be integrated by 2025 and a further 64 GW by 2030 [3].

The distribution grids, on the other hand, are taking on new functions and must meet a series of new challenges. They need to be transformed into intelligent grids that are digitally configured, remotely controllable and cyber secure and can be monitored in real time. In addition, about 40% of European distribution grids are older than 40 years and need modernization. According to industry estimates, investments in the distribution grids alone amounting to 375–425 billion EUR are required by 2030 [6]. According to Commission estimates, a total of 584 billion EUR needs to be invested in the electricity grids this decade. In many EU countries, there are long waiting times for the approval of grid connections for renewable energy generation projects. The waiting times for the approval of grid reinforcements are four to ten years, in the case of high voltages even eight to ten years [7]. Therefore, the EU has put the expansion of the electricity grids at the center of its agenda. With the revised Renewable Energy Directive [8], the approval of electricity grids needed for the integration of renewable energies is simplified. The regulation and the directive on the internal electricity market [9] contain rules that are relevant for the expansion of the grids with regard to planning, the grid tariffs and the role of ENTSO-E and the European Organization of Distribution Grid Operators (EU-DNO). To strengthen the European electricity grids, better connect them, increase their digitization and protect them against cyber-attacks, the Commission presents a 14-point action plan agreed with all stakeholders in its communication of 28.11.2023 [7]. Annex 1 of the communication (Fig. 3.2) lists the individual measures and the deadlines.

CATEGORY	ACTIONS	TIMELINE
Accelerating the implementation of PCIs and developing new projects	1. Commission, Member States and TSOs to strengthen support to PCI and PMI preparation, faster implementation and funding	**From 2024**
Improving long-term grid planning for a higher share of renewables and increased electrification	2. ENTSO-E to enhance top-down planning towards 2050 by integrating the identification of offshore and onshore system needs and further considering hydrogen 3. EU DSO Entity to support DSO grid planning by mapping the existence and characteristics of distribution development plans	**From Q1 2024** **Mid-2024**
Introducing regulatory incentives for forward-looking grid build-out	4. Commission to propose guiding principles identifying conditions under which anticipatory investments in grid projects should be granted 5. Commission to issue guidance on cross-border cost sharing for offshore projects	**Q1 2025** **Mid-2024**
Incentivising a better usage of the grids	6. ENTSO-E and EU DSO Entity to agree on harmonised definitions for available grid hosting capacity for system operators and establish a pan-EU overview 7. ENTSO-E and EU DSO Entity to promote uptake of smart grid, network efficiency and innovative technologies 8. ACER, in its next tariff report, to recommend best practices in relation to the promotion of smart grids and network efficiency technologies through tariff design, focusing on consideration of OPEX in addition to CAPEX and benefit sharing	**From adoption** **Q4 2024** **Q1 2025**
Improving access to finance	9. Commission to identify tailored financing models and strengthen dialogue to address financing obstacles 10. Commission to increase visibility on opportunities for EU funding programmes for smart grids and modernisation of distribution grids	**From adoption** **From Q1 2024**
Accelerating deployment through faster permitting and public engagement	11. Commission to support permitting acceleration providing guidance and technical support on how to implement existing legislative tools and Member States to implement acceleration measures 12. Commission to launch a Pact for Engagement for early, regular and meaningful stakeholder engagement and regulatory support	**2024-25** **From adoption**
Strengthening grid supply chains	13. ENTSO-E and EU DSO Entity to collaborate with technology providers to develop common technology specifications and improve visibility of grid project pipelines, to facilitate investments in manufacturing capacity and secure supply chains 14. Commission to promote common technical requirements for generation and demand connection	**Q4 2024** **By 2025**

Fig. 3.2 Annex 1 from *The action plan for the European electricity grids* (short version) [7]

3.3 Approval Procedures in Austria

A central component of approval procedures for high and extra high voltage lines in Austria are Environmental Impact Assessment Procedures (EIA). These are single-stage and concentrated approval procedures, are carried out by a state authority, which applies all relevant subject laws (laws that regulate a specific area). Most of the thresholds for triggering an EIA procedure for power lines have remained unchanged for many years. Recently added are the EIA criteria "deforestation" and "line clearings", where 20 ha or 50 ha are trigger criteria, as well as the complex summary of both. This applies not only to line projects, but to all projects, i.e. also for industrial plants, supermarkets, etc.

A very good overview of the environmental impact assessment in Austria is provided by the brochure "Environmental Impact Assessment" [10]. The Ministry responsible for EIA for Climate Protection, Environment, Energy, Mobility, Innovation and Technology defines the objectives of the EIA summarized as follows:

- to avoid environmental damage from the outset according to the precautionary principle,
- to consider the environmental impacts holistically and comprehensively, not just sectoral and in part,
- to achieve better preparation of the projects and the approval procedures,
- to bring environmental concerns into the balance and decision with the same value as other concerns and
- to make the approval procedures for projects more transparent and understandable with the participation of the public.

Projects of the energy transition have been granted enhanced legal status in the EIA Act since the amendment 2023 for facilities for the generation of renewables, for power lines and railways. For such projects, for example, the granting of the suspensive effect of an appeal in the procedure can be more easily rejected and thus delays reduced, and faster procedural processes can be achieved. From the perspective of supporters of the energy transition, such legislative changes are not far-reaching enough. At EU level, further measures are being discussed in this regard. Figure 3.3 shows the submission documents for the EIA as part of an approval application using the example of a 380 kV overhead line. The process depends of course on the current legal situation and changes with newly issued regulations and laws.

Procedures that are not subject to the EIA Act, and extend across two or more federal states, are conducted according to the High Voltage Transmission Line Act (*Starkstromwegegesetz).* In this case, the individual matters are considered separately. The responsibility lies with the affected ministries.

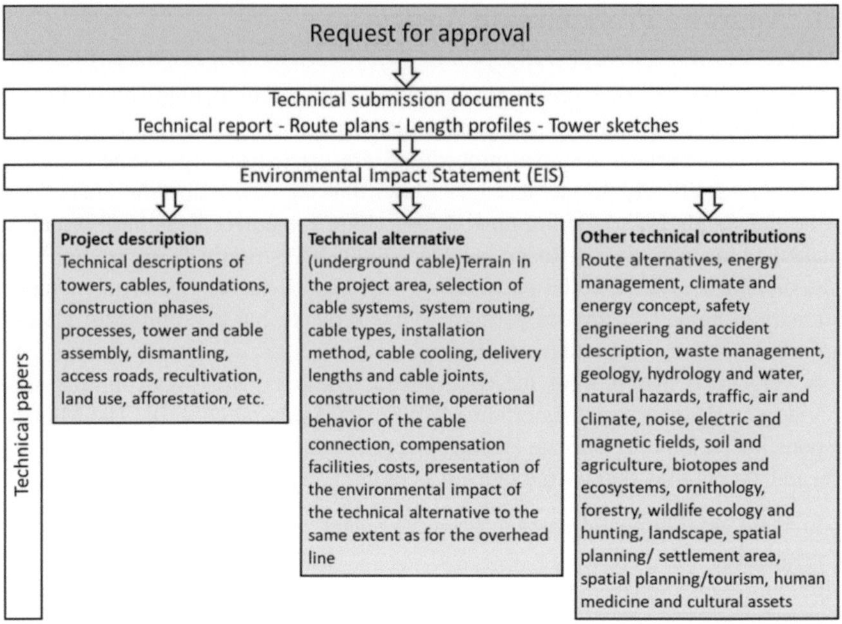

Fig. 3.3 Approval application and submission documents for Environmental Impact Statement (EIS) for a 380 kV overhead line in Austria

3.4 Approval Procedures in Germany

Basic regulations for grid expansion in Germany are contained in the *Energy Industry Act* (in German: *Energiewirtschaftsgesetz EnWG*), especially in its §§ 12a–12e and § 43. At the same time, it contains authorization for regulations and specific references to procedures to be carried out.

For the construction of high and extra high voltage lines in Germany, both public law assessment and approval procedures as well as private law agreements for the use of privately owned land and spaces are required. The legal framework for approval distinguishes between the voltage levels and also between the types of implementations. It depends on whether the planned project is an overhead line or an underground cable. Since new projects fundamentally represent an intervention in nature and landscape with possible devaluations and usage restrictions, the proof of their economic necessity is of substantial importance. However, this proof is provided differently. For projects classified as particularly or priority at the highest voltage level, the legislator has established economic necessity. This can no longer be questioned in the subsequent procedures. For the remaining projects, the proof of necessity must be provided on a case-by-case basis. This can be based on an overall view of the nationwide supply structure taking into account the expected supply requirements in the future. Figure 3.4 shows the basic implementation path for high and highest voltage lines with the individual planning, assessment and approval stages in Germany.

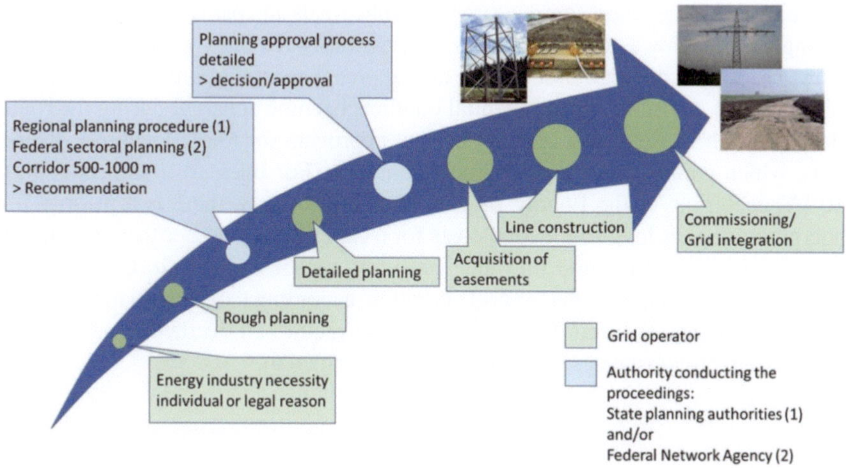

Fig. 3.4 Implementation path of line projects (overhead lines and cables) in the voltage levels 110–380 kV

The approval and environmental assessment procedures to be carried out in Germany are explained in detail in this chapter and illustrated using flow charts and diagrams. The distinction is made between procedures for overhead lines in the various voltage levels and the procedures for high and highest voltage cables. It also sheds light on the compensation practice when private land is used.

3.4.1 Procedure for High Voltage Lines

For 110 kV overhead lines, which are to be erected on a new route, a spatial planning procedure is mandatory. Excluded are 110 kV railway power lines, for which the Railway Act applies. In the spatial planning procedure, based on rough planning, line corridors with variants are examined for their spatial compatibility. The procedure usually ends with a "spatial planning assessment" or "state planning determination" for the route variant favored after careful consideration of all relevant aspects. This is followed by the implementation of a planning approval procedure. In this procedure, the project with the detailed route is "planned" and thus approved. After the decision becomes legally binding, the project can be realized in any case, if required, through expropriation. For cables at this voltage level, both spatial planning and planning approval procedures are optional (Fig. 3.8). For voltage levels below 110 kV, there are no public law procedures required for either cables or overhead lines. Here, the operator has the option to negotiate and realize lines with both local authorities and private landowners. It may be necessary to obtain approvals or permits related to nature conservation, water law, or forestry regulations. In the medium voltage level, which usually serves local supply, the grid operators in the municipal areas have the right, based on concession contracts, to use public roads and paths for the installation of supply lines. However, there is also the possibility for lines with voltage levels below 110 kV to undergo assessment and approval procedures upon application. This is always advisable when potential conflicts become apparent during the route selection, such as the refusal of private law approvals for tower locations or line spans, which could hinder project implementation. For lines up to and including 110 kV, according to the Energy Industry Act in Germany, there is a priority for undergrounding, as long as the total costs for a cable are not higher than 2.75 times that of an overhead line and no other important reasons oppose it. Conservation reasons or compelling reasons for grid operation, according to which no longer cable lines can be integrated into the grid, would also require an overhead line solution under certain circumstances, even if the factor is undercut.

3.4.2 Procedure for Non-priority High Voltage Lines

For overhead lines with a voltage of 110 kV and more, a spatial planning procedure is generally prescribed. This is followed by the mandatory planning approval procedure. If the economic necessity has not already been determined by the legislator, it must be specifically demonstrated and justified in the procedures. For lines in voltage levels above 110 kV, an underground cable is only considered in certain legally provided cases. In this case, a planning approval procedure is also required, a spatial planning procedure, however not in all cases. For lines up to 110 kV, a planning approval procedure can be carried out for an underground cable at the request of the grid operator. This may be advisable due to the comprehensive approval effect of the planning approval procedure.

3.4.3 Procedure and Laws for Priority High Voltage Lines

The federal legislator recognized at the end of the first decade of the twenty-first century, that the urgently needed and through various protests and delays endangered expansion of the transmission grid cannot be left to the free play of forces. The grid operators and the authorities were—despite greatest efforts without effective, legal support—not able to realize the necessary line connections identified on the basis of relevant grid studies. In addition, the European Union's guidelines for trans-European energy networks—in which the German transmission grid is integrated through multiple cross-border connections—recommend legislative measures aimed at accelerating implementation.

Energy Line Expansion Act (in German: Energieleitungsausbaugesetz EnLAG)

In 2009, the federal legislator, for the first time, defined in the Energy Line Expansion Act certain high-voltage line connections whose energy-economic necessity and urgent need were determined a priori. Their implementation is necessary for reasons of "predominantly public interest and in the interest of public safety". The appendix to the law currently lists 22 line projects (formerly 24) in the demand plan, which are to be built as high voltage three phase overhead lines. If the proximity to residential buildings is too close—less than 400 m in the case of closed construction and 200 m in the outdoor area, or for reasons of nature and protection of species, the grid operators can be mandated to construct parts of the line as underground cables. Six of the 22 projects are explicitly designated as pilot projects, intended to test the use of underground cables at the high voltage level in the form of partial cabling.

For the first time in the history of the energy sector, a law was enacted that legally mandates specific line connections. With the legally established planning justification, the subsequent planning procedures were to be accelerated. This also applied to necessary expansions and changes at the grid connection points through which the new lines are connected to the existing transmission grid. However, the law only defined the start and end points of the lines. It does not contain any guidance on the specific line routing. Determining the exact route was left to subsequent spatial planning and planning approval procedures.

Network Expansion Acceleration Act Transmission Grid (in German: Netzausbaubeschleunigungsgesetz NABEG)

Already in 2011, the need for further acceleration of the procedures for implementing line projects had become apparent. This was followed in the same year by the *Network Expansion Acceleration Act Transmission Grid*. As the name suggests, it served to accelerate the expansion of the high voltage grid and focused on lines that cross borders to other federal states or to other countries and therefore is of essential importance. The projects listed in the EnLAG are expressly excluded from the scope of application of the NABEG.

The planning and approval procedures for high voltage lines generally fall within the jurisdiction of federal state authorities. In the case of lines crossing federal borders or country borders, this law transferred jurisdiction to the Federal Network Agency (in German: *Bundesnetzagentur BNetzA*). It assesses the environmental and spatial compatibility of the projects within a nationally standardized, multi-stage planning procedure.

The so-called *Federal Sectoral Planning* forms the first stage of the procedure. This is followed by a planning approval procedure for the line in the previously defined line corridor, which is also carried out by the Federal Network Agency.

Federal Requirements Plan Act (in German: Bundesbedarfsplangesetz BBPlG)

The phase-out of nuclear energy and the already anticipated end of coal-fired power generation in connection with the forced expansion of generation from renewable energy sources required both an expansion and a further acceleration of power grid expansion.

As previously mentioned, this leads to fundamental changes in load flow and a fundamental shift in feed in points, which urgently require new and efficient transmission line connections. Just two years after NABEG, the German Bundestag and

Bundesrat passed the *Federal Requirements Plan Act*. The appendix to the law—originally enacted in 2013 and revised several times since—was designed to further accelerate grid expansion and includes a table of specific projects known as the 'Federal Requirements Plan' (Fig. 3.5).

It currently includes 80-line projects, each annotated with their status and implementation method. It is indicated whether they cross federal borders (A1) or country-borders (A2) or are to be constructed as overhead lines with high temperature conductors (D), as three phase underground cables (F) or as high voltage direct current transmission underground cables (E) or as underground lines initially built only as empty cable conduits (H) are to be laid.

With the 2015 amendment to the law, the German legislator ordered that the high performance north–south connections be primarily executed in the form of HVDC underground cables. An exception to the mandatory use of underground cables is only permitted for reasons of nature conservation, at the request of local authorities, or when the new electrical systems can be installed on the structures of existing overhead lines or parallel to them. In such cases, partial sections may be constructed as overhead lines.

For all projects, the law establishes energy-economic necessity and priority need. As with the lines according to EnLAG, there is therefore an *overriding public interest* in their realization, and they are necessary for public safety. Here too, their necessity cannot be questioned in later procedures, as they result from the legally regulated grid

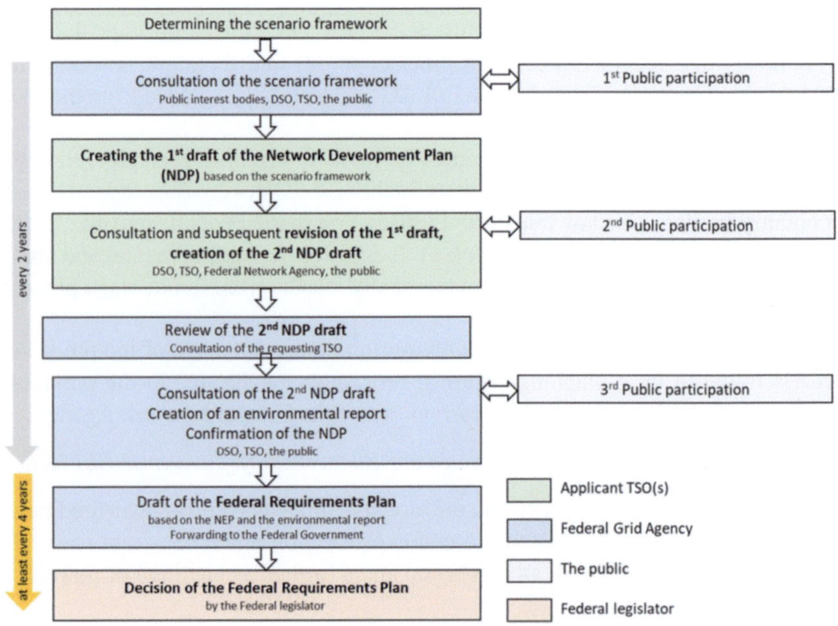

Fig. 3.5 Schematic representation of the process flow for the Federal Requirements Plan

development planning. This begins with the so-called scenario framework, which the four transmission grid operators taking into account future developments and submit it to the Federal Network Agency. It describes the expected developments in the areas of renewable energy, conventional energy, energy consumption and load. Particular consideration is given to the Federal Government's energy policy goals, to increase the share of renewable energy in electricity demand to 65% by 2030, the foreseeable shifts in generation- and load- centers, and the presumed exchange with neighboring countries in the future. Three scenarios are created for the next 15 years and one for the next 20 years. Figure 3.6 shows the forecast of generation and demand in the individual federal states for the medium scenario B 2035 (version 2021). According to this, a strong generation gradient will emerge within Germany, with the southern federal states expected to have a significant generation deficit and the northern federal states a similarly significant generation surplus. According to this medium scenario, Lower Saxony and Schleswig–Holstein alone will have an annual surplus of up to 150 TWh, which will be lacking in the southwest of the republic. At the same time, a slight net electricity imports from the neighboring countries of around 3% of the annual amount is expected. Renewable energies dominate almost all federal states in terms of generation technologies. The balance and the apparent generation gradient provide the justification for the high transmission requirement in the north–south direction.

As part of the subsequent consultation of the scenario framework, the general public as well as the distribution grid operators have the opportunity to comment. Based on the results of the scenario framework, which has been finally checked and confirmed by the Federal Network Agency, the transmission grid operators develop the first draft of the *grid development plan (in German: Netzentwicklungsplan NEP).* This plan specifies which grid connections are necessary to achieve the supply targets in the scenario framework. The transmission grid operators once again submit this draft available for consultation by the general public, the distribution grid operators and the Federal Network Agency. Based on the feedback received of this second round, the transmission grid operators create the second draft. The Federal Network Agency then reviews this version and conducts a final round of consultation with both the public and the distribution system operators. Following this, the Agency compiles an environmental report and formally approves the Grid Development Plan. This procedure is repeated every two years. At least every four years, the current version of the grid development plan is incorporated into the federal requirements plan, which is decided by the federal legislator and thus becomes law.

While the Energy Line Expansion Act (EnLAG) primarily lists projects that are within the borders of federal states, the Federal Requirements Plan Act contains also projects that cross both federal state and national borders. For projects that are within one federal state, the planning instruments of that state apply, such as spatial planning and planning approval procedures, and are carried out by federal state authorities.

If a project affects several federal states and/or crosses state borders and is designated as such in the Federal Requirements Plan Act (marked with A1 or A2), the Federal Network Agency is responsible for federal sectoral planning and the plan approval procedure; otherwise, the state authorities are responsible.

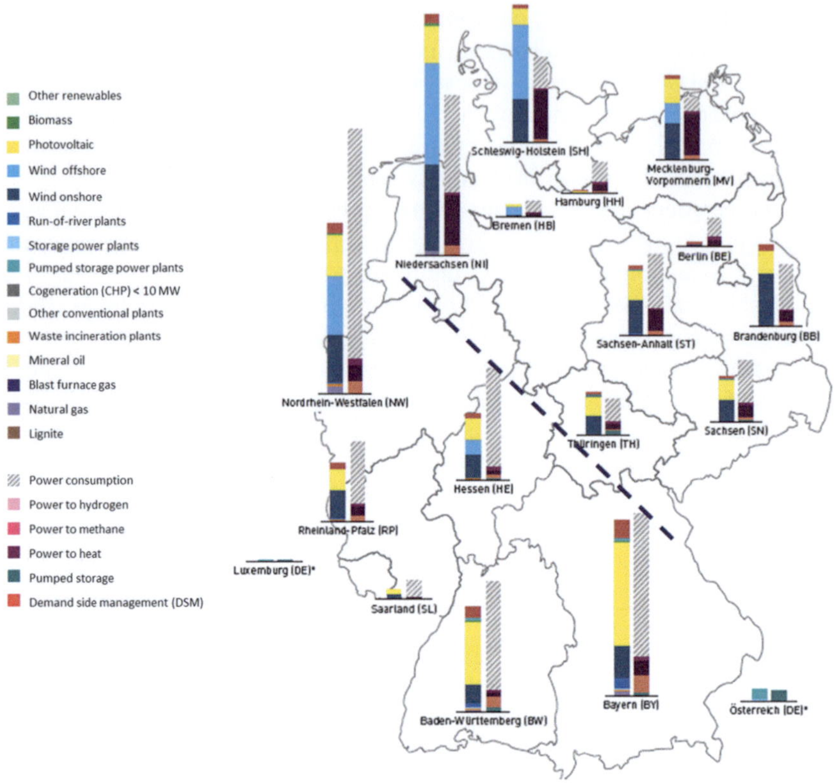

Fig. 3.6 Forecast of generation and demand in Germany according to scenario Grid Development Plan B 2035 [11]

Figure 3.7 shows the status of the projects from the Federal Requirements Plan Act (BBPlG) and the Energy Line Expansion Act (EnLAG) after the third quarter of 2020 [12].

Figure 3.8 provides an overview of the procedures to be carried out in the three categories explained above.

3.4.4 *Spatial Planning Procedure (in German:* **Raumverträglichkeitsprüfung***)*

Comparing the activities in the industrialized countries of Europe, Germany ranks undoubtedly among those with a very high spatial and environmental demand due to its intensive economic and technical development. As a result of the increasing demand for settlements and infrastructure, the multifunctionality of space has been

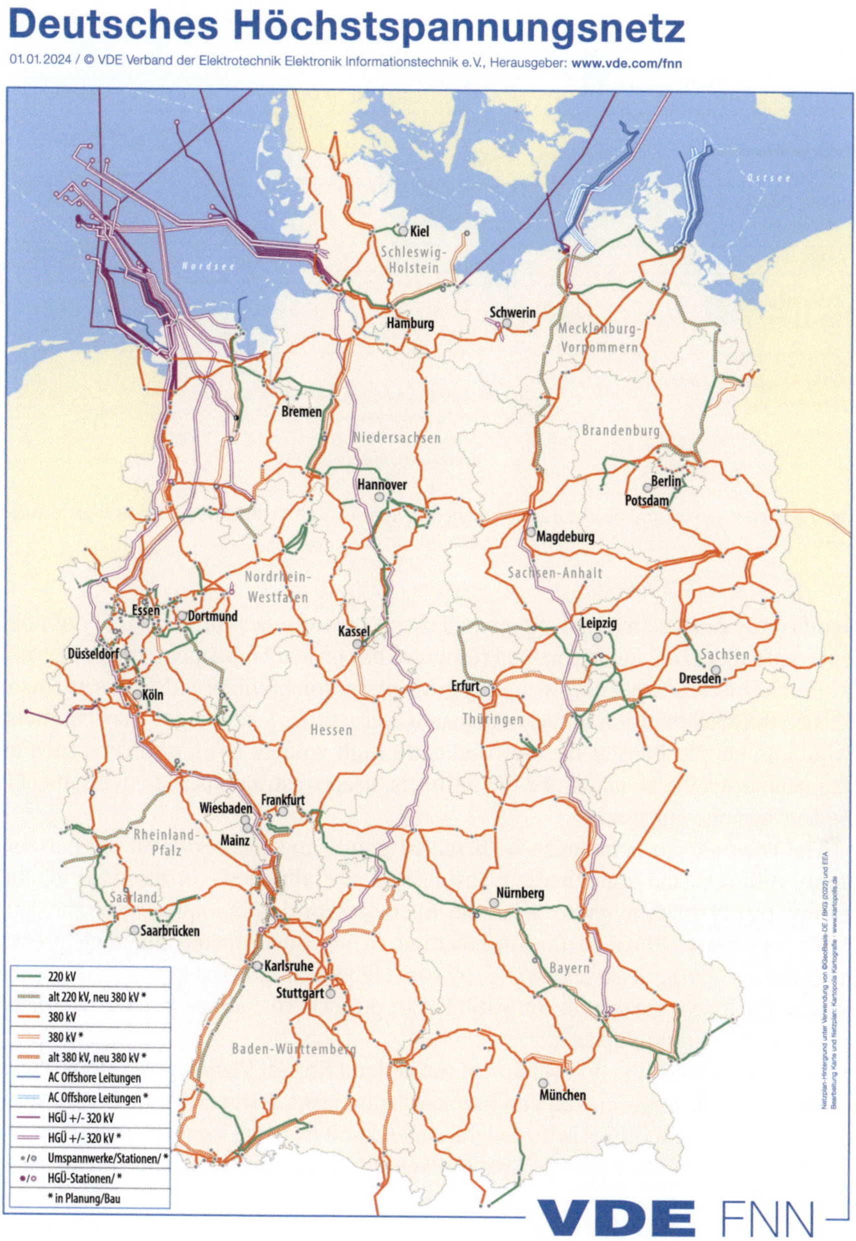

Fig. 3.7 Line routes and status EnLAG and BBPlG 12-2024

	Lower than110 kV UGC and OHL	110 kV and 220+380 kV, not urgent		380 kV, urgent UGC and OHL	
		UGC	OHL	EnLAG	BBPlG
Proof of the energy-economic necessity	individual	individual		legally determined	
Regional planning procedure	optional	optional	mandatory 1)	mandatory 1)	-
Federal sectoral planning	-	-	-	-	mandatory 2)
Planning approval procedure	optional	optional	mandatory 1)	mandatory 1)	mandatory 2)

1) Regional planning authority
2) Federal Grid Agency

UGC = Underground cable, OHL = Overhead line
EnLAG = Energieleitungsausbaugesetz (Energy line expansion act)
BBPlG = Bundesbedarfsplangesetz (Law of the Federal Republic requirements plan)

Fig. 3.8 Overview of the need for assessment and approval procedures for lines in high voltage grids

significantly limited in terms of potential uses, and awareness of the scarcity of non-renewable space and environmental resources has grown both among the public and in political circles. Especially in rapidly growing agglomerations with high settlement density, the concentration of spatial demands can at times lead to significant conflicts. Even with supply lines at the high and extra high voltage level, problems such as fragmentation effects, landscape impairments, usage restrictions and environmental impacts became apparent.

The necessity to economize with scarce goods and to coordinate the current, highly different and sometimes opposing spatial claims led—in the wake of the stormy reconstruction phase after the Second World War—already at the late 1950s and early 1960s to the introduction of so-called spatial planning procedures (*Raumverträglichkeitsprüfung*) by some federal states. These procedures were designed to assess the spatial compatibility of projects considered to have "spatially significant territorial impact".

As early as 1965, the first version of the Federal Spatial Planning Act (ROG) was passed [13]. The coordination and harmonization tasks assigned to the higher state spatial planning authorities in the federal states result from the legally defined responsibilities set out in the National Spatial Planning Regulation (*Raumverträglichkeitsprüfung*).

The names of these authorities vary from federal state to federal state. In some of them, they are the state ministries of building themselves, In Baden-Württemberg and Hesse, it is the regional councils, in North Rhine-Westphalia the district governments, and in some of the new federal states the state administrative offices.

According to § 1, para. 14 of the currently valid Federal Spatial Planning Procedure (ROV), as already mentioned, a spatial impact assessment according to state law can be carried out for the construction of high voltage overhead lines with a nominal voltage of 110 kV or more. Excluded is the construction in and immediately next to existing routes or their predominant use. However, experience over the past decades has shown that a spatial planning procedure should be pursued for line projects in the categories in question, where there is a change, e.g. enlargement of the protection strip or even just a minor change in the route. This is always advisable when spatial resistances and concerns are recognizable in the planned route, which are expected to cause problems in implementation.

Task of Spatial Planning

The task of spatial planning is to reconcile the interests of the applicant project sponsor with regard to the spatially significant aspects of his project with the interests of all affected parties and institutions. This includes the totality of spatially determined living and working conditions, which largely influence each other and thus determine and affect the structure of the space.

The aim is a balanced order of living and working space, which offers the people living there, in addition to the preservation of nature and landscape, the most favorable conditions for the free development of their personality [14].

Thus, spatial planning controls and limits economic and infrastructural activities and thus, in addition to building planning law, also makes an important contribution to the functioning of the social market economy. The spatial impact assessment is, alongside the spatial planning plans, a key instrument of spatial planning. It serves to examine specific projects for their possible effects and to bring their spatially compatible design into line with the spatial planning plans of the states and regions [13]. Specifically, the spatial impact assessment examines the spatial compatibility of a project, including various site and route alternatives. It forms a prerequisite for subsequent approval or licensing procedures, in this case for the planning approval procedure. However, the route found here is only a non-binding recommendation, on the basis of which the planning impact assessment is carried out. If new findings emerge within this planning step that justify route deviations, these can be taken into account in the procedure.

Procedure of Spatial Planning

The procedure of a spatial planning procedure follows a given structure (Fig. 3.9).

First, the project sponsor contacts the state planning authority responsible for spatial planning procedures and expresses his interest in carrying out the procedure based on a project description of the line. The authority then formally checks whether a spatial planning procedure is necessary.

If the answer is yes, it will hold an application conference, if necessary, to clarify the necessity, subject matter, scope, and procedure. The focus is on the type and scope of the investigations that the project sponsor must carry out in preparation for the procedure. The most important authorities, associations and organizations to be involved in the procedure are already involved in the application conference.

In the next step, the authority sets the Framework for the investigation (in German: *Untersuchungsrahmen*). This contains the application documents of the project

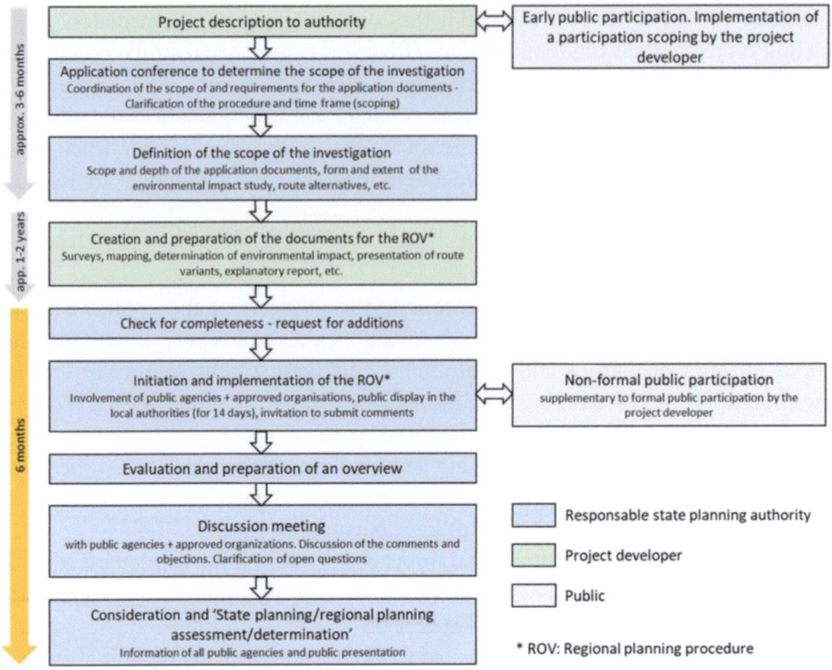

Fig. 3.9 Schematic representation of the spatial planning procedure [15, 16]

sponsor as well as the scope and depth of the environmental impact study (EIS), which serves as the basis for the integrated environmental impact assessment.

The project sponsor himself—or planning offices commissioned by him—prepare the application documents and carry out the necessary cartographic and surveying work for this. This phase of the procedure is not time-limited and usually lasts one to two years depending on the size and scope of.

Before the official initiation of the spatial planning procedure, the authority reviews the application documents and, if necessary, requests amendments.

With the start of the procedure, for which the spatial planning authority has a maximum period of six months, it carries out the involvement of public authorities (in German: *Träger öffentlicher Belange*) and other authorized organizations and arranges the display of the application documents for the public in the affected municipalities as well as on the internet. At this stage, everyone has the opportunity to submit their comments, suggestions, and objections.

With the help of the project sponsor, the authority evaluates all received statements and hints and, if necessary, arranges for their discussion in an official consultation with all participants.

Spatial Impact Assessment

Afterward, the spatial planning authority compiles all arguments and statements submitted in writing or orally. In a next and central step for the entire process, it weighs all interests put forward against each other and summarizes the result of the participation process in a spatial or state planning decision.

In this, it assesses whether the project is in line with the requirements of spatial planning, which spatially significant impacts the project has, what the examination of route alternatives has revealed and which route is to be considered the most suitable. It also contains specifications for the further planning and implementation steps, such as the subsequent planning approval procedure [17]. However, the spatial impact assessment is merely a recommendation. For justified reasons, it may be deviated from during the planning approval procedure.

3.4.5 *Federal Sectoral Planning (in German:* **Bundesfachplanung BFP***)*

Federal sectoral planning corresponds in its nature to a state-crossing spatial planning procedure with binding character [18–21]. For projects in this category, the transmission system operators must submit an application for federal sectoral planning to the Federal Network Agency—individually or jointly—based on a detailed project description and a rough route corridor for the line with alternative corridors. In a first step, identified corridors are determined, which can have a width of up to 1000 m. The aim of this planning step is to identify large scale spatial resistances and to find areas for the route corridors that are as low in conflict as possible the conflict between the search for the shortest possible route and a connection avoiding spatial resistances is often at the center of this process. Thus, the identification of rough corridors is based on a spatial resistance analysis. The main spatial resistances include designated protected areas and spatial planning determinations in the spatial planning plans of the federal states or regions. An important planning principle is the requirement of bundling of band infrastructure facilities, for example the bundling of high- and extra high voltage lines with existing or planned linear infrastructures. This is intended to reduce new environmental impacts. However, the bundling requirement also has its limits. Especially when in individual cases new routes prove to be more spatially and environmentally compatible or through a further bundling an excessive additional burden would arise. An example of the determination of a rough corridor is shown in Fig. 3.10.

Even before the start of the formal procedure, the transmission system operators inform the public in the affected regions about the project, for example through events with direct dialogues and others in the form of "information markets" After that, the application documents are checked by the Federal Network Agency and serve as the basis for the subsequent application conferences. At these events, the Federal

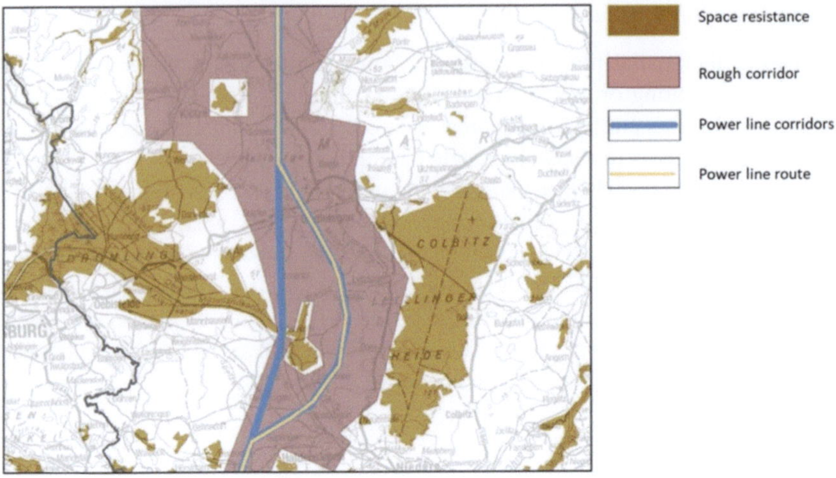

Fig. 3.10 Example of the determination of a rough corridor [4]

Network Agency collects information on the environmental and spatial compatibility of the proposed route corridor and possible alternatives. Involved are the state planning authorities, the holder of public interests, associations and the interested public. Relevant hints are incorporated into the investigation framework, which the Federal Network Agency then sets. This also determines which documents and expert opinions the transmission system operator must submit in total. A special focus is on the documents for the spatial planning assessment and for the strategic environmental assessment (SEA), which is carried out alongside the procedure. After submission of the complete documents, these are displayed for one month at the location of the Federal Network Agency in Bonn and at other suitable regional locations for inspection and comment. At the same time, they are also published on the websites of the Federal Network Agency. Subsequently, all participants have the opportunity, within one month after the end of the publication period, to submit their comments and statements in writing, orally or for the records.

Subsequently, the Federal Network Agency examines both the documents of the transmission system operator, and the comments received. It invites—if necessary—all those who have expressed themselves in time to a hearing to discuss the comments together with the transmission system operators in the context of a technical discussion and to find as far as possible consensual solutions.

Following the hearing, the Federal Network Agency weighs all arguments presented and determines the most suitable corridor based on all relevant criteria.

Following the hearing, the Federal Network Agency weighs all the arguments put forward and determines the most suitable corridor under all conditions. An important goal is to determine a corridor that is as spatially and environmentally compatible as possible, which is also technically and economically sensible. This decision should be made by the Federal Network Agency no later than six months after the submission

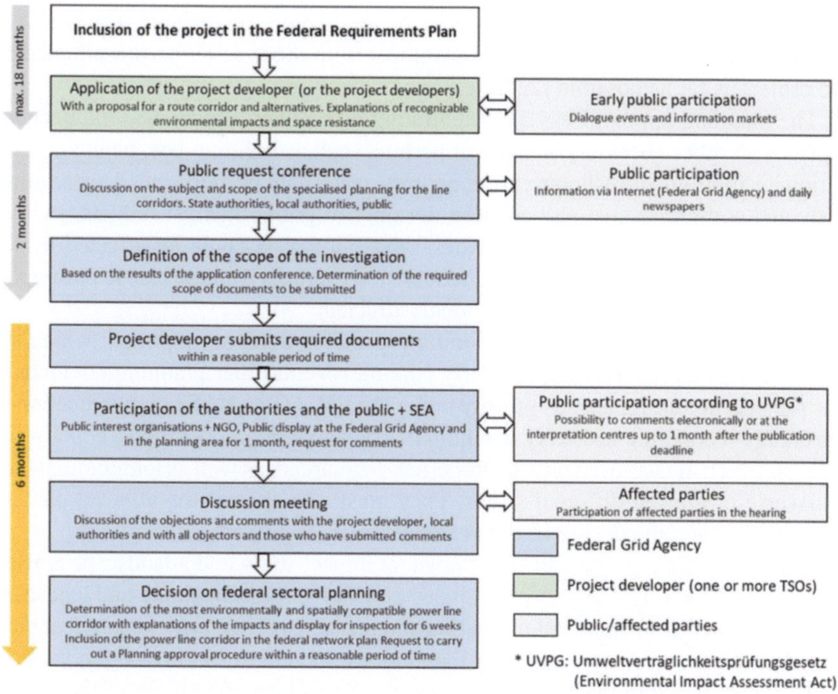

Fig. 3.11 Schematic representation of the process of federal sectoral planning (BFP)

of the complete application documents by the project sponsor(s). The decision is made public by the Federal Network Agency.

For the corridor determined in this way, a planning approval procedure follows, just as after the execution of a spatial planning procedure, which in this case is carried out by the Federal Network Agency. The schematic representation of the process of federal sectoral planning is shown on Fig. 3.11.

3.4.6 *Planning Approval Procedure (in German:* **Planfeststellungsverfahren PFV***)*

After the completion of a spatial planning procedure or federal sectoral planning, the approval procedure follows as the approval procedure for a high and extra high voltage line project. It applies to infrastructure projects that affect a multitude of public and private interests. It is required not only for high and extra-high voltage power lines, but also for spatially significant infrastructure facilities such as railways, roads, power plants, and much more. In the procedure itself and in the final decision—the planning approval decision—a careful and comprehensive balancing of public and private interests takes place, just as in the preliminary procedures. The goal is to reconcile the interests of the project developer with those of all parties affected by the project as far as possible [22, 23].

The planning approval has a concentrating effect. This means that the decision replaces all other necessary individual permits such as building law, nature conservation exemptions, forest conversion permits, water law permits, etc. This requires early and comprehensive participation not only of public interest holders, such as municipalities and specialist authorities, whose interests could be affected by the project, but also of organizations and associations approved authorized to participate in the proceedings, and private individuals affected.

The application documents, in addition to the explanatory report in which the project is described in detail, regularly contain several other planning documents. For projects not designated as priority under the EnLAG or NABEG, justification of the energy- economic necessity is required. The application documents contain site plans in which the required private properties are marked, as well as longitudinal- and transverse-profiles at different scales. They must also specify the affected parcels, their owners, and the degree of use or impact on each property.

In addition, they contain an environmental impact study and a landscape conservation accompanying plan, which includes the interventions in nature and landscape and the planned compensation mitigation and measures. They also contain technical details about the structures and their foundations, including soil investigations.

The exact scope of the planning documents depends on the effects of the project on all affected interests and can also include other documents not mentioned here.

Procedure of the Planning Approval Procedure

After the application documents have been prepared, the project sponsor submits the "Application for the implementation of the planning approval procedure". The leading authority checks the documents and verifies their completeness. For line projects within the borders of a federal state, these are regularly the responsible state planning authorities. For so identified cross-state and cross-border projects, as already mentioned, it is the Federal Network Agency.

If necessary, the project sponsor must submit any missing documents. With the complete documents and the necessary duplicates, the planning approval authority conducts a comprehensive hearing. All public interest holders, i.e. municipalities and specialist authorities, whose areas of responsibility could be affected by the project, as well as associations, possibly other transmission and distribution grid operators and other relevant bodies, are asked to comment. At the same time, the authority ensures that the documents are available online. The public is informed of these changes through customary local announcements. The customary local announcements are based on the practices of the respective municipalities, i.e., in the official gazette, municipal bulletins, and/or regional daily newspapers. Anyone whose interests are affected by the measure has the opportunity to raise objections or suggest changes within two weeks after the end of the display period. These can be made in writing to the authority or for the record at the display location.

After timely receipt of all comments, objections, and suggestions, the project sponsor is usually asked to comment on them. After that, the planning approval authority sets the so-called “discussion date” or hearing and invites all those who have submitted a statement or raised objections to the project. If the number of participants to be notified exceeds 50, a public announcement is sufficient. This is published in the local daily newspapers and in the state gazette. In this case, no personal invitation is issued.

In the hearing, the statements of the authorities and associations as well as the objections and suggestions from private persons are discussed with the project sponsor, the authorities, the affected parties, and those who have raised objections or submitted statements. The project sponsor has the task of presenting the arguments for his planning and checking whether the individual objections can be taken into account. The planning approval authority has the task of moderating the discussion neutrally and without preconceived outcome and leading to a balance of interests. In particular, this meeting has the purpose of hearing and weighing all arguments in order to come to a balanced decision. For large projects, the discussion may require large rooms and may take several days.

Following the hearing, the responsible authority checks whether a balance of interests was possible after weighing all arguments and a planning approval decision can be issued. If changes to the plans arise from the hearing, a supplementary consultation with the affected authorities and individuals may be required, which can consequently extend the duration of the procedure.

Planning Approval Decision

The following final planning approval decision is the decision that concludes the planning approval process and corresponds to a building permit for the project. The planning approval decision and all related documents will be published on the Internet.

It is possible to appeal the decision to the Federal Administrative Court, the sole court of appeal. If no appeals are filed or are unsuccessful, the project developer can proceed with the project. However, the planning approval decision does not entitle him to actually use the required properties. It only determines that the areas may be used for the project because the public interest in the measure outweighs the private interests of the owner. Nor does the decision contain any information on the amount of compensation that the project sponsor must pay to the owner. Matters concerning land acquisition and easement are excluded from the decision. They are to be resolved in subsequent negotiations between the project sponsor and the owners. If no agreement is reached here, the project sponsor has the option of initiating expropriation [15, 17]. Figure 3.12 shows the procedure schematically.

Detailed information on the respective status of the procedures to be carried out according to EnLAG and BBPlG can be found on the website of the German Federal Network Agency.

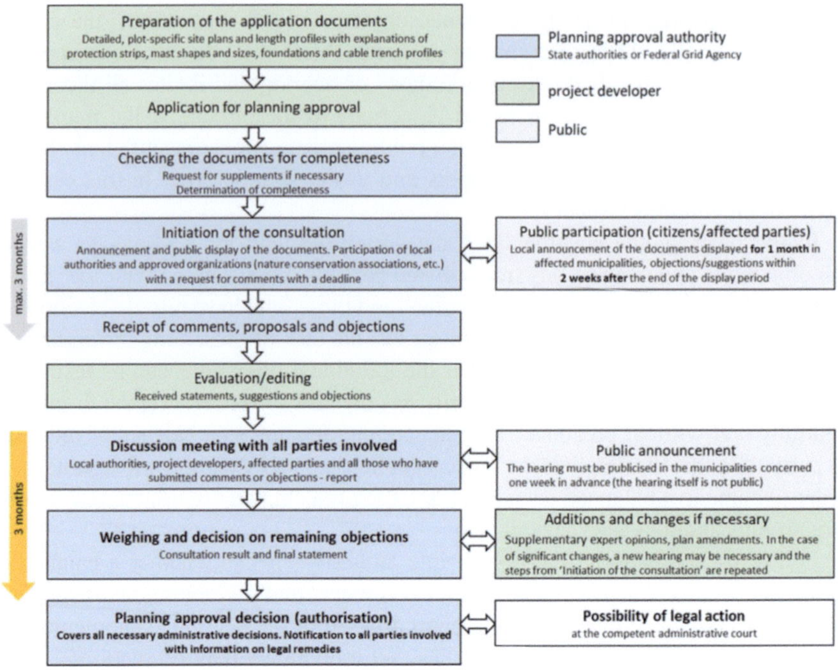

Fig. 3.12 Schematic representation of the procedure of a planning approval procedure according to the Energy Industry Act, as it is applied, for example, for EnLAG projects

3.4.7 *The NOVA Principle*

In all grid development tasks, the grid operators usually follow the NOVA principle, regardless of the type of line (in German: "**N**etz-**O**ptimierung vor Netz-**V**erstärkung vor Netz-**A**usbau"). It means: Grid Optimization before Grid Strengthening before Grid Expansion [24] (Fig. 3.13).

Grid Optimization

Through the already mentioned overhead line monitoring, high and extra high voltage overhead lines can be used more intensively under appropriate weather conditions. At low outside temperatures or with stronger wind, the current carrying capacity can be increased through natural cooling effects.

Overhead lines, for example, which are heavily loaded with the transport of wind-generated energy, the wind cools the conductors and can thus under certain circumstances reduce or prevent redispatch measures. Another example is high grid loads in winter, where the conductors are cooled by low outside temperatures. This also includes the corresponding adaptation and reinforcement measures in the associated facilities.

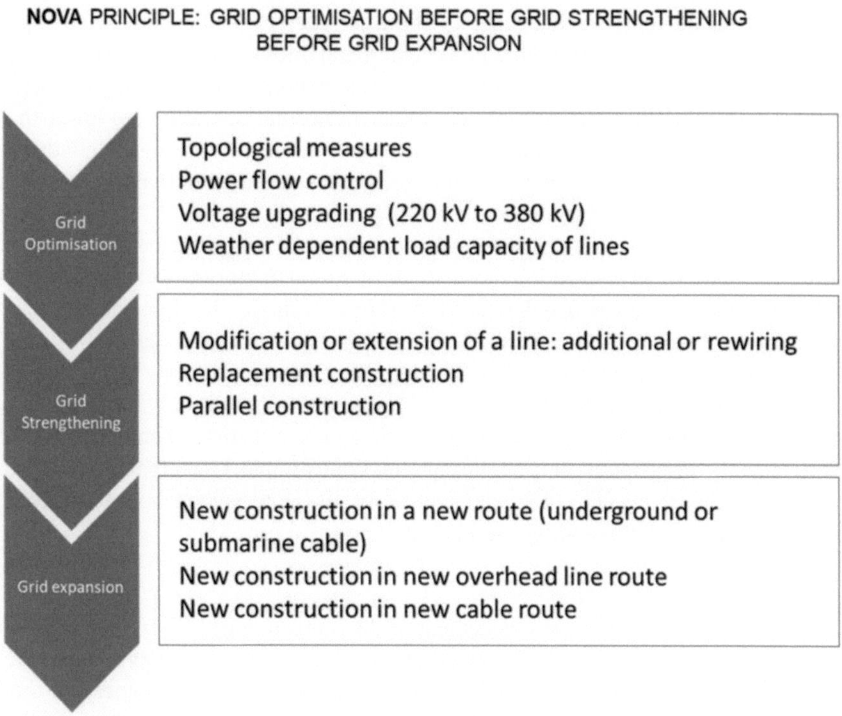

Fig. 3.13 The NOVA principle with the 2021 specification [25]

Some monitoring systems are based on direct temperature measurements on the conductors and underground cables, which are taken into account during line operation. Others use continuous measurements of meteorological data and enable a variable operation adapted to the weather via calculation models. In this way, bottlenecks can be eliminated at many sections without major modifications to existing lines or new constructions. For overhead lines, at strong wind and very low outside temperatures, the transmission capacity can be increased by up to 20%. However, with changing weather conditions, the permanently possible increase in performance is only in the order of 5–10% [11]. The load capacities for conventional conductors are limited by the maximum conductor temperature, which is usually 80 °C. Measures such as cross-zone grid congestion management or load flow control via cross regulators can also be used to avoid transmission bottlenecks (Chap. 6).

Grid Strengthening

When the optimization possibilities reach their limits for overhead lines, high temperature conductors come into question, which resemble conventional conductors in terms of their static requirements, but even at temperatures of 200 °C and more, they do not have a larger passage than the conventional ones at their maximum operating temperature. They are referred to as HTLS conductors (**H**igh **T**emperature **L**ow **S**ag).

With them, the current carrying capacity of the conductors can be increased by up to 100%. This can be achieved both with special metal alloys with low thermal expansion coefficients [26] and with conductor cores made of high-strength carbon and glass fibers [27]. While the current heat losses in normal operation are lower than those of conventional conductors, they increase significantly at higher conductor temperatures. The latter has meant that their use has so far been mainly limited to lines with temporary load increases.

As another future-proof reinforcement that allows for a permanent increase in power transmission in conventional construction, re-conductoring with a larger cross section is considered, provided the statics of the towers and foundations can withstand the additional load. If necessary, reinforcement measures can also help.

Alternatively, a replacement of the line along the existing route may be considered, where the route or the profile of the line remains unchanged or is only insignificantly changed. Deviations from the existing route are always sensible when this results in improvements for residents or the environment. For example, by increasing distances to residential buildings or to protected natural spaces. However, changes to the route can also trigger spatial planning and planning approval procedure.

Expansion

For new connections, where a replacement of the line along the existing route is not sufficient e.g. because the desired power transmission requires a higher voltage level and results in larger towers, an expansion is considered. Existing routes are used again if possible. In the case of overhead lines this often leads to the construction of multiple circuit lines on existing routes, where the circuits of the existing line are added to the circuits of the other new voltage level. If the new superregional line connections cannot use existing routes or only in sections, as is the case with the necessary high performance north–south connections in Germany, new constructions on new routes are inevitable. This step-by-step approach allows grid adjustments to be implemented as cost-effectively, spatially and environmentally friendly as possible. Figure 3.14 shows the scheme of the graduated procedure as it was updated in the German Grid Development Plan 2035 from the year 2021.

3.4.8 Environmental Compatibility of Overhead Lines and Cables

The environmental assessment is usually an integrated part of assessment and approval procedures, such as federal planning or the spatial planning procedure. However, in different forms. While the so-called strategic environmental assessment (SEA) is used in federal planning, an environmental impact assessment (EIA) must be carried out for projects that require a spatial planning procedure, depending on the route length of the line. An exception are the HVDC connections. Although they

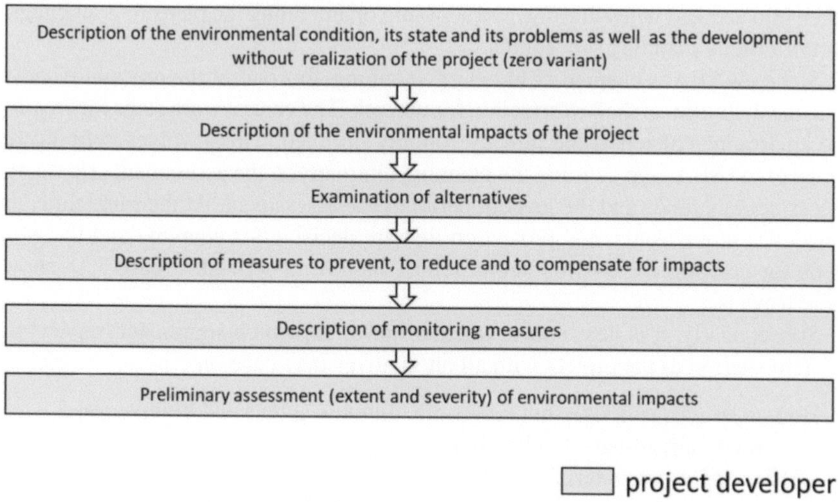

Fig. 3.14 Scheme of the graduated procedure according to the German Grid Development Plan 2035

are subject to federal planning, they are subject to EIA. The principles of both environmental assessment procedures are largely the same. In both cases, the impacts on the environment and humans are determined and described. The legal basis for conducting environmental assessments is the Environmental Impact Assessment Act EIA (in German: *Umweltverträglichkeitsprüfungsgesetz UVPG*). It sets out all principles and regulations on how the projects must be checked for their environmental impacts. In both procedural variants, the public and the authorities responsible for environmental matters are consulted on the plans or projects based on suitable documentation. Subsequently, the authorities leading the procedure evaluate the information collected in the consultation procedures and take them into account in their decision on the admissibility of a project. The strategic environmental assessment refers to a plan with a sometimes very broad and long line corridor. In contrast, in the environmental impact assessment, the implementation plans are more concrete, and they refer to routes and route alternatives that already include the rough route and possible angle points, the latter in the case of overhead lines.

Strategic Environmental Assessment (SEA)

As demonstrated in the procedure description for federal planning, the SEA is conducted in parallel with the procedure. There are continual links and feedback mechanisms to the lead procedure—in this case to the federal planning.

The aim of the SEA is the early, systematic and transparent determination, description and evaluation of the environmental impacts of the planned project including the planning alternatives as well as the participation of the public and the authorities responsible for environmental and health matters. The results of the SEA must be taken into account when drafting and adopting or amending the plans. It is an integral part of official planning procedures.

Since the SEA is mandatory in federal planning, this part of the procedure begins with the definition of the investigation framework. The central work steps for creating the environmental report are already roughly outlined. This involves defining the scope of the SEA, and defining the planning alternatives, the assessment criteria, the assessment methods and the level of detail of the assessment. At the same time, the objectives of environmental protection that are decisive for planning and the work steps for creating the environmental report should be defined. Figure 3.15 shows these aspects.

Subsequently, it is determined whether and if so, which significant impacts the implementation of the project with all alternatives may have on

- Human beings, including human health, animals, plants and biodiversity,
- Soil, water, air, climate and landscape,
- Culture and other material goods, and
- The interactions between the aforementioned protected assets.

In principle, the subject of the assessment includes all aspects of the plan, including the alternatives considered, that could have significant environmental impacts. An additional assessment of cumulative effects arising from the overlapping of impacts on two or more protected interests may also be required.

The environmental report to be created thereafter serves to describe and evaluate the anticipated significant environmental impacts of the project including any variants. It forms the substantive basis for the participation of the public and the specialist authorities. Responsible for the creation of the environmental report is always the project sponsor. As part of the authority and public participation process,

No.	Project	col. 1	col. 2
19.	Line systems and other systems:		
19. 1	Construction and operation of a high-voltage overhead line within the meaning of the Energy Act with ...		
19. 1.1	a length of more than 15 km, rated voltage 220 kV or more,	X	
19. 1.2	a length of more than 15 km, rated voltage 110 kV to 220 kV,		A
19. 1.3	a length of 5 km to 15 km, rated voltage 110 kV or more,		A
19. 1.4	a length of less than 5 km, rated voltage 110 kV or more		S
19. 11	Construction and operation of an underground cable in accordance with § 2 (5) BBPIG*	X	

Keywords:

No.	=	Number of the project
Project	=	Type of project, if applicable with size and power according to § 6, sentence 2 as well as test values for size and power according to § 7, subsection 5, sentence 3
X in column 1	=	Project is subject to Environmental Impact Assessment Act (EIA)
A in column 2	=	General preliminary assessment of the individual case according to § 7
S in column 2	=	Site-specific preliminary assessment of the individual case according to § 7

* BBPIG: German Federal Requirements Plan Act

Fig. 3.15 Work steps for creating the environmental report [28] those authorities are generally to be heard, which were already involved in the scoping appointment. Authorities whose interests are not recognizably affected, do not need to be consulted. Others who were not previously involved, but whose interests could be affected, must be consulted. If the plans have significant impacts on the protected assets in one of the neighboring states, the responsible authorities there must be involved.

The participation of the public takes place on the basis of the display of the documents. Mandatory component is the planning draft, which the environmental assessment refers to, as well as the environmental report. The affected public can comment on the planning draft and the environmental report within a reasonable period, which is at least one month.

After the completion of the participations, the project sponsor must review and revise the environmental report in light of the knowledge gained. Ideally, the presentations and evaluations in the environmental report are confirmed. Otherwise, changes or additions are necessary. The final assessment of the environmental impacts is carried out simultaneously with the review. The results of the SEA must be duly appreciated and considered in the lead procedure [28]. An overview of all SEA-required line projects, which are led by a federal agency, in this case by the Federal Network Agency, is contained in the EIA portal of the federal government.

Environmental Impact Assessment (EIA)

The construction and operation of high voltage overhead lines with a length over 15 km is generally subject to an EIA. The same applies for the construction and operation of the HVDC cables contained in the Federal Requirements Plan. High voltage lines with 110 kV and more are subjected to a general or a site-specific preliminary examination depending on their length. Figure 3.16 shows an excerpt from Annex 1 of the Environmental Impact Assessment Act (EIA/UVPG), in which the framework conditions for EIA-required lines are listed.

Since it is an important task of assessment and approval procedures for high and extra high voltage lines to avoid or minimize impairments of protectable areas, this—usually dependent—part of the procedure has a special significance to. The EIA, like the SEA, should help to avoid disturbances and environmental impacts or to reduce unavoidable disturbances and environmental impacts as far as possible, and thus contribute to an environmentally compatible development of space. Figure 3.17 shows the schematic procedure of an EIA.

The official EIA process begins with so-called screening, i.e. the examination of whether an environmental impact assessment is to be carried out for the project or not. For projects listed in Fig. 3.16 in the columns 9.1.1 and 9.11, EIA is mandatory. The actual screening process is therefore only to be carried out for the projects with the numbers 19.1.2–19.1.4.

These are marked in Fig. 3.15 in column 2 with A and S. The preliminary examination of the individual case always applies when adverse environmental impacts are possible but not necessarily expected. In such cases, the objective is to provide a preliminary assessment of the potential environmental effects of the project and to determine whether a formal Environmental Impact Assessment (EIA) is necessary.

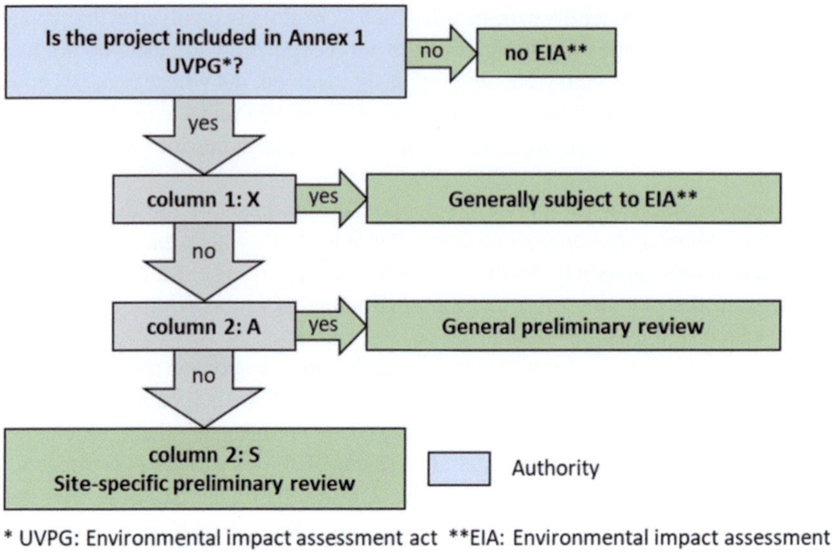

Fig. 3.16 Excerpt from Annex 1 of the Environmental Impact Assessment Act (EIA/UVPG)

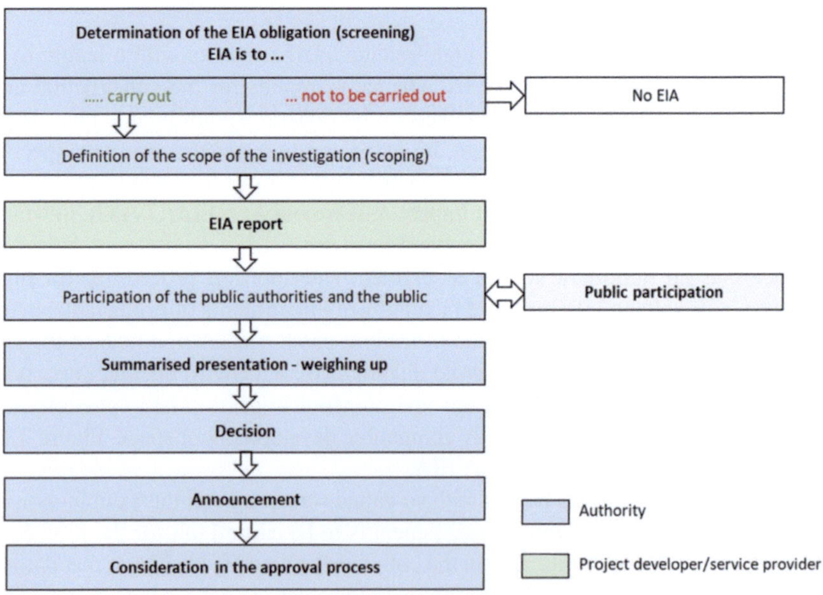

Fig. 3.17 Schematic representation of the process of an EIA as an example from Germany

Here, a distinction is made between general and site-specific preliminary examinations. In the *general preliminary examination (in German: allgemeine Vorprüfung)*, in Fig. 3.15, column 2 marked with A, the authority determines whether the line with its characteristics, for example, due to their severity, duration, and frequency could have environmental impacts that require an EIA. This includes, for example, 110 kV high voltage overhead lines with a length or partial length between 5 and 15 km. For overhead lines in the high voltage level, which are shorter than 5 km, the local situation is particularly considered. Here can the *site-specific preliminary examination (in German: standortbezogene Vorprüfung)* according to Fig. 3.15, column 2, marked with S, be considered. The conversion of overhead lines in existing routes, for example, can become subject to EIA, even if it does not require a spatial planning procedure. This is particularly possible when the line profile and/or tower locations differ from the existing line and other, potentially greater, environmental impacts are to be expected.

If the screening results in the need for an EIA, the scope of investigation for the so-called EIA report is determined in the next step. It is advisable to determine the scope of investigation in a so-called scoping meeting together with the authorities, experts from other specialist authorities and environmental associations. This serves as the basis for the EIA report. Before the public authorities and the public are involved in the respective assessment or approval procedure, the authority checks the EIA report for completeness. The documents are made available to the public for inspection, so that the affected public can comment on the project as part of the participation process. The incoming comments are documented and taken into consideration. The authority then presents a summary of the environmental impacts of the project, the comments and the statements of the specialist authorities. On this basis, it assesses the environmental impacts of the project in a weighing process and justifies its assessment. After the procedure is completed, the public authorities and the public are informed about the decision. The result is taken into account in the higher-level lead procedure [29].

Protected Assets in Environmental Assessments

Both the strategic environmental assessment and the environmental impact assessment focus on the so-called *protected assets* [30]. Approval of grid expansion projects requires, in particular, assessments concerning the protection of the coherent European Natura 2000 network as well as specific species protection. The aim of these examinations is to protect Fauna-Flora-Habitat (FFH) and bird-protection areas as well as certain, particularly worthy of protection animal and plant species from impairments. Strict legal and nature conservation requirements apply here [30]. The environmental assessments examine and document the influence of the intervention of the considered project on the individual protected assets, and which possibilities can be taken to avoid or minimize them. Be it through a change in design or through compensation. In a first step, the current state in the planned line corridor and its possible variants—including all previous environmental impacts—are recorded and evaluated. By considering planned changes in the areas under review—resulting from planning requirements, such as those arising from state planning, as well as sufficiently concrete future construction projects—the so-called zero option is defined. It documents the development of the environmental situation in the considered area, without realizing the project [31]. In a second step, the estimation and prognosis of the possible impairment of the protected assets by the construction of the overhead line or the underground cable including possible interactions are assessed and forecast.

The following protected assets and their influence or impairment are considered.

Humans and Human Health

The condition of the investigation area with all its burdens affects the health, well-being and quality of life of the people living there permanently or temporarily. Pre-existing environmental burdens include, for example, impacts from settlement infrastructure as well as disruptive elements in the landscape, such as industrial facilities, railways or existing overhead lines in the immediate vicinity. The addition of overhead lines in particular in the higher voltage levels changes the visual landscape. This aspect is perceived by most people as the most serious consequence of line construction. From this subjective assessment, many overhead line projects are rejected by the population. Since the subjective landscape experience influences the recreational value, overhead lines were particularly critically viewed in recreational regions. To assess the visibility and the spatial effects, several studies and publications were created in the early 90s of the last century, which attempted to objectify the subjective human perception [32, 33]. However, in the subsequent approval practice, the individual case was always evaluated and the assessment results mostly remained subjective. As a result, as already described in Chap. 2, routing aspects were developed, which with tower shapes adapted to the landscape and the best possible use of the topology should promote acceptance. In addition, parallel guidance of the new lines to existing band infrastructure facilities, especially other overhead lines, were always sought, or carried on newly erected shared towers in the existing route.

Other impairing aspects are emissions caused by the operation of the lines. These include the electric and magnetic fields that occur in the vicinity of overhead lines and cables (Sect. 6.1.8) as well as noise emissions and ozone formations that occur due to so-called corona phenomena on overhead lines. While the noise emissions, depending on the type of tower and type of conductor (single or bundle), can be at a sound level of 40–50 dB (A), which is well above the permissible level in residential areas. The ozone production at 380 kV overhead lines with 2–3 ppm is harmless [31]. Audible noise from wind depends on the direction and generally decreases with increasing distance from the line.

Animals, Plants and Biological Diversity

Animals and plants are an important part of biological diversity. For example, if the soil condition or the groundwater level have changed by an intervention, this also has effects on the population of animal and plant species as well as on the species composition at a location. Potential impacts on the protected assets animals, plants, and biological diversity caused by high- and extra-high-voltage overhead lines and cables include, among other things, the destruction or damage of plants and the killing or injury of animals. The same applies to site alterations affecting plants and animals, the impairment or destruction of habitats, the damage or loss of developmental stages, the obstruction or prevention of biotope connectivity, as well as other significant disturbances. The effects are generally divided into three categories:

– *Construction-Related Effects*

The temporary effects during the construction phase mainly concern the removal of vegetation at the locations of the towers and along a cable corridor. For cables in open construction it should also be taken into account that in addition to the cable trench, areas for the temporary storage of the excavated soil are needed, which is subsequently refilled in the order of its removal for the most part. Similarly, construction roads are created, which after the completion of the construction work are to be dismantled again. At the same time, a construction pit water retention may be necessary, which can result in a temporary groundwater lowering. Driving on the route corridor with heavy machines, can cause soil compaction in both types of lines which can have lasting effects on the flora.

– *Facility-Related Effects*

While it may take several vegetation periods after the restoration of the original state until the former yield situation is reached on agriculturally used areas, a permanent habitat loss for fauna and flora is created in the tower area of overhead lines due to the partial sealing of the tower area with tower foundations. The extent of the effects here is very much dependent on the biotope type present. While outside sealed parts of tower foundations many biotopes can re-emerge after the construction phase, this is in the overhead line open span below the conductors only limitedly permitted due to a possible height restriction for vegetation.

Overhead lines at different voltage levels pose considerable dangers to the free-living birds. The grids are particularly dense in metropolitan areas. On the other hand, towers and conductors of medium, high and extra high voltage lines are welcome resting and perching places for a large number of bird species. Some of them even use the towers as nesting sites. Overhead lines can cause serious losses, especially among large birds. They pose essentially three risks to birdlife [14, 34]:

- Certain tower types, especially of medium voltage overhead lines pose dangerous perching and resting places, on which primarily large birds are threatened by electric shock.
- Meadow breeders avoid the area under and around overhead lines as breeding place, because towers and conductors serve as perching places for birds of prey. This can potentially affect breeding and feeding habitats.
- Conductors and earth wires of high and extra high voltage overhead lines can pose considerable obstacles to bird flight. Primarily large birds can be seriously injured or killed by collision.

Medium voltage towers with post insulators are among the most dangerous perching places due to their short insulator designs, because large birds—when resting on a crossarm—can easily reach live parts. Therefore, the German legislator already decreed in 2002 in § 41 of the Federal Nature Conservation Act that medium voltage overhead lines must be constructed in such a way that birds are protected against electric shock. Within a ten-year transition period, existing lines also had to be retrofitted accordingly. In 2011, bird protection on medium voltage overhead lines was made mandatory in a VDE application rule [35]. Among other things, all conductive towers with post insulators must be covered with plastic hoods at least 1.3 m long (Fig. 3.18, left). For tension towers, the insulating bodies must have a minimum length of 60 cm to maintain the minimum distances to avoid conductor-earth contacts (Fig. 3.18, right). By now, medium voltage overhead lines with post insulators should have been defused nationwide using plastic cover hoods [36].

Although overhead line corridors in forested areas can have high biodiversity—when properly maintained—they are avoided as nesting sites by meadow-breeding birds due to the presence of perching birds of prey. Another reason for avoidance is likely to be the regular presence of predators such as the fox, which roams the area in search of victims of a collision. This brings us to the last immediate danger to birdlife.

The collision with conductors and ground wires as a cause of death for birds affects a much larger number of bird species than electrocution. In the high and ultra high voltage range, where electrocution as a cause of death hardly plays a role, collision with conductors or earth wires is the bigger problem [14]. Due to the poor visibility of the wires, which optically blend with the horizon primarily at dusk and in unfavorable weather conditions, they are either not recognized at all or too late. This primarily affects bird species that see significantly worse than birds of prey and, due to their size, can avoid obstacles with difficulty. Thus, the risk of collision increases by decreasing distance to the obstacle [37].

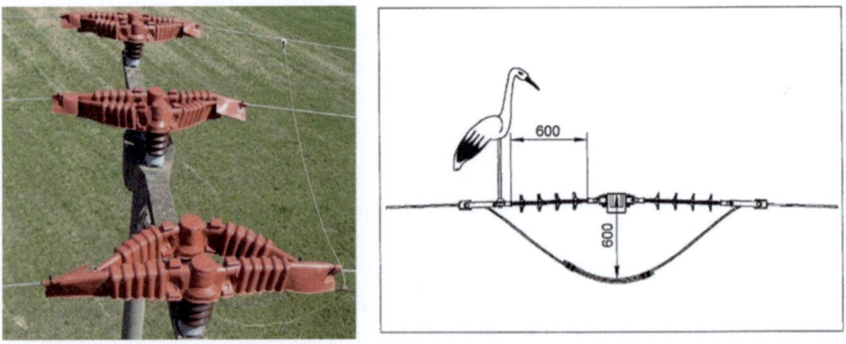

Fig. 3.18 Bird protection hood on a 20 kV overhead line with post insulators (left, picture TE Connectivity) and (right) distance specifications for tension towers from the VDE-AR-N 4210-11 [35]

To reduce the number of collisions, the horizontal single level arrangement of the conductors is recommended as a constructive measure in affected areas to avoid further vertical dimensions of the obstacles. In addition, highly visible bird protection markers with the highest possible attention-grabbing potential should be mounted on the usually highest located conductors or ground wires. They increase their visibility and reduce the risk of collision [38]. At the same time, they must meet technical and mechanical requirements so that they do not damage the wires and do not affect the statics of them or the structures. The markers should be effective for all bird species occurring in the affected area, i.e., universally applicable. They are divided into passive and active marker types. Passive markers, such as warning balls and bird protection spirals, have no moving components. Active markers, on the other hand, have movable parts that can rotate or swing, thus attracting more attention [39]. Figure 3.19 shows two examples of bird protection markers of the active category as they are used today.

The so-called Zebra markers with hanging, movable black and white stripes (Fig. 3.19, left) achieve a reduction in the collision rate, especially for large birds, of over 90% as determined in studies [39]. The white stripes are phosphorescent and visible at dusk and at night. They are installed by helicopter or line car.

The Firefly marker in Fig. 3.19, right, is rotatably mounted and shaped so that it rotates vertically in windy weather. It achieves high visibility as well as UV reflection and glows at dusk and night. The deterrent reflects all colors perceptible by birds and achieves a reduction in the collision rate of over 90% as proven by studies [41]. The installation and removal can be done here with a drone. For both types of markers, the manufacturers recommend a mounting distance of approx. 20–25 m and a lifespan of at least 20 years.

Due to the underground course, the installation-related effects of cables differ significantly from those of overhead lines. Here, the route area is generally available

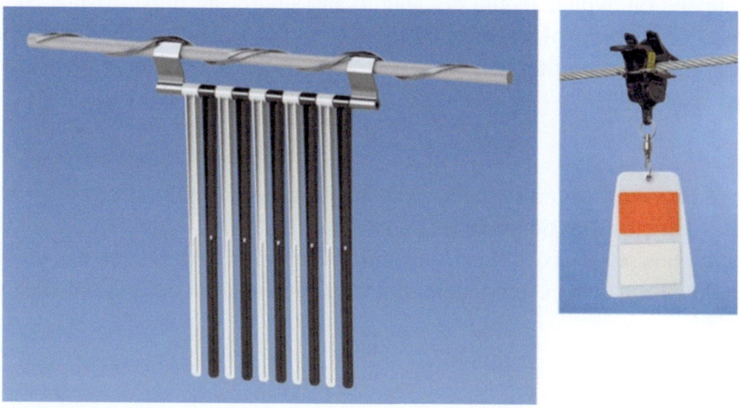

Fig. 3.19 Examples of bird protection markers, left the Zebra marker model from the German manufacturer Richard Bergner (factory photo Richard Bergner) [40], right the FIREFLY model from the Swedish manufacturer Hammarprodukter [41]

again as a planting site after the cable has been laid. Possible restrictions exist when planting deep rooted vegetation that can cause damage to the cable systems. Due to the planting restrictions, the effects of underground cables on shrub biotopes can be significant. The emergence of new biotopes at moisture-influenced sites is uncertain. Many ecological factors are susceptible to disturbance here. The change in the water balance affects the growth of plants, which can lead to a shift in the spectrum of the species. Agricultural areas usually regenerate well, provided civil engineering work was carried out professionally and excessive soil compaction was avoided [30].

– *Operational Impacts*

The operation of overhead lines is associated with increased traffic during maintenance, servicing, and repair work, as well as the possible pruning of vegetation that could endanger the line operation. This can lead to disturbances in the habitat of wild animals and flora in the immediate vicinity of the line. However, significant disturbances are not to be expected. The operating temperature of current-loaded conductors of lines could lead to damage in birds resting on them, especially in the case of high temperature wires (Chap. 4). Since ground wires are mainly preferred as resting and perching places in high and ultra high voltage overhead lines, the risk of injury from these reasons is relatively low.

With underground cables, maintenance, servicing, and repair work on the cable itself is not necessary. The operational monitoring usually protects the cable from damage or destruction. However, the grid operators are responsible for operational safety and must regularly check the route and, if necessary, remove growth (deep-rooted) that could endanger operations, which can lead to temporary disturbances of the flora and fauna in the route area. They also have to ensure that the cable corridor remains accessible in case of a fault. However, disturbances only occur sporadically throughout the year, so that sensitive disturbances remain an exception. On agricultural land, slight heat emissions into the cable surrounding soil can be expected during normal operation, which should not lead to any significant impairments. However, the effects of operational warming of the soil and cumulative effects pose the risk of drying out and impairing soil life. These are currently being investigated in some research projects [30].

Air and Climate

The protected asset *air and climate* is temporarily and locally affected by exhaust emissions from construction vehicles and machinery, as well as dust emissions during the construction phase of overhead lines and underground cables. In case of prolonged drought, localized dust developments and wind erosions can occur from such works [30].

Landscape

For the protected asset "landscape," both the ecological functions of the landscape and its visual appearance are considered.

For the protected asset *landscape*, the visual impression as well as the balance of landscape and the balance of nature are considered. The most important aspects regarding overhead lines and cables have already been described for the protected assets "humans, human health" and "animals, plants, biological diversity". Changes in the landscape are primarily changes in the visual landscape caused by humans or by nature itself. It is also an expression of the intensity of the use. The visual landscape is perceived subjectively, so it always depends on the interpretation by those people affected. Characteristics such as diversity, uniqueness, beauty, and recreational value are assessed, which, as previously described, are difficult to objectify. During the construction phase of both types of lines, the operation dominates with a number of disturbing factors such as emissions from exhaust and noise, the creation of extensive storage sites, access roads, and all the associated vegetation damage. This can temporarily affect the experience of the landscape. However, in the case of spectacular constructions in the high voltage range, especially for technology enthusiasts, it can also encourage lingering and observing the construction work.

With overhead lines, visible aisles are created when crossing closed tree stands, or gaps in linear tree stands (tree rows, avenues, tree hedges) that can permanently negatively affect the appearance of these structures and are perceived in the visual landscape.

For the laying of underground cables in an open trench, the required areas for work must be completely cleared of vegetation. Forest areas can thus be visibly impaired in their appearance. Especially in areas with few forests, the possible loss of shaping tree structures noticeably affects the visual appearance of the landscape [30].

Soil

The construction, existence, and operation of high and ultra high voltage lines can have effects on the protected asset *soil*, depending on the technical design (both, overhead line or underground cable). These can be temporary, but they can also result in permanent impairments and soil changes (for example, sealing of soils by foundations). The sensitivity and condition of the soil play a special role. Particularly moist, compaction-sensitive, and erosion-prone soils deserve special attention. When constructing overhead lines, the effects on the protected asset soil mainly occur during the construction phase. The soil is usually driven on with heavy construction vehicles in the area of the construction sites and on the access roads. This can lead to compactions that negatively affect the natural soil functions. The compression of the soil can lead to a reduction in pore volume and an interruption of vertical pore channels. This can reduce the infiltration rates, which can lead to waterlogging and impair growth of plants.

The impact on the soil is higher during the construction phase of underground cables than during the construction of overhead lines. With open laying, the entire cable route is excavated and must be accessible for transport. Excavation, sailing, soil compaction, and soil relocation can temporarily, but also permanently, disturb the soil structure and the water balance of the soil. To enable recultivation of the extensive earth movements when laying underground cables, the soil layers and soil horizons, as already mentioned, must be carefully stored separately during soil excavation and then reinstalled in their natural layering.

Area

The protection of the *area* (of the project), which originally formed a common protected asset with *soil*, was included as an independent protected asset in the catalog of protected assets of the environmental assessment by an EU EIA amendment directive. Therefore, the protected asset area is considered in the environmental assessment with a focus on land use, i.e., the quantitative aspects of change of land use and intensity of use. A distinction is made between temporary and permanent land use. The "land use by potential abandonment of use" arises, for example, from sealing by foundations and is permanent. The "land use due to potential usage restrictions" can be permanent (e.g., in the protection strip of overhead lines due to the height restriction in effect there), or temporary (e.g., due to areas that are used for construction work or as access routes in the short to medium term during the construction works).

The construction and existence of high and extra high voltage lines can have effects on the protected asset of land, which can vary depending on the technical design (overhead line or underground cable). In addition, maintenance during the operation of overhead lines (including the restoration of old towers) can lead to temporary and also permanent land use. Far greater effects can result from repair work on underground cables, for example after a cable fault. Here, the defective cable must be excavated at the fault location and replaced with a new cable section with joints or joint structures (Chap. 5).

Water

Water is filled with diverse life and subjected to a constant cycle, which is referred to as the water balance. Surface waters are in diverse contact with groundwater and form complex ecosystems that can be altered or impaired by construction measures. In the case of overhead lines and cables, this mainly affects groundwater or retention zones that can be temporarily flooded. While localized interventions for tower construction are necessary during overhead line construction, linear and significantly more extensive civil engineering measures are necessary for underground cables, which generally also have stronger effects on the protected asset of water (for example temporary groundwater maintenance, which among other things can lead to a temporary lowering of the groundwater level). During the construction phase of overhead lines, there is a risk that the installation of foundations can permanently change the groundwater aquifer and the cover layer. It is possible that when excavating the pits for the tower foundations, an exposure of the near surface groundwater occurs. In these cases, temporary groundwater maintenance is necessary, which can have effects on the groundwater balance and its flow direction and can lead to a temporary lowering of the groundwater level. For foundations made of driven piles, groundwater maintenance is generally not necessary, as excavation of the construction pit can be dispensed with here. The work could be particularly risky if water endangering substances such as lubricants and cleaning agents as well as paints and fuels are handled at tower locations during the construction phase and these leaks out. Such damage can be countered by defining extensive protective measures, which are written down in ancillary provisions of a planning approval decision. When founding towers, the provisions of § 78 paragraph 1 Water Management Act (WHG) and the respective legal framework from state water laws for water protection must be observed.

The effects of the construction phase of underground cables are generally greater due to the significantly more extensive civil engineering works. Thus, water maintenances in temporary pits with the effects and protective measures described above are also necessary to a greater extent. Irreversible damage caused by laying underground cables in areas of groundwater is largely ruled out due to the short duration of the construction works. With proper implementation of the construction measures, water contamination by pollutant input can be avoided [30].

Cultural Heritage and Other Material Goods

For the protected assets *Cultural heritage and other material goods*, significant testimonies of human action that can be located and defined as places, as spatial dispositions or as things in the cultural landscape are particularly relevant. These cultural assets include both individual objects and groups of objects, together with their environmental context, as well as extensive structures and spatial relationships, up to and including cultural landscapes under historical protection. Parts of landscapes, as well as evidence of prehistoric developments such as archaeological monuments and finds, are also included.

Work on overhead line corridors can endanger soil, cultural and architectural monuments as well as archaeological sites during the construction phase. With underground cables, the linear extent of the construction sites and the more extensive earthworks pose a far greater threat to these protected assets [30]. To avoid damage or destruction, careful surveying and mapping of possible finds is necessary in both types of lines already during the planning stage.

Landscape Conservation Plan and Impact Compensation

In the case of new lines at the high and extra high voltage level, the examination of environmental compatibility in the EIA report regularly comes to the conclusion that the projects represent an intervention in the sense of the Federal Nature Conservation Act. In this case, the project sponsors are obliged to draw up a "Landscape Conservation Plan", which is often combined with the EIA report to form an environmental study in the case of large projects during the planning approval procedure.

Under the Federal Nature Conservation Act, the party responsible for an intervention is generally obliged to refrain from avoidable impairments of nature and the landscape if reasonable alternatives are available that can prevent the intervention or reduce its impacts.

The party responsible for an intervention is generally obliged under the Federal Nature Conservation Act to refrain from avoidable impairments of nature and landscape if reasonable alternatives are available that can avoid the intervention or reduce its effects. Unavoidable impairments must primarily be compensated or replaced with measures of nature conservation and landscape maintenance on site or at least near the intervention. Impairments are always considered compensated when the impaired functions of the natural balance can be directly restored. A replacement for the impairment is provided when the impaired functions of the ecological balance can be restored in the same way in the affected natural area. If both are not possible, but the project is essential for energy-economic reasons, the perpetrator is obliged to provide monetary compensation. The form and amount of compensation is regulated in the Federal Compensation Regulation or in regulations of the respective federal state and differ from one state to another.

3.4.9 Compensations

With their consent to the use of their property on a voluntary basis or after expropriation (in this case the compulsory granting of a servitude) on the basis of a planning approval decision, owners of properties affected by line projects receive compensation for restrictions they permanently incur from an overhead line or a cable. This can be a usage restriction under an overhead line as well as restrictions for overbuilding and growth with a cable. In addition to compensation for the tower location and the overcrossing in the protection strip or for the protection strip of an underground cable and any buildings for cable joints, the owners receive compensation for land damage. For agricultural and forestry used areas they receive additional compensation for potential loss of harvest during the construction and possibly recultivation phase. This usually takes into account far more than the protection strips of overhead lines or cables. It also includes compensation for temporary installed lines and construction roads, which are removed after the construction work, temporary storage places for the building materials as well as for the transport, construction and laying vehicles (Fig. 3.20).

The framework for compensations of property owners or property users is formed by constitutional requirements in connection with compensation and expropriation rules of the federal states. The basis for the claim is always a public law permit, which is regularly available in the form of a planning approval decision for line projects at the high and highest voltage level.

Subsequently, the grid operators benefiting from the decision, following the principle of proportionality, seek to obtain servitudes from the owners of the properties affected by the route, through negotiation. Servitudes do not establish ownership of the areas needed by the grid operator.

They generally consist solely of a compensable right of construction and access for the purpose of erecting, operating, and maintaining the line within the protection strip and any permanent access roads, which right is notarized and entered in the land register.

They generally only include a compensable right of construction and access for the installation, operation and maintenance of the line within the protection strip and any permanent access routes, which is notarized and entered in the land register. The compensation primarily serves to compensate all disadvantages resulting from the claim. In practice, negotiations already take place during the planning approval

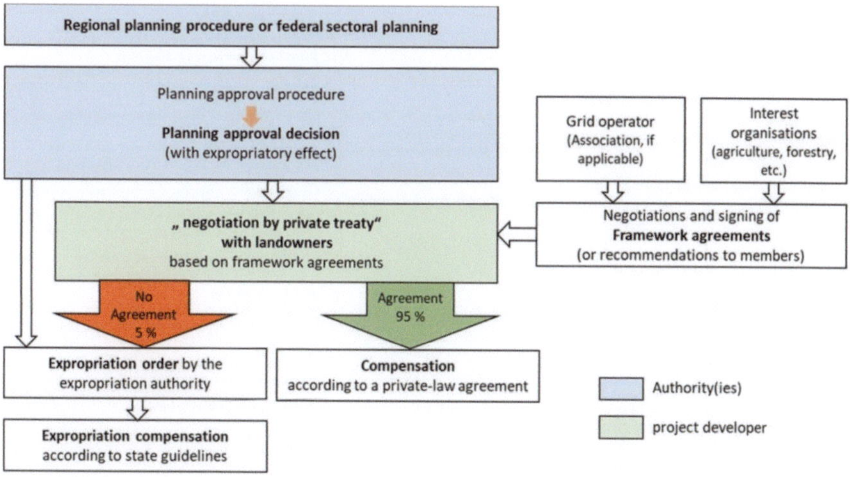

Fig. 3.20 Scheme of the process of regulations for compensation [42]

procedure. This is permissible, but not without risk, as route corrections could occur during the ongoing procedure.

If the project sponsors, despite a reasonable offer, cannot come to an agreement with the owners on a compensation regulation, they must initiate expropriation after the planning approval decision. The offers made on the negotiation path are usually based on current framework agreements, which the interest groups of those affected (agricultural associations, forest owner associations etc.) and the grid operators determine together. In over 95% of cases, the grid operators receive the rights to use properties to be claimed within the framework of private agreements [42].

One reason for the high "signing rate" in the case of private agreements is likely to be that there is a right of expropriation for areas that have to be claimed for high and highest voltage lines. The main reason, however, is likely to be that the amounts of compensation in the framework agreements are higher than the alternative expropriation compensations according to the state regulations. However, the grid operators are not completely free in their offers. Higher compensations than those shown in the framework agreements violate the efficiency requirement and could lead to them not being fully recognized in a cost audit when determining the revenue caps. Since the revenue caps determine the income opportunities of the grid operators via the so-called grid usage fees, non-recognized costs have a direct impact on the result. Figure 3.20 shows the typical process in the compensation regulation.

The right of use—or way right—for the grid operator established with the compensation is secured by a perpetual "limited personal servitude" registered in the land register. The devaluation associated with the burden compared to comparable and unencumbered properties is compensated by one-time servitude compensation, oriented towards the current market value. In addition to the compensation for the servitude, the property owners or users, as mentioned, also receive compensations

Component	Purpose	Form/Type	Amount of compensation
Compensation for restrictions (easement)	Compensation for the loss in value of the land due to restrictions on use as a result of the easement	**One-off payment** to the site **owner**	▪ **Overhead line: 20%** of the market value for protective strips (approx. 40-70m) ▪ **Underground cable: 25-30%** (incl. surcharge for amicable agreement), protection strip approx. 6-20m
Acceleration surcharge	Additional payment to accelerate grid construction	**One-off payment** to the **owner** if the offer is accepted within a certain period (usually 8 weeks)	▪ **0.3 to 0.5 €/sqm** of the protection strip area used ▪ **may not exceed 50%** of the market value of the property according to the Federal Grid Agency
Compensation for permanent use	Compensation for yield losses due to permanent land use by structures (e.g. pylons, cable sleeve buildings) or overhead line corridors in the forest	**One-off payment** to the **owner**	▪ Compensation for pylons, based on **expert opinions** ▪ Compensation amount usually depends on the length of the pylon edge and the gross yield of the area ▪ Loss of use due to management in forestry is vovered by so-called 'ground gross rent' depending on the type of tree
Compensation for construction damages andfor consequent damages	Compensation for crop yield reductions (e.g. destroyed/premature harvest)	Payment to **user** (land damage regulation)	▪ Amount of payment based on guidelines values from the Chamber of Agriculture/farmers' associations depending on the kind of crop ▪ In forestry, usually a parcel-specific assessment based on the applicable guidelines by experts
Compensation for expenses	Expenses for conclusion of contract & notarisation or land damage settlement	**Separate one-off** payment to the owner/user	▪ **50-300 € each for owners & users** ▪ For HVDC cable pilot projects € 250 - 1.000 for owners and € 500 - 1.500 for users

Fig. 3.21 Overview of compensation practice and the amount for compensations in Germany [42]

for the permanent use of land, for example by towers and buildings for joints of cables, compensations for consequential damage caused during the construction, such as field and crop damage, as well as further compensation components, for example expense allowances or compensations for used private roads. An overview of the compensation practice and the determination of the one-time compensation payments is shown on Fig. 3.21.

3.5 Approval Procedures in Brazil

3.5.1 Environmental License for Underground Transmission Lines in Brazil [43]

The environmental license process for underground transmission lines in Brazil is based on the same procedures applied to overhead lines, but with specific requirements related to underground installation. The requirements may vary depending on the state/municipality where the line will be installed.

The main steps are:

Preliminary License (LP)

A preliminary study is conducted to evaluate the potential environmental impacts of installing the underground transmission line. At this stage, factors such as proximity to sensitive areas (conservation zones, water bodies, densely populated urban areas), the type of soil to be excavated, and the characteristics of the project are identified.

Studies such as the *Environmental Impact Study (EIA)* and the *Environmental Impact Report (RIMA)* may be required to detail the project's effects and propose mitigation measures. After analyzing these studies, if the project is considered environmentally viable, the LP is issued.

Installation License (LI)

With the LP in hand, the developer can apply for the Installation License, which authorizes the start of construction for the underground transmission lines. At this stage, it is essential to follow the requirements outlined in the Preliminary License. This includes the implementation of mitigation measures, noise control plans, emission control, and, most importantly, care during excavation to prevent soil and groundwater contamination. A detailed waste management plan must also be provided for the excavation materials resulting from the construction of the line's infrastructure.

Operation License (LO)

After completing the construction and verifying compliance with all the conditions imposed by the previous licenses, the developer applies for the Operation License. This license authorizes the transmission line to begin operation. To obtain the LO, it is necessary to demonstrate that all mitigation measures were effective, and that the operation will not cause significant environmental harm, as well as to verify the levels of magnetic field emissions. The required level of magnetic fields may vary depending on the State/City—not all accept the exposure limits established for both the public and workers, as recommended by the *International Commission on Non-Ionizing Radiation Protection (ICNIRP)*, adopted by the World Health Organization, and endorsed by Brazilian National Law No. 11,934 of May 5, 2009.

Special Considerations for Underground Cables

- Excavation: underground lines require the removal and movement of large volumes of soil. Licensing requires strict measures to prevent erosion, sedimentation, soil contamination, and the submission of the residual waste management plan.
- Interference: often, these lines are installed in urban areas, which may require special measures to minimize the impact on traffic and local infrastructure.
- Environmental Risks: The possibility of interference with underground watercourses or contamination of aquifers is a significant environmental concern, and the disposal of excavation materials must be done in certified and authorized landfills.

The license approval process typically takes around six months in Rio de Janeiro City.

Additionally, licensing from other local public authorities (such as Drainage, Urban Planning, Traffic, and Public Road Works and Repairs Commissions) is required to carry out the construction.

3.6 Approval Procedures in Slovenia [44]

As in many other countries, the Authorization Procedures in Slovenia take several years, and can exceed 10 years. Base for the line planning are European standards, e.g. EN 50341 and their National Normative Annexes for Slovenia, also European Environmental Directives are considered which are sometimes implemented more strictly in the national laws. Slovenia has defied other nature conservation protected areas in addition to the ones given in the Natura 2000 regulations.

In the planning state of a project, a strategic environmental assessment (SEA) has to be carried out in addition to the environmental impact assessment (EIA).

The procedures of spatial planning and line routing are among the most complicated and most time consuming. Figure 3.22 shows two possibilities for the routing. The left procedure has been used so far. In the first phase, several possible parallel corridors are investigated. They consider the different importance of protected assets (e.g. nature, archaeology, settlement, rivers, etc.). According to the SEA procedure variants need to be considered and the one with the smallest impacts will be proposed at the end. The results and the variants which led to the decision are presented in public hearings and the preferred one is carried out now more in detail. For all transmission lines of 110 kV and above and lengths of more than 5 km, a preliminary assessment is carried out. An EIA procedure is carried out for all projects of 220 kV and above and lengths exceeding 15 km, as well as all projects that previously have been assessed as causing impacts. The EIA leads to a second hearing.

The Ministry of Natural Resources and Spatial Planning will grant the permission for the project, based on the completed documentation, contracts with landowners in the-right-of way and positive notifications from the concerned authorities (ministries, local communities, and others).

3.7 Duration of Approval Procedures in Different Countries

The trend towards electrification of the economy continues. For instance, the International Energy Agency (IEA) affirms that the electric grid would need to double by 2050 to reach the Net Zero Emission scenario [45]. The timeframes for the realization of energy infrastructure projects differ from country to country. It comes out that some "advanced economies" require significantly longer realization times than, for example, India or China. The most striking differences are evident in the realization of high voltage connections. Here, the planning periods are, for example, up to seven years in Europe and just one and a half years in China [45, 46]. Figure 3.23 considers the period from 2023 to 2035 and shows this relation.

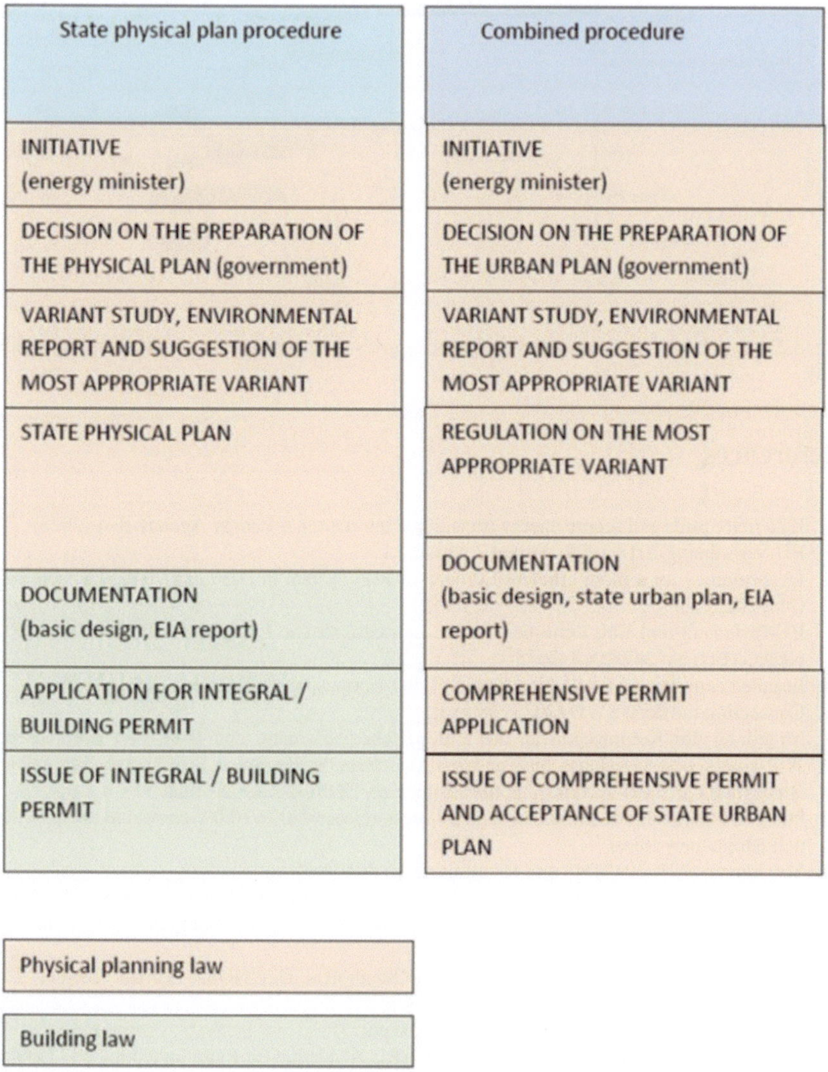

Fig. 3.22 Process for spatial planning and line routing in in Slovenia

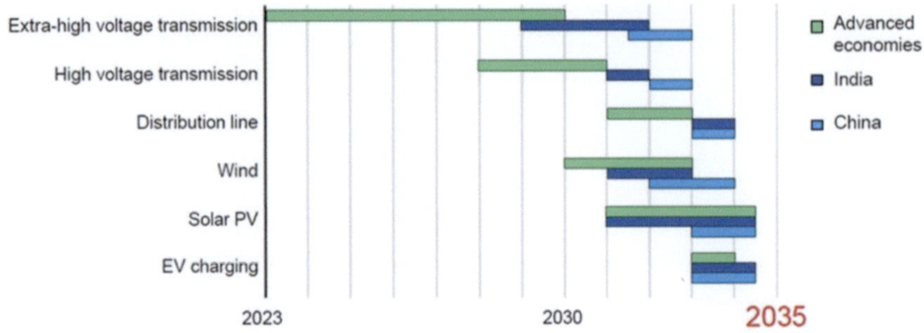

Fig. 3.23 Typical deployment time for electricity grids, solar PV, wind and EV charging stations [45]

References

1. Electricity grids and secure energy transitions. International Energy Agency IEA (2023)
2. EU-Verordnung 2018/1999, Artikel 2 Absatz 11
3. Opportunities for a more efficient European power system in 2030 and 2040, TYNDP 2022 (Mai 2023)
4. EU-Parlament und Rat, Leitlinien für die transeuropäische Energieinfrastruktur, VERORD-NUNG (EU) Nr. 347/2013, 2013
5. Installed capacity per production type, ENTSO-E, transparency platform (2024+2025)
6. Connecting the dots, SWD (2023) 68 final
7. Mitteilung der Kommission an das Europäische Parlament, den RAT, den Europäischen Wirtschafts- und Sozialausschuss und den Ausschuss der Regionen, Stromnetze, das fehlende Bindeglied – Ein EU-Aktionsplan für Stromnetze, COM (2023) 757 final
8. https://www.consilium.europa.eu/de/press/press-releases/2023/10/09/renewable-energy-council-adopts-new-rules/
9. Verordnung (EU) 2019/943 und Richtlinie (EU) 2019/944
10. Broschüre „Umweltverträglichkeitsprüfung", Österreichisches Bundesministerium für Nachhaltigkeit und Tourismus, Wien, 2019, im www: Anwendungen und Ziele der Umweltverträglichkeitsprüfung (bmk.gv.at)
11. TenneT (2019) Ersatzneubau Oberbachern-Ottenhofen, Das NOVA-Prinzip. TenneT TSO GmbH, Bayreuth
12. ENTSO-E, Über das TYNDP, TYNDP (entsoe.eu)
13. Jannsen G (2018) Raumordnungsverfahren. ARL-Akademie für Raumforschung und Landesplanung, Hannover, S. 1919–1929
14. Palic M et al (1992) Kabel und Freileitungen in Überregionalen Versorgungsnetzen. expert Verlag, Ehingen
15. Panebianco S et al (2019) Rumordnungsverfahren Grundlagen, Beispiele, Empfehlungen. ARL Akademie für Raumforschung und Landesplanung, Hannover
16. Regierungspräsidien Baden-Württemberg. Durchführung von Raumordnungsverfahren. Regierungspräsidien Baden-Württemberg (baden-wuerttemberg.de)
17. Regierungspräsidien Baden-Württemberg. Planfeststellung (baden-wuerttemberg.de)
18. Bundesnetzagentur, Netzausbau, Online-Informationsplattform der Bundesnetzagentur
19. Bundesnetzagentur (2020) Jahresbericht 2020. Bundesnetzagentur, Bonn
20. Bundesnetzagentur (2012) Leitfaden zur Bundesfachplanung. Bundesnetzagentur, Bonn
21. Bundesnetzagentur (2015) Ablauf und Fristen der Bundesfachplanung. Bundesnetzagentur, Bonn

22. de Witt S, Durinke P, Krause H (2019) Höchstspannungsleitungen Planung, Genehmigung und Enteignung, 2nd edn. Erich Schmidt Verlag, Berlin
23. Transnet BW. Genehmigungsverfahren, Dialog Netzbau. Transnet BW, Stuttgart
24. Transnet BW. Das NOVA-Prinzip - Verantwortung im Netzbau, Welt der Energie. Transnet BW, Stuttgart
25. Drees T et al. Netzentwicklungsplan Strom 2035, Version 2021. www.netzentwicklungsplan.de
26. Lumpi Berndorf. Hochtemperaturbeständige Freileitungsseile belastbar bis 210 °C (HTLS). Lumpi-Berndorf
27. CTC Global (2018) Hoch belastbare ACCC-Leiterseile, 1. Auflage, deutsche Version. CTC Global, Wiesbaden
28. Richter M et al (2010) Leitfaden zur Strategischen Umweltprüfung (Langfassung). Umweltbundesamt
29. UVP-Portal des Bundes, Wie läuft eine Umweltverträglichkeitsprüfung ab, Startseite | UVP-Portal
30. Anon., Plattform zu Umweltthemen beim Stromnetzausbau Bundesnetzagentur 2018
31. Kießling F, Nefzger P, Kaintzyk U (2001) Freileitungen - Planung, Berechnung, Ausführung, 5. Auflage. Springer
32. Fleckenstein K, Palic M, Rhiem W (1992) Räumliche Auswirkungen von Freileitungen. etz 113(1)
33. Groß M (1990) Graphische Datenverarbeitung in der Freileitungsplanung - Innovative Methoden mittel Sichtbarkeitsanalyse, Elektrizitätswirtschaft, Jahrgang 89 m, Heft 6
34. Haas D et al. Vogelschutz an Freileitungen, Studie. NABU-Naturschutzbund Deutschland e.V., Bonn
35. Anon., Vogelschutz an Mittelspannungsfreileitungen, VDE-AR-N 4210-11, Anwendungsregel: 2011-08 (2011)
36. Richarz K, Grohs A, Bohn T (2010) Vogelschutz an Mittelspannungsfreileitungen. netzpraxis 49(12)
37. Jödicke K (2018) Wirksamkeit von Vogelschutzmarkierungen an Erdseilen von Höchstspannungsfreileitungen. NATURSCHUTZ und Landschaftsplanung 50(8)
38. Anon., Vogelschutzmarkierung an Hoch- und Höchstspannungsfreileitungen, FNN-Hinweis, Berlin (2014) 3.23
39. Liesenjohann M et al (2019) Artspezifische Wirksamkeiten von Vogelschutzmarkern an Freileitungen, BfN-Skripten 537, Bundesamt für Naturschutz
40. Olbrich A (2018) Vogelschutz - Verpflichtung für Netzbetreiber, Vortrag, EW-Fachtagung Freileitungen, Erfurt, EW Medien und Kongresse, Berlin
41. Schmitt K (2021) Neues zum Vogelschutz, Vortrag, Forum Netzbau und Netzbetrieb Dresden. MVV Regioplan, Mannheim
42. Anon., Entschädigung von Grundstückeigentümern und Nutzern beim Stromnetzausbau, Gemeinschaftsstudie, frontier economics und WHITE&CASE (2016)
43. Damasceno PC (2024) HV Underground Lines Consultant, "Environmental license for underground transmission lines in Brazil", October 2024, Brazil
44. Kregar A. Personal information. ELES, Slovenia
45. Iglesias J (2024) Challenges for electric grid development. CIGRE ELECTRA 335. Figure Source: IEA 2024
46. Palic M (2025) 33. Ausgabe des Forums Netzbau und Netzbetrieb. Electrosuisse Bulletin.ch, Mai 2025

Chapter 4
Overhead Lines

Summary

High-voltage overhead lines have been the backbone of power grids for more than 125 years. This chapter begins with a discussion of the external loads on overhead lines and their components, from material selection to manufacturing processes to dimensioning. This is followed by the sub-chapter on conductors, including sag calculation, state equation, monitoring, conductor vibrations, and high-temperature conductors. The focus for insulators is on porcelain, glass, and composite insulators, while for fittings it is on conductor clamps and vibration dampers. Also, overhead line structures, from classic steel lattice towers to compact lines often used today, are discussed in detail along with their foundations. The chapter concludes with a discussion of line construction and maintenance, as well as the upgrading and uprating of lines. Many calculation examples and an extensive bibliography round off this chapter.

4.1 Introduction

Overhead lines are and will remain the most important medium for the transmission of electrical energy for a long time to come, despite serious objections from the public in many countries and recent progress in high-voltage underground cable technology. This development has been massively intensified in recent years by the energy transition and the global mega-trend towards renewable energies as well as the liberalization of the electricity market in Europe in the 1990s. The result is the need for a considerable number of new power transmission corridors. Overhead lines as an essential component of the worldwide high voltage networks will remain the backbone of a secure and reliable power supply for the foreseeable future.

From a technical point of view, there is broad agreement that the reinforcement and upgrading of existing lines and the construction of new ones are urgently needed to

K. O. Papailiou et al., *High-Voltage Overhead Lines and Underground Cables*, Power Systems, https://doi.org/10.1007/978-3-032-07498-0_4

maintain network stability and reliability. The big problem with network expansion, however, is the rapidly growing resistance from the public that has emerged in recent decades. This has led to approval procedures that—if they are successful at all—take quite long, with endless discussions and mostly emotional arguments from the parties (Chap. 3).

As determined in a recent survey by the Study Committee for Overhead Lines (SC B2) of CIGRE, the following topics are of particular importance for the overhead line community worldwide:

- Increasing the capacity of existing lines,
- New materials and technologies for use in overhead lines,
- Public acceptance of overhead lines.

Other topics of interest identified in the survey include:

- Condition assessment and estimated remaining lifetime,
- Methods for optimizing the design,
- Best practice for construction and maintenance procedures,
- Foundations for difficult soil conditions.

These topics are examined in greater detail below, based on the *Springer Handbook of Power Systems* [1] and the CIGRE Green Book *Overhead Lines* [2], authored by leading experts within CIGRE.

4.1.1 CIGRE

CIGRE (*Conseil International des Grands Réseaux Électriques—in English: International Council on Large Electric Systems*), is the largest international non-governmental, non-profit organization in the field of electrical power engineering that has been active since 1921 and has its headquarters in Paris. Its goal is the exchange of experience and the sharing of knowledge among its thousands of members and the public. CIGRE publishes comprehensive technical reports on emerging developments, best practices, and procedures in the power engineering sector, while also actively contributing to international standardization efforts. Its publications are widely used as authoritative references across a broad range of projects and technical domains. CIGRE's global membership spans academia, industry, research institutions, and utilities, reflecting the organization's multidisciplinary reach. Recognized worldwide as a leading authority in power engineering, CIGRE has played a pivotal role in the development and advancement of overhead line design and construction. Its work has provided the technical foundation for numerous international standards, including:

- Vibration behavior of conductors, measurements and damping
- Dynamic line thermal rating
- Application of high-temperature conductors
- Probabilistic line design acc. to IEC 826/80826, EN 50341 (CENELEC).

CIGRE has its head office in Paris, but over the years four regional subgroups have emerged. They deal with aspects of the power community within CIGRE which are of particular importance for the respective regions. These are listed according to their year of establishment:

2000: AORC—Asia-Oceania Regional Council of CIGRE (Australia, China, Gulf Cooperation Council—Saudi Arabia, Kuwait, United Arab Emirates, Qatar, Bahrain, Oman—India, Indonesia, Iran, Japan, Jordan, Korea, Malaysia, New Zealand, Thailand)
2001: NRCC—Nordic Regional Council of CIGRE (Denmark, Estonia, Latvia, Lithuania, Finland, Iceland, Norway, Sweden)
2006: RIAC—Regional Ibero-American Council of CIGRE (Argentina, Brazil, Chile, Colombia, Mexico, Paraguay, Peru, Portugal, Spain)
2013: SEERC—South-East European Region of CIGRE (Austria, Bosnia/Herzegovina, Croatia, Czech Republic and Slovakia, Georgia, Greece, Hungary, Israel, Italy, Kosovo, Montenegro, North Macedonia, Romania, Serbia, Slovenia, Turkey, Ukraine).

Through worldwide symposia, colloquia, and tutorials—including online formats—CIGRE provides its members with up-to-date insights into the latest advancements in their respective technical fields. These events also serve as valuable entry points for young professionals, offering them the opportunity to engage with cutting-edge developments, build foundational knowledge, and deepen their expertise in this dynamic and impactful domain.

4.1.2 Brief History of Overhead Lines

The first alternating current high-voltage overhead line was built in 1891 in Germany, from Lauffen am Neckar to Frankfurt am Main (Sect. 1.2). The voltage was only 15 kV at the time, although the line was later operated at 25 kV. It is interesting that the industrial transmission of electrical energy in direct current was already developed and used in Germany already in 1882.

Overhead line engineers realized early that current and voltage had to be increased to transmit large amounts of electrical energy. This led to the intensive development of conductors for higher currents and insulators for higher voltages.

Since the originally used copper wires were heavy and expensive, aluminum wires were soon introduced. As early as 1897, a seven-wire stranded aluminum conductor was put into operation in Connecticut, USA. Another important milestone was the development of composite conductors, i.e., steel-reinforced aluminum conductors (ACSR, *Aluminum Conductor Steel Reinforced*), in 1907. Today's conductors with diameters of 50 mm or more can transmit up to 1500 A, and modern high temperature conductors (HTLS, *High Temperature Low Sag Conductors*) even more (Sect. 4.3.9).

Because ever greater distances had to be covered from the generation to the load centers and due to the empirical rule for the optimal transmission voltage of 1 kV/km (Fig. 4.1), higher voltages were required. These could only be realized through the parallel development of suitable insulators. The invention of the cap-and-pin insulators made of porcelain or glass allowed long insulator strings and thus higher voltages by simply adding further elements. The next development led—via the motor insulator, the predecessor of the long rod insulator, made of porcelain, so-called, as invented by Motor-Columbus in Baden/Switzerland—to today's composite insulators.

With higher voltages and larger size conductors, line supports also had to evolve to meet the new electrical and mechanical demands. This can be seen in the development of the towers, the most visible elements of an overhead line. Here the range extends from the wooden poles of the early days to the impressive steel lattice towers of the 1200 kV AC lines of the present (Fig. 4.2).

Direct current lines are normally used to cover greater line lengths; they are covered in Sect. 4.10 and in Chaps. 1 and 7.

In recent years, in addition to cost-optimized design [3, 4], the appearance of overhead lines has gained in importance, which has promoted the development of so-called compact lines and alternative tower designs (Sect. 4.6.9).

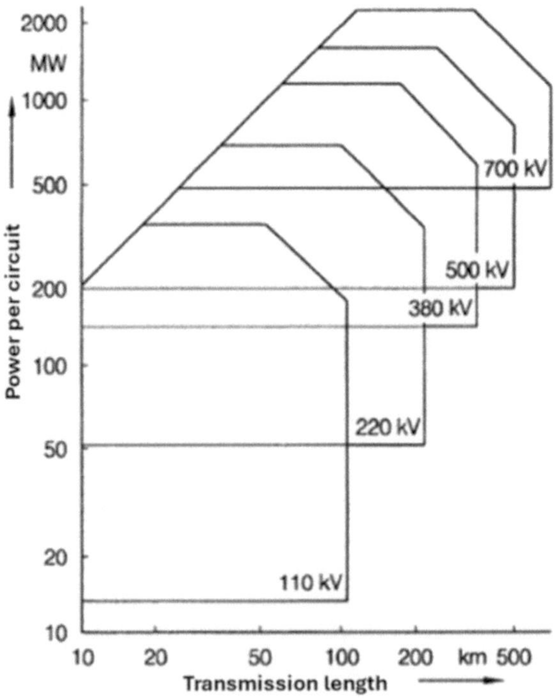

Fig. 4.1 Optimal transmission voltage for high voltage overhead lines

Fig. 4.2 Development of overhead line structures: From the wooden pole of the first 15 kV overhead line in 1891 in Germany (*left*) to the steel lattice towers of the first 1200 kV AC double circuit line in India (*right*)

4.2 Line Design

An overhead line must be planned, constructed, and operated to reliably fulfill its primary function—the transmission of electric power—while maintaining economic efficiency throughout its expected service life. Safety is also paramount; the design must ensure that risks of injury or loss of life during both construction and operation are minimized.

To achieve these goals, all line components must be carefully designed, manufactured, and installed to ensure uninterrupted operation under prevailing climatic conditions, nominal voltage and current levels, and anticipated short-circuit loads. In addition, appropriate measures must be taken to address the impacts of overvoltages—both atmospheric and system-generated, including switching overvoltages.

These conditions are best met when a line is designed according to a generally recognized national or international standard. In this context, the European Standard EN 50341-1 "Overhead electrical lines exceeding AC 1 kV—Part 1: General requirements—Common specifications", CENELEC, Brussels (2012) [5] is a good example. It is widely used in Europe and it applies to European CENELEC-member-countries in combination with National Normative Annexes [6]. About CENELEC and the related standards, see Chap. 1, Sect. 1.2.3.

4.2.1 Loading Assumptions

An overhead line is subject to a variety of different mechanical loads, such as:

- Conductor self-weight,
- Ice weight,
- Wind pressure,
- Conductor tension,
- Conductor vibrations,
- Ambient and conductor temperature.

With the exception of conductor vibrations, these are static or quasi-static loads.

In addition, there are extraordinary loads, which occur during construction and operation. These include short circuits, differential settlements of foundations, or—very rarely—loads due to conductor rupture, avalanches and earthquakes.

However, the most important loads for the design of an overhead line, besides the conductor weight, are the so-called climatic loads, i.e. wind and ice loads, which will be examined in the following.

Conductor Self-Weight

The self-weight of the conductors and the ice on the conductors are the main causes of the vertical load on a line. These loads are transferred via the fittings and the insulators to the structures and finally to the foundations.

The weight per unit length of the conductor is typically provided in the data sheets of the suppliers. Typical values for conductors of high-voltage lines are between 10 and 30 N/m, with usual spans at 380 kV lines of 400 m.

To calculate the associated load, one must know the so-called *weight span* L_G, which is defined as the horizontal distance between the apex points of the conductor curve on both sides of the tower. If there is no difference in height between the conductor attachment points of the adjacent spans, the weight span is identical to the wind span L_w, i.e. the arithmetic mean of the length of the two adjacent spans (Fig. 4.3):

$$L_W = \frac{L_1 + L_2}{2} = L_G \quad (h_1 = h_2 = 0) \tag{4.1}$$

If there is a height difference, the weight span must be calculated with the so-called *supplementary span* L_E [7, p. 481]:

$$L_E = L + 2Z \frac{\pm h}{w_c\, L} \tag{4.2}$$

Here L is the span length between two adjacent towers, which have a difference in height h (+ sign: adjacent tower is lower, − sign: adjacent tower is higher) and Z is the tension in the conductor. Herewith, the vertical load V_c acting on the support from

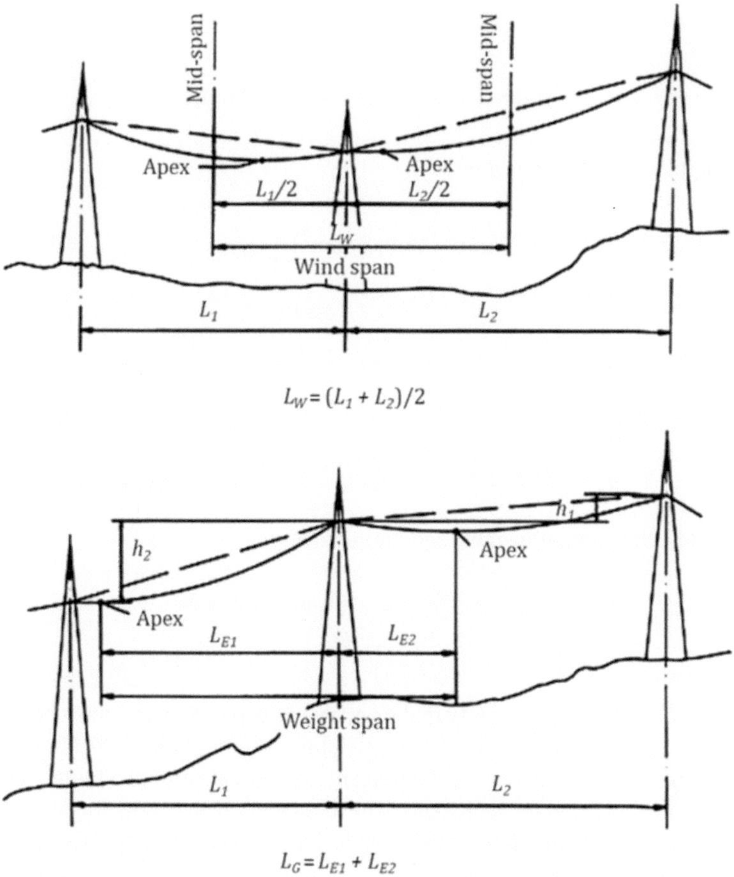

Fig. 4.3 Wind span (top) and weight span (bottom) (*Source* Omexom)

the conductor self-weight w_c of the two adjacent spans (Fig. 4.3) can be calculated to:

$$V_{Gc} = w_c\ (L_{G1} + L_{G2}) = w_c\ \frac{L_{E1} + L_{E2}}{2} \tag{4.3}$$

Example

$$L_1 = 350 \text{ m}, L_2 = 350 \text{ m}, h_1 = 50 \text{ m}, h_1 = 0 \text{ m}$$

Conductor: 231-AL1/39-ST1A (ACSR 230/30) with $\sigma = 85$ N/mm^2, $A_c = 260.7$ mm^2, $w_c = 8.77$ N/m.

The wind span L_W is:

$$L_w = \frac{350\ \text{m} + 350\ \text{m}}{2} = 350\ \text{m}$$

The supplementary spans are (4.2):

$$L_{E1} = 350\ \text{m} + 2\left(85\ \text{N/mm}^2 \cdot 260,7\ \text{mm}^2\right)\frac{50\ \text{m}}{8.77\ \text{N/m} \cdot 350\ \text{m}} = 1072\ \text{m}$$

$$L_{E2} = 350\,\text{m} + 2\,Z\frac{h\ (=0)}{w_c\ L} = 350\,\text{m}$$

And the weight span:

$$L_G = \frac{L_{E1} + L_{E2}}{2} = \frac{1072\ \text{m} + 350\ \text{m}}{2} = 536\ \text{m}$$

As well as the vertical weight load on the tower:

$$V_{Gc} = 8.77\ \text{N/m} \cdot 536\ \text{m} = 6235\ \text{N}$$

Ice Loads

Ice on the conductor acts evidently in the vertical direction and thus must be added to the weight of the conductor. Under extreme climatic conditions, which are increasingly occurring today, heavy ice loads can lead to catastrophic failure and line outages—if not designed for such loads. The ice accretion on conductors is a complex physical process. A distinction is made between the direct conversion of water vapor into ice in fog or clouds, the so-called *rime*, and ice accretion from precipitation, which occurs as *clear ice from freezing rain*, *wet snow* and *dry snow*.

These vertical loads result in increased tensile stress in the conductors. The associated loads on clamps, insulators, and structures can be calculated in a similar way to the conductor's weight, i.e. using the corresponding weight span. The values for ice loads under average climatic conditions depend heavily on the region and its climate and are defined in the project specifications. They can change along the route of a line. The standards determine minimum values. For Germany, for example, the ice load for a 30 mm conductor in the ice load zone 3 (see list below) is approx. 25 N/m.

Basically, based on weather services, statistical evaluations, ice load measurements, observations from the network operators and their experience regarding damage to overhead lines, an ice load map is created. Germany, for example, is divided into four ice load zones [8]:

Ice load zone 1: Areas where, due to the climatic conditions and confirmed by many years of experience, only low ice loads occur that have not led to any damage.

Ice load zone 2: Areas where, due to the climatic conditions, the geographical location and confirmed by many years of experience, high ice loads are to be expected, which have led to damage.

Ice load zone 3: Areas where, due to the climatic conditions, the geographical location and confirmed by many years of experience, frequently high ice loads are to be expected, which have led to significant damage.
Ice load zone 4: Areas where, due to the climatic conditions, the geographical location and the exposure, very high ice loads are to be expected. The magnitude of ice loads in this area should be determined based on the experience of the line operator or through expert assessment.

The associated ice loads per unit length q_i depend on the conductor diameter d [mm] and are [N/m]:

E1: $g_i = 5 + 0.1\ d$
E2: $g_\mathrm{i} = 10 + 0.2\ d$
E3: $g_\mathrm{i} = 15 + 0.3\ d$
E4: $g_i = 20 + 0.4\ d$

For insulators, the ice load is dependent on the length of the insulator string. The following values should be considered for each ice load zone: 50 N/m in zone E1, 100 N/m in zone E2, 150 N/m in zone E3, and at least 200 N/m in zone E4.

Wind Loads

Wind can also lead to a strong loading on overhead lines. In addition to increased conductor tensile stresses, high winds can cause massive conductor deflections, which may violate the required safety clearances of the live conductors to the grounded structures and therefore must be considered in the design of the *tower top geometry.*

When the wind hits a surface, be it the surface of a conductor, an insulator or a structure, it generates a dynamic pressure q, which depends on the wind velocity, which in turn is primarily determined by the climate and terrain. In this sense as an example Germany is divided into four wind zones according to [8], with the corresponding dynamic pressures q_0:

W1: 320 N/m^2,
W2: 390 N/m^2,
W3: 470 N/m^2,
W4: 560 N/m^2.

Wind velocity—and consequently wind pressure—increases with height h above ground level, due to decreasing frictional resistance. It shall be calculated in accordance with Sect. 4.3 DE.1 [8] as follows:

$$q_p(h) = 1.5 q_0 \text{ for } h \le 7\,\mathrm{m}$$

$$q_p(h) = 1.7 q_0 \left(\frac{h}{10}\right)^{0.37} \quad \text{for } 7\,\mathrm{m}\, h \le 50\,\mathrm{m}$$

$$q_p(h) = 2.1 q_0 \left(\frac{h}{10}\right)^{0.24} \quad \text{for } 50\,\mathrm{m}\, h \le 300\,\mathrm{m}$$

Wind load on conductors

The wind load on the conductors of an overhead line can be calculated as follows:

$$Q_{wc}(\vartheta) = q(h)Q_{wc}G_{xc}C_c d_c L_w \cos^2(\vartheta) \tag{4.4}$$

With d_c, the conductor diameter, L_w, the wind span defined above, *h,* the height of the suspension point of the conductor at the insulator and ϑ, the angle between the wind direction and the conductor. In the case that the wind acts perpendicular to the conductor, $\vartheta = 0$ and $\cos^2(\vartheta) = 1$. For typical conductors for high voltage overhead lines ($d_c > 15.8$ mm), the drag coefficient $C_c = 1$. The span coefficient G_{xc} takes into account—here in the example according to the German standard DIN EN 50341-3-4 [8] —the fact that the wind does not act uniformly on the entire wind span L_w.

In the wind zones W_1 and W_2 the following applies:

$G_c = 0.75$ for spans up to 200 m

$G_c = 0.45 + 60/L$ for spans over 200 m

In the wind zone W3 the following applies:

$G_c = 0.67$ for spans up to 200 m

$G_c = 0.40 + 54/L$ for spans over 200 m

In wind zone W4 the following applies:

$G_c = 0.60$ for spans up to 200 m

$G_c = 0.36 + 48/L$ for spans over 200 m

For a conductor bundle, the total wind load is defined as the sum of the wind loads on the individual sub-conductors and, depending on the specification, with or without consideration of shielding effects on leeward conductors.

The wind also leads to a deflection of the conductor, which influences its distance to adjacent conductors, to the line support and to objects in the vicinity and must be appropriately considered to maintain the mandatory clearances. The swing-out angle φ_c of the conductor in midspan results from the ratio of the horizontal wind load Q_{wc} according to (4.4) to the vertical self-weight of the conductor W_c based on the weight span L_W (4.1). Here, according to [8, Sect. 5.6.3.2/DE.1] the conductor temperature is assumed to be constant at +40 °C and the impact of the wind load on the sag of the conductor is neglected. As the design value of the wind load, 58% of the wind pressure specified in the section *Wind load* must be assumed. In general:

$$\phi_c = \arctan\left(\frac{Q_{wc}}{W_c}\right) = \arctan\left(\frac{q(h)\ G_c\ C_c\ d_c\ L_W}{w_c\ L_G}\right) \tag{4.5}$$

where w_c is the weight per unit length of the conductor.

Combination of Ice Loads and Wind Loads on Conductors

When wind and ice act simultaneously, the wind pressure on a conductor with diameter d_c [m] with ice accretion according to the considered ice zone, may be reduced to 50%. This pressure acts on the so-called equivalent diameter d_i [m] of the conductor with the ice weight per unit length g_i [N/m], with the specific weight (density) of the ice $\rho_i = 7500$ N/m^3, which can be calculated as follows [8, Sect. 4.6.6.1 DE.1]:

$$d_i = \sqrt{d_c^2 + \frac{4g}{(\pi \rho_i)}} = \sqrt{d_c^2 + 0.00017 g_i} \tag{4.6}$$

This value can then be substituted into (4.4) instead of d_c, where the drag coefficient C_c is set to 1.0.

Wind Load on Insulators

The wind load $Q_{w,is}$ (h) on the insulator depends on the height h of its suspension point above ground, and can be calculated as follows:

$$Q_{w,is}\,(h) = q(h)\ \ G_{is}\ C_{is}\ A_{is} \tag{4.7}$$

The terms mean:

$q\,(h)$ Wind pressure on the insulator
h Height above ground of the insulator's suspension point on the structure
G_{is} Structure coefficient of the insulator; a value of 1.0 is recommended
C_{is} Drag coefficient of the insulator; a value of 1.2 is recommended
A_{is} Outline area of the insulator string that is exposed to the wind.

Thus, using (4.5) and (4.7), the swing-out angle ϕ of the suspension string (Fig. 4.4), as well as the critical distance of the conductor from the structure (Fig. 4.102), can be calculated:

$$\varphi = \arctan\left(\frac{nq(h)G_cC_cd_cL_W + Q_{w,is}(h)}{nw_cL_G + w_{is}/2}\right) \tag{4.8}$$

With:

n Number of sub-conductors.
$q(h)$ Wind pressure (N/mm^2) on conductor and insulator string
C_c Drag coefficient of the conductor (1.0 for typical conductors)
G_c Span coefficient
d_c Conductor diameter (m)
L_W Wind span (m)
L_G Weight span (m)
w_c Weight per unit length of the conductor (N/m)
w_{is} Weight of the insulator string (N).

Fig. 4.4 Forces and swing-out angle of a suspension string

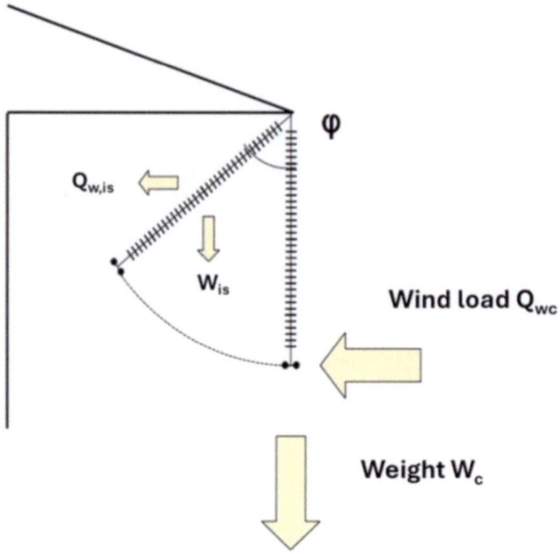

Example

Conductor: Quad bundle 264-AL1/34-ST1A (ACSR 265/35), diameter 22.4 mm, weight per unit length 10 N/m.

Suspension height: 50 m, wind span: 350 m, weight span: 536 m, wind perpendicular to the direction of the line: $\vartheta = 0$, wind zone 2, ice load zone 2.

Insulators: Porcelain long rods, $G_{is} = 3$ kN, $A_{is} = 1.2$ m ^{2}

Wind pressure (wind zone 2) on conductor and insulator, suspension height of the conductor ($h = 50$ m):

$$q(50m) = 1.7 \cdot 390\left(\frac{50}{10}\right)^{0.37} \frac{N}{m^2} = 1203\frac{N}{m^2}$$

Wind load per sub-conductor (4.1):

$$Q_{wc}(0^0) = 1203 \cdot (0.45 + 60/400) \cdot 1.0 \cdot 0.0224 \cdot 350 = 5688\,\text{N}$$

Wind load on suspension string (4.4):

$$Q_{w,is}(50\,\text{m}) = 1203\,\text{N/m}^2 \cdot 1.0 \cdot 1.2 \cdot 1.2\,\text{m}^2 = 1732\,\text{N}$$

Weight per sub-conductor at the suspension point:

$$V_c = w_c L_G = 10\,\text{N/m} \cdot 536\,\text{m} = 5360\,\text{N}.$$

Swing-out angle of the suspension string (Fig. 4.4):

$$\phi_c = arctan\left(\frac{4 \cdot 5688N + 3000N}{4 \cdot 5360N + 1732N}\right) = arctan(1.11) \approx 48^0$$

Ice load per unit length for ice load zone 2:

$$E_2 : g_i = 10 + 0.2d = 10 + 0.2 \cdot 22.4 = 15\,\text{N/m}$$

Conductor diameter with ice accretion (4.3):

$$d_i = \sqrt{0.0224^2\,\text{m}^2 + \frac{4 \cdot 15\,\text{N/m}}{3.14 \cdot 7500\,\text{N/m}^3}} = 55.2\,\text{mm}$$

Wind load on conductor with ice (4.1 and 4.3):

$$Q_{wc,i}(0^\circ) = 0.5 \cdot 1203 \cdot (0.45 + 60/400) \cdot 1.0 \cdot 0.0552 \cdot 400 = 8\text{kN}$$

Wind load on structures

There are two basic types of structures for overhead lines

The classic steel lattice towers, and the single-piece tubular *poles* increasingly used for compact lines. The latter are made of wood, steel, concrete, and recently also of glass fiber reinforced composites. The calculation of wind loads is relatively simple for poles, as they essentially have a cylindrical shape, and is given by:

$$Q_{w,p}(h) = q(h) G_p C_p A_p \tag{4.9}$$

With:

- $q(h)$ Wind pressure
- h Reference height
- G_P Structure coefficient; a value of 1 is recommended
- C_P Drag coefficient with values between 0.7 for round and 1.4 for hexagonal poles; for wooden poles, $C_p = 0.9$ applies
- A_P Area filled by tower profiles (steel lattice towers).

The calculation of the wind load on steel lattice towers is also done according to Eq. (4.6). The drag coefficient C_P can be assumed to be 2.8 in the case of the usual steel lattice towers with angle profiles [6].

Safety Loads

These include one-sided acting tensile forces in the conductor, e.g. due to conductor failure or unequal ice loads in adjacent spans. They cause longitudinal and torsional stresses on the structures and must be considered accordingly.

Construction Loads

The structures of an overhead line must be able to withstand all loads, which arise from line construction work, temporary guy anchors, lifting devices, etc., as well as loads from the weight of the linemen. The latter are usually assumed to be 1.0 kN per person.

Short-Circuit Loads

The major concern during a short circuit is that the swinging motion of the conductors, as caused by the short circuit, can lead to flashovers, and thus to power interruptions. In addition, short-circuit forces generate high mechanical loads on conductor fittings, such as spacers [9] (Sect. 4.5.3).

4.2.2 Design

The *external loads* explained in the previous section, which are mainly mechanical loads, cause *mechanical stresses* in the individual components of an overhead line. In general, components or their materials can only be stressed up to a certain limit, i.e., they have a *load-bearing capacity*, which resists unacceptable deformations or material failure, when they reach their *ultimate strength.*

The basic principle of any *design* or *dimensioning* process is to choose the material and dimensions of a component in such a way that its stress is less than its load-bearing capacity.

The determination of the various loads involves inherent uncertainties, as most of them exhibit stochastic behavior. Moreover, the models used to represent these loads are often subject to simplifications and inaccuracies. To account for these factors, *partial safety factors* have been introduced. These are applied multiplicatively to the *characteristic value* of a given force to derive its corresponding *design value* (see Table 4.1). This approach ensures a sufficient level of safety in the design process.

Similarly, uncertainties also exist regarding component dimensions and the variability of material properties, particularly strength. Also, in this case partial safety factors are applied to account for these effects. In accordance with the "safe side" principle, the characteristic material strength is divided by the respective safety factors to obtain the design strength (see Table 4.2).

Electrical Requirements

In addition to the mechanical aspects, electrical requirements must also be considered when designing overhead lines. Essentially, this involves questions of insulation and

Table 4.1 Partial safety factors and combination values ([6], Table 4.7)

Action	Symbol	Reliability levels		
		1	2	3
Variable actions (climate loads)				
Extreme wind load	γ_{W}	1.0	1.2	1.4
Nominal wind load	Ψ_{W}	0.4	0.4	0.4
Extreme ice load	γ_{i}	1.0	1.25	1.5
Nominal ice load	Ψ_{i}	0.35	0.35	0.35
Permanent actions				
Self-weight	γ_{G}	1.0		
Security loads (accidental actions)				
Torsional loads due to conductor tension	γ_{A1}	1.0		
Longitudinal loads due to conductor tension	γ_{A2}	1.0		
Safety loads				
Construction and maintenance loads	γ_{P}	1.5		

Table 4.2 Partial safety factors γ_M according to EN 50341-1 for the strength of the individual components of an overhead line. The *ultimate load* refers to the failure load, whereas the *nominal load* denotes the required minimum bearing capacity as specified in [7], Table 6.9

Component	Type of stress	Type of loading	γ_{M}
Conductors	Tension	Limit load	1.25
Insulators	Tension, bending	Limit load	2.00
Fittings	All types of loading	Minimum failing load	1.90
Lattice towers			
– Members	Tension, compression, buckling, bearing	Limit load	1.10
– Bolted connections	Shearing, bearing	Limit load	1.25
– Welded connections	Shearing	Limit load	1.25
Poles			
– Made of steel	Tension, compression, bending, bearing	Limit load	1.10
– Concrete	Compression	Limit load	1.50
– Reinforcement	Tension	Limit load	1.15
Guyed towers, anchors	Tension	Limit load	1.60
Foundations	Uplift loading	Limit load	National requirements

dielectric strength. In an overhead line, on the one hand, solid insulating materials are used (Sect. 4.4), on the other hand, the atmospheric air takes over the insulation task of the live conductors against earth, as well as between the conductors themselves. In this context a distinction is also made between *internal* and *external clearances*.

Internal Clearances

The geometry of the structures of an overhead line and in particular of the tower top must be designed (Sect. 4.6) so that the so-called internal clearances, i.e. the distances between live parts, i.e. the conductors and to the grounded towers [10] or the ground wires, as well as between the conductors in the span, especially in mid-span [11] are maintained. With large spans, there can be different swinging angles between horizontally arranged adjacent conductors, so that one conductor swings more than a neighboring one, which may lead to a reduction of clearances.

External Clearances

Also, between the live parts of an overhead line and the environment sufficient distances must be maintained, so that danger to persons, animals and objects, which are in the vicinity of the line, can be avoided. The following situations must be considered:

- Distance to the ground in remote and regularly accessed areas
- Distance to residential and other buildings
- Distance to sports fields and recreational areas
- Distance to wind turbines and energy plants
- Distance when crossing or near roads, railways and other tracks and navigable waterways
- Distance to cable cars, lifts, and similar
- Distance when crossing or running parallel to other power or telecommunication lines.

These are vertical *and* horizontal distances, at the maximum conductor temperature, but also the sag of conductors with ice accretion, as well as the deflection of conductors by wind have to be considered (Sect. 4.2.1).

The external clearances consist of the electrically required clearance D_{el} and an additional safety distance, so that neither persons nor objects come dangerously close to live parts of the line. Their minimum values are to be taken from the relevant regulations, e.g. [6].

4.3 Conductors

The conductors are considered the most important part of an overhead line, as they provide the transmission of electric power and their share of the total costs of the line is considerable. The costs of conductors—material and installation—can make up to 40% of the total investment costs of the line. In addition, conductor loads, and conductor sag, determine the shape and dimensions and thus also the costs for the supports and foundations of the line. Therefore, careful attention is always given to the proper selection of conductors [12].

4.3.1 Types and Basic Properties of Conductors

Conductor Structure

Conductors consist of several layers of concentrically stranded wires. The layers are *counter-rotating*, i.e., the direction of rotation of each layer is opposite to the layer below, with the outer layer always being right-handed (Fig. 4.5). This reduces the torsional moment that occurs in the conductor during stringing which can cause problems during installation. This also improves the electrical properties of the conductor, especially their self-inductance and their alternating current resistance [13]. In conductors, each layer has six more wires than the layer below, which allows a high filling factor of the conductor cross-section (Fig. 4.6).

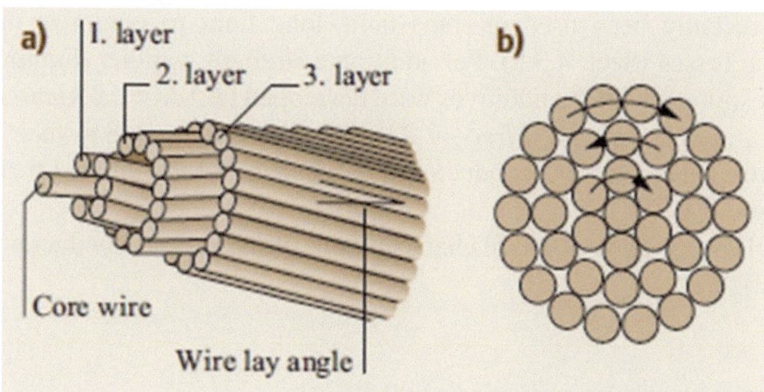

Fig. 4.5 Stranded conductor structure and designations **a** conductor geometry, **b** direction of lay

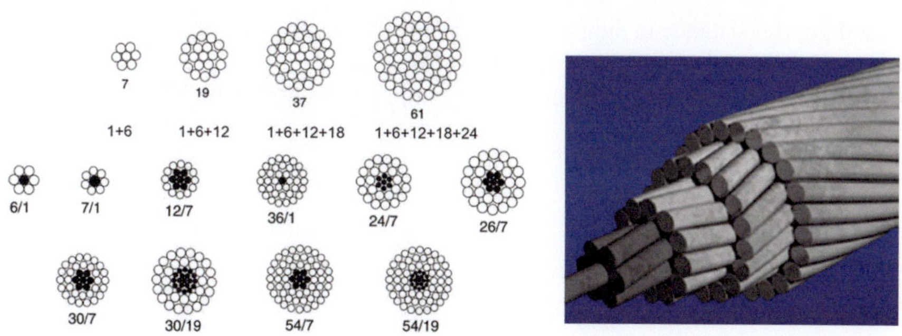

Fig. 4.6 Structure of typical stranded conductors (steel wires in black, aluminum wires in gray)

Already at the end of the nineteenth century, aluminum conductors began to replace the originally used copper conductors. Aluminum wires are produced using the so-called *Properzi process*, which is a rolling/casting process [14]. Most conductors nowadays consist of several layers of aluminum wires, which are helically wound around a core of steel wires, these primarily provide mechanical strength. Such conductors are known as ACSR (*Aluminum Conductor Steel Reinforced*). The wires of the steel core consist of medium-strength galvanized steel with a carbon content of 0.50 to 0.85 weight-% and a nominal tensile strength of 1400 MPa. They are greased for corrosion protection [15]. The diameter of the conductor's wires is generally between 2 and 5 mm and the *lay angle*, which is the angle of a wire with the conductor axis (Fig. 4.5 left), is between 10° and 15°. In recent years, a trend has emerged to increase the aluminum content of the conductor. Nowadays, conductors with a ratio of the aluminum to the steel cross-section of 7:1 are increasingly used—previously 6:1 conductors were used more often—and sometimes up to 11:1. This reduces the Ohmic losses. Table 4.3 summarizes the properties of selected standard ACSR conductors.

Overhead line conductors made of pure aluminum are referred to as AAC (*All Aluminum Conductor*). Due to their lower strength-weight ratio, these conductor types are more suitable for short spans in distribution networks and for areas where ice and wind loads are limited, as well as for flexible busbars in substations. However, they have recently been used in some quite long lines to minimize the Ohmic transmission losses (Sect. 4.3.5). For additional strength, various aluminum alloys with magnesium and silicon additives were developed (AAAC: *All Aluminum Alloy Conductor*), also known as *Aldrey*- or *Almelec-conductors*. Due to their excellent corrosion resistance, AAAC conductors are the preferred solution for the use in coastal areas.

In the following some essential characteristics of the various conductor materials are summarized:

Aluminum

- Hard drawn aluminum consisting of 99.5% Al
- Conductivity approx. 35 m/Ωmm^2 (corresponds to 61% of copper)
- Tensile strength (before stranding): 160–200 N/mm^2
- In air formation of an oxide layer, thus protecting against further corrosion attack and good resistance in aggressive environments, e.g. coastal areas
- Coefficient of thermal expansion: 23×10^{-6} /°C
- Strength reduction begins at conductor temperatures above 80 °C.

Aluminum-Magnesium-Silizium Alloy

- Trade names: Aldrey, Almelec
- Tensile strength (before stranding) approx. 300 N/mm ^{2} (corresponds to 160% of aluminum)
- Conductor without steel core possible
- 86–93% of the aluminum conductivity (corresponds to 52.5–57.5% of copper)
- Continuous current carrying capacity approx. 95% of aluminum
- Coefficient of thermal expansion and softening equal to aluminum.

Steel

- Use of galvanized steel wires in ACSR as well as in ground wires
- Labeling is according to strength
- Galvanization dependent on wire diameter (EN 10244-2)
- Diameter: 1.25–4.75 mm
- Tensile strength: 1200–1800 N/mm^2
- Coefficient of thermal expansion 11.5×10^{-6}/°C (50% of aluminum)
- Conductivity: approx. 5 m/Ωmm^2
- Max. continuous operating temperature >180 °C.

Table 4.3 Characteristic data of ACSR conductors [7]

Designation	Former code	Cross section (mm^2)	Number of strands	Number of strands	Conductor diam. (mm)	Mass per unit length (kg/km)	Rated strength (kN)	DC resistance (Ω/km)	Current-carrying capacity (A)
	ACSR	Total (mm^2)	Al	St					
15-AL1/3-ST1A	16/2.5	17.8	6	1	5.40	61.6	5.80	1.8769	105
24-AL1/4-ST1A	25/4	27.8	6	1	6.75	96.3	8.95	1.2012	140
34-AL1/6-ST1A	35/6	40.1	6	1	8.10	138.7	12.37	0.8342	170
44-AL1/32-ST1A	44/32	75.6	14	7	11.2	369.3	44.24	0.6574	200
48-AL1/8-ST1A	50/8	56.3	6	1	9.60	194.8	16.81	0.5939	210
51-AL1/30-ST1A	50/30	81.0	12	7	11.7	374.7	42.98	0.5644	230
70-AL1/11-ST1A	70/12	81.3	26	7	11.7	282.2	26.27	0.4132	290
94-AL1/15-ST1A	95/15	109.7	26	7	13.6	380.6	35.93	0.3060	350
97-AL1/56-ST1A	95/55	152.8	12	7	16.0	706.8	77.85	0.2992	370
106-AL1/76-ST1A	105/75	181.2	14	19	17.5	885.3	105.82	0.2742	400

(continued)

Table 4.3 (continued)

Designation	Former code	Cross section (mm^2)	Number of strands	Number of strands	Conductor diam. (mm)	Mass per unit length (kg/km)	Rated strength (kN)	DC resistance (Ω/km)	Current-carrying capacity (A)
	ACSR	Total (mm^2)	Al	St					
122-AL1/20-ST1A	120/20	141.4	26	7	15.5	491.0	44.50	0.2376	410
122-AL1/71-ST1A	120/70	193.4	12	7	18.0	894.5	97.92	0.2364	415
128-AL1/30-ST1A	125/30	157.8	30	7	16.3	587.0	56.41	0.2260	425
149-AL1/24-ST1A	150/25	173.1	26	7	17.1	600.8	53.67	0.1940	470
172-AL1/40-ST1A	170/40	211.8	30	7	18.9	788.2	74.89	0.1683	520
184-AL1/30-ST1A	185/30	213.6	26	7	19.0	741.0	65.27	0.1571	535
209-AL1/34-ST1A	210/35	243.2	26	7	20.3	844.1	73.36	0.1381	590
212-AL1/49-ST1A	210/50	261.5	30	7	21.0	973.1	92.46	0.1363	620
231-AL1/30-ST1A	230/30	260.8	24	7	21.0	870.9	72.13	0.1250	630
243-AL1/39-ST1A	240/40	282.5	26	7	21.8	980.1	85.12	0.1188	645
264-AL1/34-ST1A	265/35	297.7	24	7	22.4	994.4	81.04	0.1095	680

(continued)

Table 4.3 (continued)

Designation	Former code	Cross section (mm²)	Number of strands	Number of strands	Conductor diam. (mm)	Mass per unit length (kg/km)	Rated strength (kN)	DC resistance (Ω/km)	Current-carrying capacity (A)
	ACSR	Total (mm²)	Al	St					
304-AL1/49-ST1A	300/50	353.7	26	7	24.4	1227.3	105.09	0.0949	740
305-AL1/39-ST1A	305/40	344.1	54	7	24.1	1151.2	96.80	0.0949	740
339-AL1/30-ST1A	340/30	369.1	48	7	25.0	1171.2	91.71	0.0852	790
382-AL1/49-ST1A	380/50	431.2	54	7	27.0	1442.5	121.30	0.0758	840
389-AL1/34-ST1A	385/35	420.1	48	7	26.7	1333.6	102.56	0.0749	850
434-AL1/56-ST1A	435/55	490.6	54	7	28.8	1641.3	133.59	0.0666	900
449-AL1/39-ST1A	450/40	488.2	48	7	28.7	1549.1	119.05	0.0644	920
490-AL1/64-ST1A	490/65	553.8	54	7	30.6	1852.9	150.61	0.0590	960
494-AL1/34-ST1A	495/35	528.4	45	7	29.9	1632.6	117.96	0.0584	985

(continued)

Table 4.3 (continued)

Designation	Former code	Cross section (mm^2)	Number of strands	Number of strands	Conductor diam. (mm)	Mass per unit length (kg/km)	Rated strength (kN)	DC resistance (Ω/km)	Current-carrying capacity (A)
	ACSR	Total (mm^2)	Al	St					
511-AL1/45-ST1A	510/45	585.8	48	7	30.7	1765.3	133.31	0.0566	995
550-AL1/71-ST1A	550/70	620.9	54	7	32.4	2077.2	166.32	0.0526	1020
562-AL1/49-ST1A	560/50	611.2	48	7	32.2	1939.5	146.28	0.0515	1040
571-AL1/39-ST1A	570/40	610.6	45	7	32.3	1887.1	136.40	0.0506	1050
653-AL1/45-ST1A	650/45	698.8	45	7	34.4	2159.9	156.18	0.0442	1120
679-AL1/86-ST1A	680/85	764.5	54	19	36.0	2549.7	206.56	0.0426	1150
1046-AL1/45-ST1A	1045/45	1090.9	72	7	43.0	3249.2	218.92	0.0277	1580

In addition to the standard conductor designs, there are special constructions, such as conductors with a high steel content or with high strength steel for very long spans, like river crossings, or with trapezoidal wires to better utilize the conductor cross-section than with round wires. Recently, conductors made of copper and its alloys are also offered to exploit the lower resistance of this material. The mechanical properties of such conductors limit their suitability for all applications and present an optimization challenge [16].

In Europe, the designation of conductors according to European standards includes the abbreviation of the conductor material type and the respective cross-section in mm^2. For example, a frequently used ACSR conductor with an aluminum cross-section of 243 mm^2 and a steel cross-section of 39 mm^2, previously called Al/St 240/40, is now referred to as 243-AL1/39-ST1A. In other parts of the world, the North American practice is often followed, where the conductors bear names of flowers, cities, and animals, e.g. Tulip, Rose and Magnolia for AAC, Akron, Montreal and Halifax for AAAC and Pelican, Hawk and Drake for ACSR. MCM and kcmil are further abbreviations from the English-speaking world. They stand for "thousands of circular mils" and also denote cross-sections, where a circular mill corresponds to 0.0005067 mm^2. A conductor with a 1000 mm^2 cross-section would therefore have 1973 kcmil.

Manufacturing and Testing

The production of conductors uses so-called *stranding machines* (Fig. 4.7). This is a quite complex process with high quality requirements, in which even small details, like the unwinding of the wires, play a major role [17, 18]. Additional important information on the design and manufacturing of conductors can be found in [12, 16].

The individual wires and the finished conductor are both stringently tested. As a rule, samples are taken from 10% of the wire length pending for stranding and from wires of the finished conductors. The following tests are performed:

- Diameter: max. 1% deviation from the value in the standard
- Wire surface: smooth, free of cracks, jumps and other damage, no inclusions of copper particles
- Tensile strength: is tested on a wire sample of 250 mm length
- Wrapping test for aluminum wires: A sample wire is wound eight times around a mandrel of the same diameter as the wire and unwound six times; it is considered passed, if the wire remains intact
- Galvanization for steel wires: Determination of the zinc mass and the uniformity of the zinc layer of the steel wires of the core as well as wrapping test: A sample wire is wound eight times around a mandrel of the same diameter as the wire; the galvanization must not peel off

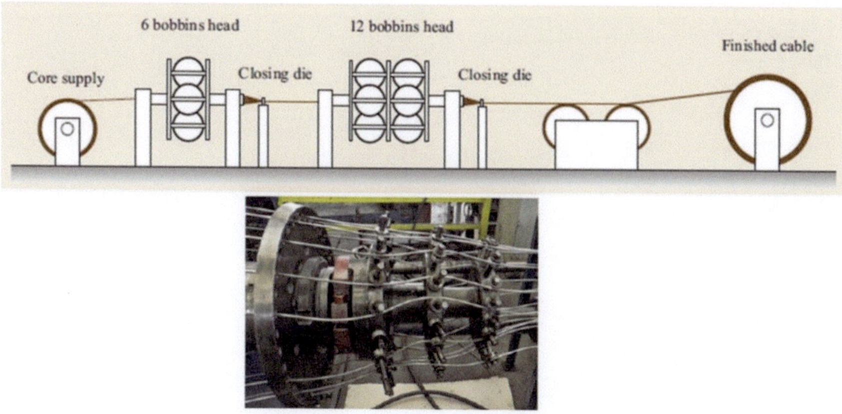

Fig. 4.7 Schematic of orbital strander (above), preforming head (below)

- Specific electric resistance: Measured between 10 °C and 30 °C and adjusted to the standard value at 20 °C
- Welding spots on wires made of aluminum and aluminum alloy are limited in number and should, if present in the outer layer, be marked.

For the finished conductor, on the other hand, the following tests are carried out (Table 4.4) [19].

Table 4.4 Conductor tests

Conductor	Type test	Sample test
– Surface condition	×	×
– Diameter	×	×
– Inertness	×	×
– Lay ratio and direction of lay	×	×
– Number and type of wires	×	×
– Mass per unit length	×	×
– Stress–strain curve	a	–
– Tensile breaking strength	a	–
– Stringing test	a	–
– Creep test[b]	a	–
Grease		
– Mass per unit length	×	×
– Drop point	×	×

[a] According to agreement between manufacturer and purchaser
[b] According to [20]

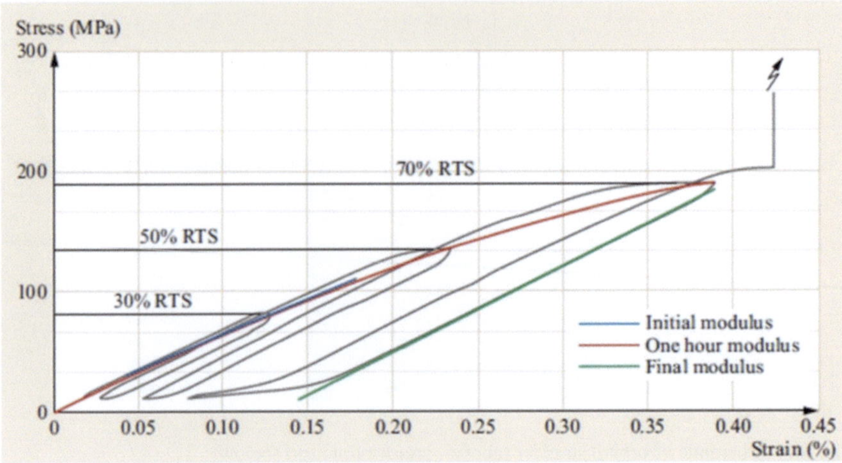

Fig. 4.8 Typical stress–strain curve of a conductor [21]

The most important test is the *tensile test*, described in more detail below, as it provides both the tensile strength of the conductor and its stress–strain curve (Fig. 4.8) [21]; the latter is essential for the sag calculation (Sect. 4.3.3).

This test is carried out on a conductor sample with a length of at least 400 times the conductor diameter, but not less than 10 m, in the following steps, with the conductor tensile force being completely relieved after each step:

- Load of 5% of the nominal tensile strength F_B (RTS: *Rated Tensile Strength*) for straightening and setting the conductor
- Load to 30% F_B and hold for 0.5 h
- Load to 50% F_B and hold for 1 h
- Load to 70% F_B and hold for 1 h
- Load to 85% F_B and hold for 1 h.

Subsequently, the conductor is again loaded with a steadily increasing tensile force until failure. This is achieved—and the test is considered passed—when one or more wires break at a load equal to or greater than 95% of the nominal tensile strength F_B—this is the calculated failure load—of the conductor. The latter is calculated for ACSR conductors according to [19] from the minimum tensile strength $\sigma_{B,Al}$ of the aluminum wires before stranding, multiplied by their nominal cross-section A_{Al}, plus the tensile stress of the steel wires at 1% elongation $\sigma_{1\%\ St}$ on a measuring length of 250 mm, also before stranding, multiplied by their nominal cross-section A_{St}, i.e.:

$$F_B = \sigma_{B,Al}\, A_{Al} \;+ \sigma_{1\%,St}\, A_{St} \tag{4.10}$$

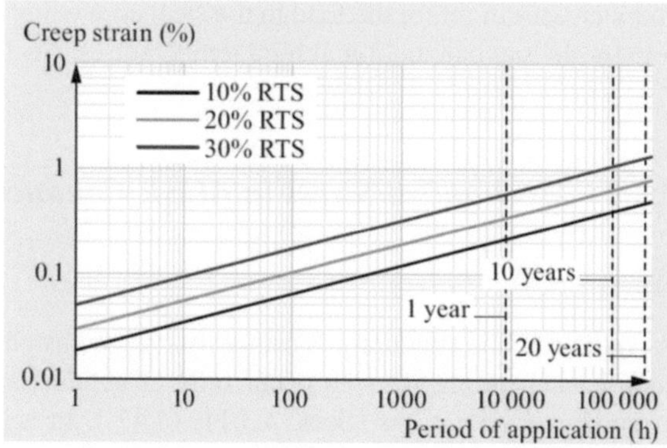

Fig. 4.9 Creep curves for ACSR Cardinal (483-AL1/63-ST1A) [7]

Creep

The time-dependent plastic strain or deformation of a material under sustained load is known as creep. Aas explained in detail in [22], creep occurs in crystalline materials through the thermal self-diffusion of atoms within the lattice structure. Atoms migrate from regions of higher to regions of lower density, leading to a more uniform distribution and a reduction in the material's strain stiffness.

Since the activation energy required for this atomic movement increases with the bonding energy between atoms, materials with higher bonding energy also tend to have higher melting points. As a result, creep in aluminum—whose melting temperature is considerably lower than that of steel—initiates at lower temperatures and progresses at a faster rate than in steel. The metallurgical creep behavior of aluminum under sustained elevated temperatures has been extensively studied in [17].

Creep strain, measured in long-term tests, is typically plotted on a double-logarithmic scale (Fig. 4.9). The data points generally align closely with a straight line (4.11), allowing for reliable extrapolation to practical timeframes—for example, up to 100000 h (over 10 years) [23]

$$e = e_{1h} t^n \tag{4.11}$$

In this formula, e and n are parameters, which depend on the type of conductor and its load; guideline values for these are given in [24]. In practice, the creep strain ε_{kr} is converted to an equivalent temperature increase ΔT_{kr}:

$$\Delta T_{kr} = \frac{\varepsilon_{kr}}{\alpha} \tag{4.12}$$

with α is the linear thermal expansion coefficient of the conductor (among others see section *Temperature behavior*).

The creep of steel wires is much less and is usually neglected. As a result, the higher creep in the aluminum wires of an ACSR conductor reduces their tensile load and increases in return the load in the steel core wires [25], which plays a significant role for the sag calculation at high temperatures (Sect. 4.3.4).

4.3.2 Mechanical Stress in ACSR Conductors

Tensile Stress

An overhead line conductor is primarily stressed by an axial tensile load. In addition, significant bending stresses occur only at conductor suspensions and play a role in conductor vibrations (Sect. 4.3.11, (4.47)). In a tension-loaded conductor, the load Z is distributed to individual wires according to the basic rules of engineering mechanics, i.e. in a monometallic conductor like aluminum or aluminum alloy with n wires of the same diameter d and with the modulus of elasticity E, the mechanical tension σ of a single wire is given by:

$$\sigma = \frac{Z}{n\pi d^2/4} \tag{4.13}$$

This is related to the wire strain ε via the so-called *Hooke's Law* (Fig. 4.10):

$$\sigma = E\varepsilon \tag{4.14}$$

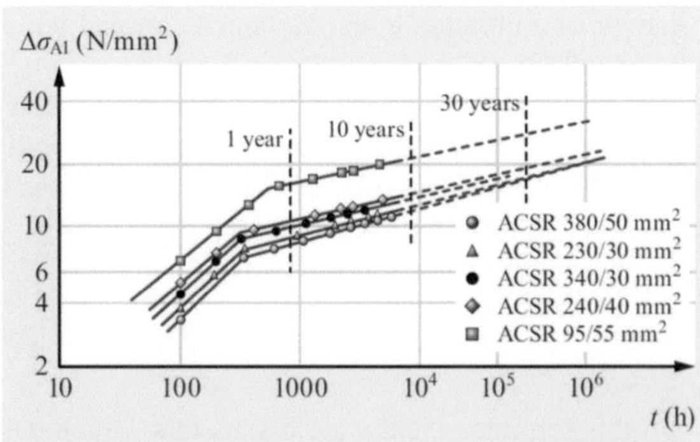

Fig. 4.10 Tensile stress decrease $\Delta\sigma_{Al}$ due to creep in the aluminum wires of different ACSR with a load of 25% RTS at 20 °C [25]

The horizontal component H of the conductor tension Z (in the direction of the conductor axis) used in (4.9) differs minimally (typically by 0.1%) from Z [22].

The situation is more complex for ACSR conductors. Since the conductor tension causes a radial compression on the individual wires due to the transverse contraction, (Fig. 4.11), the conductor behaves mechanically like a homogeneous round rod, i.e. the strains of the steel and the aluminum wires are equal:

$$\varepsilon_{St} = \varepsilon_{Al} \tag{4.15}$$

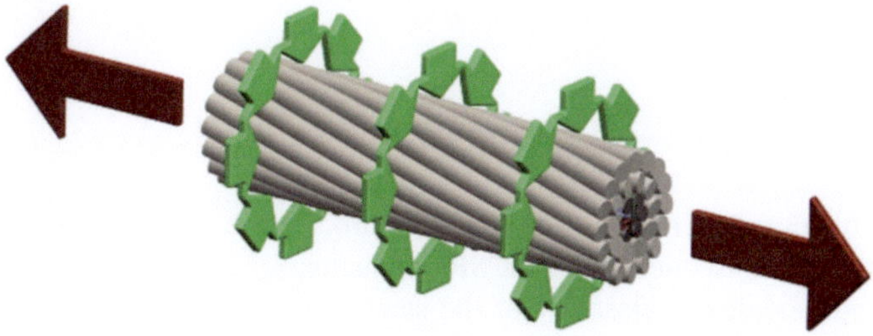

Fig. 4.11 Transverse contraction and radial compression of a tension-loaded conductor

By applying (4.14) to the aluminum and the steel wires follows:

$$\sigma_{Al} = E_{Al}\ \varepsilon_{Al} \tag{4.16}$$

and:

$$\sigma_{St} = E_{St}\varepsilon_{St} \tag{4.17}$$

And by equating the strains of aluminum and steel per (4.15):

$$\sigma_{St} = \frac{E_{St}}{E_{Al}}\ \sigma_{Al} \tag{4.18}$$

and:

$$\sigma_{Al} = \frac{E_{Al}}{E_{St}}\ \sigma_{St} \tag{4.19}$$

In addition, the force equilibrium over the conductor cross-section applies:

$$Z_{Al/St} = Z_{Al} + Z_{St} = \sigma_{Al}\ A_{Al} + \sigma_{St}\ A_{St} \tag{4.20}$$

Elementary conversions yield the modulus of elasticity of an ACSR conductor:

$$E_{Al/St} = E_{Al}\ \frac{A_{Al}}{A_{St}} + E_{St}\ \frac{A_{St}}{A_{Al}} \tag{4.21}$$

as well as the stresses in the aluminum and the steel wires:

$$\sigma_{Al} = \frac{Z}{A_{Al/St}}\ \frac{E_{Al}}{E_{Al/St}} \tag{4.22}$$

and:

$$\sigma_{St} = \frac{Z}{A_{Al/St}} \frac{E_{St}}{E_{Al/St}} \tag{4.23}$$

where A_{Al} denotes the cross-section of aluminum, A_{St} the cross-section of steel and $A_{Al/St}$ the total cross-section of the ACSR conductor.

Since the modulus of elasticity of the steel wires at approx. 210000 N/mm^2 is significantly higher than the modulus of elasticity of the aluminum wires at approx. 70000 N/mm^2 [26, Table A1-2], it follows from (4.23) that the tensile stresses in the steel wires are approx. three times larger than the tensile stresses in the aluminum wires, i.e. even a modest steel content in the conductor of e.g. 11% of the conductor cross-section, increases its tensile strength by almost 25% [22].

Temperature Behavior

Conductors not only elongate when they are under a tensile load but also under the influence of temperature. For individual wires with the original length *L*, the change of length *ΔL* can be calculated as:

$$\Delta L = \alpha\ \Delta T\ L \tag{4.24}$$

with α the *linear coefficient of thermal expansion* and ΔT is the temperature difference.

With similar considerations as with the modulus of elasticity, the coefficient of thermal expansion of ACSR conductors can also be calculated:

$$\alpha_{Al/St} = \alpha_{Al} \frac{E_{Al}}{E_{Al/St}} \frac{A_{Al}}{A_{Al/St}} + \alpha_{St} \frac{E_{St}}{E_{Al/St}} \frac{A_{St}}{A_{Al/St}} \tag{4.25}$$

Since in ACSR conductors the expansion coefficients of aluminum and steel have a ratio of 2:1 ($\alpha_{Al} = 23 \times 10^{-6}$ /°C vs. $\alpha_{St} = 11.5 \times 10^{-6}$/°C), a load transposition between the aluminum and the steel wires occurs during temperature changes: When the temperature rises, the load shifts from aluminum to steel, while, when the temperature drops, the opposite is the case. Subsequently, the aluminum wires must bear a higher mechanical load [27], which can be critical in winter, as then the conductor is more heavily tensioned due to the prevailing low temperatures. This load transposition is also important for the investigation of conductor vibrations [25].

4.3.3 Conductor Sag

The shape of a conductor, which is suspended at two points at the same height above the ground with a distance *L*, the span length, between these points, and with a distributed load *w* from its self-weight, plus possibly an additional load, such as an

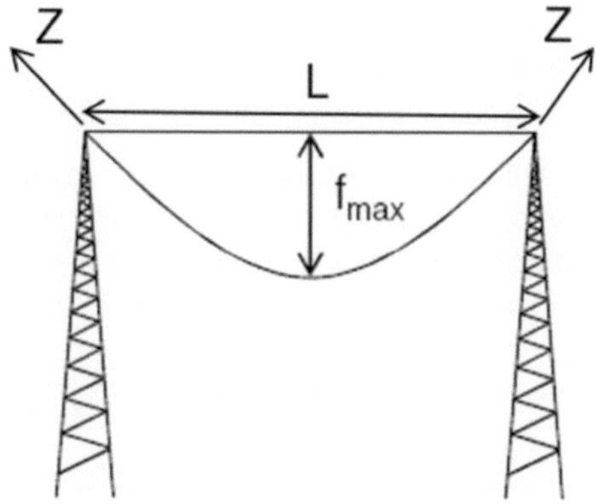

Fig. 4.12 Conductor sag

ice load, under a tensile load Z, forms a curve, which in mathematics is referred to as *catenary* (Fig. 4.12) (*catenary* from Latin *catena*: chain). This is – for most cases – represented with sufficient accuracy by a parabola.

The highest deflection of the conductor curve occurs in the middle of the span L and is referred to as maximum *sag* f_{max} (Fig. 4.12) and is given by:

$$f_{\max} = \frac{wL^2}{8Z} \tag{4.26}$$

As expected, the sag increases with the conductor's weight w and decreases with the tensile load Z in the conductor. In addition, it is quadratic proportional to the square of span length L, i.e., when the span length is doubled, the sag quadruples, which is significant for large spans, e.g. at river crossings, fjords or deep valleys, and is an important factor in determining the height of the adjacent towers.

Both, conductor length S in the span, (4.28), as well as the so-called *slack* $S - L$, (4.29), i.e. the difference between conductor length and span length, can be calculated under the condition (4.27), that applies with sufficient accuracy to almost all overhead lines.

$$\frac{w^2L^2}{48Z^2} \ll 1 \tag{4.27}$$

$$S = L + \frac{8f^2}{3L} = L\left(1 + \frac{S^2w^2}{24H^2}\right) \tag{4.28}$$

$$S - L = \frac{8f^2}{3L} \tag{4.29}$$

However, the following situations cannot be adequately addressed with these simple formulas:

- Position and length of the insulator strings
- Differences in height of the conductor's suspension points

- Different span lengths in a tension-section (this is the line section between two tension towers, see Sect. 4.6.2)
- Unequal ice loads in the individual spans of a section.

Although special computer programs are available for the above [22], it is instructive to calculate the sag for a reference span, referred to as *ruling span,* for a section with n spans, [28, 29]. Its length L_{id} is calculated as:

$$L_{id} = \sqrt{\frac{{L_1}^3 + {L_2}^3 + \cdots + {L_n}^3}{L_1 + L_2 + \ldots L_n}} \tag{4.30}$$

Since the numerator in (4.30) depends on the cube of the individual span lengths, the value of the ruling span strongly depends on the longest span. And since the conductor tension in all spans is approximately constant, after calculating the sag for the longest span $f_{\max,i}$, the sag f_i in each span of the section can be calculated as follows:

$$f_i = f_{\max,i}\left(\frac{L_{id}}{L_{\max,i}}\right)^2 \tag{4.31}$$

For clearance purposes, it is often necessary to determine the conductor's sag at any point in the span (Fig. 4.13). This is done with:

$$f_c = 4\ \frac{L_l\ L_r}{L^2}\ f_m \tag{4.32}$$

By substituting the numerical values from Fig. 4.13, we get:

$$f_c = 4 \cdot \frac{82.6 \cdot 272.4}{355^2} \cdot 13.92\ \text{m} = 9.94\ \text{m}$$

4.3.4 State Equation

The state of a conductor is defined by its temperature T and its tensile force Z, or its tensile stress σ, (Fig. 4.14).

The so-called *state equation* defines the relationship between two conductor states, denoted as state "1" and state "2", with the respective parameters identified by corresponding subscripts in the following expressions. This is accomplished by calculating the changes of the conductor length S in a span of length L, ΔS_{Temp}, due to conductor temperature T, and ΔS_{Spng}, due to tensile stress σ, (4.28):

$$S_2 = S_1 + \Delta S_{1Temp} + \Delta S_{1Spng} = S_1\left[1 + \alpha\ (T_2 - T_1)\right]\left[1 + \frac{\sigma_2 - \sigma_1}{E}\right] \tag{4.33}$$

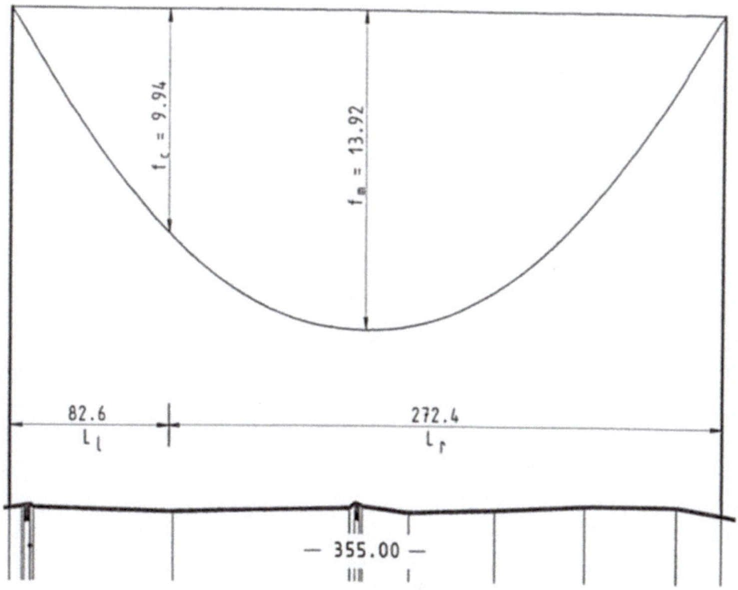

Fig. 4.13 Calculation of the sag at any point in the span (*Source* Omexom)

Fig. 4.14 Conductor sag for different conductor temperatures, ice, wind loads, and elapsed time after installation compared with the initial sag during stringing (top curve)

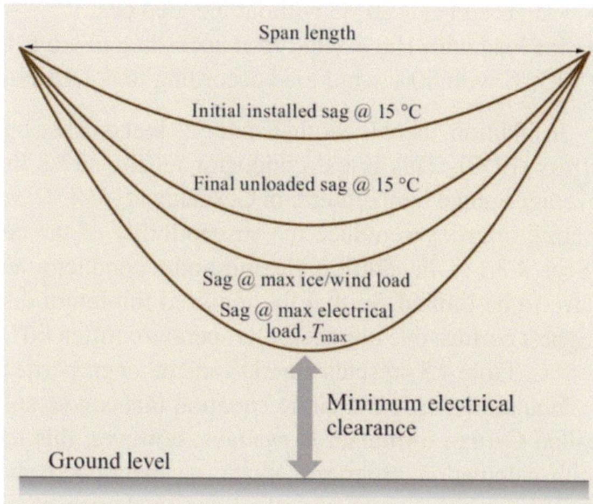

By substituting (4.28) into (4.33) we get the state equation:

$$\sigma_2^2\left(\sigma_2 - \sigma_1 + \frac{E(w_1L)^2}{24A^2\sigma_1^2} + E\alpha(\mathrm{T}_2 - \mathrm{T}_1)\right) = \frac{E(w_2L)^2}{24A^2} \tag{4.34}$$

Or in the more common form with the conductor tensile force Z_2 as the unknown:

$$Z_2^2\left(Z_2 - Z_1 + \frac{EA(w_1L)^2}{24H_1^2} + EA\alpha(\mathrm{T}_2 - \mathrm{T}_1)\right) = \frac{EA(w_2L)^2}{24} \tag{4.35}$$

In (4.34) and (4.35) mean:

σ_1, σ_2 Tension stress in state "1" or "2".
Z_1, Z_2 Tension force in state "1" or "2".
w_1, w_2 Weight of the conductor in state "1" or "2".
T_1, T_2 Conductor temperature in state "1" or "2".
L Span length (the ruling span according to (4.30) can also be used).
A Cross-sectional area of the conductor.
E Modulus of elasticity of the conductor.
α Coefficient of thermal expansion of the conductor.

State "1" is characterized by the ambient temperature during conductor stringing (Sect. 4.7.5), whereby the conductor is only loaded by its own weight. The tensile load Z_1 corresponding to a conductor tensile stress σ_1, must be set in such a way that for state "2", the tensile load and the corresponding tensile stress of the conductor Z_2, respectively σ_2, multiplied by the partial safety factor $\gamma_C = 1.35$ do not exceed a certain value. In Germany this value corresponds to 95% of the *rated tensile strength* (RTS) of the conductor divided by the material partial safety factor $\gamma_M = 1.25$, in accordance with [8, Sect. 6.2/EN.1]. In the same standard the critical conductor loads and temperatures are specified for state "2":

- −5 °C and ice load according to 4.5.2/EN.1 or
- −5 °C and at −20 °C without ice load or
- Ice load with 100% wind load according to 4.6.6.1/EN.1
- +5 °C with 50% wind load according to 4.4.1/EN.1.

In addition, there is another state "2" that must be observed, the so-called *everyday stress* (EDS). This is the conductor tensile stress that must be maintained at the average annual temperature, in Germany at +10 °C, without wind load [8, Para. 6.2/EN.2], in order to reduce the susceptibility of the conductor to aeolian vibrations (Sect. 4.3.11). In addition, the minimum conductor tensile force and the related sag have to be limited, so that the required minimum distances are not violated at the highest permissible conductor temperature (often 60° or 80°), or in the case of ice at −5 °C. Table 4.5 presents the EDS and other properties of some common conductors.

Equation (4.35) is a cubic equation that can be solved analytically using the so-called Cardan formulae. Nowadays, however, this task is conveniently performed with calculation programs, since, as mentioned above, several combinations of conductor states must be calculated to determine which one is decisive for the conductor installation load. With normal ice and wind loads, the other criteria mentioned above are usually fulfilled if the EDS limit is observed [7, p. 218].

Example

An ACSR conductor 300/50 ($A = 353.7\ mm^2$, $w = 12.115$ N/m, $E = 77{,}000\ N/mm^2$, $\alpha = 18.9\ 10^{-6}$ K) is strung at 10 °C with a conductor tensile stress $\sigma_1 = 50\ N/mm^2$, which roughly corresponds to its EDS (Table 4.5). The sag at the maximum design temperature 60 °C in a 390 m long span is required.

Table 4.5 Properties and everyday stress for common conductors [8, Table 9/EN.1]

Conductor type and reference standard	Cross-sectional ratio	Number of strands	Coefficient of thermal expansion α (× 10^{-6} K^{-1})	Effective modulus of elasticity E (kN/mm^2)	Everyday stress (N/mm^2)
AL1/ST1A and AL3/ST1A	1.4:1	14/7	15.0	110	90
		14/19			
	1.7:1	12/7	15.3	107	84
	4.3:1	30/7	17.8	82	57
	6.0:1	6/1	19.2	81	56
		26/7	18.9	77	
	7.7:1	24/7	19.6	74	52
		54/7	19.3	70	
		54/19	19.4	68	
	11.3:1	48/7	20.5	62	44
	14.5:1	45/7	20.9	61	40
	23.1:1	72/7	21.7	60	35
AL1	/	7	23.0	60	30
		19		57	
		37			
		61		55	
		91			
AL3 (E-AlMgSi)	/	7	23.0	60	44
		19		57	
		37			
		61		55	
		91			

The tensile force during stringing is:

$$Z_1 = \sigma_1/A = 50\,\text{N/mm}^2/352.7\,\text{mm}^2 = 17685\,\text{N}$$

By inserting into (4.35) follows:

$$\left[Z_2 - 17685 + \frac{77000 \cdot 353{,}7 \cdot (12{,}115)^2}{24 \cdot 17685^2} \cdot 390^2 \right.$$
$$\left. + 77000 \cdot 353{,}7 \cdot 18{,}9 \cdot 10^{-6}(60 - 10)\right] Z_2^2$$
$$= \frac{77000 \cdot 353{,}7 \cdot (12{,}115)^2}{24} \cdot 390^2$$

Summing equal terms together, provides:

$$Z_2^2[Z_2 + 89472] = 2555 \cdot 10^{10}$$

By iteration, it follows:
$Z_2 = 15{,}575$ N and from this $\sigma_2 = 44$ N/mm^2
With (4.26), the sag f at 60 °C can now be calculated:

$$f = \frac{12.115 \cdot 390^2\,\mathrm{m}}{8 \cdot 15575} = 14.8\,\mathrm{m}$$

In (4.22), it is assumed that the conductor behaves linearly elastic. This is a reasonable approximation, especially for conductors with a maximum operating temperature of up to 60 °C. Therefore, it is referred to as LE model (*Linear Elongation Model*). However, conductors do not behave purely elastically. They stretch under a load and, when this is reduced back to zero, they do not return to their original length, i.e., the conductors behave plastically, and their stress–strain curves are nonlinear (Fig. 4.8). These curves are obtained through laboratory testing and integrated into dedicated computational tools. Based on the so-called EPE model (*Experimental Plastic Elongation Model*), these tools are used to resolve the state equation [30]. Conductor creep (Fig. 4.9) can also be considered by an equivalent temperature increase [22, 31].

The situation becomes more complicated when ACSR conductors are operated at elevated temperatures [32]. With increasing temperature, the tensile stress carried by the aluminum wires shifts to the steel wires, as their coefficient of thermal expansion is twice as high as that of the steel core (Sect. 4.3.2). Therefore, above a certain conductor temperature, the mechanical load carried by the aluminum wires is completely transferred to the steel core; this is considered in the so-called SPE model (*Simplified Plastic Elongation Model*) [30]. The sag-temperature curve flattens from this so-called *knee-point temperature*, also called *transition point* [33], reducing the increase in sag with increasing temperature, as shown in Fig. 4.15. The knee-point temperature T_{KP} can be calculated as follows:

$$T_{KP} = \frac{Z_{Al/St}}{E_{St}\,A_{St}\,(\alpha_{Al} + \alpha_{St})} + T_i \tag{4.36}$$

where T_i is the temperature that prevailed during the installation of the conductor (index i for *initial*).

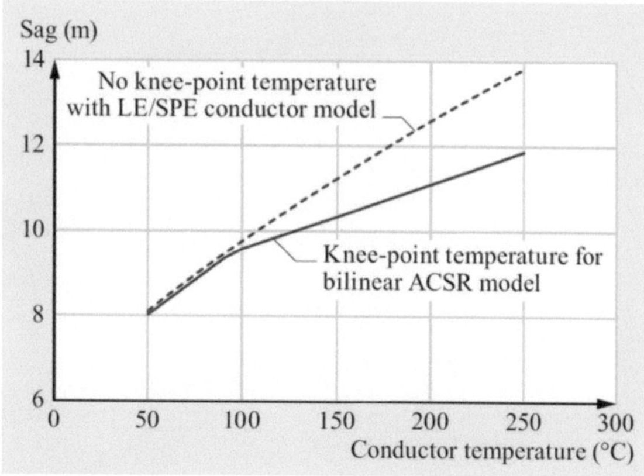

Fig. 4.15 Sag versus temperature of a 300-m span for ACSR Cardinal (54/7) using the LE/SPE and EPE conductor models [30]

4.3.5 *Current Carrying Capacity*

The flow of current through a conductor generates heat, which elevates the conductor temperature. To avoid overheating and damage, the *current carrying capacity* of the conductor must be determined.

If the conductor has reached a thermal equilibrium state under the influence of various factors, i.e., when its temperature over time is essentially constant, the so-called *steady state*, the following equilibrium condition applies, as schematically represented in Fig. 4.16 [34]:

$$P_J + P_S = P_C + P_R \tag{4.37}$$

The terms mean:

Joule Losses

The Joule losses per unit length P_J in a conductor are proportional to the square of its current I:

$$P_J = R_{AC}I^2 = k_J R_{DC}I^2 = k_J[1 + \alpha(T_m - 20)]I^2 \tag{4.38}$$

where α is the thermal expansion coefficient of the conductor (Sect. 4.3.2).

They arise predominantly from the direct current (DC) resistance R_{DC}, which opposes the movement of electrons through the conductor material. This resistance increases with the average conductor temperature T_m, which can be intuitively explained by the fact that as the temperature increases, so does the amplitude of the vibrations of the atoms in the metal lattice of the conductor, making the passage of electrons more difficult. The increase in direct current resistance R_{DC} with conductor temperature is about 4% per 10 °C. In addition, for alternating current (AC), the

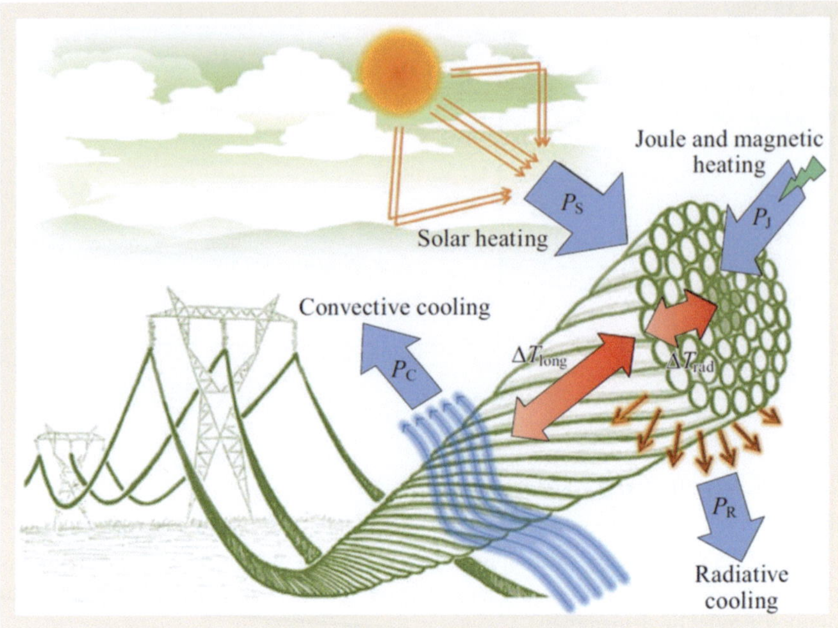

Fig. 4.16 Factors affecting the thermal equilibrium of an OHL conductor

conductor resistance R_{AC} increases due to the skin effect, and in bi-metallic ACSR conductors due to eddy currents and magnetization losses in the steel core [13, 35]. In (4.38), these effects are summarized in the factor k_J, which rarely takes values above 1.1.

Solar Radiation

The solar radiation per unit length P_S consists of direct solar radiation, diffuse sky radiation, and radiation reflected from the ground, the so-called *albedo*, and is proportional to the conductor diameter and the global radiation intensity, for which a value of 900 W/m^2 is often used in Central Europe.

Convection

Convection is the transfer of heat in the direction of a temperature gradient due to the mass movement of molecules, for example, in gases. The heat transfer from a conductor to the surrounding atmosphere is quantified by the convective cooling per unit length P_C, which depends on the thermal conductivity of the air and the temperature difference between the conductor surface and the surrounding air.

There are two types of convection: Natural convection occurs when the movement of air molecules is caused by buoyancy forces that result from density differences in the air. Forced convection occurs when wind flows over the surface of the conductor and transfers heat from the hot conductor to the cooler surrounding air. Even at fairly

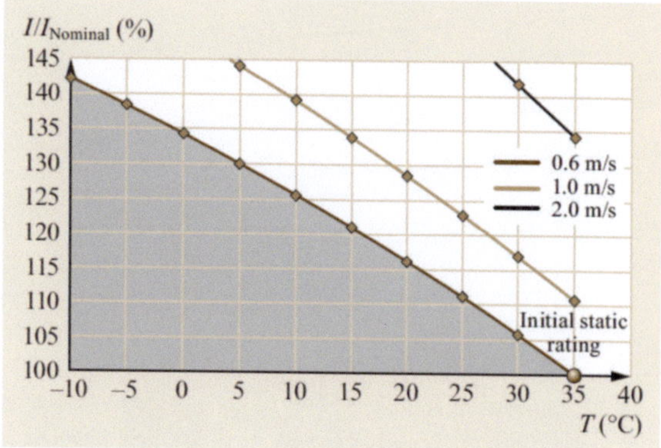

Fig. 4.17 Thermal rating of ACSR 265/35 as a function of ambient temperature, with wind velocity as a parameter [36]; $I_{Nominal}$ is the rated current

low wind speeds below 0.5 m/s, forced convection dominates and natural convection can be neglected.

Radiation

Radiation is caused by the movement of atoms and molecules in the material and occurs at any temperature above absolute zero. Since these atoms and molecules consist of charged particles (protons and electrons), their movement leads to the emission of electromagnetic radiation in the form of photons, which carry energy away from the surface of the object in question, here the conductor. For overhead lines, cooling by radiation P_R in most cases is three times smaller than convective cooling.

In summary, substituting (4.38) into (4.37) gives the maximum current $I_{AC}^{\max}$ that a conductor—in thermal equilibrium—can carry at a conductor temperature T_m:

$$I_{AC}^{\max} = \frac{\sqrt{P_C + P_R - P_S}}{k_j[1 + \alpha(T_m - 20)]} \tag{4.39}$$

Figure 4.17 illustrates the strong dependence of the current-carrying capacity of a typical ACSR 265/35 conductor on wind speed and ambient temperature, assuming a maximum permissible conductor temperature of 80 °C. Additional calculation examples are provided in [34].

4.3.6 Corona

Corona is a discharge phenomenon caused by the acceleration of electrons in an inhomogeneous electric field, e.g. on the surface of conductors or on sharp edges of fittings. When the electric field strength is high enough, the electrons gain sufficient energy to ionize neutral molecules of the surrounding air [37, 38]. This way, new

Fig. 4.18 Complex cleaning of conductors in service

free electrons are created, which, because they are exposed to the same field, ionize further air molecules, leading to so-called *partial discharges* in the air.

Weather conditions and air pressure influence the insulating properties of air. Corona effects can occur in unfavorable weather situations, e.g. in rain, when the voltage gradient on the conductor, the clamps, the arc protection or the conductor fittings, exceeds a certain value. Conductors with the largest possible conductor diameters or conductor bundles with 2, 3, 4, 6 and even more sub-conductors and well-designed fittings, can reduce the voltage gradient and thus corona. Even slight damages on the conductor surface during production, transport or stringing (scratches, abrasions) also lead to an increase in the voltage gradient and can cause corona and audible noise. Associated fittings—such as clamps, spacers, and arcing horns—also operate at line potential and must therefore be designed to avoid sharp edges, corners, or surface irregularities that could result in elevated electric field gradients and, consequently, corona discharge.

Normally such sharp imperfections should not occur with professional production and assembling of fittings and conductors. Contamination of conductors and fittings from industrial activity and traffic can also lead to corona discharge. Applying an appropriate coating can mitigate this risk [39], as cleaning conductors during operation—if feasible at all—is extremely complex (Fig. 4.18).

Requirements for conductors and fittings to reduce corona:

For conductors:

- Clean, smooth surface
- Large conductor diameter
- Increased number of sub-conductors
- Use of alternative conductor types (subject to tower static verification).

For fittings:

- Clean, smooth surface
- Generous radii and rounded contours
- Absence of protruding elements (e.g., pins).

Potential effects of corona:

- Generation of ozone (O_3)
- Emission of ultraviolet (UV) radiation
- Power losses due to corona discharge
- Audible noise
- Electromagnetic interference (EMI: *Electromagnetic Interference*)
- Radio and television interference (RI: *Radio Interference*).

It is common practice to limit the electric field strength E_C at the conductor surface to 17 kV_{rms}/cm [7]. It can be calculated with sufficient accuracy using Peek's formula [1, p. 654]:

$$E_c = m\, E_0\, \delta \left[1 + \frac{k}{\sqrt{\delta\, r}}\right] \tag{4.40}$$

where:

$E_0 = 31$ kV/cm	Breakdown field strength of the air in a quasi-homogeneous field at normal pressure (760 mm Hg) and an ambient temperature of $T_a = 25$ °C
r	Conductor radius in cm.
$K = 0.308$	Empirical factor.
δ	Temperature and height (pressure) correction factor.
m	Shape factor 0.9 due to stranding effect, 0.6 in rain.

Water droplets on conductors can be also a cause for initiation of corona and related acoustic noise (Fig. 4.19, left side). Special surface treatments and coatings can contribute to noise reduction in rainy situations of new conductors [40] (Fig. 4.19, right side).

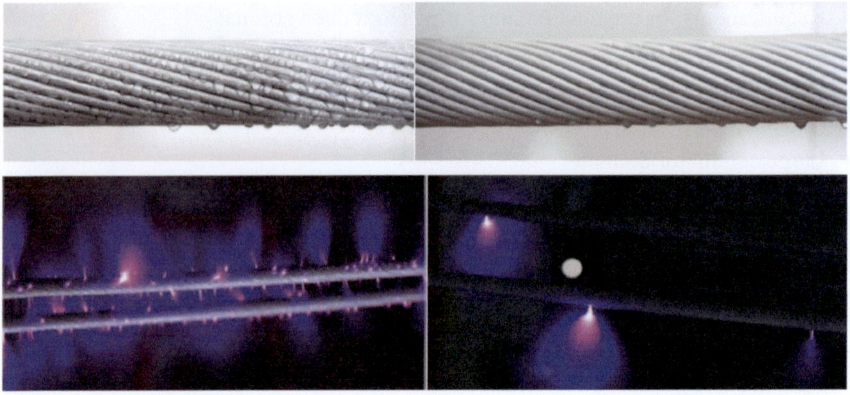

Fig. 4.19 Corona experiments in the high voltage laboratory. Top *left* and *below*: untreated conductor; top *right* and *below*: surface-treated conductor to reduce corona [40]

4.3.7 Bundle Conductors

For voltages above 220 kV, maintaining acceptable electric field strength at the conductor surface would require conductor diameters so large that they become impractical to manufacture and install. For instance, in the case of a 380 kV overhead line, using a single conductor would necessitate a diameter of approximately 4.7 cm—an engineering challenge that is difficult to realize in practice [41]. This limitation has led to the development of *bundle conductors*, which utilize multiple sub-conductors per phase to control the electric field and reduce corona effects. Typical configurations include *double-bundle conductors* for 220 kV, and *triple- or quadruple-bundles* for 380 kV systems, with a common sub-conductor spacing of around 400 mm. For ultra-high voltage applications, such as 765 kV lines and above, bundle configurations with up to eight sub-conductors per phase are employed internationally (see Fig. 4.20).

The conductor bundle establishes an effective radius considerably greater than that of a single conductor. Since the electric field strength is inversely related to the conductor radius, this leads to a notable decrease in the electric field magnitude. This is shown schematically in (Fig. 4.21) using the field images and the reduction factors of the maximum field intensity for different bundle arrangements [42].

Bundle conductors also contribute to a lower characteristic impedance of the transmission line, thereby increasing its *Surge Impedance Loading* (SIL) and improving its capacity for efficient power transfer (Sect. 6.1).

Fig. 4.20 Six-bundle with spacer dampers (Sect. 4.5.4) for a 765 kV line in South Africa (*Source* U. Cosmai, PFISTERER)

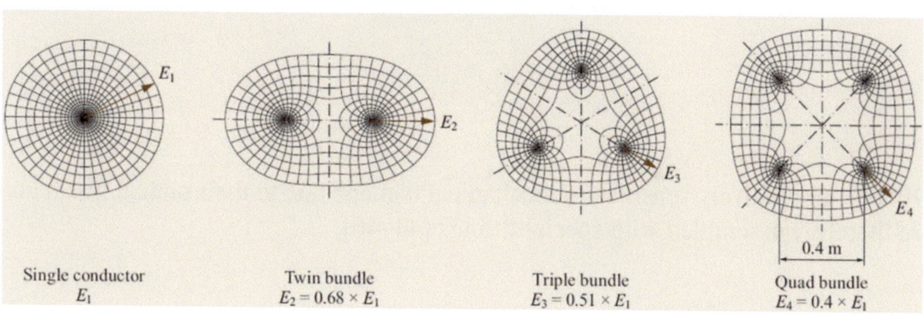

Fig. 4.21 Field images and maximum electric field strengths E_i of different bundle arrangements

4.3.8 Ground Wires

High-voltage overhead lines are practically always equipped with ground wires, also called earth wires. Their task is to protect the conductors from direct lightning strikes and thus to avoid line failures. For effective lightning protection, ground wires are connected to the ground through the tower in order to provide a low-impedance path to earth.

In the past, ground wires with low conductivity, such as made of steel strands or ACSR conductors with low aluminum content, were often used. To conduct safely the higher earth fault currents of today, ACSR conductors with larger cross-sections are increasingly being used. They often carry in the core integrated optical fibers (Fig. 4.22). These are used for communication and information transmission and are referred to as OPGW (*Optical Ground Wire*) [43].

Fig. 4.22 Optical ground wire (*Source* De Angeli Prodotti)

OPGW are relatively sensitive to mechanical damage due to their optical fibers and are therefore assembled with special fittings [44–46].

4.3.9 High-Temperature Conductors

One of the most successful and at the same time relatively simple methods to increase the power transmission capacity of an existing line without extensive modification measures, is to increase the current that it can safely transmit. With the original conductors, this would result in a corresponding increase in conductor temperature over the allowed limits. To avoid this, the conductors are replaced by so-called HTLS (*High Temperature Low Sag*) conductors. As the name suggests, two aspects need to be considered.

First, that conventional hard-drawn aluminum for normal conductors recrystallizes at temperatures above 100 °C and loses its mechanical strength over time, as shown in Fig. 4.23 [47]).

Second, the increased temperature leads to a larger sag, which could violate the required safety clearances. In this context, it is worth mentioning that the conductor

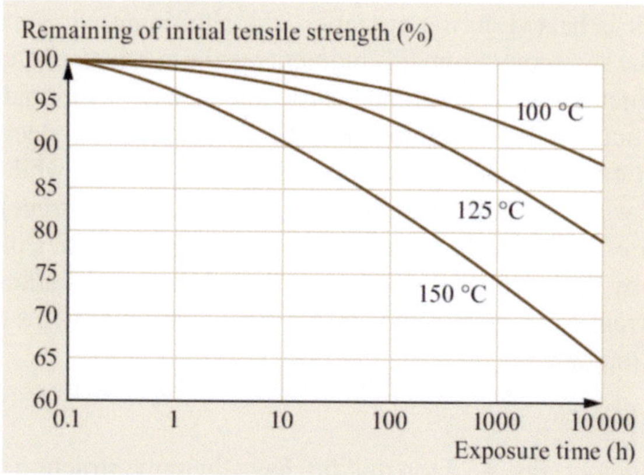

Fig. 4.23 Residual strength of aluminum depending on the temperature and the duration of exposure

sag at elevated temperatures is influenced by some additional factors that unfavorably increase the sag [23].

To solve these problems, various types of high-temperature conductors have been developed in recent years [48–50]. They can operate continuously at temperatures of at least 150 °C, in emergencies (up to a maximum of 10 h per year) even up to 250 °C. The most important ones are shown on Fig. 4.24, and their properties are briefly described below.

ACSS Conductors

ACSS conductors (*Aluminum Conductor Steel-Supported*, Fig. 4.24 top left) consist of a steel core with one or more aluminum layers. They do not differ in appearance and geometry from ACSR conductors. They can be operated continuously up to 250 °C, as the aluminum wires are annealed in the factory before stranding, causing them to lose their mechanical strength and only must carry the current. When an

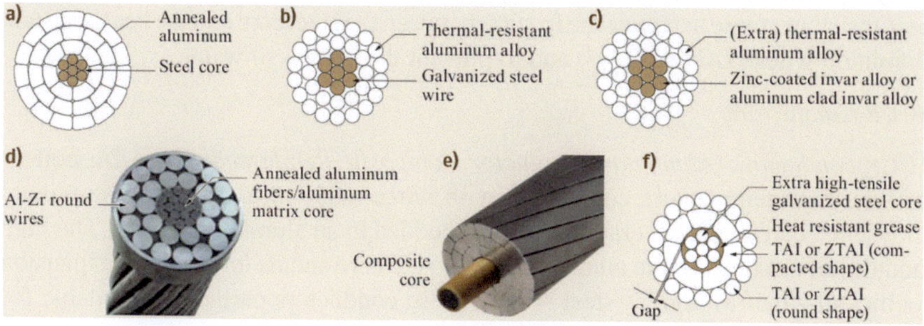

Fig. 4.24 a–f: HTLS conductors: **a** ACSS/TW, **b** (Z)TACSR, **c** (Z)TACIR, **d** ACCR, **e** ACCC, **f** G(Z)TACSR [48]

ACSS conductor is heated during operation, the aluminum wires expand and shift their load onto the steel core. From this moment on, the conductor behaves essentially like a steel conductor with an additional vertical load, i.e., its thermal expansion and modulus of elasticity of the conductor correspond to those of the steel core. Since the ultimate elongation of annealed aluminum wires is about 20%, while it is limited to 1% for the steel wires, the full strength of the steel core can be used. In addition, the high elongation of the aluminum means that the creep properties of the conductor are determined by the steel core, which has a very low creep coefficient. To further increase the current carrying capacity, trapezoidal aluminum wires (TW) are often used (Fig. 4.24 top left).

(Z)TACSR Conductors

(Z)TACSR conductors (Fig. 4.24 top middle) have the same structure as conventional ACSR conductors, but use heat-resistant aluminum wires, which are referred to as *(Z)TAL*. These are made of a thermally resistant zirconium-aluminum alloy, which can be operated up to 210 °C without loss of strength. However, these conductors have the same thermal expansion behavior as normal Al/St conductors, i.e., they show large sag at high temperatures, which must be considered when determining the required safety clearances. Therefore, such conductors are mainly used in new line constructions that expect high current transport and where a large sag can be taken into account in the design stage.

The problem of increased sag can be solved by using *(Z)TACIR conductors* (Fig. 4.24, top right), which instead of the steel strands of (Z)TACSR conductors use galvanized or aluminum-coated Invar steel wires. *Invar* is an iron-nickel alloy with a coefficient of thermal expansion of about one third of that of steel. These conductors can be operated with low sags at temperatures up to 210 °C.

G(Z)TACSR, Gap Conductors

G(Z)TACSR—or *Gap conductors* (Fig. 4.24, bottom right) have a small gap between the steel core and the adjacent trapezoidal aluminum wires, so that during the stringing the conductor can only be tensioned by the steel core, thus its advantageous mechanical properties can be used over a larger temperature range. The gap is filled with heat-resistant grease to reduce the friction between the steel core and the heat-resistant aluminum wires (*TAl* or *(Z)TAl*) and to prevent the ingress of water.

ACCR Conductors

ACCR conductors (*Aluminum Conductor Composite Reinforced*, Fig. 4.24, bottom left) are high temperature conductors with wires made of a zirconium-aluminum alloy and a core made of ceramic fibers embedded in an aluminum matrix. The high modulus of elasticity of the aluminum composite core and its low thermal expansion coefficient—half as high as steel—make these conductors particularly suitable for existing lines with high ice loads and restrictive high temperature sags.

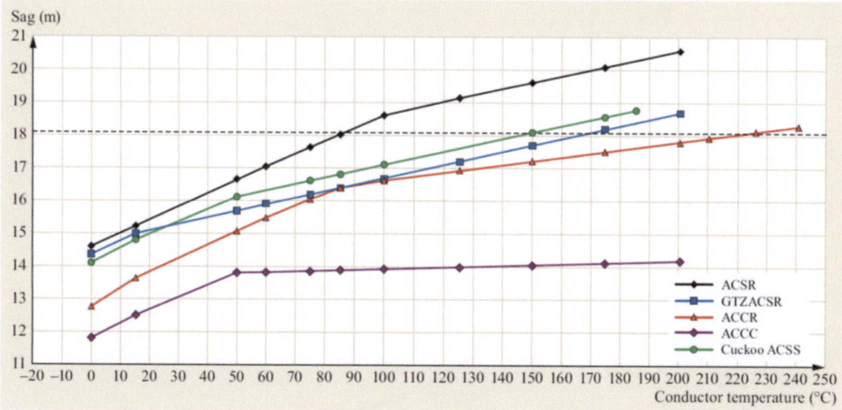

Fig. 4.25 Temperature-dependent sag of ACSR 400/70 (Drake) and various HTLS conductors of the same size for a 420 m span [22] (Black: ACSR, Blue: GTZACSR, Red: ACCR, Magenta: ACCC, Green: ACCC)

ACCC Conductors

ACCC conductors (*Aluminum Conductor Composite Core,* Fig. 4.24 bottom middle) have a carbon-reinforced high-temperature thermoset core and annealed aluminum wires, often in trapezoidal shape. Due to the core's very low coefficient of thermal expansion, the sag increases only marginally beyond the knee-point (Fig. 4.25). However, for these conductors a maximum continuous operating temperature of 180 °C is recommended due to the thermosetting resin in the core wire.

Figure 4.25 [22] compares the temperature-dependent sag of a 420 m span for a conventional ACSR 400/70 conductor (code name: Drake) with those of the various HTLS conductor types discussed above.

It should be though noted that the use of high-temperature conductors for line lengths up to approx. 80–100 km (area "1" in Fig. 4.26) has advantages because their high thermal load capacity can be fully utilized. In area "2", however, the usual maximum voltage drop of 5% limits the highest possible transferable power $P_{ü}$. The latter is further limited in area "3" by system stability limits, for which in practice a maximum phase angle between input and output voltage of the line of 30° to 36° has to be observed. This fact is graphically represented in Fig. 4.26 where the transferable power $P_{ü}$ is shown in relation to P_{nat}, the SIL of the line (Sect. 6.1) [51].

HTLS conductors are usually chosen and installed in such a way that when replacing a conductor, no or only a few changes to the line are required (e.g. increase of tower height).

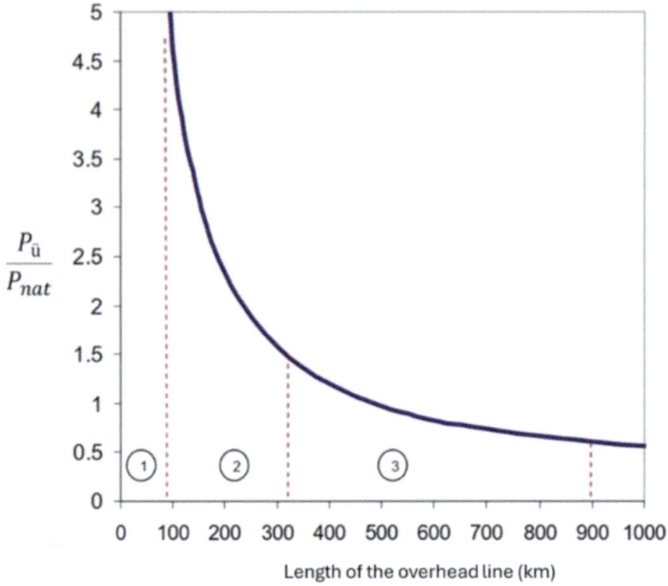

Fig. 4.26 Transmission capacity of an overhead line

4.3.10 Monitoring

Alternatively, or in addition to HTLS conductors, *real-time monitoring systems* (RTM) can be used to determine in real time the maximum power that the line can transmit depending on the permissible conductor temperature. The related devices measure directly or indirectly the temperature or the sag of the conductor, as well as the environmental conditions (temperature, solar radiation, wind). Based on these measurements, the current carrying capacity of the conductor can be determined (Sect. 4.3.5). This so-called *Dynamic Line Rating* (DLR) enables the identification of potential capacity reserves in the transmission line and serves as an important tool for network operation. A key prerequisite is the continuous and reliable transmission of measurement data from the line-mounted monitoring devices to the load dispatch center, followed by evaluation using dedicated software.

Aspects of Dynamic Line Rating (as e.g. methods, weather conditions, risks, system operation, application examples) can be found in detail in [52].

There are several types of monitoring devices [53], which work according to the following methods:

Temperature Measurement

Donut devices, so called because of their shape, which reminds of donut pastry, were among the first monitoring systems to be used. They work with temperature sensors and are mounted directly onto the conductor (Fig. 4.27). They can simultaneously measure the surface temperature of the conductor and the current and transmit this

Fig. 4.27 Power-Donut [53]

data for evaluation via a base station to the control room. They are not so expensive and easy to install, but their relatively high mass acts like a heat sink and can influence both the temperature measurement and the vibration behavior of the conductor.

Load Cells

Monitoring the tensile force of the conductor, from which the conductor sag can be calculated, is one of the most widely used methods. It is based on the installation of a load cell on the earthed side of a tension string (Fig. 4.28). From the conductor tensile load, the sag of the conductor can be determined by formula (4.26) and transmitted to the control room to determine the current carrying capacity of the conductor and the remaining transmission reserve. These monitoring systems are robust, largely mature and have proven themselves in numerous installations worldwide.

Vibration Measurement

In this case (Fig. 4.29), the conductor tension and the corresponding sag—and from this the conductor temperature—can be derived by recording the vibration frequency of aeolian vibrations in the relevant span (Sect. 4.3.11). Such monitors are light, easy to install, autonomous—by power harvesting from the conductor—and can also transmit the measurement data to a base station or even the control room.

Fig. 4.28 Load cells (red) for monitoring the tensile force of the conductor [53]

Fig. 4.29 Monitoring by vibration analysis [53]

4.3.11 Conductor Vibrations

Conductor vibrations are a widespread phenomenon and can cause significant damage not only to the conductors themselves but also to other components of the overhead line (Fig. 4.30). These vibrations are considered dynamic loads, alongside events such as conductor rupture, short circuits, ice shedding, and, in rare cases, earthquakes. In this section, we will focus on wind-induced vibrations, as they are the most common type.

With the widespread introduction of aluminum-based conductors at the beginning of the twentieth century, wind-induced vibrations experienced often damage, as they have a tendency to vibrate due to their lower weight compared to the previously used copper conductors. In addition, aluminum is more vulnerable to material fatigue due to its metallographic structure.

We distinguish between three main types of wind-induced vibrations: *aeolian vibrations*, *sub-span oscillations* and *galloping*. Statistically, for every case of galloping, there are ten cases of sub-span oscillations and a thousand cases of aeolian vibrations.

Aeolian Vibrations

The most common type of wind-induced conductor movements is the so-called aeolian vibrations, named after *Aeolus*, the God of winds in Greek mythology. They are excited by moderate, laminar wind flows with a wind speed between approx. 1 m/s and 7 m/s, and their amplitude (peak-to-peak) rarely exceeds a conductor diameter. Physically, these vibrations are triggered by the shedding of so-called *von Kármán vortices*, which generate pressure fluctuations along the conductor surface.

Fig. 4.30 Severely damaged conductor at a suspension clamp (*Source* PFISTERER)

These fluctuations, in turn, induce alternating lift forces that act on the conductor in an upward and downward motion (Fig. 4.31).

Although commonly referred to as a single frequency, the vibration usually consists of a combination of multiple harmonic components that form a beat pattern (Fig. 4.32). The dominant frequency can be estimated using the *Strouhal formula*:

$$f_S = S\frac{v_w}{d_c} \tag{4.41}$$

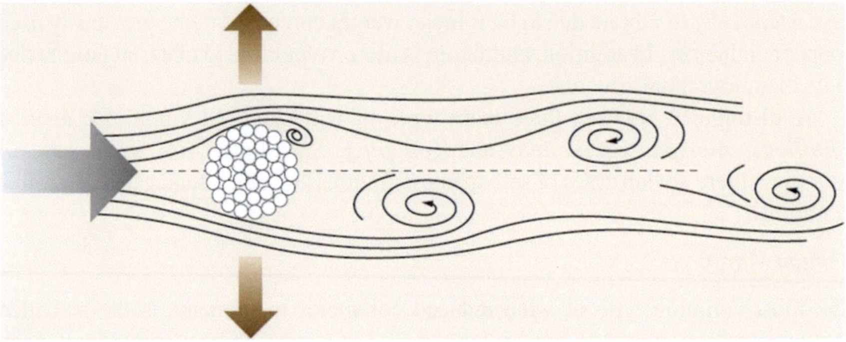

Fig. 4.31 Aeolian vibration initiated by vortex shedding (*grey arrow*: wind velocity, *brown arrows*: upward and downward lift forces)

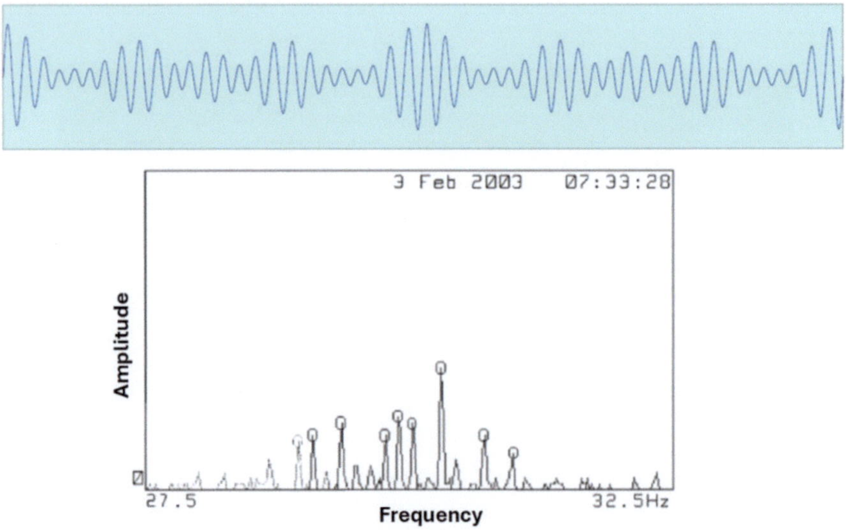

Fig. 4.32 Typical frequency spectrum for aeolian conductor vibrations; the individual frequencies have been obtained by *Fourier analysis*

Herein, v_w is the wind speed perpendicular to the conductor [m/s], d_c is the conductor diameter [m], and S is the so-called *Strouhal number*, which for aeolian vibrations assumes a dimensionless value of approx. 0.2. The conductor vibration begins when the *Strouhal frequency* approaches a natural frequency of the conductor. A lock-in effect occurs, i.e., the frequency of vortex shedding is controlled by the vibration, which causes it to be maintained, even if the wind speed fluctuates in the range between 0.9 and 1.3 of the original wind speed [26, pp. 2–9].

The movements of the conductor spread as traveling waves to both ends of the span, where they are fully or partially reflected and overlap to form stationary waves along the span. This fact is well approximated by the model of a vibrating string, which obeys the wave equation [54–56]. Its solution provides the curve $y(x, t)$ of the vibrating conductor in space x and time t:

$$y(x,t) = y_{\max}\left(n\pi \frac{x}{L}\right)\sin(\omega t) \tag{4.42}$$

The velocity of propagation of the wave is:

$$v = \sqrt{\frac{Z}{m'}} \tag{4.43}$$

And its frequency:

$$f = \frac{v}{\lambda} \tag{4.44}$$

With:

y_{max} Oscillation amplitude
n Mode number, it corresponds to the number of half-wavelengths along the span
λ Wavelength
L Span length
Z Tensile force of the conductor
m' Conductor mass per unit length.

By substituting (4.43) into (4.44) and considering

$$n\,\frac{\lambda}{2} = L \tag{4.45}$$

The vibration mode number n and its associated *eigenfrequency* f_n can be determined for a specific wind speed using the following relationships:

$$n = 2\,L f_s \sqrt{\frac{m'}{Z}} = 2\,L\,S \frac{v}{d_c}\sqrt{\frac{m'}{Z}} \tag{4.46}$$

$$f_n = \frac{n}{2\,L}\sqrt{\frac{Z}{m'}} \tag{4.47}$$

Vibration frequencies of common conductors lie between 7 and 35 Hz, the corresponding wavelengths are between 20 and 4 m.

Example

We consider an ACSR conductor 400/70 mm^2, codename Drake, 26 × 4.44 mm aluminum and 7 × 3.47 mm steel wires), with a diameter *d* of 28 mm and a mass per unit length *m*' of 1.63 kg/m) and a tensile force *Z* of 28 kN—this corresponds to 20% of its ultimate load of 140 kN—in a $L = 400$ m long span. With the above formulas, we get:

Velocity of propagation:

$$\upsilon = \sqrt{\frac{Z}{m'}} = \sqrt{\frac{28000\,\mathrm{N}}{1,63\,\mathrm{kg/m}}} = \sqrt{\frac{28000\,\mathrm{N}}{1,63\,\mathrm{N}/(\mathrm{m}^2/s^2)}} = 131\,\mathrm{m/s}$$

Fundamental frequency of the conductor:

$$f_n = \frac{n\ (=1)}{2\,L}\,v = \frac{131\ \mathrm{m/s}}{2\cdot 400\ \mathrm{m}} = 0.164\ \mathrm{Hz}$$

Since the vibration of the conductor is initiated by the wind, its vibration frequency must match the excitation frequency, here the Strouhal frequency (4.41). In this case we get for a wind speed $\upsilon_w = 1$ m/s:

$$f_{S,1m/s} = 0.2\cdot\frac{1\ m/s}{0.028\ m} \approx 7\ Hz$$

And for $\upsilon_w = 5$ m/s:

$$f_{S,5m/s} = 0.2\cdot\frac{5\ m/s}{0.028\ m} \approx 35\ Hz$$

Thus, the conductor vibrates at $\upsilon = 1$ m/s in the 43rd (= 7 Hz/0.164 Hz) and at υ = 5 m/s in the 215th (= 35 Hz/0.164 Hz) mode. The corresponding wavelengths λ result from (4.20) to 18.7 m and 3.7 m, respectively.

The law of energy conservation governs the behavior of a vibrating conductor, as it does throughout nature; in this case, the term *Energy Balance Principle* (EBP) is used. When a conductor starts to vibrate, the energy supplied to it by the wind will amplify the vibration and continuously increase its amplitude. On the other hand, energy is dissipated in the vibrating conductor—primarily due to friction at contact points (red points in Fig. 4.33 left) of the conductor wires when they mutually shift during conductor bending (Fig. 4.33 right) [20].

Fig. 4.33 Contact points of the individual wires (*left*) during mutual displacement due to bending movement (*right*)

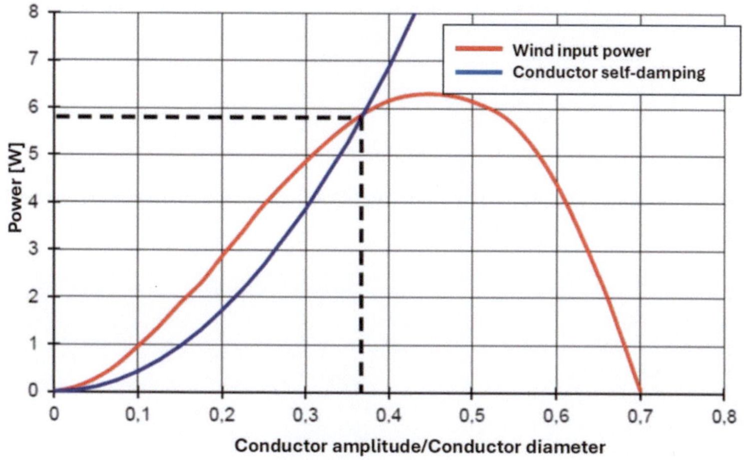

Fig. 4.34 Wind input power and conductor self-damping over the ratio of conductor amplitude *A* to conductor diameter *D* (*Source* J. L. Lilien)

As shown in Fig. 4.34, the system reaches equilibrium, and the vibration amplitude stabilizes at a level where the aerodynamic energy input from the wind is balanced by the energy dissipated through the conductor's *self-damping* [57].

The self-damping of the conductor increases nonlinearly with rising vibration frequency (Fig. 4.35), due to the increased curvature of the conductor's shape. However, it diminishes with higher tensile stress, as the internal wire displacements that generate self-damping (Fig. 4.33) are initiated only at larger bending amplitudes.

The terrain, and in particular its roughness, plays a critical role in the excitation of aeolian vibrations by strongly influencing the nature of the airflow and the level of turbulence. As an example, laminar airflow over flat desert terrain (Fig. 4.36a) is typically more favorable for the development of conductor vibrations compared to the turbulent flow conditions associated with built environments featuring vegetation and structures (Fig. 4.36b). To evaluate the vibration intensity, four terrain categories with increasing turbulence intensity were introduced, Table 4.6 [58].

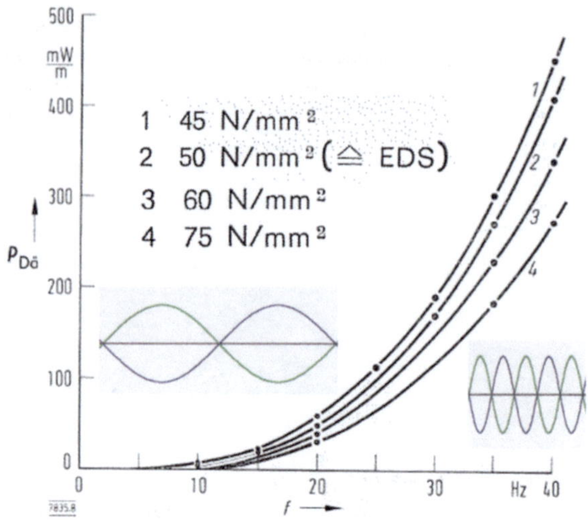

Fig. 4.35 Self-damping $P_{Dä}$ of an Al/St conductor 560/60 ($D = 28$ mm) depending on its vibration frequency f, with EDS as parameter [7]

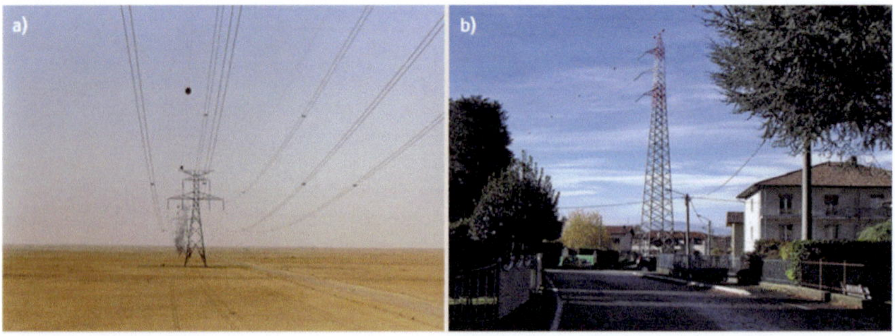

Fig. 4.36 Different terrain situations influence wind-induced conductor vibrations

Table 4.6 Terrain categories [58]

Terrain category	Terrain characteristics
1	Open, flat, no trees, no obstruction, with snow cover, or near/across large bodies of water; flat desert
2	Open, flat, no obstruction, no snow; e.g., farmland without any obstruction, summer time
3	Open, flat or undulating with very few obstacle; e.g., open grass or farmland with few trees, hedgerows and other barriers; prairie, tundra
4	Built-up with some trees and buildings, e.g., residential suburbs; small towns; woodlands and shrubs. Small fields with bushes, trees, and hedges

Material Fatigue

The internal movement of wires within a conductor during vibration contributes significantly to damping, as this mechanism—described earlier—provides the largest share of the conductor's inherent damping capacity. A smaller portion of the damping results from material-based mechanisms, such as *Coulomb damping*. However, these internal wire movements are also a primary cause of conductor failures due to so-called *fretting fatigue* [59]. This type of damage typically occurs at the interface between the conductor and the suspension clamp (see Sect. 4.5.1), where static bending stresses in the outer wires are highest. Additionally, compressive forces exerted by the clamp keeper create an unfavorable stress regime in the individual wires, further contributing to fatigue-related failures.

More specifically, relative motion between wires (Fig. 4.33) can lead to minor surface wear through abrasion, resulting in the formation of aluminum oxide. This hard, powdery black substance functions as a secondary abrasive, potentially intensifying the degradation process. Due to the pressures in the contact area of the wires (Fig. 4.33, left), in combination with the high static bending stresses, cracks can occur on the wire surface (arrows in Fig. 4.37a), which may spread across the wire cross-section, until it completely separates (circle in Fig. 4.37a). With continued vibration further wire breakages may develop, until the whole conductor fails (Fig. 4.37b). Since wire breakages predominantly occur in the inner layers of the conductor and concentrate in the lower part of the conductor near the contact area of the clamp, it is difficult to detect these failures through visual inspections.

Experiments, which recorded the strain of a Finch conductor (ACSR 565/71) from production to installation in the line, have shown that 36% of the permissible total strain for the conductor manufacturing process including testing, 11% for installation, 6.5% for conductor tension and creep and 22% for the static bending in the suspension clamp are "consumed", leaving a rather small portion for the vibration-induced alternating bending [60].

Field Measurements

Vibrating conductors are complex mechanical systems, for which field measurements are indispensable. It is nowadays state of the art to monitor conductor vibrations in the field with vibration recorders (Fig. 4.38). Most of these recorders measure the so-called *bending amplitude* Y_b—this is the conductor deflection peak-to-peak in mm at a standardized distance (normally 89 mm (=3½ inches) from the last point of contact of the conductor with the clamp body, store them in a matrix, together with the vibration frequency and the corresponding number of cycles. Modern devices can work autonomously for many months and can wirelessly transmit the measured data to a base station. The concept and evaluation of vibration measurements are described in detail in [61–63].

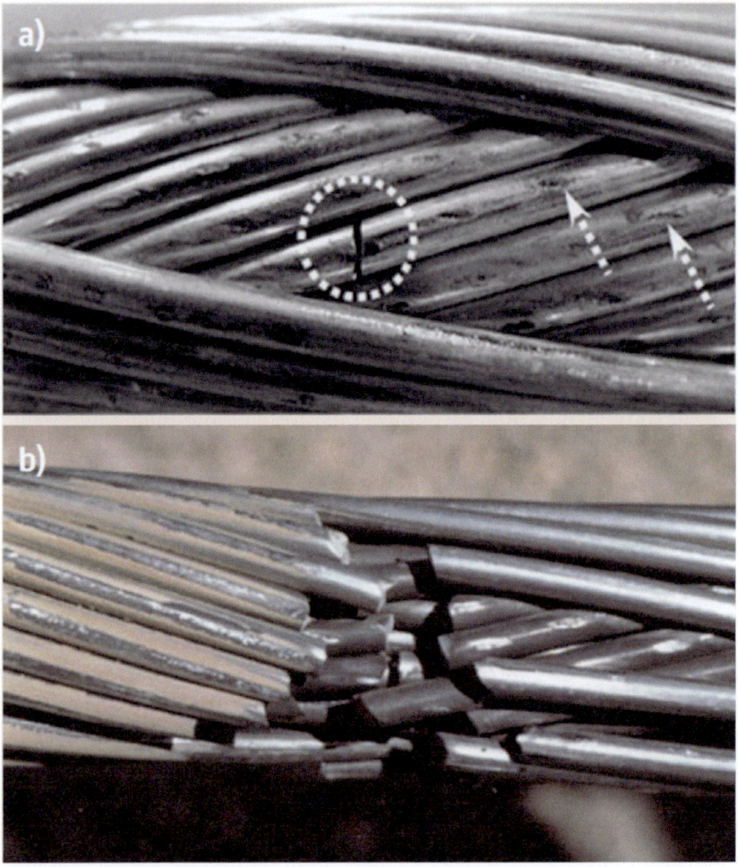

Fig. 4.37 Wire failure (**a**) leading to conductor rupture (**b**) (*Source* U. Cosmai)

The *dynamic bending stress* σ_b in N/mm^2 at the top wire of the conductor outer layer (Fig. 4.39) is calculated by the so-called *Poffenberger-Swart formula* [64]:

$$\sigma_b = \frac{E_a\, d_a\, p^2}{4\,(e^{-p\,x} - 1 + p\,x)}\; Y_b \tag{4.48}$$

With:

d_a wire diameter in the conductor's outer layer (mm)
E_a modulus of elasticity of the conductor wires of the outer layer (N/mm ^{2})
p^2 Z/EI (1/mm^2), parameter
Z tensile force in the conductor (N)
$EI_{\min}$ minimum bending stiffness of the conductor (N mm^2) under the assumption that the conductor wires bend around their own axis.

Fig. 4.38 VIBREC™ vibration recorder in use (*Source* PFISTERER)

Fig. 4.39 Definition of bending amplitude [65]

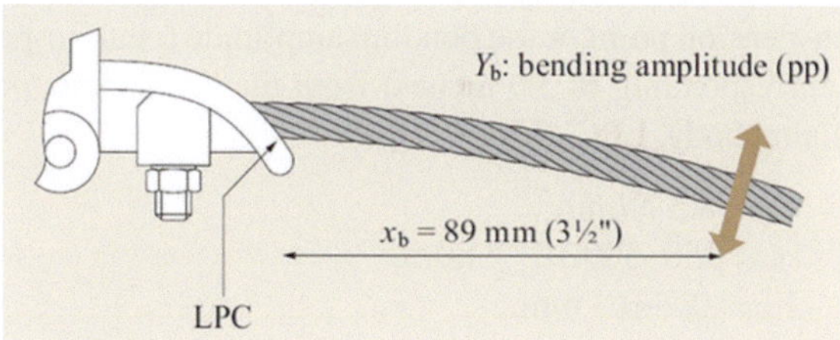

For an ACSR conductor, the following applies:

$$EI_{\min} = n_{St}\, E_{st}\, \frac{\pi\, d_{St}{}^4}{64} + n_{Al}\, E_{Al}\, \frac{\pi\, d_{Al}{}^4}{64} \tag{4.49}$$

And for a Drake conductor (see example above):

$$\begin{aligned} EI_{min,Drake} &= 7 \cdot 210\,\text{kN/mm}^2 \frac{3.14 \cdot 3.47^4}{64}\,\text{mm}^4 + 26 \cdot 70\,\frac{\text{kN}}{\text{mm}^2}\,\frac{3.14 \cdot 4.44^4}{64}\,\text{mm}^4 \\ &= 10469 + 34707\,\text{kNmm}^2 = 45.2\,\text{Nm}^2 \end{aligned}$$

x = Distance of the sensor from the last contact point between conductor and clamp (mm) (Fig. 4.39).

Thus, with (4.48) for the above-mentioned conductor at a maximum permissible bending amplitude Y_b of 0.25 mm (see below *Permissible limit values*) a bending stress of:

$$\sigma_{b,Drake} = \frac{70{,}000\,\text{N/mm}^2 \cdot 4.44\,\text{mm} \cdot 0.00062\,\text{mm}^{-2}}{4 \cdot \left(e^{-0.025\times 89} - 1 + 0.025 \times 89\right)} 0.25\,\text{mm} = 9.0\,\text{N/mm}^2$$

The alternating bending stress σ_b calculated from the measured bending amplitude Y_b is primarily suitable for comparison purposes, e.g. with different clamp constructions (Fig. 4.73), or when changing the conductor tension. The stress regime prevailing in the conductor wires is much more complex and cannot be calculated with simple formulas, since among other things the bending stiffness *EI* of the conductor during the bending process does not change linearly [66].

Permissible Limit Values

To ensure that aeolian vibrations remain within levels that the conductor can withstand without sustaining damage, limit values have been established. One possibility is to limit the conductor's tension. Long-term field experience has shown that if the conductor tension without wind load at the annual average temperature, for example in Germany at +10 °C, the so- called *Every Day Stress,* EDS, (Sect. 4.3.4), does not exceed the values mentioned in Table 4.5, the tendency of the conductors to vibrate is massively reduced and damage is practically avoided [67]. Alternatively, one can also limit the bending stress σ_b or the bending strain ε_b (single peak) at the conductor suspension point or the bending amplitude (peak-to-peak) Y_b at a distance of 89 mm (corresponding to 3½ inches) from the last contact point of the conductor with the clamp body, LPC, (Fig. 4.39) to:

$\sigma_b = 8.5$ N/mm ^{2}
$\varepsilon_b = 150$–$300\ 10^{-6}$ m/m.
$Y_b = 0.2$–0.3 mm.

It has been shown that these limit values in certain cases, such as AAAC conductors, are rather high and can lead to conductor damage. For this reason, in 2005 the CIGRE proposed new recommendations [58], which refer to *H/w*, the ratio between the horizontal tensile force in the conductor *H* (N)—which deviates minimally from its axial tensile force *Z* [22, Sect. 2.4]—and the conductor's self-weight per unit length, *w* (N/m). For undamped single conductors, the value for *H/w* = 1000, for bundle conductors 2500 are recommended. Both values are considered conservative.

Subspan Oscillations

Subspan oscillations can occur in bundle conductors and are caused by the wake created by the windward conductors of a horizontal bundle on the leeward ones. In this process, the individual conductors of the bundle move against each other in cyclic or elliptical orbitals (Fig. 4.40). Subspan oscillations occur between two adjacent spacers at wind speeds in the range of 8 m/s to 20 m/s, and in most cases in the fundamental mode [65]. Their frequencies are between approx. 0.7 Hz and 2 Hz,

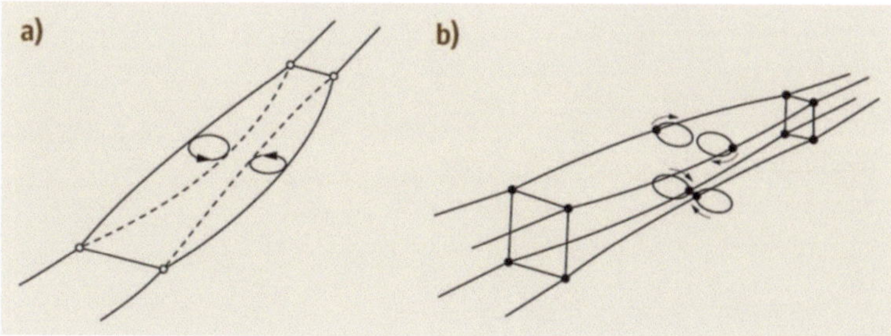

Fig. 4.40 Schematic presentation of a subspan oscillation of a twin-(**a**) and a quad bundle (**b**) [65]

and their amplitudes are often high enough to cause the sub-conductors to clash. The consequences are overstretching of the conductor wires and high compressive forces on the spacers, which can lead to damage to the conductors and loosening of the bolts of the spacer's clamps [26].

Vertical twin bundles are less vulnerable to subspan oscillations, if the sag of the lower sub-conductor is chosen slightly larger than that of the upper one. The lower sub-conductor "hangs" a little on the upper one and dampens the bundle.

Typical values for the ratio a/d of the bundle distance a to the sub-conductor diameter d are in the range of 10–17, with ratios of 12 or below being considered critical. The subspans, i.e., the distance between two spacers should not exceed a length of approx. 80 m, with high wind speeds of 20–25 m/s a value of a maximum of 65 m is recommended [68]. Also, the use of unequal subspans has proven to be effective, where a ratio of the lengths of two adjacent subspans of about 0.85–0.9 has produced good results [65]. Accepted practice nowadays to mitigate subspan oscillations is the installation of vibration-damping spacers, so-called *spacer dampers*) (Sect. 4.5.4).

Galloping

Galloping is a form of vibration of single or bundle conductors due to wind action, when there is an asymmetric accretion of ice or wet snow on the conductors, as shown in Fig. 4.41. It is characterized by frequencies from 0.08 to 3 Hz and amplitudes of the order of the sag. It occurs at wind speeds of about 25 km/h.

Of all wind-induced vibrations, galloping is the most spectacular and can cause costly damage, such as the failure of conductors and fittings, damage to towers or even their collapse. The phenomenon is characterized by quite large amplitudes—up to the conductor's sag—and can last from a few hours to several days. Due to the risk of conductors contacting each other, or because of the possibility of conductor and/or insulator failure, the affected circuits should be switched off.

Alongside various anti-galloping devices [65], notably the use of interphase spacers (Fig. 4.68), conductor de-icing techniques have also proven to be effective preventive measures against galloping [69].

Fig. 4.41 Galloping-endangered overhead line; asymmetric ice accretion (picture in picture) favors galloping

4.4 Insulators

Overhead line insulators serve a dual function: they support the conductors and bear all associated mechanical loads, while also electrically insulating the conductor potential from the grounded line supports. To fulfill these roles, insulators must be capable of withstanding continuous operating voltage, transient overvoltages, and potential power arcs.

Although insulators typically account for only 3 to 8% of the total line cost—making them one of the least expensive components—they are critical to the operational safety and reliability of the transmission line. Electrical or mechanical failure of an insulator can not only lead to power outages but also pose significant risks to human safety and property.

4.4.1 Classification

Insulators can be divided according to their *material*, their *shape*, their *function* and their *geometry*:

Material

Currently, the most used insulating materials are *porcelain*, *toughened glass*, and *silicone rubber composites*. The first overhead line insulators were made of porcelain. A few decades later, glass followed and only many years later (in the 1960s) *composite insulators*, also called *non-ceramic* or *polymeric insulators* (NCI), were introduced. At both ends of the *insulating body*, so-called *end fittings* are attached, which transmit the mechanical loads of the conductors, such as wind and/or ice loads, to the insulator and from there to the supports. End fittings are made of metal for all insulator types.

Shape

Over the years, three basic insulator shapes have emerged: *bell-*, *cap-and-pin-* and *long rod*-insulators. The initially used bell-shaped porcelain insulators were mechanically too weak to carry the conductors with larger cross-sections and the bundle conductors necessary for higher transmission voltages. For this reason, so-called *cap-and-pin insulators* made of porcelain, and from the 1920s made of glass, were developed. Bell insulators are still used as post insulators for low and medium voltage lines.

Advances in the manufacturing technology of large, one-piece porcelain bodies, without air inclusions and cracks from the burning process, led to the so-called *long rod insulators* in 1936. These are puncture-proof, i.e. in case of an overvoltage, a flashover will occur in air, along the surface of the insulator and not through the dielectric.

As a consistent development of long rod insulators made of porcelain, the first *composite insulators* appeared around 1960. These are insulators with a core made of *glass fiber reinforced composite* (GFRP) and a shell made of a *polymeric* material, nowadays almost always *silicone rubber* (Fig. 4.42).

Function

The function of an insulator in an overhead line is primarily determined by the mechanical loads which it must carry. Thus, *suspension-* and *tension insulators* (Fig. 4.43) are subjected to *tension*, whereas *post insulators* (Fig. 4.44 left) to *bending* and *compression*. Insulating crossarms (Fig. 4.44, right) are subjected to *tension* as well as to *compression*.

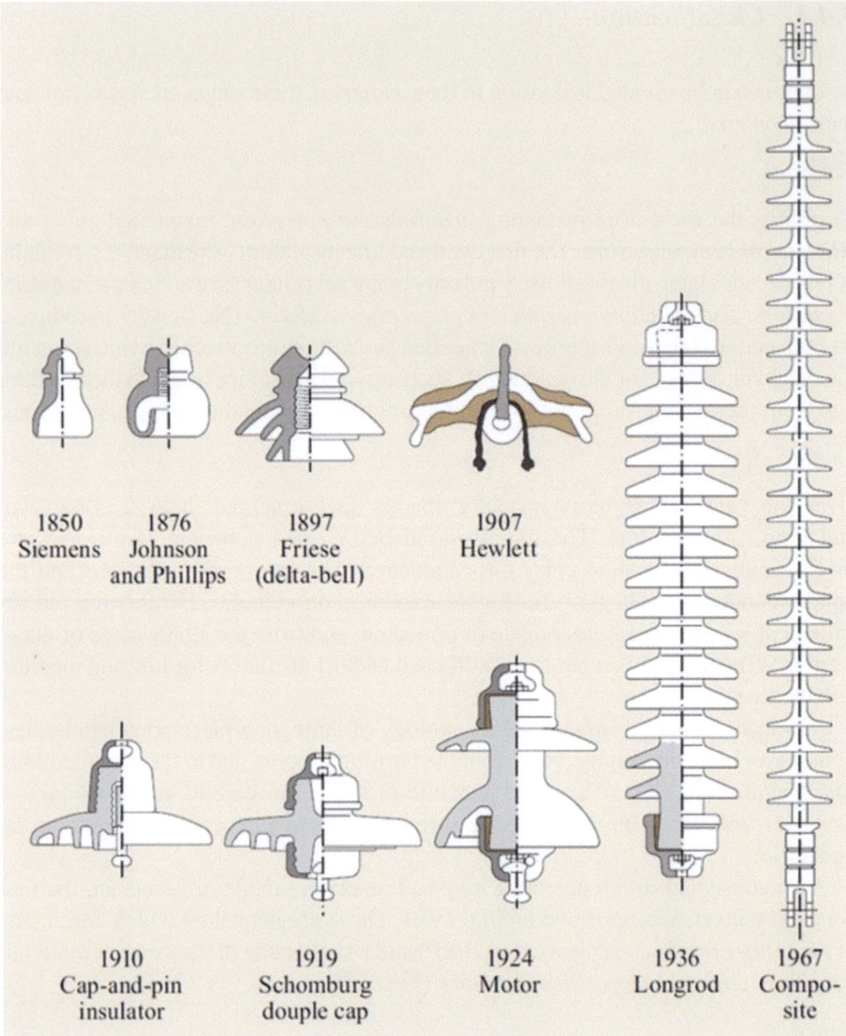

Fig. 4.42 Historical development of overhead line insulators (*Source* Pfisterer Lapp)

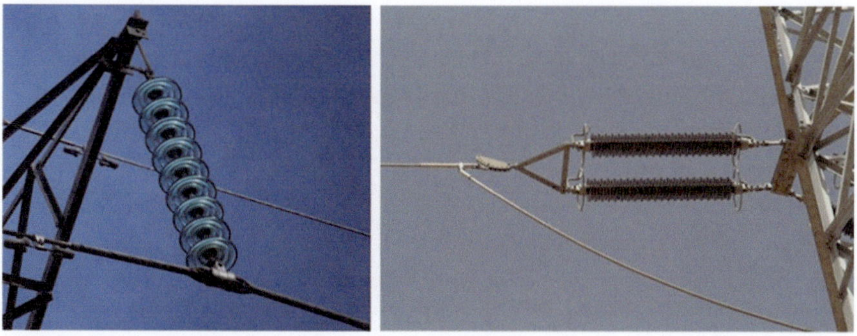

Fig. 4.43 Suspension insulator made of glass caps (*left* part of the picture), long rod insulators as tension insulators (*right* part of the picture) (*Source* U. Cosmai)

Fig. 4.44 Post insulators (*left* part of the picture) and insulating crossarm (*right* part of the picture) (*Source* PFISTERER)

Geometry

All types of insulators share the following geometric parameters (Fig. 4.45).

Connection length: This is the shortest *direct* path between the end fittings of the insulator.

Arcing distance: This is the shortest path *in the air* between the end fittings and is decisive for the behavior of the insulator during electrical flashover due to overvoltages.

Creepage distance: This is the shortest path *along the surface* of the insulator between the end fittings. This measure determines the electrical behavior of the insulator when contaminated by a *pollution layer*.

The shape of the insulator's surface is independent of the insulating material used and is formed with *sheds. This* is necessary to ensure the insulator's functionality when contaminated by a pollution layer, e.g., marine pollution. In this case low-resistance paths (*creepage paths*) can be formed along the surface of the insulator, which can lead to a *pollution layer flashover* (Fig. 4.46). The power arc typically comprises the arc root points, the arc column, and the arc flare, which forms as the

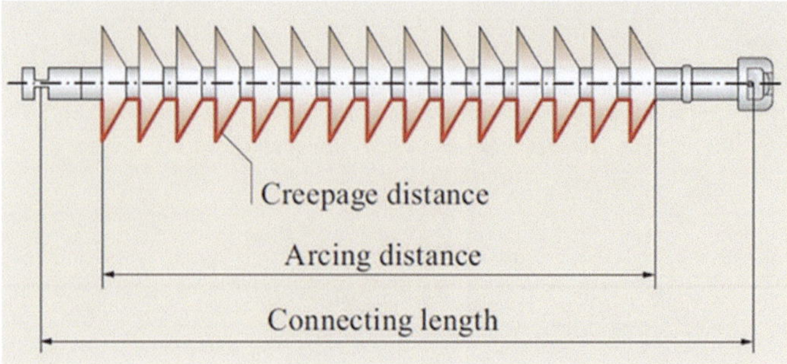

Fig. 4.45 Dimensional properties of long rod insulators (porcelain/composite)

arc expands into the surrounding environment. The temperatures that occur depend on the value of the fault current. The highest values occur in the foot points with up to 18000 K.

Design of insulators regarding *contamination* is done according to IEC [70], which has also been adopted as DIN IEC 60815 [71]. This standard qualitatively defines five classes of *site pollution severity*:

a—very light
b—light
c—medium
d—heavy
e—very heavy.

Fig. 4.46 Pollution layer flashover

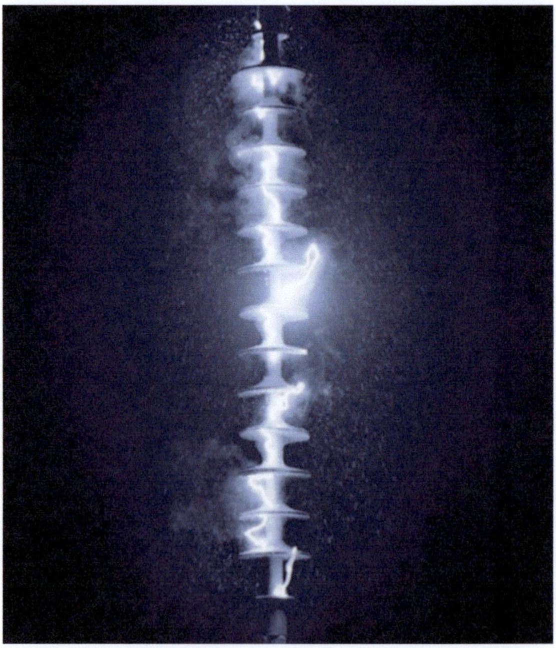

Table 4.7 Correspondence table of SCD and USCD—values for AC

Specific creepage distance (SCD) for AC (mm/kV)	USCD (mm/kV)
12.70	**22.0**
14.40	**25.0**
16.00	27.7
16.17	**28.0**
20.00	34.7
20.21	**35.0**
25.00	43.3
25.40	44.0
31.00	53.7
31.75	**55.0**
38.70	**67.0**

It should be noted that these letter assignments to the classes does not directly correspond to the earlier number assignment in IEC 60815 (1986).

Typical pollution environments are also given in the new standard [70], with E2 corresponding to a *light*, E3 to a *medium*, E5 to a *heavy* and E7 to a *very heavy* pollution.

The *specific creepage distance* (SCD) used in the previous edition of [70] was based on the system voltage. In AC systems, this is the phase-to-phase voltage. The *unified specified creepage distance* (USCD) in the new edition of the standard refers to the voltage across the insulator, i.e. in AC systems the phase-to-earth voltage. In the standard it is defined as the *creepage distance* of an insulator divided by the highest voltage at the insulator $\frac{U_m}{\sqrt{3}}$. It is generally given in mm/kV. In this, U_m means the highest operating voltage.

Table 4.7 provides the attribution between commonly used values of SCD and USCD. It includes the values for the different pollution classes according to old and new standards, highlighted in bold.

For DC the determination of the required creepage distance is more complex, as DC insulators have a different pollution layer behavior than insulators for AC. This is mainly due to an uneven accumulation of the pollution layer on the insulator surface due to electrostatic effects, which is why they require a higher creepage distance under the same pollution conditions. For this reason, composite insulators were used early on in *high-voltage direct current* (HVDC) transmission lines [72], as they require up to 30% less creepage distance in heavy pollution than porcelain or glass insulators [70].

Profile

Over the years, various insulator shed profiles have been developed to maximize creepage distance while maintaining the specified arcing distance of the insulator.

Fig. 4.47 Profiles for different types of insulators

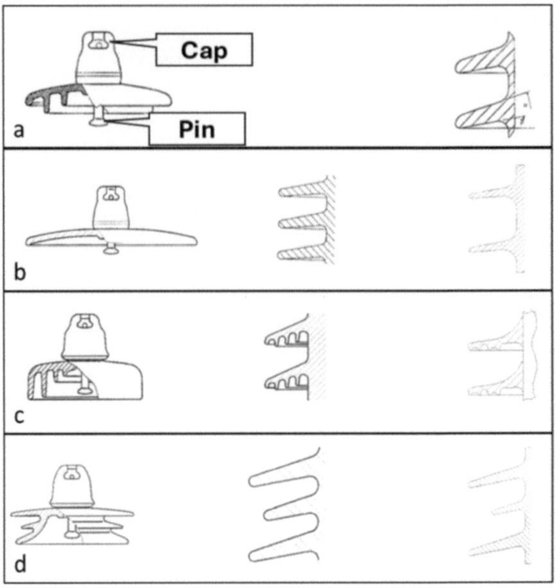

Over the years, various insulator shed profiles have been developed to accommodate the required creepage distance within the arcing distance of the insulator (Fig. 4.45), the latter being mainly determined under insulation coordination aspects. Experience has shown that the ratio between creepage distance and arcing distance, the so-called *creepage factor*, is limited to values between 3.5 and 5.5, a value of 4 being often used.

The use of *deep under-ribs* (Fig. 4.47c) allows for the extension of the creepage distance required in polluted environments. Additionally, employing *alternating sheds* of different diameters (Fig. 4.47d) further increases the creepage distance and enhances insulation performance under heavy rain or icing. Profiles with *under-ribs* (Fig. 4.47a) are effective in areas with light to medium pollution. Aerodynamic or open profiles (Fig. 4.47b) on the other hand are favorable in areas with heavy pollution, as they have good self-cleaning properties. More information on insulator profiles can be found in the corresponding sections of [70].

Corrosion

Another problem arising from pollution is the corrosion of the metallic end fittings, such as the balls and sockets of cap-and-pin insulators. These can be damaged by corrosion so severely that they may mechanically fail. Corrosion debris can cause cracks in the cement used to attach the end fitting to the dielectric or in the dielectric itself and reduce its dielectric strength. This is particularly critical for DC.

The corrosion resistance of an insulator's fittings can be enhanced through heavy-duty galvanizing—that is, by increasing the thickness of the protective zinc coating from 85 to 140 μm in areas with high corrosion exposure, such as coastal regions [73, 74]. For HVDC insulators, the use of a so-called sacrificial zinc sleeve is indispensable [75].

4.4.2 Design

Overhead line insulators must meet various requirements and must be dimensioned accordingly. These are:

- Mechanical requirements (self-weight, wind, additional load)
- Electrical requirements (insulation level)
- Radio interference resistance, corona behavior (EMI aspects)
- Behavior under pollution (pollution layer)
- Power arc behavior (short circuit)
- Audible noise behavior (wind induced)
- Durability (vandalism, corrosion, aging).

Mechanical Design

According to [8], for suspension strings (Fig. 4.48) and post insulators, the highest load is comprised of:

- Conductor weight, ice load, simultaneous wind load and possibly the resultant of the horizontal conductor tensile forces (load case D, Sect. 4.6.5); or
- Conductor weight, wind load and possibly the resultant of the horizontal conductor tensile forces; or
- Conductor weight, ice load and differential load of the horizontal conductor tensile forces (exceptional load).

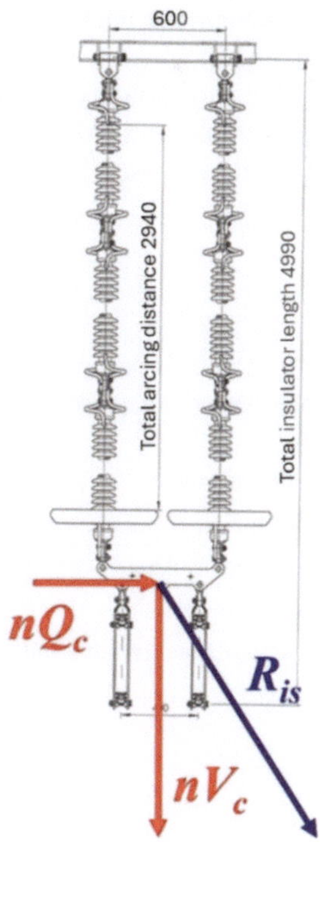

Fig. 4.48 Design of a suspension string for load case D (*Source* J. Seifert)

The resultant R_{is} on the suspension string is with (4.3) and (4.4):

$$R_{is} = \sqrt{(nV_c)^2 + (nQ_c)^2}$$
$$= \sqrt{\left(nL_G\left(w_c + w_{c,i}\right)\right)^2 + \left(n\left(0.5q(h)G_cC_cd_iL_w\cos^2(v)\right)\right)^2} \quad (4.50)$$

And with the numerical values from the example in Sect. 4.2.1:

$$R_{is} = \sqrt{\begin{array}{l}(4 \cdot 536\text{ m }(10\text{ N/m} + 15\text{ N/m}))^2 \\ \quad +(4 \cdot 1.203\ (0.45 + 60/400) \cdot 1.0 \cdot 0.0224 \cdot 350)^2\end{array}}$$
$$= \sqrt{(53.6\ kN)^2 + (22.8\ kN)^2} = 58.3\ kN$$

For tension strings, however, the highest load Z_i is the maximum conductor tensile force

$Z_i = \sigma_{max} \cdot A_c$, with A_c, the total cross-section of the conductor(s) attached, under the following conditions:

- −20 °C without ice load
- −5 °C and ice load according to 4.3.3/DE.1
- −5 °C and ice and wind load according to 4.3.4/DE.1
- +5 °C with wind load according to 4.3.2/DE.1.

Therein, the tensile stress $\sigma_{\max}$ in the conductor can be calculated from the state Eq. (4.13).

The partial safety factors for the loads are uniformly set at 1.35. In addition, the material partial safety factors listed in Table 4.8 should be applied to the rated load of the insulators.

Table 4.8 Material partial safety factors for various types of insulators [8], excerpt from Table 10/DE.1

Type of insulators	Material partial safety factor
Cap-and-pin insulators	2.3
Long rod insulators	2.3
Composite insulators	2.3
Post insulators (ceramic and composite)	1.85

Table 4.9 Standard insulation levels from DIN EN 60072-1 (VDE 0111 Part 1)

U_m [kV]	Lightning impulse voltage [kV] BIL = U_{10}	Switching impulse voltage [kV] BSL = U_{10}	Min. arcing distance D_{el} [mm]
123	550	–	1000
145	650	–	1200
170	750	–	1300
245	1050	(750)	1700
300	1175	850	2100
362	1300	950	2300
420	**1425**	**1050**	**2800...3000**
525	1550	1175	3300...3500
765	1800	1550	4900...5500

Electrical Design

For electrical design, the minimum arcing distance D_{el} (Fig. 4.45) must first be determined. This depends on the maximum system voltage (phase-to-phase) U_m and the associated lighting impulse and switching surge withstand voltages. Typical values are given in Table 4.8. It must then be verified that the minimum clearance distance does not fall below the values required in the standards for D_{el} (for example, for a 380 kV insulator, marked in red in Table 4.9). Very often for HVDC, the required creepage distance according to the pollution class (see Sect. 4.4.1) determines the length of the insulator string.

4.4.3 Porcelain Insulators

Porcelain insulators have the longest service history of all insulating materials, with well over 100 years of use, and are associated with excellent operational performance. Over this long period of time, two types have emerged for use in high-voltage overhead lines: *cap-and-pin* insulators and *long rod* insulators (Fig. 4.49).

While cap-and-pin insulators are used worldwide, long rods are mainly used in Central Europe as well as in the Arabian Peninsula and other countries with strong desert pollution. Due to their advantageous aerodynamic profile, the build-up of contamination is significantly reduced. For both types, the creepage distance can easily be adapted to the pollution regime by simply adding further units for the cap-and-pin insulators and by length extension for the long rods.

Cap-and-pin insulators, unlike long rods, maintain their integrity when a cap breaks, e.g., due to gunshots or a production error. They also react less sensitively to bending stresses, such as those caused by load transposition, e.g., when one of the two insulator strings in a double string fails. This can lead to catastrophic failure of

Fig. 4.49 Cap-and-pin insulator (*left*) and long rod insulator (*right*) made of porcelain

the healthy long rod, if there are no trapezoidal-shaped or triangular-shaped damping elements installed (Fig. 4.77) [76].

On the other hand, long rod insulators also have advantages over cap-and-pin insulators. They are puncture-proof and can be more easily protected against power arcs. They show better behavior in heavy pollution, are not vulnerable to cascade flashovers due to the lack of intermediate metallic fittings and show quite low radio interference levels. Long rod insulators are standardized in [77].

Porcelain insulators are made from *quartz porcelain* C-110 for low strength or from *alumina porcelain* C-120 or C-130 for high strength requirements [78, 79]. The production of porcelain insulators is a complex process, which has been continuously developed over the years. It is based on extensive manufacturer-specific know-how and requires strict quality control in all phases to obtain a low-porosity and thus mechanically strong porcelain body. The glazing process is also of great importance, as the glaze not only forms a smooth and easily washable surface, but also seals it, thus practically excluding the formation of cracks due to surface defects.

Their end fittings are thoroughly standardized [80, 81], so the insulators can easily be replaced—often by composite insulators. For cap-and-pin insulators, ball and socket connections are increasingly used, while for long rods, it is usually clevis and tongue fittings.

4.4.4 Glass Insulators

In the first low-voltage overhead lines built in the 1880s, glass insulators were used, which evolved from the insulators used for telegraph lines. However, the glass quality at that time was not sufficient to withstand the higher loads that came with increasing voltage. This changed with the introduction of so-called pre-stressed *toughened glass* around 1930. Since then, glass insulators have been used worldwide in overhead lines up to the highest voltages. They are practically always cap-and-pin insulators (Figs. 4.50 and 4.51).

As with porcelain cap-and-pin insulators, arcing distance and creepage distance of glass cap-and-pin insulators can be adapted to requirements by simply adding

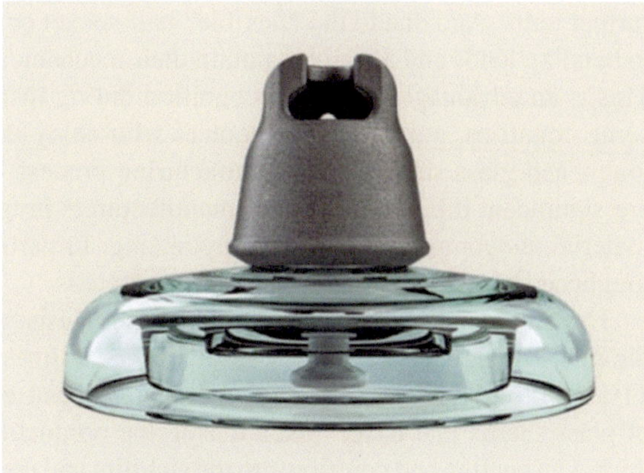

Fig. 4.50 Glass insulator (*Source* MacLean Power Systems)

Fig. 4.51 Glass cap-and-pin insulators in a UHV line

further units. And due to the "flexible" ball-socket connections, they are insensitive to bending loads and usually maintain their mechanical integrity when a cap breaks. This is an advantage for easy recognition during inspections but, unfortunately, in some countries, attractive for shooters who enjoy the explosive shattering of the toughened glass shell. Their manufacturing process is energy-intensive, and there are significant differences among manufacturers in terms of the quality of the raw material, its composition, and its processing. In particular, the pre-stressing of the toughened glass is a know-how-intensive process.

The raw materials for the glass dielectric consist roughly of 57% silica (silicon dioxide, which occurs naturally as quartz), 9% limestone, 14% soda, 4% feldspar, 11% dolomite, and 6% other mineral salts. About one third of the mass consists of glass shards that have broken during the production process or were sorted out during inspection and contribute to the stability and consistency of the manufacturing process [82].

The connections of glass insulators—essentially the metallic caps and balls—are very similar to those of porcelain cap-and-pin insulators. They are usually forged parts made of steel, as this process can achieve higher ductile strength than can be achieved by casting. They are hot-dip galvanized to protect against corrosion, like the end fittings of porcelain cap-and-pin insulators.

4.4.5 Composite Insulators

As early as the 1960s, composite insulators were developed in Germany and used in high-voltage overhead lines for initial field trials [83]. In the early 1970s, manufacturers from Germany, France, Great Britain, the USA, and Switzerland brought the first generation of commercial composite insulators for overhead lines to the market. For almost 60 years millions of units have been in operation with very good service experience and as various CIGRE and IEEE studies have shown, they have very low failure rates, comparable to those of porcelain and glass [84–88].

They came into general use in the 1980s and today make up about 40% of the global insulator market. Since then, the number of composite insulators installed has increased exponentially and counts nowadays more than 20 million units for AC lines [84]. In addition, more than 100000 units are installed in HVDC lines—due to the aforementioned advantages in terms of pollution (Sect. 4.4.1) [72].

This remarkable—and for many, unexpected—development has been made possible by the following advantageous properties of composite insulators:

- Hydrophobic properties of the insulating housing, which are transferred to the pollution layer and improve the insulator performance under pollution conditions,
- Low weight (e.g., 10% of the weight of a 380 kV porcelain string),
- Easy adjustment of the arcing distance to the insulation coordination requirements of the network and of the creepage distance to the pollution level of the

environment. This can happen by adjusting the geometry of the insulator, such as its length and the shape and number of sheds,
- Largely insensitive to vandalism by firearms and to dynamic bending stresses by load transposition,
- Development of numerous international standards based on considerable research effort by manufacturers, system operators, and academia,
- With the increasing spread of high-voltage lines up to 1200 kV AC and 1000 kV DC, composite insulators with single lengths of 10 m and more and a rated load up to 1000 kN, offer significant cost savings compared to porcelain or glass insulators,
- Growing environmental concerns of the public against the construction of new lines can often be met by so-called compact lines, which are visually less noticeable. Such lines have become feasible using composite insulators [89] and
- Due to high quantities (*economies of scale*) and automated industrial production, their prices have continuously fallen while simultaneously improving their quality.

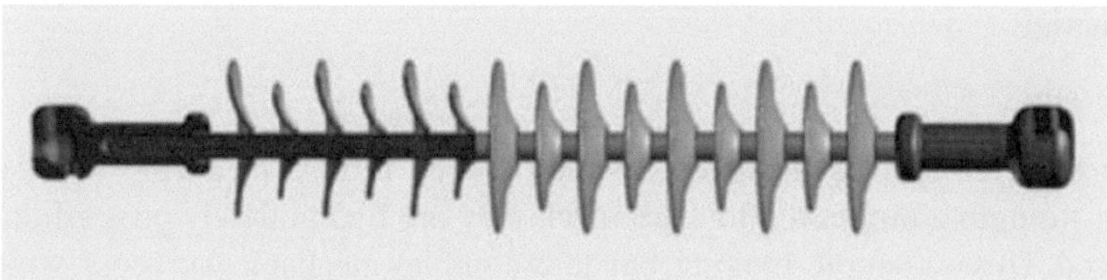

Fig. 4.52 Structure of a silicone rubber composite insulator

The basic structure of a modern composite insulator is shown in Fig. 4.52.

GFRP Rod

The heart of a composite insulator is the GFRP rod (*Glass Fiber Reinforced Plastic*). It consists of thousands of glass fibers (Ø 10–30 μm), which are embedded in an epoxy resin matrix. A glass content of 60 to 70 w% allows a high axial tensile strength of the rod of about 1000 N/mm^2. On the other hand, the GFRP rod, as so-called orthotropic material, is mechanically weak perpendicular to the fiber axis. This must be considered in the design and attachment process of the metal fittings to the rod.

GFRP rods are produced in the so-called *pultrusion process*, [90], with diameters of up to 300 mm and lengths well over 10 m. The glass fibers used in the GFRP rods of modern insulators are in most cases made of so-called *E-CR glass* (*E-Glass Corrosion Resistant*). This is an acid-resistant glass that contains no boron and replaces the previously used *E-glass* to avoid so-called brittle fractures which occasionally occurred in the early days [90].

Fig. 4.53 End fittings of composite insulators

Sealing of the end fitting is of crucial importance for the long-term reliability and operational safety of composite insulators. Because if water and especially acid rain were to penetrate to the so-called triple point of the insulator, i.e., to the point where the GFRP rod, the silicone housing, and the metal end fitting meet, partial discharges could produce oxalic acid, which would lead to a brittle fracture in inferior GFRP rod material.

End Fittings

For long rod composite insulators, mainly forged fittings made of steel, seldom of high-strength aluminum alloy, are used; they are hydraulically pressed onto the GFRP rod. This is a simple-looking, but quite complex mechanical process, which has been extensively studied in [91]. To ensure the compatibility of composite insulators with porcelain and glass insulators, the dimensions of their fittings are based on the standards for conventional insulators. Typical end fittings are shown in Fig. 4.53 [92]. They are hot dip galvanized with a typical zinc layer thickness of 85 μm.

Housing

The housing of a composite insulator has the dual task of protecting the GFRP rod, which becomes brittle if not protected from ultraviolet radiation (UV) and pollution, but also to ensure the required creepage distance to avoid pollution flashovers. This is the reason why the housing of polymer insulators uses basically the same shed shapes of porcelain long rod insulators.

For the housing, various polymers such as ethylene-propylene-diene monomer (EPDM), ethylene-propylene monomer (EPM), liquid silicone rubber (LSR), room temperature volcanizing (RTV) and high temperature vulcanizing (HTV) silicone rubber (SR), but also polytetrafluoroethylene (PTFE, often called Teflon) have been used. Meanwhile, HTV silicone, due to its superior properties and positive service experiences, is the material of choice.

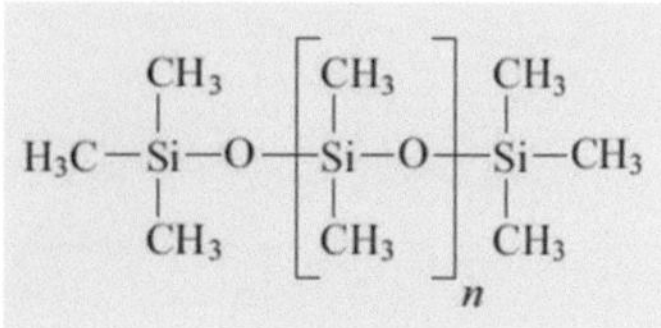

Fig. 4.54 Chemical structure of Polydimethylsiloxane (PDMS), n is the number of repeating methyl groups [93]

Properties of Silicone Polymer

The main components of silicone polymer are:

- Base polymer
- Fillers (silica, quartz flour)
- Vulcanizing agents (peroxides, H-siloxane)
- Catalysts (platinum compounds, tin compounds)
- Additives (adhesives, plasticizers, pigments)
- Aluminum trihydrate (ATH), i.e. Al_2O_3.

The base polymer is a three-dimensional cross-linked *Polydimethylsiloxane* (PDMS). To improve its mechanical properties, silica is incorporated. The creep and erosion properties are improved by adding aluminum trihydrate (ATH). The basic structure of a silicone polymer consists of an inorganic main string of *Si–O bonds* with each silicon atom bound to two methyl groups (Fig. 4.54) [93].

This structure explains some of the remarkable chemical and physical properties of the silicone polymer. Its high flexibility, even at low temperatures, is a result of the low intermolecular binding forces. It also has good thermal stability and excellent resistance to ozone, corona discharges, and moisture. But its essential advantage over the classic insulating materials porcelain and glass, is its *hydrophobicity* (from Greek: *Hydro* = water and *Phobos* = fear) (Fig. 4.55). This is caused by the low surface tension of PDMS and leads to water droplets on the silicone surface maintaining their quasi-round shape and not wetting the surface, which combined with a possibly existing pollution layer, would lead to conductive paths along the surface and ultimately to flashovers (Fig. 4.46). In addition, the diffusion of *low molecular weight siloxanes* to the surface, which encase the existing pollution particles there, also makes the pollution layer hydrophobic (Fig. 4.56); this is called *hydrophobicity transfer*.

In case of heavy pollution, the hydrophobicity of the silicone sheds can temporarily be lost, and surface discharges may occur, which would cause tracking and erosion of the silicone housing. To avoid this, adding of ATH filler of more than 50 wt% is required. However, it should be noted that the sole addition is not sufficient per se. Rather, the whole system (polymer, fillers, treatment of the fillers, etc.) must be optimized in a balanced manner [94].

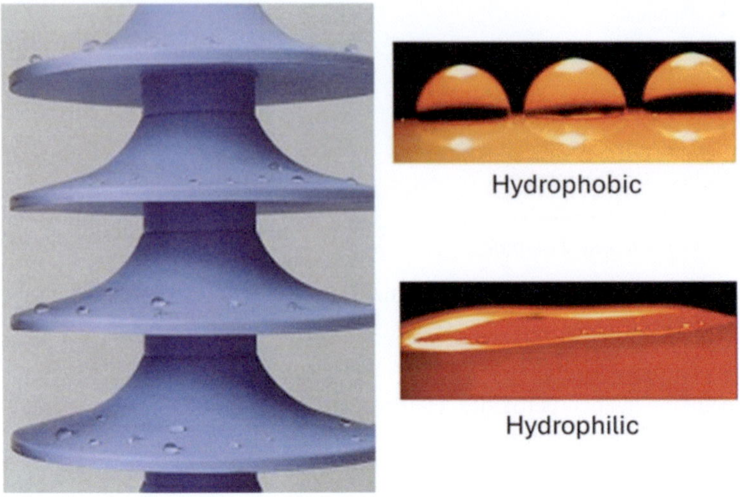

Fig. 4.55 Hydrophobic surface of silicone (*left*) versus hydrophilic surface of porcelain and glass insulators (*right*)

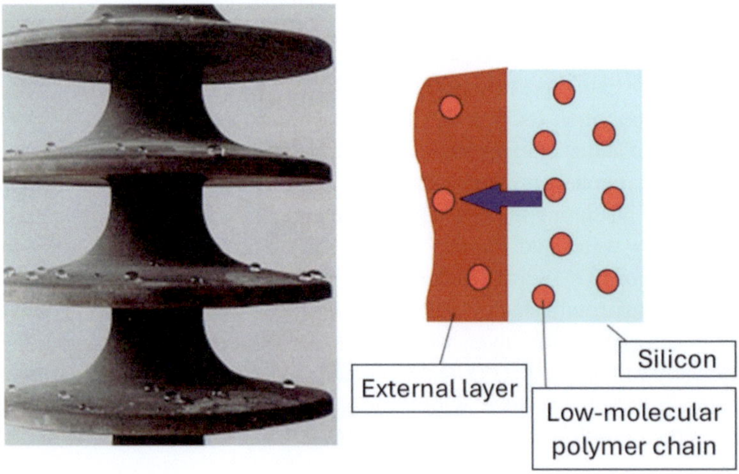

Fig. 4.56 Hydrophobicity transfer on composite insulators

Manufacture of Housing

Traditional methods for processing the silicone polymer for the insulator housing are casting (still used for hollow core insulators for electrical apparatus), compression molding or injection molding. The latter is the most advanced method and is

Fig. 4.57 *Left*: Injection molding equipment with operator. *Right*: flexibility of modular design of composite insulators (*Source* PFISTERER)

used by the majority of established manufacturers of composite insulators, however, it is demanding and requires significant investments in production machinery and molds (Fig. 4.57, left), extensive process know-how, as working pressures of up to 1500 bar and temperatures of up to 200 °C are required in order to produce a cavity-free housing integrally bonded to the GFRP rod. Injection molding is also suitable for a high degree of automation, such as in the ACIM process (Automated Continuous Injection Molding), thereby minimizing human errors [90].

A simpler but well-developed alternative is the modular method. In this method, a silicone sheath with a minimum thickness of 3mm is first extruded on the GFRP rod, after which the individual HTV-SR sheds, which have been separately produced by compression molding, are placed and volcanized on the silicon sheath of the rod with the help of a special silicone adhesive based on RTV. The modular method offers a high degree of flexibility regarding the required creepage distance at a given connection distance (Fig. 4.57, right) [90], but at the cost of an additional interface between the sheds and the sheath of the GFRP rod, which can lead to problems in service if improperly manufactured.

An important common feature of HTV injection molding and the modular method, is the interface between silicone and the GFRP rod. This must be absolutely tight, as any penetration of moisture through air inclusions at the interface can lead to partial discharges and ultimately to catastrophic failure of the insulator through a so-called

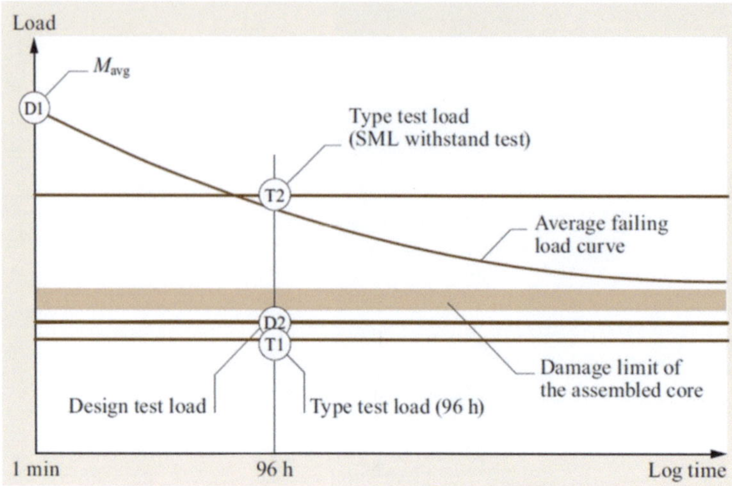

Fig. 4.58 Load-time curve for composite insulators according to [96]

«*flushunder*». Experienced manufacturers use automated spraying and UV detection techniques for the priming of the GFRP rod, which ensures a seamless and permanent application of silicone housing on it.

Testing of Composite Long Rod Insulators

Most composite insulators for high voltage overhead lines are built into suspension- or tension strings. They are primarily loaded in tension and are referred to as *composite long rods* with typical GFRP rod diameters between 16 and 32 mm. Their most important mechanical characteristic is the *specified mechanical load,* SML, defined as the load that the insulator must withstand in a 1-min tensile test after a 96-h test at 70% of the SML; it is determined in a type test. In addition, each insulator must, before leaving the factory, pass a routine test, at the so-called *routine test load,* RML, which amounts to 50% of the SML. A third important mechanical parameter is the *damage limit load,* DLL. This is the threshold for the tensile load, below which hardly any fibers fail when the load is permanently applied to the insulator. The DLL of high-quality insulators can be assumed to be up to 60% of the *average failing load* M_{avg}. The load-time behavior of a composite long rod insulator was described in detail in [95] and is shown in Fig. 4.58 [96].

Common SML ratings are:

- Suspension insulators: 120 and 160 kN
- Tension insulators: 210 and 300 kN.

The ultimate load F_{max} of a composite long rod insulator, can be determined with following formula:

$$F_{max} = 2\pi \rho \ell \tau_{max} \tag{4.51}$$

With r, the diameter of the GFRP rod, ℓ the crimping length and τ_{max} the shear strength of the GFRP rod, which is about 50 N/mm^2 [90, Table 2.21].

For a typical insulator with $r = 9.285$ mm and $= 50$ mm, a failure load $F_{max} = 155$ kN is obtained with [23], which corresponds well to the measured value of 148 kN [91].

The preferred failure mode of a composite long rod is the peeling of a surface layer of the FRP rod (a) and not the broom-type failure (b) or the lateral separation (c), as the latter indicate poor compression or pre-damage of the rod due to over-compression (Fig. 4.59).

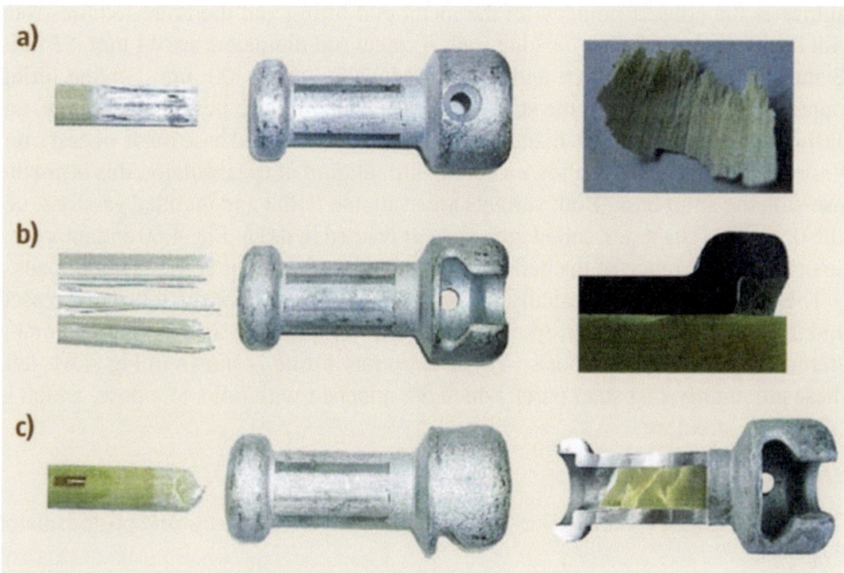

Fig. 4.59 Failure modes of long rod composite insulators during tensile testing: top—correct assembly, middle—under-compression, bottom—over-compression [91]

Fig. 4.60 220 kV composite line post insulator with gain base (*Source* INMR)

Composite Line Post Insulators

Composite line post insulators are predominantly loaded perpendicular to their axis, i.e., they are subjected to bending stresses (Fig. 4.60). As the resulting linear stress distribution over the GFRP rod cross-section, [91], may cause fiber and resin matrix failures at the contact points with the metal end fitting and therefore requires rods with larger diameters than for long rods. Typical rod diameters are 44 mm, 51 mm, 63 mm, 76 mm and 88 mm and can nowadays be up to 300 mm. For the fitting connecting the insulator to the structure, one can distinguish between a flexible, i.e. plastically deformable base, made of formed steel, and a rigid base made of cast iron. While the flexible version allows a certain articulation of the insulator, this is not the case with the solid base. Both variants are available in flat and inclined versions, i.e. with 0° or 12°. The latter, called *gain base, is* framed in red in Fig. 4.60 and improves the optical appearance of the deflection shape of the insulator at high lateral loads.

The top end fittings are usually, as for composite long rod insulators, compressed onto the GFRP rod, however, epoxy adhesives or temperature shrinkage are common alternatives for composite posts. A typical bottom fitting is shown in Fig. 4.61, left. These are mainly cast steel parts, which are attached with bolts to a base, which is fixed to the structure.

Mechanical Characteristics

The following loads characterize the bending behavior of a composite post insulator [97]:

- Minimum bending failing load, called *Specified Cantilever Load*, SCL, is the bending load that an insulator must withstand during testing.
- Ultimate load, called *Cantilever Failing Load,* CFL, is the maximum load reached during the testing of the insulator.

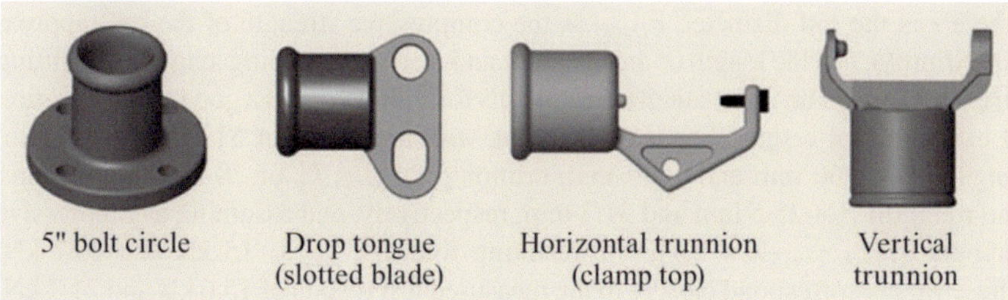

Fig. 4.61 Typical fittings for composite posts (*Source* MacLean Power Systems)

- Maximum bending load in service is called *Maximum Design Cantilever Load,* MDCL; this is usually about 25% below the so-called *Damage Limit Load*, DLL. Exceeding this load may cause damage to the FRP rod.

A simple test to determine the DLL, and subsequently the MDCL, is the measurement of the load–deflection curve of the post insulator in the same test. Such a curve is shown on Fig. 4.62 [91]. The load at which the slope of the curve begins to change corresponds to the DLL of the insulator, above which cracks occur inside the rod, these are practically invisible from the outside. This fast-track test is very practical compared to the MDCL test prescribed in [97], which is a lengthy 96-h test and is therefore often omitted during inspection and acceptance tests.

The mechanical behavior of composite post insulators was extensively studied in [91, 98] where the following simple analytical formula for their ultimate load (CFL) was developed:

$$F_{\max} = \frac{\pi\, r^3\, \sigma_{D,\max}}{4\left(L - L_p\right)} \tag{4.52}$$

Fig. 4.62 Load–deflection curve of a 38-mm composite post insulator [91]

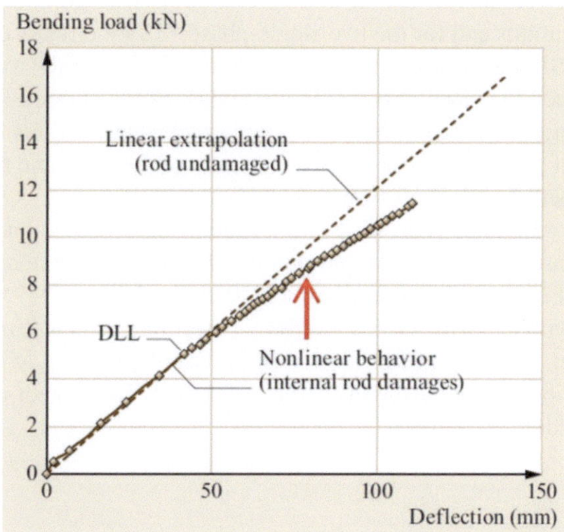

where r is the rod diameter, $\sigma_{D,\max}$ is the compressive strength of the rod (approx. 800 N/mm^2), L is the length of the insulator and L_p is the crimping length of the fitting on the FRP rod. The force calculated with this formula provides a good approximation of experimental results. For two insulators with diameters of 51 mm and 63 mm, lengths L of 760 mm and 1030 mm, crimping lengths L_p of 70 mm and 90 mm, and rod radii r of 25.5 mm and 31.7 mm, respectively, and assuming a compressive strength $\sigma_{Druck} = -794$ MPa, the resulting forces, $F_{\max} = 15$ kN and 20.7 kN. These values correspond closely to the measured test results of 15.6 kN and 21.7 kN, respectively.

Applications in Compact Lines

Compact lines were first developed in the 1970s but gained popularity only with the widespread introduction of composite insulators, as the insulating crossarms, which are often employed for their construction, are mainly subjected to compression and bending loads and thus may experience large deformations. These can be better withstood by the flexible composite materials than by the brittle ceramic and glass materials. Compact lines and their properties are extensively discussed in a recently published CIGRE Green Book [99]. The examples below demonstrate typical uses of composite insulators in compact line configurations.

Compact Lines in Switzerland

A prominent example is the world's first 380 kV compact line with composite insulators, which was erected in Switzerland in 1998 [100]. The line was necessary because the existing 125 kV line (Fig. 4.63 left) was no longer sufficient to cover the increased electricity demand in the Lake Geneva region and had to be replaced by a 380 kV line. The conventional steel lattice towers, which the local transmission system operator, EOS Lausanne (today Alpiq), would use for this type of line (Fig. 4.63 right), required a relatively wide corridor, which at certain locations would pass unacceptably close to residential areas. The solution was to construct very slender two-dimensional (2D) towers and to use insulating crossarms with composite insulators for the 380 kV circuits and for the two single-phase 132 kV circuits of the Swiss Federal Railways (SBB) on the top (Fig. 4.63 middle). Since the FRP rods of maximum 76 mm diameter, then available, could not withstand the compression forces, composite hollow insulators had to be used. Another advantage of this design was that the reduction of phase distances led to a reduction of the magnetic field at ground level below the permissible limits (Sect. 6.1.8).

Another interesting application of line compaction by V-strings with composite insulators is the multiple circuit line with slender concrete poles in central Switzerland (Fig. 4.64). In this case, a 380/220 kV line built in 1949 was replaced on the same poles with the SBB's 132 kV railway power line, which also needed renewal. As common in Switzerland for 380 kV lines, twin-bundle AAAC with 30.6 mm diameter and silicone composite insulators were used. The project was successfully completed in 2018.

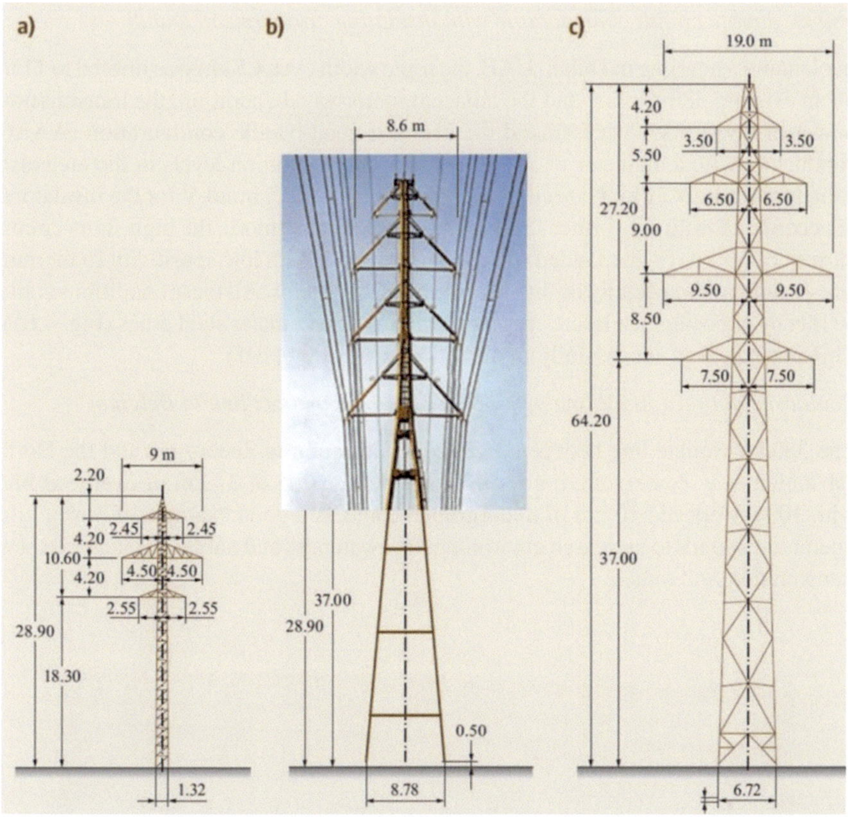

Fig. 4.63 Steel lattice tower of an existing 125 kV line (**a**), Swiss compact line (bipod) for 400 kV/132 kV (**b**), standard lattice design 400 kV/132 kV (**c**) (*Source* INMR)

Fig. 4.64 Swissgrid's 380/220 kV line Amsteg–Mettlen and SBB 132 kV Amsteg–Steinen in the Reuss valley in Uri, Switzerland (*Source* Bouygues E&S EnerTrans AG)

380 kV double circuit compact line with insulating crossarms in Dubai

In planning a new line in Dubai, UAE, the route width over 4.5 km was limited to 11 m by an existing 380 kV line and the adjacent motorway. In addition, the transmission power of 2600 MVA necessitated the use of a quad-bundle configuration (AAAC conductors with a diameter of 28.4 mm). The high pollution levels in the area also called for a relatively high specific creepage distance of 55 mm/kV for the insulators, in accordance with the former IEC 815 standard. Furthermore, the high short-circuit current of 63 kA/1 s demanded the use of insulators with low sensitivity to thermal stress, along with specially designed arc protection fittings. All these conditions could be met by choosing insulating crossarms, mounted on tubular steel poles (Fig. 4.65). This line has been successfully in operation since 2004 [101].

Replacement of a 150 kV line by a 380 kV double compact line in Belgium

The 380 kV double line between the Stevin substation in Zeebrugge and the Horta substation near Zomergem, both in Belgium, consisting of 37 km of overhead line with 100 towers and 10 km of underground cable, is a good example of how a grid operator responds to increased environmental awareness and public resistance to new transmission line routes.

Fig. 4.65 380 kV compact line in Dubai during conductor stringing; on the *left* in the picture the existing conventional 380 kV line, on the *right* in the picture the new 380 kV compact line and the city motorway, in the background the skyline of Dubai (*Source* PFISTERER)

One of the main reasons for the Stevin project was the connection of 800 MVA to the Belgian electricity grid, which comes from existing wind farms and photovoltaic plants in the north of the country, as well as the connection of 1200 MVA from new wind turbines. Further considerations concerned the connection to the 1000 MW HVDC submarine cable with England (Nemo-Link), as well as covering the increasing electricity demand in the rapidly developing city of Zeebrugge.

A widely accepted solution involved reusing existing towers along a 16 km corridor, the laying of 10 km XLPE cables and the construction of a 21 km long compact line with composite insulators. In addition, 53 km of the existing 150 kV line, which runs through the same corridor, were dismantled and 35 km of it were laid underground [102].

While glass cap-and-pin insulators are traditionally used in Belgium, silicone composite insulators were used for this project. One of the main factors for this decision was the investment costs, and indeed the new insulator strings proved to be about half as expensive as the glass alternative, although they had a relatively large creepage distance of 13,000 mm or 54 mm/kV USCD.

As can be seen from Fig. 4.66, the new 380 kV line (right in the picture) is similar in height and visual effect to the 150 kV line that has been dismantled. However, it transmits 10 times more power through the same corridor.

Special attention was paid to the assembling in the construction phase and the accessibility of the insulating crossarms during operation. This was solved by the innovative development of a working platform, Fig. 4.67. It allows all installation and maintenance work to be carried out safely and without the use of cranes.

Fig. 4.66 Comparison of the visual effect of the existing 150 kV line (*left*) with the new 380 kV line (*right*) in Belgium (*Source* [96])

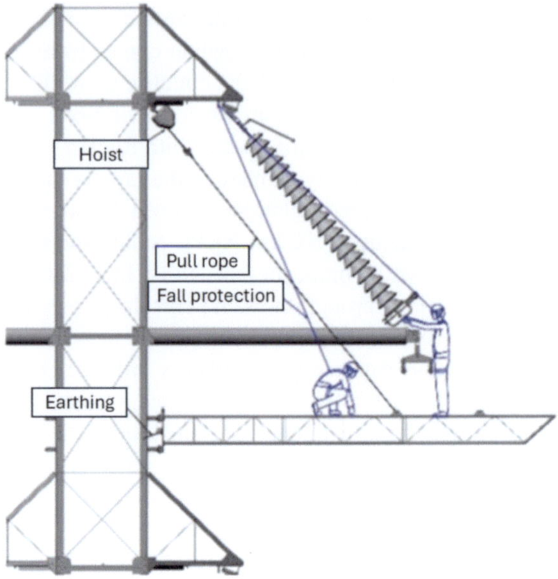

Fig. 4.67 Working platform for the new 380 kV compact line in Belgium (*Source* [102])

Interphase Spacers

Interphase spacers (IPS) are primarily used to prevent conductors from clashing under dynamic loads, such as strong winds, ice shedding or short circuits, and thus to avoid outages. They are also used worldwide to prevent galloping (Sect. 4.3.11). Their properties and design are discussed in detail in [90]. Although some applications with IPS made of porcelain were known in the past, composite insulators have been used almost exclusively in recent years due to their low weight and high flexibility.

The use of interphase spacers for line compaction is shown in Fig. 4.68. In this case, a second 12 kV circuit could be installed on an existing 50/12 kV line on concrete poles using IPS with silicone composite insulators. These IPSs are installed at intervals of 40 m along the span and ensure that the necessary in-span phase clearances are observed. Thousands of them e.g. have been successfully used in Switzerland for many years.

Interphase spacers equipped with composite insulators are also used to control conductor galloping and to suppress upward conductor motion caused by ice shedding, as implemented on one of Europe's highest-altitude 380 kV lines, in the *Vorab Glacier* in Switzerland. Due to the harsh weather conditions that prevail there in winter, there were frequent outages. These were caused especially in one span by two-phase short circuits, when the conductors clashed after ice shedding. Besides

Fig. 4.68 Line compaction in medium voltage lines with interphase spacers (*Source* CKW, today Axpo)

that, the outer wire layers were severely damaged by physical contact but also by the high short-circuit currents, which made short-term repairs under particularly difficult environmental conditions necessary. For this reason, it was decided to install interphase spacers and to design them mechanically in such a way that they could maintain the in-span phase clearances even under extremely dynamic loads. This resulted in IPSs between 10 and 12 m length weighting approx. 100 kg. They consist of four to five silicone composite insulators with a rod diameter of 76 mm and have been successfully in operation since 1995 (Fig. 4.69).

Fig. 4.69 380 kV interphase spacers in the Vorab line in the Glarner Alps of Switzerland (*Source* NOK, today Axpo)

4.5 Fittings

Fittings or *hardware* provide the mechanical attachment, the electrical connection and the protection of conductors and insulators (Fig. 4.70). They are also referred to as *accessories*. A distinction is made between *conductor fittings* and *string fittings*.

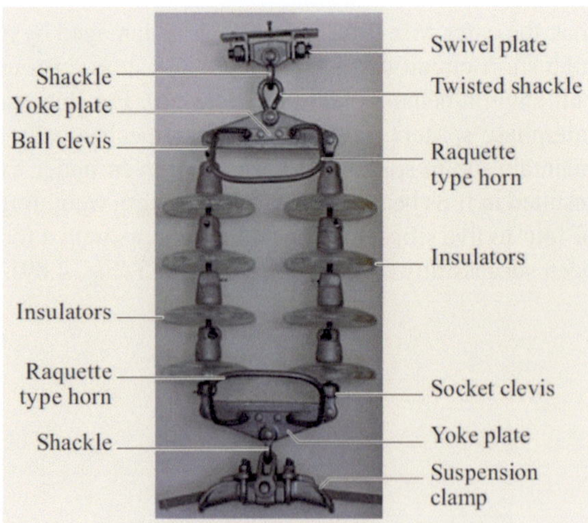

Fig. 4.70 Suspension string with string fittings (*Source* U. Cosmai)

Most fittings are relatively simple metal components however, they must withstand all types of mechanical, electrical, and environmental stresses.

String fittings are components that are used to connect the insulators with the line supports, and, in the case of longer strings, the individual insulators with each other.

Basically, fittings should be designed so that they:

- Avoid damage to the conductor
- Withstand mechanical stress during installation, operation, maintenance and repair,
- Are secured against loosening of individual components while in service,
- Exhibit better corona behavior than the conductors on which they are installed.

Furthermore, where applicable, fittings should be designed with live-line maintenance in mind (Sect. 4.8.3), ensuring safe and straightforward handling in accordance with established work procedures.

Figure 4.71 illustrates the components of a 110 kV tension insulator string in an exploded assembly drawing, showing the individual fittings and their configuration within the string.

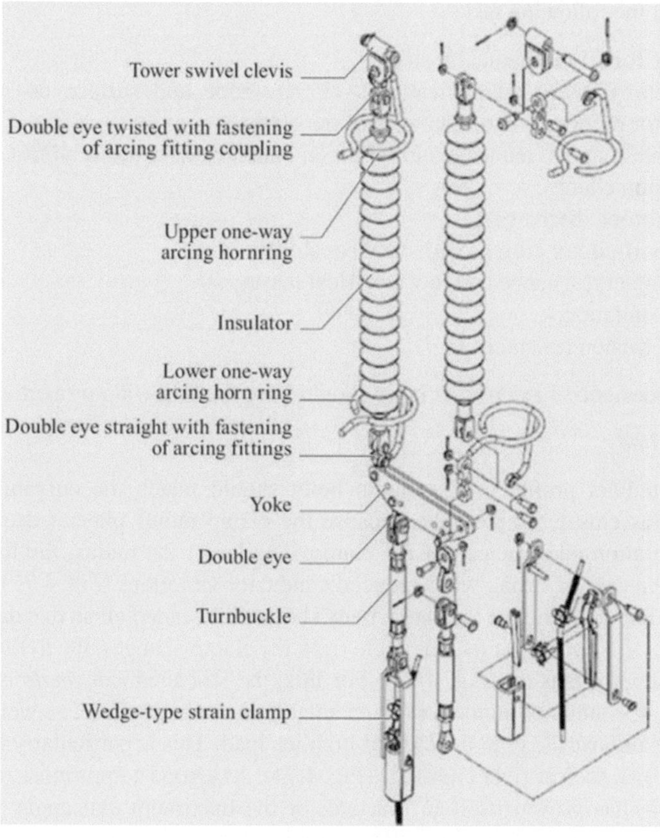

Fig. 4.71 Structure of a 110 kV double tension string for a twin bundle

4.5.1 Conductor Fittings

Conductor fittings are components that are directly attached to the conductors and include damping devices (Sect. 4.5.4). There are a variety of conductor fittings, such as:

- Suspension clamps
- Tension clamps
- Joints
- Repair sleeves
- Preformed fittings
- Tie-off or T-clamps
- Warning devices (spheres, flags for bird warning, radar markers, etc.)
- Vibration dampers
- Spacers and Spacer Dampers.

Suspension Clamps

Suspension clamps are used to attach the conductor to the suspension string. They must fulfill the following tasks:

- Suitable for all mechanical loads,
- Conductor-friendly attachment, i.e. clamp shape and surface do not cause conductor deformation or damages of the clamp,
- Favorable shape to reduce/avoid vibration damage; these occur almost always at suspension clamps,
- Avoid corona discharges,
- High short-circuit current withstand capability,
- Low contact resistance and low electrical losses,
- Easy installation,
- High corrosion resistance.

A well-designed suspension clamp should have the following properties [103].

Clamp Body

The longitudinal profile of the clamp body should match the curvature of the conductor as closely as possible, because the clamp radius plays a decisive role for the vibration performance of the clamp: The larger the radius, the fewer wire breakages in case of strong wind-induced conductor vibrations (Fig. 4.72) [104].

In addition, both ends of the clamp body should be rounded off so that damages to the conductor is avoided if its exit angle from the clamp changes due to fluctuations in the conductor tension (Fig. 4.73). For this, the so-called *exit angle* α must be large at high conductor temperature and small at low temperature, as well as large at high tensile loads, e.g. in the case of high ice load. This is particularly important for long spans, such as river crossings (Fig. 4.74). As a good compromise, a value of $\alpha_m/2 = 30°$ has proven itself in practice for the maximum exit angle. To avoid damaging the conductor's surface, all parts of the clamp, including the *clamp keeper*, that come into contact with the conductor should be smooth.

Fig. 4.72 Suspension clamp: number of wire breakages in a laboratory fatigue test vs. the suspension clamp radius

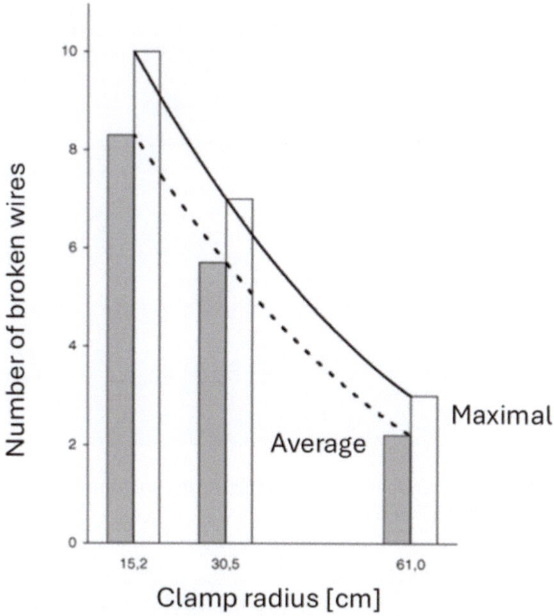

Fig. 4.73 Technical terms related to suspension clamps (α: exit angle; α_m: maximum exit angle; γ: turning angle; γ_m: maximum turning angle) (*Source* [103])

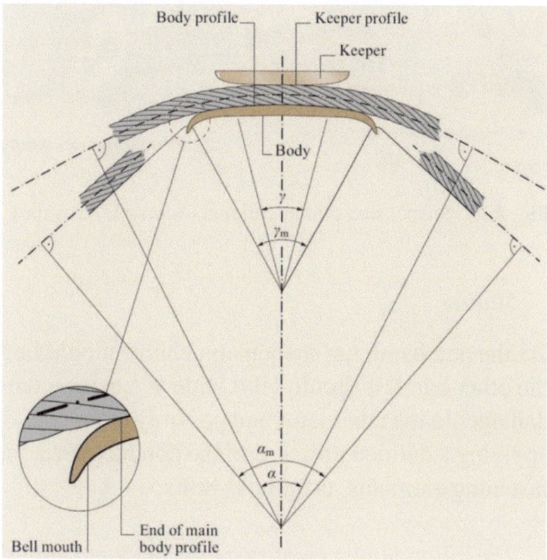

Mobility

The suspension clamp should allow rotation in the vertical longitudinal plane to be able to absorb asymmetric loads, e.g. from unequal adjacent spans (Sect. 4.2.1). The rotation axis of the clamp should be as close as possible to the conductor axis to reduce additional bending stress on the conductor. In this context, it is advantageous to keep the mass of the clamp low and thus reduce its moment of inertia.

Fig. 4.74 Special suspension clamp for the crossing of the Blue Nile River in Egypt (*Source* W. Huiber, PFISTERER)

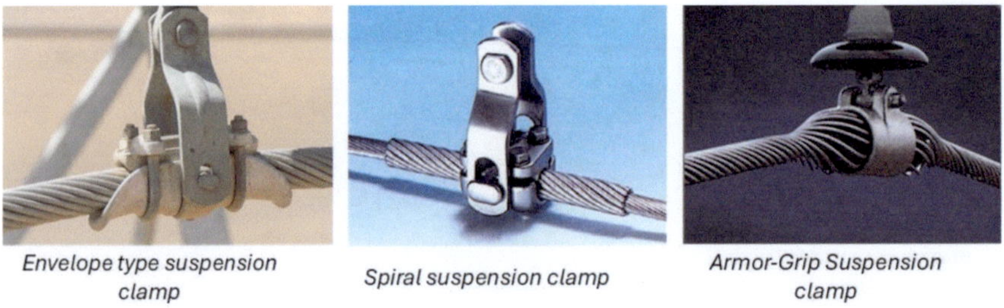

Fig. 4.75 Suspension clamp variants (*Source* U. Cosmai)

Sliding

On the one hand, the suspension clamp should hold the conductor in its position; on the other hand, it should let it slide at a certain tensile load to avoid overloading and damages to the conductor and possibly to the line support. This is mainly achieved by applying a defined pressure to the conductor via the clamp keeper and the associated fastening elements, usually U-bolts.

The following types of suspension clamps are distinguished:

Envelope type suspension clamps (Fig. 4.75 left) are the most used with about 50% of all installations. They come in various forms, are easy to install, inexpensive and provide, when properly designed and manufactured adequate grip and protection for the conductor.
Spiral suspension clamps (Fig. 4.75 middle) are used in about 25% of all installations. They consist of spiral wires that are wound around the conductor and reinforce it at the suspension. The usually forged clamp body is then mounted over the spirals.

Helically Attached Elastomer Suspension, also known under the trade name *Armor-Grip Suspension* (AGS) (Fig. 4.75 right), have proven themselves in situations with strong conductor vibrations [105], and are used in about 10% of installations, often for optical fiber ground wires, OPGW (Sect. 4.3.8). On the other hand, their costs are higher and their installation more complex than the other types.

Tension Clamps

Tension clamps, also *strain clamps* or *dead-end clamps,* are used to attach the conductors via the tension strings to line supports, such as at tension-, angle- and terminal towers (Sect. 4.6.1), and must withstand the full tensile load of the conductors. There are four main types:

Compression Clamps

Compression clamps are the most used tension clamps worldwide. This clamp and the compression process with hydraulic presses, are shown on Fig. 4.76. The steel core is first compressed and then an aluminum tube is slipped over the aluminum layers and compressed. Adjusting the conductor sag or moving the clamp after installation is no longer possible. For this purpose, turnbuckles or adjustable extension links, so-called *sag adjusters,* are used (Fig. 4.77). The same figure shows a so-called *triangular frame yoke*, which is used to protect porcelain long rod insulators from failure during load transposition (Sect. 4.4.3). In case of failure of an insulator of a double string, it deforms and thus reduces the bending stresses which may damage the healthy insulator of the string. This situation applies mainly to long rod porcelain insulators [106].

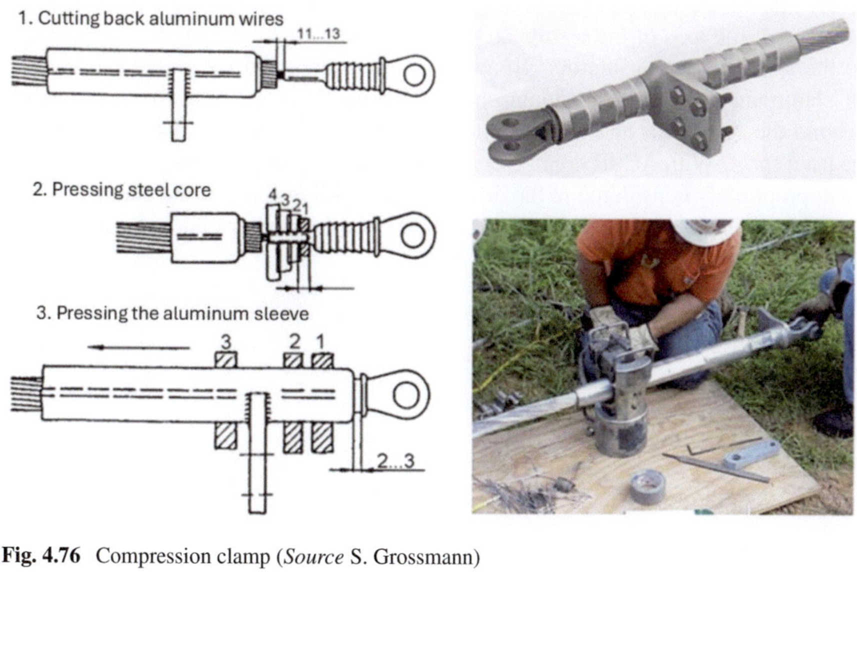

Fig. 4.76 Compression clamp (*Source* S. Grossmann)

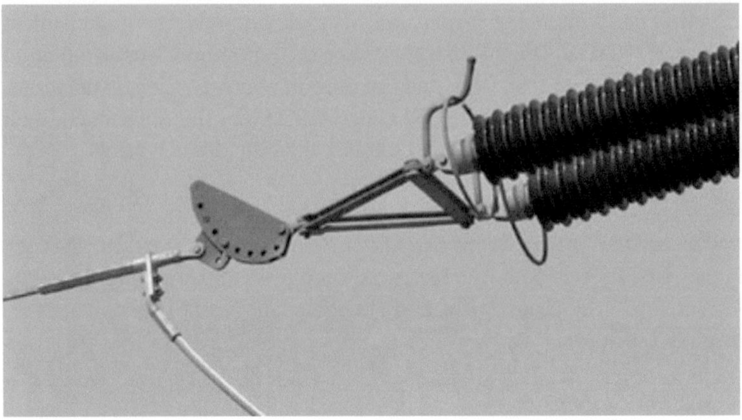

Fig. 4.77 Compression clamp with sag adjusting plates (*Source* PFISTERER)

Wedge Type Clamps

Wedge type clamps do not require special tools for installation and facilitate conductor sag adjustments. They consist of a pair of wedges and a two-piece clamp body (Fig. 4.78). When the tensile force of the conductor increases, the pair of wedges slides in the clamp body and provides a steadily increasing lateral force. Due to friction, the conductor—which becomes thereby slightly oval—but without significant plastic deformations of its outer layer, as is the case with compression clamps, cannot slide. The attachment of the clamp at the tension string is done with straps that are attached at the axis of the clamp, so that any conductor vibrations are "passed on" to the less sensitive insulators. To avoid eddy currents and the associated heating of clamp and conductor, the clamp body must not form a closed magnetic circuit around the conductor and only paramagnetic materials, such as aluminum, should be used for it. With ACSR conductors, the presence of grease in the steel core must be appropriately considered in the design of the holding force of the clamp [107].

With the help of Coulomb's friction law (4.53), the operating principle of the wedge type clamp can be explained [108]:

$$R = \mu N \tag{4.53}$$

With:

R: Frictional force
N: Normal force
μ: Coefficient of friction.

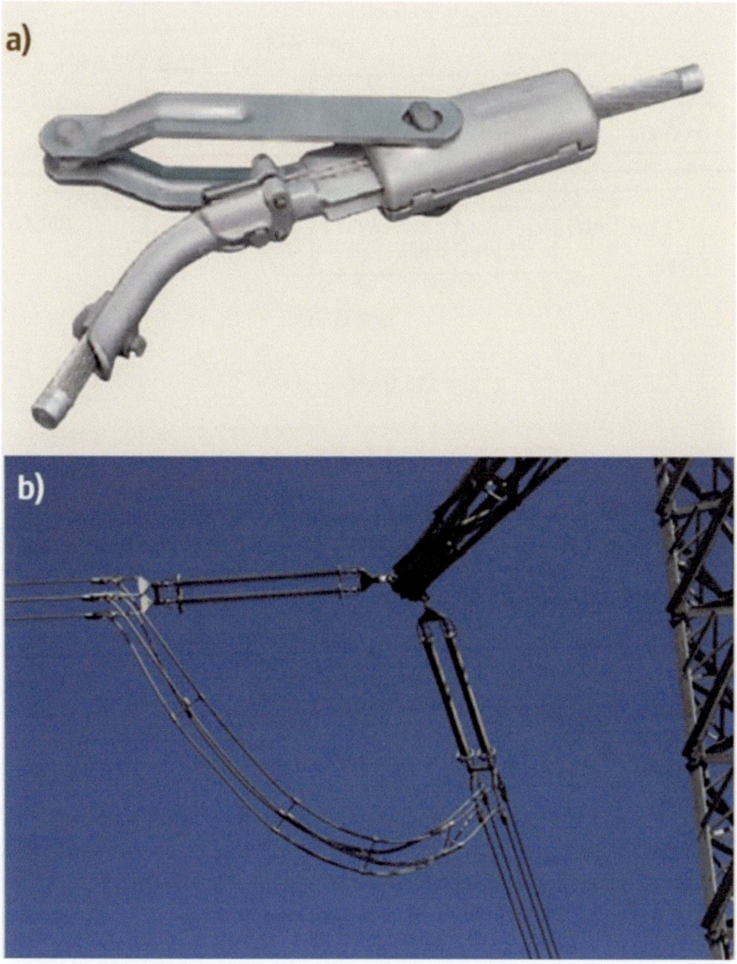

Fig. 4.78 Wedge type clamp (**a**) picture (**b**) at an angle tower with jumper (*Source* PFISTERER)

With the values for the friction coefficients given in Fig. 4.79, the tangential forces at the respective boundary surfaces of the wedge type clamp can be determined.

In the case shown in Fig. 4.79, the ACSR 240/40 conductor at 20 °C operates under a tensile stress of 40N/mm^2, corresponding to a tensile force of $Z = 11.3$ kN (Fig. 4.79 top). From this according to (4.22) and (4.23) the aluminum layers carry $Z_A = 7.2$ kN and the steel core $Z_S = 4.1$ kN. Due to the high coefficient of friction between the outer aluminum layer and the wedge, the load transfer from the conductor to the wedge and subsequently to the clamp body is ensured. The

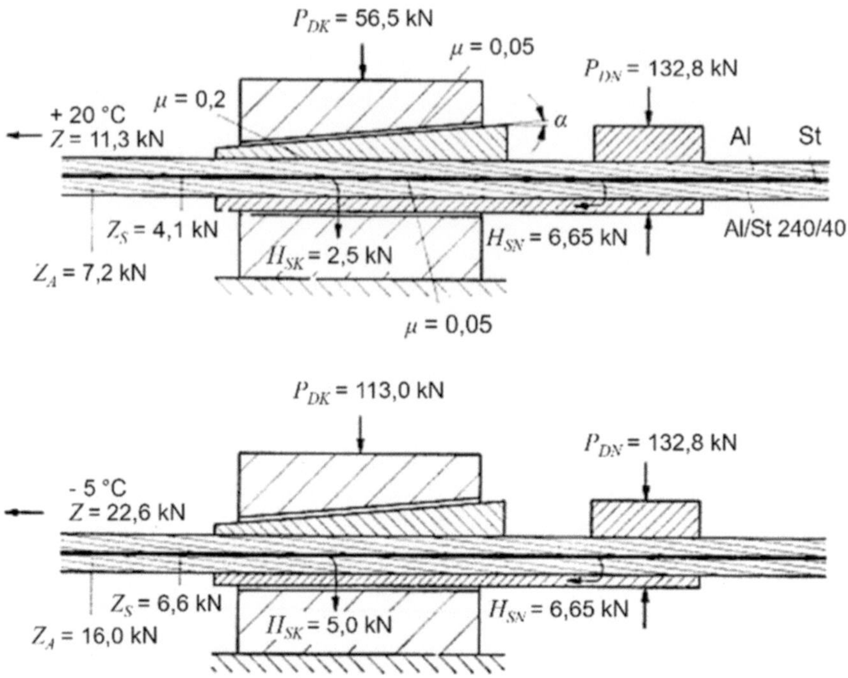

Fig. 4.79 Function and force balance of the wedge type clamp at +20 °C (top) and at −5 °C (bottom) (*Source* [108])

limited compressive strength of the aluminum material is thereby considered. The conductor tension Z of 11.3 kN generates via the frictional connection between the outer aluminum layer and the wedge on one side and the inclined plane between the wedge and the clamp body on the other, a compressive force of $P_{DK} = 56.5$ kN. From this, about 10% will not reach the steel core because the surrounding aluminum layer acts partly like a cage. With the remaining compressive force of $0.9 \cdot 56{,}500$ N = 50.85 kN and the, due to the greasing of the steel core, quite low friction coefficient between steel core and aluminum layer of 0.02 to 0.05, the contribution of the steel core to the retention load of the clamp amounts to $H_{SK} = 2.54$ kN. In addition, the four M12 bolts in the so-called *wedge follower* generate pressure on the conductor of $P_{DN} = 132.8$ kN, resulting, because of friction, to a contribution of $H_{DN} = 6.65$ kN.

The resulting total holding force on the steel core, of $H_S = H_{SK} + H_{DN} = 2.54$ kN + 6.65 kN = 9.15 kN clearly exceeds its axial tensile load of $Z_S = 4.1$ kN. offering a sufficient safety margin to prevent axial displacement of the steel core within the aluminum strands. This margin is only then ensured if the wedge follower is properly mounted. If the tensile stress increases due to ice load at −5 °C to 80 N/mm^2 corresponding to $Z = 22.6$ kN (Fig. 4.79), the clamp body pressure P_{DK} and thus also the steel core holding force H_S increases. A steel core tensile force $Z_S = 6.6$ kN is now countered by a holding force $H_S = 11.65$ kN. The safety of the steel core against slipping is therefore particularly high at low temperatures and high conductor loads.

If the assumed friction conditions in the clamp are maintained, the balance of forces is ensured. In practice, however, these conditions change over time. The shear force in the wedge follower is reduced by creep phenomena in the aluminum layers. Also, the lubricity of the sliding grease in the clamp decreases over time. To safeguard long-time operation, the clamp must have considerable reserves, which is ensured by a sophisticated design and manufacturer know-how.

Tension Spiral Fittings

Tension spiral fittings consist of spirally preformed aluminum wires evenly distributed on the outer layer of the conductor. They can be installed by hand without tools. Since the inner diameter of the spirals is slightly smaller than the conductor diameter, lateral pressure and a frictional connection between the spiral and the conductor is generated when the spirals are attached to the conductor. As a result, the tensile force is distributed smoothly by the helical grip over a longer conductor length—relative to other clamp types (Fig. 4.80)—which helps prevent damage to the outer wires. For this reason, tension spiral fittings are often used for OPGW which, because of the sensitive optical fibers they carry, cannot withstand large lateral stresses (Fig. 4.81). Spiral fittings are also used in case of damage to the outer wire layer of a conductor, to ensure electrical continuity.

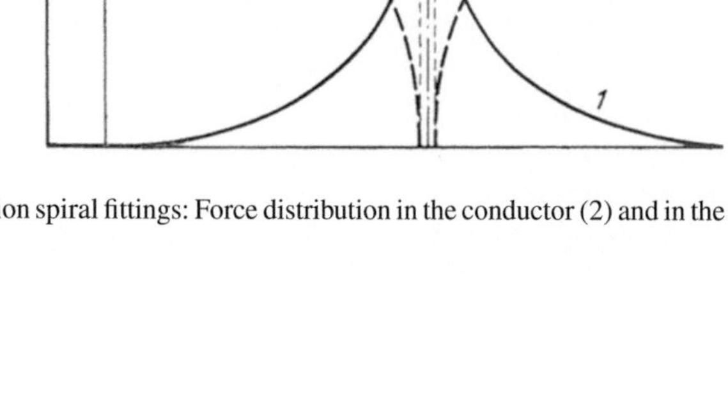

Fig. 4.80 Tension spiral fittings: Force distribution in the conductor (2) and in the spiral (1) (*Source* RIBE)

Fig. 4.81 Helical tension fitting with adjustable extension links at the Suez crossing; inset: rod ends with so-called parrot bill shape to avoid corona (*Source* RIBE)

Joints

Joints are conductor fittings used to connect the two cut ends of a conductor.

Experience has shown that the failure of a connection is often caused by increased electrical resistance of the connector due to corrosion, which can lead to unacceptable temperature increase in the conductor and unstable mechanical conditions (Sect. 4.8.1).

Because of the increasing tendency of network operators to operate lines with conventional conductors at higher temperatures, sometimes up to 100 °C, a better understanding of the behavior of conductor joints at higher temperatures is required. In this regard, extensive laboratory tests lasting a number of years have shown that properly designed and installed joints exhibit stable long-term behavior up to temperatures of about 150 °C [109].

There are two types of conductor joints:

Full-Tension Joints

These are joints that establish the electrical connection of the conductors and are at the same time subjected to tensile loads. They are suitable for use in spans.

Full-tension joints must be capable of safely carrying the conductor up to 1.55 times its maximum service load, or 0.85 times 95% of its rated tensile load (RTL), whichever is lower. Additionally, under both continuous current and maximum short-circuit conditions, the joint's temperature must not exceed that of the conductor itself. Furthermore, the voltage drop across the joint must not exceed the voltage drop across an equivalent length of conductor.

Full-tension joints for high-voltage overhead lines are predominantly compression joints. These consist for monometallic conductors of a sleeve made of aluminum or an aluminum alloy. Joints for ACSR conductors additionally have a steel sleeve

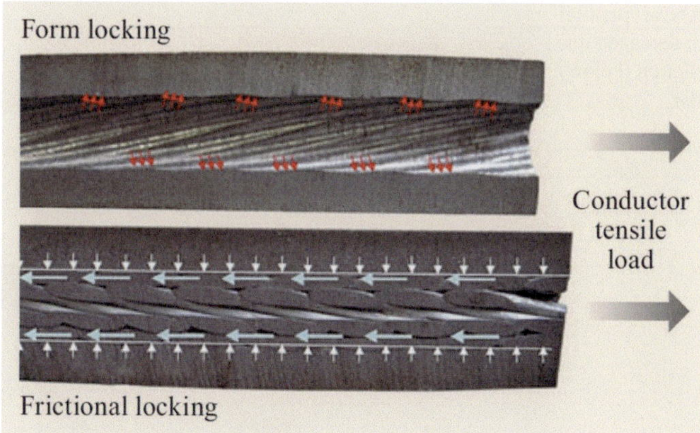

Fig. 4.82 Force equilibrium in a compression joint under traction via form and frictional locking [110]

for connecting the steel core. The compression is carried out, as described for the compression tension clamps, with hydraulic presses with mostly hexagonal tools (Fig. 4.76). The compression process may appear simple, but in reality, it is a complex plasticity problem, as the joint and the conductor, during the compression, experience high plastic deformations, which affect the quality of the connection. The force transmission between conductor and joint consists of a form fitted and a frictional component (Fig. 4.82). It should be noted that due to the residual elasticity of the metal sleeve (*spring-back*) the permanent pressure force F_{Kres} on the conductor is only 20% of the initial pressure force F_{Kpress} (Fig. 4.83) [109].

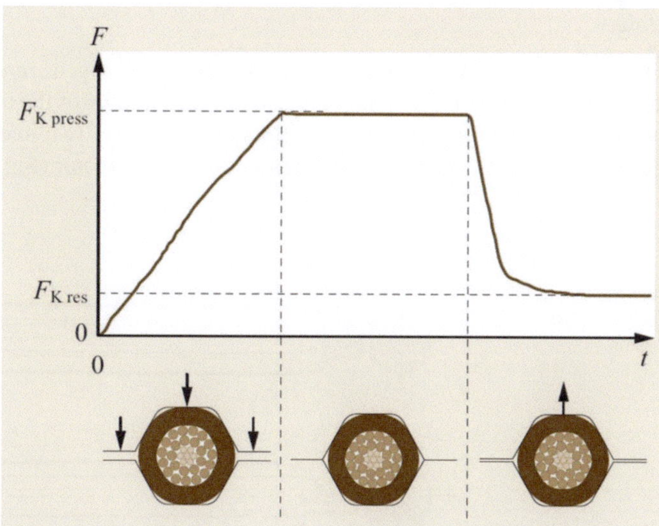

Fig. 4.83 Compression force variation during the compression process in a compression joint (schematic) (*Source* [109])

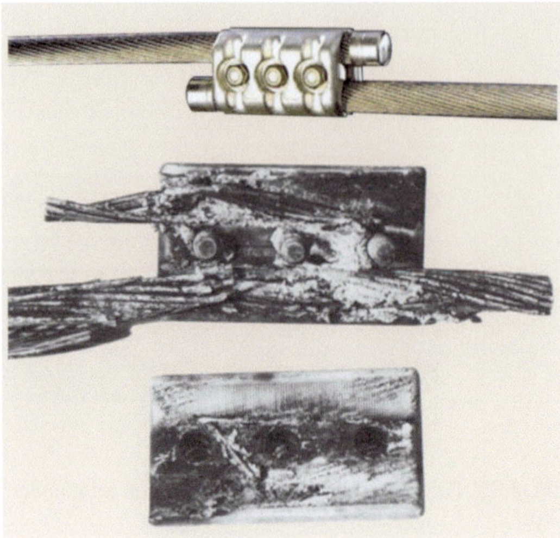

Fig. 4.84 (Above) parallel groove clamp damaged due to poor installation (below) (*Source* RIBE)

Non-Tension Joints

This includes all connectors that only establish the electrical connection of the conductors but are not or only partially subjected to tensile loads. They are preferably used for connecting conductors in the jumpers of tension strings (Fig. 4.78).

Even with *non-tension joints,* like the so-called *parallel groove clamps*, it can happen due to improperly cleaned surfaces of the conductors and clamps, that unacceptably high contact resistances occur which will impair the clamp operation and may lead to outages (Fig. 4.84).

Repair Joints

If individual wires of the outer layer of the conductor are damaged, *repair joints* can be used to restore the mechanical strength and the current flow of the conductor. Repair joints belong to the non-tension connectors and consist of two interlocking, semi-circular tube segments, which are compressed on the conductor (Fig. 4.85).

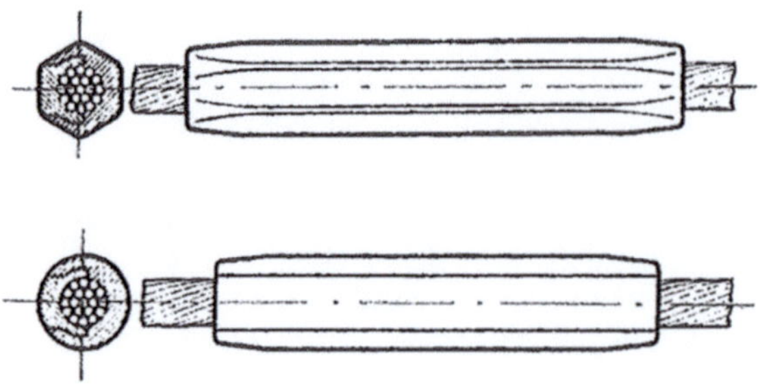

Fig. 4.85 Repair joint before (bottom) and after (top) compression

Warning Devices

Warning devices on conductors and ground wires are intended to make them visible to prevent collisions with aircrafts, ships, birds, parachutes and gliders [111]. This is especially important for valley crossings. They are also installed at line crossings with railways and roads when these are designated as flight routes for rescue operations with helicopters or as bad weather emergency flight routes. This visibility contradicts the efforts of planners to make overhead lines rather invisible to achieve greater acceptance in the public.

Although warning devices only make up a very small proportion of the construction costs of an overhead line, they are very important and sensitive components for its function. If they fail, e.g. due to color fading or clamp breakages, or if unsuitable clamping methods damage the outer layer of the conductors, they must be replaced, which is often not possible when the line is live and requires a shutdown. The type of warning devices and where they are to be installed is usually stipulated by official regulations. The same loads from ice, snow, and wind as for the conductors are assumed for their mechanical design; internal company standards may define other values.

Aircraft Warning Spheres

Aircraft warning spheres (AWM: *Aircraft Warning Markers*), Fig. 4.86, are primarily used on the ground wires of overhead lines. They are made of plastic or aluminum and have diameters from 30 to 100 cm and are clamped onto the conductors by linemen from the line car. Recently, systems are available that are placed on the conductor by a helicopter and automatically clamped there. They can also be dismantled in a similar way, i.e. without human interaction at the conductors.

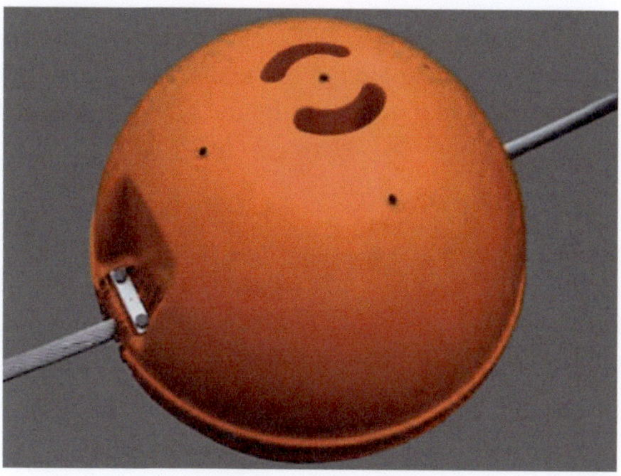

Fig. 4.86 Warning sphere (note the intrusions on the top for easy handling) (*Source* PFISTERER)

Fig. 4.87 Bird warning devices: double balls with 30 cm diameter (top *left*); warning flags (top *right*, *Source* RIBE), insulator cover (bottom *left*), bird broom (bottom *right*)

There are considerations and investigations in Europe to replace aircraft warning devices on overhead lines with satellite-based systems. As soon as acceptance for this is given, it requires a coordinated procedure for the marking, introduction, and maintenance of the system, as well as the corresponding technical equipment in the aircrafts.

Signal Lamps

In some countries near airports or flight paths, self-illuminating lights are attached to the conductors to warn low-flying aircraft of the obstacle. They are powered either by induction from the conductor current or by the line potential and often contain buffer batteries [111].

Bird Protection Warning Devices

For bird protection, there exists a wide range of warning devices (Fig. 4.87). The birds are encouraged by the markings to fly over or under the line. The selection of the most effective marking depends on the bird species that are to be protected since the width of their viewing angles varies between species.

For example, black and white balls are mounted side by side or at intervals, or black and white bird warning panels between the sub-conductors of a conductor bundle, similar to spacers, or bird warning flags. Recently, movable bird deflectors are available, which can be mounted by a drone. Also, the covering of the insulators on poles of medium voltage lines with plastic covers (Fig. 4.87, bottom left) or the mounting of broom-shaped diverters, so-called *bird brooms* at the fixing point of the insulator on the crossarm (Fig. 4.87, bottom right) have proven to provide adequate bird protection against short circuits in case the birds touch the live parts of the line (electric shock) (Sect. 3.8).

Radar Markers

Radar markers on the conductors are primarily used at the crossings of waterways. They often resemble the usual red or orange warning spheres, and contain metallic elements inside, which can be seen by the radar systems of the ships. This helps the detection of line crossings during day and night trips.

4.5.2 Insulator String Fittings

Insulator *string fittings* connect the conductor fittings with the insulators, the insulators with each other and these with the line supports. We distinguish between *connection fittings* and *protection fittings*.

String fittings must have the following properties:

- Withstand loads of 40 to 800 kN (does not apply to protection fittings)
- Short-circuit current capacity
- Mobility: ball-socket; clevis-tongue
- Compatibility: IEC 60120; IEC 60471
- Easy assembling: no special tools
- Corrosion resistance: hot dip galvanized or made of A2 steel.

Especially in the case of a short circuit and the associated temperature increase, their mechanical integrity must not be impaired, and their protective zinc layer must not be damaged. For this purpose, their short-circuit current density in mechanically loaded fittings should be limited to 70 A/mm^2 (1 s) and in unloaded ones, such as arc protection fittings (see below), to 80 A/mm^2 (1 s) [8].

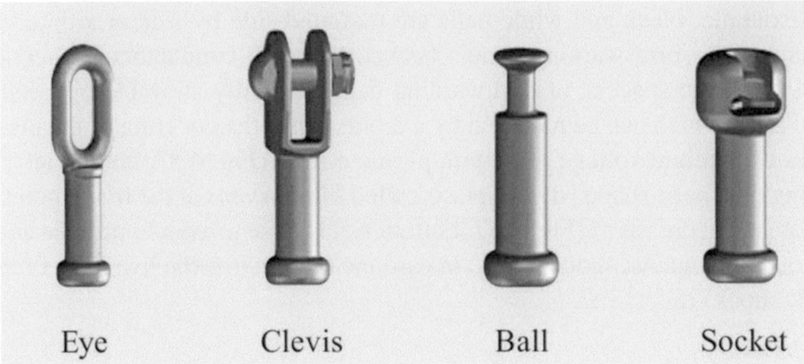

Fig. 4.88 Connection fittings

Figure 4.88 shows different types of connection fittings for insulator strings.

Insulator protection fittings, such as arcing horns and rings, are important components of insulator strings, especially at high voltages. In Europe, they are usually used from 50 kV upwards, overseas often only from 150 kV or even from 220 kV and above. They are intended to protect the insulators from the power arc in the event of flashovers. These can be caused by:

- Overvoltages due to lightning strikes and switching operations,
- Conductive pollution layers on the surface of the insulator,
- Reduced clearances to grounded structures of the line due to strong wind, ice shedding or conductor galloping,
- Bridging the clearance of the insulator, e.g. by animals or bird droppings, which are released when birds are flying off the top of an insulator,
- Reduced electrical strength of the air in case of fires below a power line, e.g. in the case of devastating forest fires nowadays.

Since arcs can reach temperatures of over 18000 K, the main task of arc protection fittings is to quickly capture the arc and lead it away from the insulator so that it is not thermally overloaded, with porcelain and glass being more sensitive in this respect than composite insulators. This is done by providing the arc with a conductive path and a suitable focal point in the form of a sphere with enough *sacrificial material* (Fig. 4.89). The material loss at the end focal point can be calculated for steel with 0.4 cm^3 /kAs and for aluminum with 1.4 cm^3 /kAs; for 380 kV strings, for example, a sphere diameter of 80 mm is common.

Steel is the preferred material for insulator protection fittings over aluminum, as the arc moves almost 50% faster with steel, thus reaching the sacrificial electrode faster. In addition, the combustion heat of aluminum is five times higher than that of steel, which increases the risk of thermal damage to the insulator. Furthermore, arc protection fittings should be aligned so that the electromagnetic forces generated by the arc current support the movement of the arc away from the insulator [112].

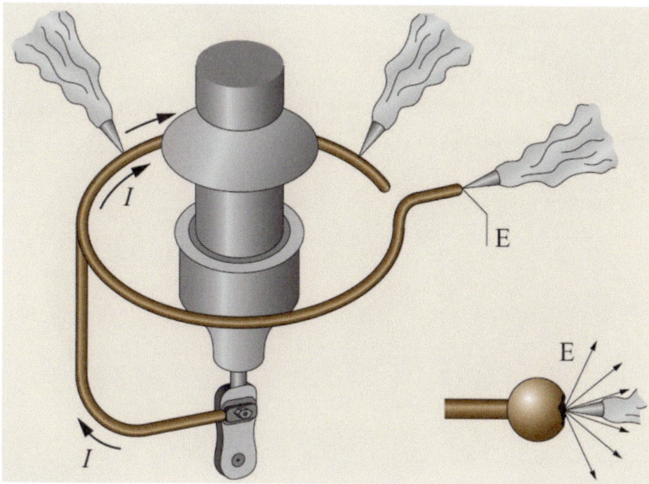

Fig. 4.89 Schematic movement of the power arc (current *I*) to the focal point of the fitting with correct shape of its extremity E (sacrificial sphere)

At higher operating voltages, the electric field strength can reach values at the live end of the insulator that are above the corona onset voltage (Sect. 4.3.6). This is more critical with composite insulators, as their silicone sheath can be damaged by the so-called *water drop corona*. For this reason, arc protection fittings should be dimensioned so that the electric field strength on the shield sheath of composite insulators does not exceed the recommended limit of 4.2 kV/cm [113]. However, this can only be proven by Finite Element simulations. The following limits generally apply:

Metallic string components:	17…21 kV/cm (test and simulation)
Insulator trunk:	≤4.2 kV/cm for 10 mm thickness (simulation)
Seal:	≤3.5 kV/cm (simulation)
Insulating material:	≤30 kV/cm (simulation)

There are also arc protection fittings that fulfil both tasks, i.e. the control of the arc and of the field strength, such as the so-called Combi-protection ring (*C-ring*) (Fig. 4.90).

For composite insulators, it is recommended not to attach the arc protection fittings directly to the end fittings (Fig. 4.91), as in the event of an arc, the full short-circuit current would flow over the end fitting and cause it to overheat. The resulting radial expansion of the fitting could cause the GRP-rod to slip out of the end fitting and the insulator to lose its mechanical integrity. For this reason, weights are attached to the

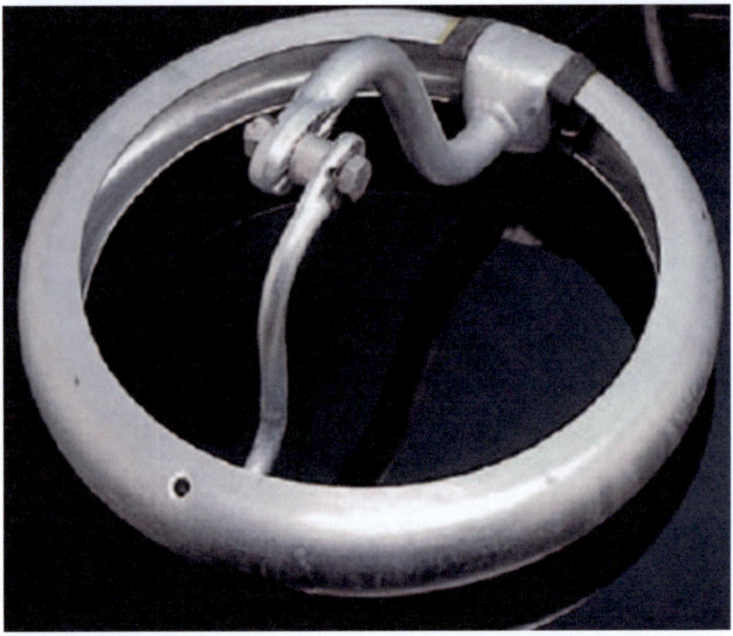

Fig. 4.90 C-ring for combined power arc and corona protection (*Source* PFISTERER)

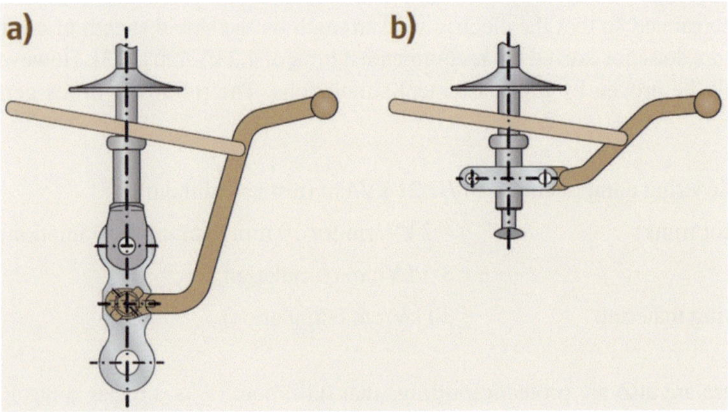

Fig. 4.91 Best practice for positioning an arc protection fitting on a composite insulator end fitting. **a** Configuration for high-current power arc, **b** configuration for low-current power arc (*Source* PFISTERER)

insulator during a specific short-circuit test to simulate this load case. Recommendations for the use of arc protection fittings for composite insulators were published in [114, 115].

4.5.3 Fitting Design

Suspension clamps must be designed to hold 1.35 times the value of the forces acting on the conductor according to [8, Sect. 11.6/DE.1]. Suspension clamps must also reliably hold the conductor under the differential tensile forces according to [8, Sect. 4.12.2/DE.1].

Full-tension joints and *tension clamps* must according to [8, Sect. 11.6/DE.1] hold the conductor with at least 1.55 times the tensile force at maximum load [8, Sect. 9.6.2/DE.1] with 0.85 times 95% of the calculated breaking force (nominal tensile force) according to (4.10). The lower value is always decisive.

String fittings are designed per [8, Sect. 10.7/DE.1] with a *partial safety factor* $\gamma_c = 1.35$ for the load, while for their material strength, the partial safety factors γ_M listed in Table 4.10 must be applied (Sect. 4.4.2).

Table 4.10 Material partial safety factors γ_M [8]

Material	Material partial safety factor γM
Structural steel according to DIN EN 10025	2.4
Cast steel according to DIN 1681	
Malleable cast iron according to DIN EN 1562	3
Cast iron with spheroidal graphite according to DIN EN 1563	3
Al wrought alloy according to DIN EN 573-4	2.4
Al cast alloy according to DIN EN 1706	3.3

4.5.4 Vibration Protection Fittings

Vibration Dampers—Stockbridge Type

Wind-induced vibrations (Sect. 4.3.11) can, if not controlled, lead to wire breakages due to material fatigue and in extreme cases to conductor failures. An early demand of the industry was therefore to develop suitable vibration dampers. The most common damper is the so-called *Stockbridge damper*, named after its inventor George H. Stockbridge (Fig. 4.92), [116].

The Stockbridge damper is a so-called resonance damper (TMD: *Tuned Mass Damper*). It is attached to the conductor by a clamp which is connected with a steel strand, so-called *messenger*, to the two damper weights, also called *dog-bones* because of their shape. When the conductor vibrates, a part of the vibration energy is transferred to the two weights, which move and cause a bending of the messenger wire. Through the friction between the wires of the messenger the vibration energy is converted into heat and thus reduces the vibration amplitude of the conductor.

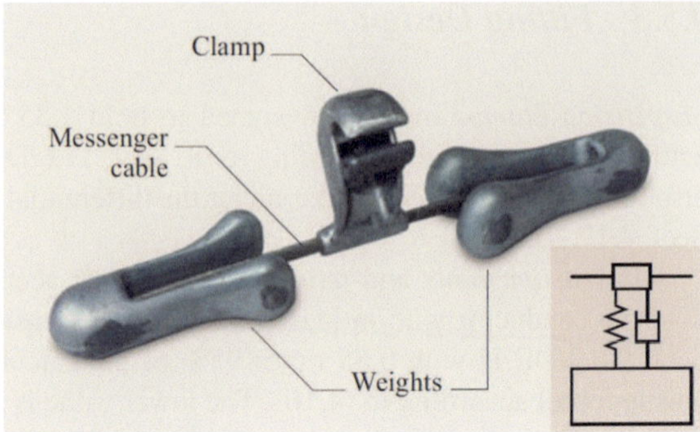

Fig. 4.92 Stockbridge damper (*Source* PFISTERER)

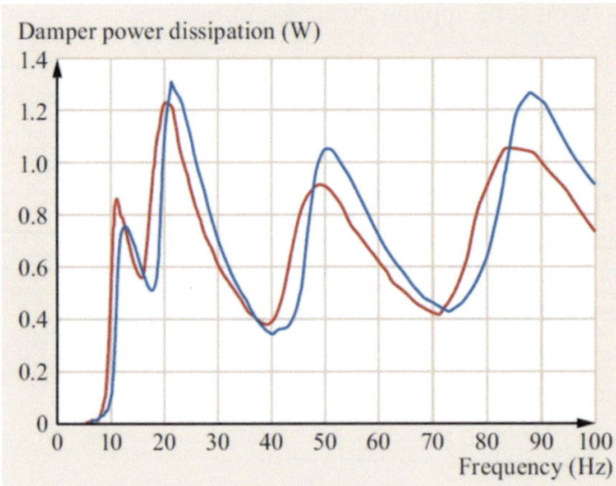

Fig. 4.93 Power dissipation of a Stockbridge damper before (blue) and after (red) the fatigue test (*Source* RIBE)

In Fig. 4.93, the power dissipation over the vibration frequency is shown in blue. In the same diagram, the power dissipation of the same damper after a fatigue test with 10^6 load cycles is plotted in red. The curve shift is within the tolerance specified in the Standard for damper tests [117]. The peaks in the curve correspond to the *eigenfrequencies* of the damper, which is usually designed as a *4R damper* (4R: four resonances). For this reason, the structure of a damper is asymmetric, i.e. its weights are of different sizes and the messenger wires on which they are attached, which are usually 19-wire galvanized steel strands, are not of equal length. As a result, the first and second (lower) eigenfrequency occur at the maximum deflection of the two damper weights, while the third and fourth (upper) eigenfrequencies are caused by the rotation of the weights around their respective center of gravity. This is illustrated in Fig. 4.94, which presents the mechanical impedance of the damper, i.e. the relation between the applied dynamic force and the resulting velocity at the damper attachment point on the conductor.

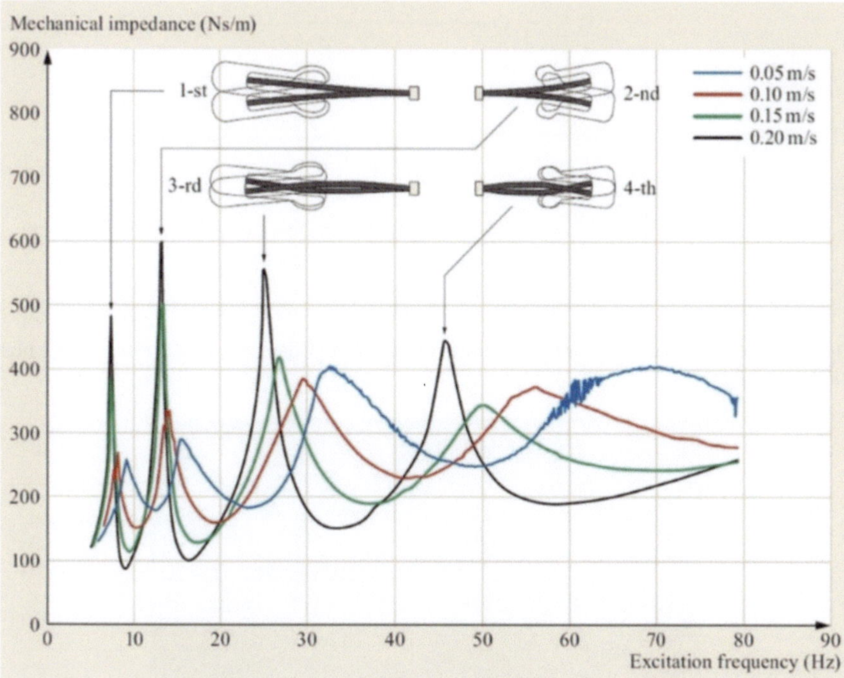

Fig. 4.94 Mechanical impedance over frequency of a four-resonance (4-R) Stockbridge damper at four different clamp velocities

The correct position of a damper on the conductor is often determined by special calculation programs [118]. This is particularly necessary in critical cases, such as long spans with high conductor tension. In standard situations, damper placement is based on experience. It has proven effective if the damper is placed at a distance S_d in front of the clamp, which corresponds to approximately 80% of half the wavelength (*loop*) of the aeolian vibration, (Fig. 4.95b), at the highest wind speed of 7 m/s causing aeolian vibrations (Sect. 4.3.11). The damper is then located approximately in the vibration's maximum amplitude, the so-called *antinode*, at the lowest relevant wind speed of 1 m/s (Fig. 4.95a) and can generate sufficient damping through the bending movement of its weights. A rule of thumb for the distance of the first damper from the clamp is approximately 2 m (Fig. 4.96).

Generally, it is common to attach a damper at both ends of the span. However, if a second damper per suspension point is deemed necessary, for example in long spans over 400 m, it should be located approximately at the same distance (e.g. 2 m) from the first damper towards mid-span.

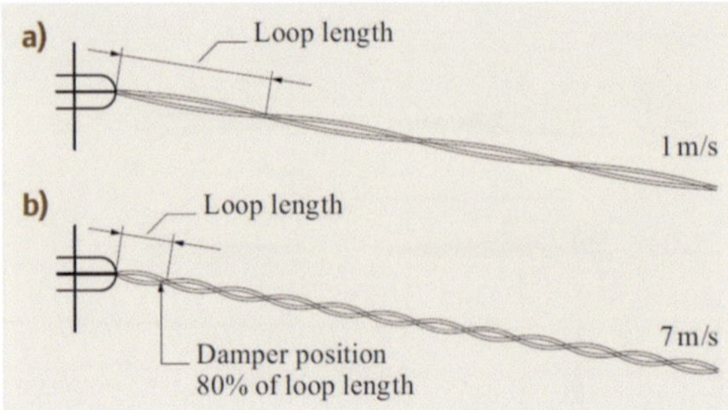

Fig. 4.95 Recommendations for the placement of Stockbridge dampers for **a** low wind and **b** high wind

Fig. 4.96 Stockbridge damper in service; *remark:* the metallic oval weight on the string, so-called *counterweight,* serves to suppress the upward conductor pull at suspension towers when there is a large difference in height between two adjacent towers

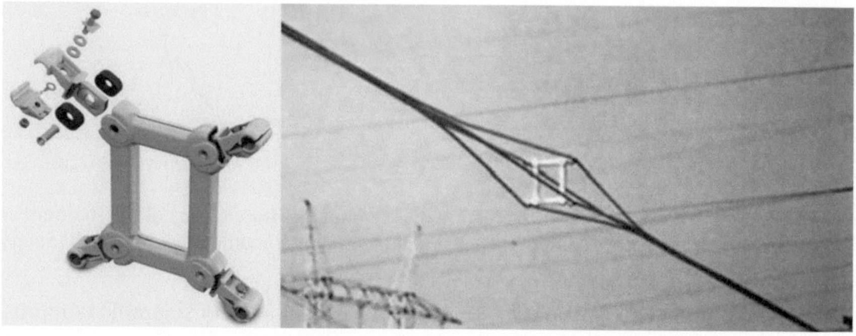

Fig. 4.97 Typical spacer damper for quad bundles (*left*); Sub-conductor clashing during short-circuit test (*right*) (*Source* PFISTERER)

Spacer Dampers

At the beginning of the 20th century, when network voltages began to rise dramatically to allow the transmission of higher power, restrictions due to corona and the wave impedance of single conductors soon came up. The solution to both problems was the introduction of bundle conductors [119–121] (Sect. 4.3.7). This required a new type of line hardware, the spacer. Their main task is to maintain the distance between the sub-conductors, often 400 mm, while preventing the sub-conductors from clashing together due to sub-span oscillations, gusts or short circuits, or the bundle from tipping, for example due to unequal ice loads in the sub-conductors. At a later stage, damping properties were also integrated into the spacers to suppress disturbing aeolian vibrations of the sub-conductors. This led to the development of damping spacers, so-called *spacer dampers*, which are used practically in all bundle configurations except for twin bundles, for which flexible, articulated or semi-rigid spacers combined with Stockbridge dampers are sometimes used as an alternative. An exploded view of a typical spacer for quad bundles is shown in Fig. 4.97 (left part of the picture). The damping is provided by the elastomeric disc elements which are built into the articulated joints and are cyclically loaded and unloaded by the movement of the arms. This creates so-called Coulomb damping in the elastomer discs, which generates frictional energy that is converted into heat energy and counteracts the vibration energy.

In general, spacers and spacer dampers must withstand the mechanical, electrical and environmental stress during installation, operation and maintenance of the line, without failing and without damaging the conductors. In [122], the various types, materials and operational experiences of spacers and spacer dampers are presented in detail.

The dynamic behavior of spacer dampers under wind-induced vibrations is a complex matter and is beyond the scope of this book. Various calculation methods are available, which are described in [118].

A specific load on spacers arises from electromagnetic forces generated by short-circuit currents flowing through the sub-conductors of the bundle. Short-circuit currents cause strong lateral forces between the sub-conductors, leading to large deflections and even collisions of the sub-conductors (Fig. 4.97, right) [123]. The

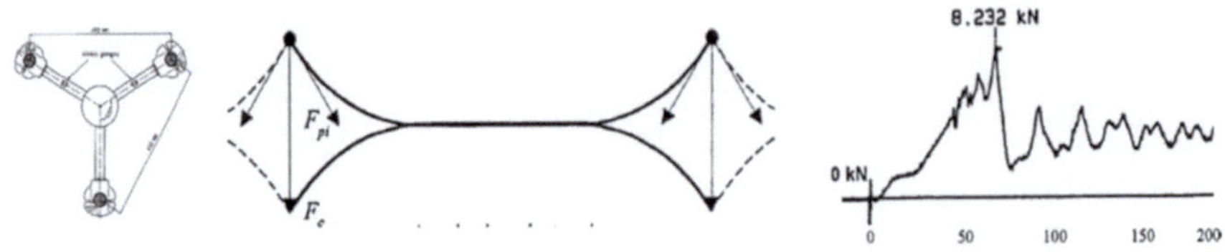

Fig. 4.98 Schematic representation (*left* and middle) and time history (*right*) of the short-circuit force acting on a triple spacer damper under a continuous short-circuit current of 40 kA (*Source* [124])

massive increase in tensile force in the conductors results in considerable compression forces on the arms of the spacer, which can reach up to 15 kN and more with short-circuit currents common today of 50 kA and above (Fig. 4.98), [124]. Their calculation is carried approximately according to the so-called *Manuzio formula* [125], and if necessary, more accurate, according to the method presented in [124].

4.6 Supports

Line supports are certainly the most noticeable components (Fig. 4.99), and, along with the conductors, the most expensive components of an overhead line. The supports of high voltage overhead lines are in their great majority steel lattice towers, although in recent years *aesthetic towers*, also called designer towers, which are often

Fig. 4.99 Example of guyed towers (*Source* U. Cosmai)

Fig. 4.100 From *left*: wooden poles, cast concrete, tubular poles made of steel or concrete, steel lattice tower and composite tower

tubular poles made of concrete or steel, are playing an increasingly important role in improving public acceptance of new lines. Structures made of composite materials are still in the development phase [126]. Wooden poles are used mostly for medium voltage and lines up to 110 kV. Only a few old lines with towers made of cast concrete profiles are still in operation (Fig. 4.100).

4.6.1 *Tower Types*

Line supports are being differentiated according to their function and deflection angle in the line route (Fig. 4.101).

- Suspension tower T: In straight route; tensile forces act on the related suspension strings, they hang vertically
- Light angle tower WT: for line deviations up to 3°; suspension strings inclined

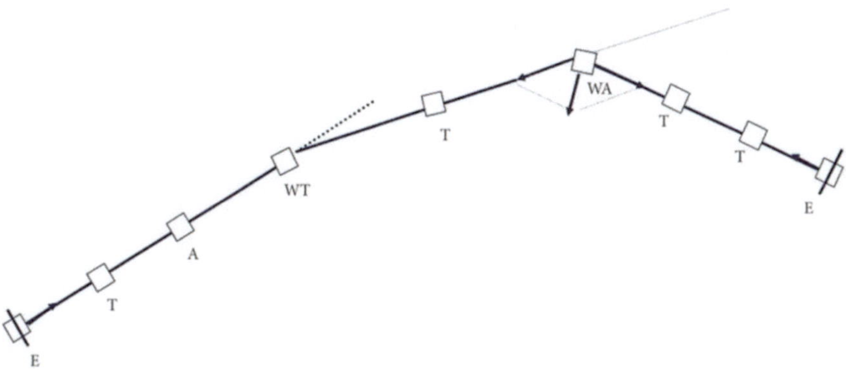

Fig. 4.101 Tower types in a line section, E: terminal tower, T: suspension tower, A: tension tower, WT: light angle tower, WA: heavy angle tower (*Source* Y. Voyatzakis)

- Heavy angle tower WA: Takes the resulting conductor tensile forces at the angle point; equipped with tension strings
- Strain tower A: Designed for differential conductor tensile forces in line direction; equipped with tension strings
- Terminal tower E: Takes the one-sided conductor tensile forces; equipped with tension strings
- Special towers (portals, transition towers from overhead lines to underground cables, transposition towers, tie-off towers).

4.6.2 *Tower Top Geometry*

Overhead line towers must be designed so that clearances between the conductors in the field and between grounded parts on the tower are sufficiently safe under climatic and electrical related loads. The necessary distance between the conductors depends on the sag and thus according to [31] on the span, while the distance to grounded parts in the tower is determined by the type and arrangement of the insulators and the wind (Fig. 4.102, Table 4.11). As stated in [6, Sect. 5.6.5, DE.1.2], the applicable reduction factor K_1 is 0.75.

The maximum conductor sag at the highest permissible conductor temperature or under additional loads from ice and snow, determines the height of the tower.

Table 4.11 Safety clearances at the tower [2]

Nominal voltage (kV)	Max. operating voltage (kV)	Del	Dpp (m)	K1 × Del (m)	K1 × Dpp
110	123	1.00	1.15	0.75	0.86
220	242	1.70	2.00	1.28	1.50
380	420	2.80	3.20	2.10	2.40

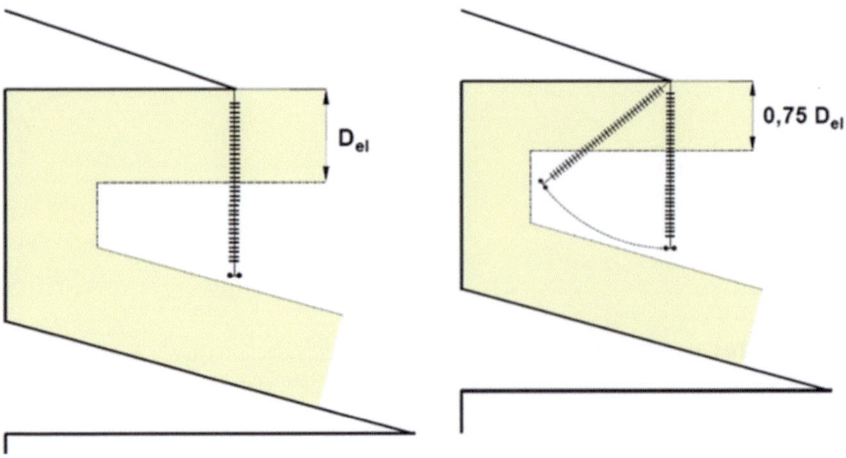

Fig. 4.102 Electrical safety clearances at the tower top

4.6.3 Design

The design of a steel lattice tower is carried out in the following steps:

- Geometrical layout and configuration
- Determination of loads
- Calculation of member forces
- Structural strength verification
- Cross-section verification
- Stability analysis
- Verification of connecting elements.

4.6.4 Loads

External Loads

The external loads acting on a line support are:

- Permanent loads: Self-weight of conductors and insulators
- Meteorological loads: Wind and ice loads (typical values)
  - max. wind at +5 °C
  - max. ice at −5 °C
  - Combination max. ice −5 °C and 50% wind
- Assembling loads
  - Assembly method/stringing
  - Anchors during assembly
  - Lifting devices
- Loads for operational safety
  - One-sided reduced conductor tensions (torsion)
  - Different ice loads on adjacent spans cause differential tension in line direction
  - Risk of tower cascade faults
- Special loads
  - Short circuit
  - Avalanches.

Load Cases

The next step is to summarize the loads into load cases, which consider probable scenarios of load combinations. It is common and helpful to represent the forces for each load case schematically in the form of so-called *load- trees* (Fig. 4.103).

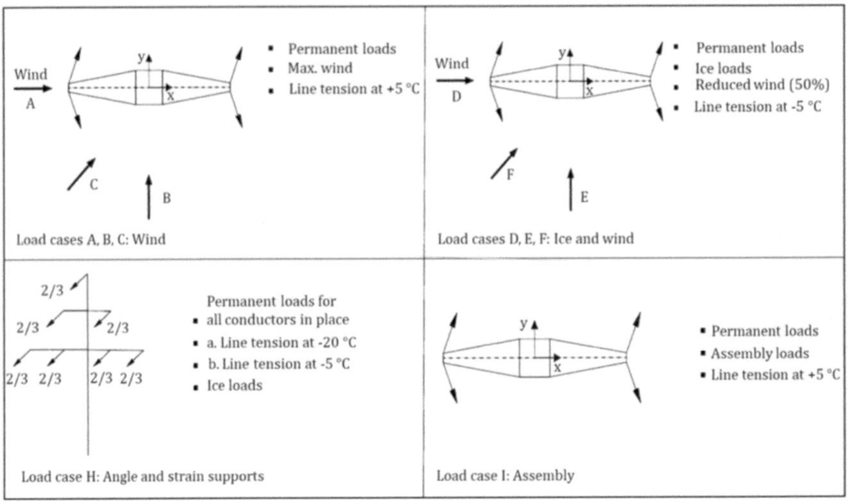

Fig. 4.103 Load trees for different load cases

4.6.5 *Steel Lattice Towers*

Steel lattice towers are generally used from 220 kV upwards, as they allow for economic implementation of large span lengths. In addition, when properly maintained, they offer a long service life. Due to their modular design, they can be easily transported from the production facility to the construction site. This modular concept also proves advantageous when upgrades or repairs to individual tower components are needed, and it facilitates straightforward adaptation to the terrain—for example, by extending the main legs.

Over the years, various types of towers have been developed and used (Sect. 2.1.1), with guyed towers often used mainly outside Europe (Fig. 4.99).

The designation of the various components of a steel lattice tower—in this case it is a so-called *Danube tower* design—are shown in Fig. 4.104.

Materials and Design of Steel Lattice Towers

In [127] there is a good overview of the industry practice for materials, manufacturing and design of steel lattice towers. For their members, angle profiles made of high-strength steel of the quality class S355 (old designation St 52) and regular steel of the quality class S235 (St 37) are used [128, 129]. They come in sizes from 40 × 3 mm to 250 × 28 mm and are connected with bolts according to [130].

To ensure a long service life, it is important that they have sufficient protection against corrosion. Presently the best corrosion protection is achieved by the so-called *In-Factory-Duplex-System*, i.e. by the additional coating of the steel profiles immediately after their galvanizing in a coating plant.

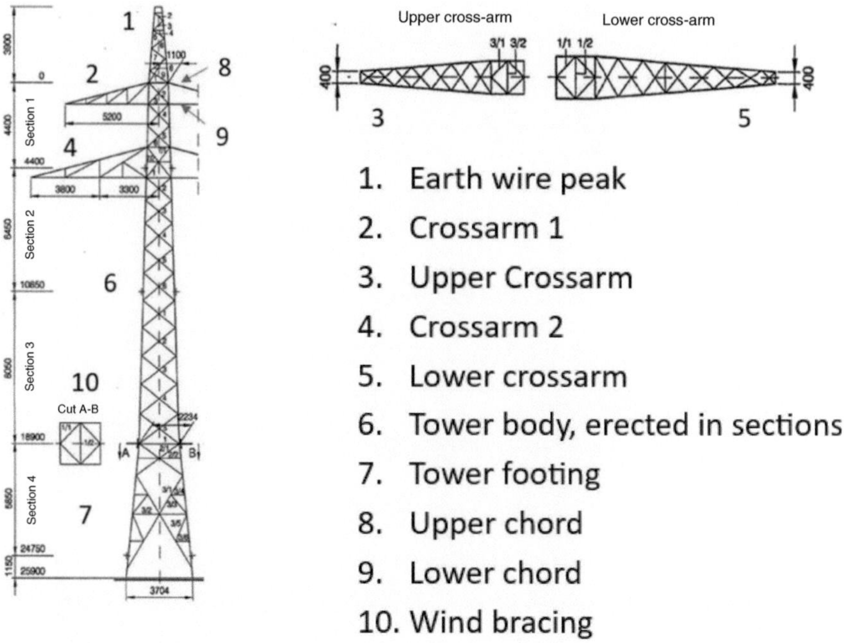

Fig. 4.104 Designation of a steel lattice tower (*Source* Omexom)

This process has the following advantages over an on-site coating of an already erected tower:

- Independent of the weather (temperature, humidity, etc.)
- Independent of the season (can be done all year round)
- Uniform layer thickness
- Uniform quality (can be checked at any time and easily)
- If necessary, it can be coated 24 h a day
- Minimal environmental impact
- No additional outages of the line for applying the coating
- Significant increase in the service life of the coating and thus reduction of the effort for subsequent maintenance coatings.

In addition to providing corrosion protection, coatings may be applied to visually blend towers into their surroundings (Chap. 7), or, in contrast, to enhance their visibility for aircraft warning (Fig. 4.105).

One quite curious example is a pink-colored tower in Great Britain that plays a prominent role in a film. Its large fan base strongly opposed any attempt to repaint it (Fig. 4.106).

Other rarities are towers with rectangular or triangular cross-sections of the tower bodies (Fig. 4.107).

Fig. 4.105 380 kV suspension tower with red–white–red coating for aircraft warning

Guyed (Anchored) Towers

Guyed towers are advantageous due to their structural simplicity, ease of transport—including aerial delivery by helicopter—and uncomplicated assembly. They can also be faster replaced in case of outages. Disadvantages compared to self-supporting towers are the large area required to place the guy wires, their large width, and that they can normally carry only one circuit. Guyed towers are mainly suspension towers, tension towers in such a line are usually designed as self-supporting lattice structures (Figs. 4.108 and 4.109).

Brazil stretches over 4000 km from its northern to its southern borders, spanning tropical and subtropical regions as well as savanna landscapes. Even for very long

Fig. 4.106 The pink tower has a large fan base

Fig. 4.107 *Left*: towers with rectangular body; *right*: triangular tower body [131]

Fig. 4.108 Guyed towers. *Left*: iceland, *right*: the line corridor in Canada also functions as a fire barrier, and is therefore wider than would be electrically necessary

Fig. 4.109 Comparison between self-supporting and guyed towers in South Africa, with a rough estimation of the relative construction costs (*Source* CIGRE Capetown Symposium 2015)

lines that cross these diverse climatic zones, guyed towers are used due to the advantages outlined above. An example is shown in Fig. 4.110. It is a 653 km long 500 kV AC line with a transmission capacity of 1670 MW, which went into operation in 2016. The suspension tower bodies with heights of 38m have an articulated connection with the foundation and are stabilized by guy wires. The tension towers are designed as self-supporting steel lattice towers. A special feature of this line is its use of so-called *extended conductor bundles*, [132], consisting of six sub-conductors, each 27 mm in diameter—rather than the standard quad bundles of 32 mm conductors—with a bundle diameter of up to 2.5 m. This configuration reduces the surge impedance and thus increases the power transfer capability (SIL) of the line—an advantage that is especially significant for long transmission lines. Another notable aspect is the variation in bundle dimensions, the middle bundle has smaller dimensions.

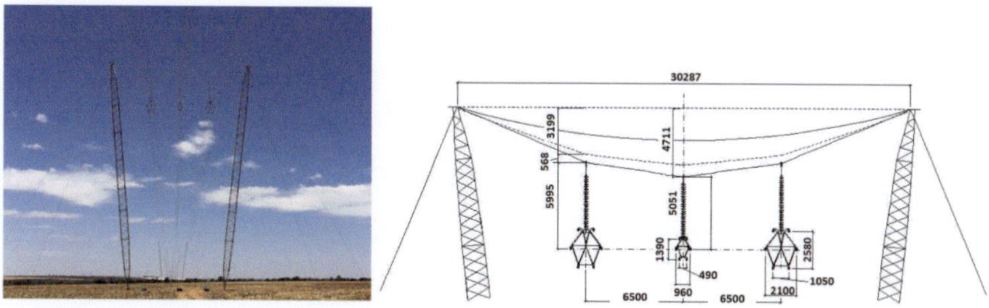

Fig. 4.110 500 kV line in Brazil with extended conductor bundles (*Source* CIGRE Session Paris 2024, Paper 10766)

Fig. 4.111 DC Bipole lines, *left* and middle with guyed towers, *right* with self-supporting towers

Guyed towers are also used for DC lines. Figure 4.111 shows a comparison.

Calculation of Member Forces in Steel Lattice Towers

Overhead line towers are statically determinate frameworks. In the early days of transmission line design, engineers often relied on manual methods such as the *Cremona diagram* or *Ritter's method* to determine member forces. With the advent of computer technology, modern analysis increasingly employs numerical approaches—typically using the *Finite Element Method* (FEM)—based on linear-elastic, three-dimensional framework models. For line supports with greater deformation potential, such as tension towers, nonlinear analysis methods are required. A comprehensive overview of available software tools can be found in [133].

Nevertheless, it is instructive to demonstrate the analytical calculation of member forces using a simplified example. For this purpose, we consider a steel lattice tower with single diagonals (Fig. 4.112), assuming that external loads are symmetrically distributed across the two opposing tower faces.

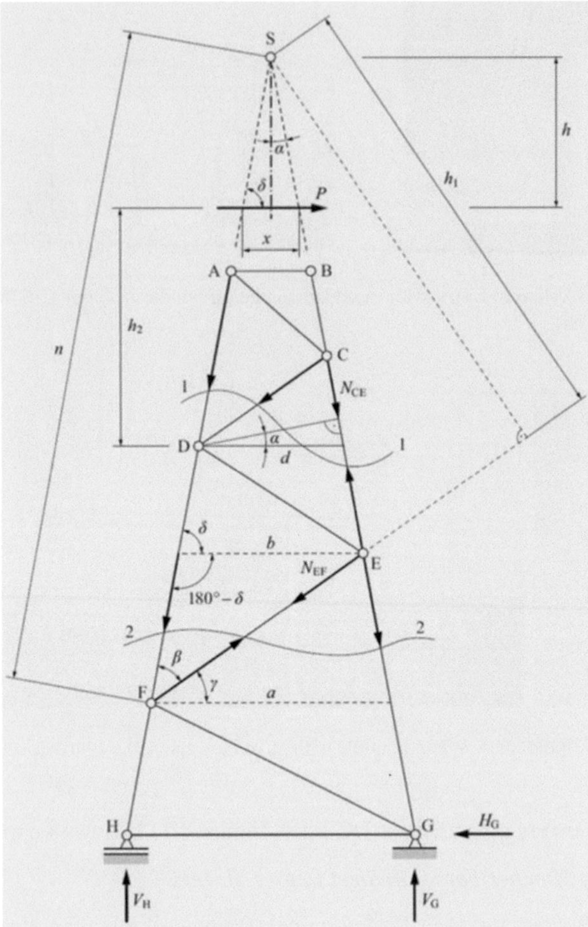

Fig. 4.112 Equivalent static system, geometry, loads and member forces of a tower face

We perform two cuts according to *Ritter*, whereby these must only separate three members each, so that with the conditions of equilibrium of plane statics, the three cut-free member forces can be determined. From the equilibrium of moments around the point D at cut 1–1 we get the main leg's force N_{CE} to:

$$N_{CE} = \frac{P\ h_2}{2\ d\ \cos\ \alpha} \tag{4.54}$$

Similarly, we can determine the diagonal force N_{EF} with cut 2–2 and the moment of equilibrium around the point S to:

$$N_{EF} = \frac{P\,x}{2\,a\,b\,l_{EF}} \tag{4.55}$$

Here, l_{EF} is the length of the relevant diagonal:

$$l_{EF} = b\,\frac{\sin\ \delta}{\sin\ \beta} \tag{4.56}$$

Factor 2 in the denominator considers the above-mentioned assumption that the members in the "parallel" tower faces share the external loads equally.

This process can be repeated with all external loads, be they horizontal (wind loads) or vertical (conductor weight and ice loads) and by simple addition (*superposition principle*) the respective total forces in the tower members are determined.

With the same considerations, the member forces in the tower crossarms, the cross bracings, the main leg bracings, as well as torsional loads, even with crossed diagonals in the tower faces, can be determined [7, p. 346 ff].

The deflection of the tower peak caused by conductor tension is a critical design consideration and must be kept within the specified limits; otherwise, the resulting deformation may negatively affect the tower's visual appearance.

Strength Calculation of Steel Lattice Towers

The following failure modes can occur not only in steel lattice towers but also in towers made of other materials:

Stability failure: component buckles laterally under compressive load.

- Flexural buckling
- Lateral–torsional buckling
- Torsional buckling.

Failure of the net cross-section: member fractures due to tensile overload.
Tensile strength of the member is exceeded
Connection failure: bolted connection deforms or shears off.

- Bearing failure
- Shear failure.

The latter are shown schematically in Fig. 4.113. In contrast, Fig. 4.114 presents a real-world example of a *bearing failure*: in this case, the material of a diagonal around the bolt—rather than the bolt itself—tore due to excessive stress. On the other hand, Fig. 4.115 illustrates a *stability failure* of tower diagonals; some members buckled about their weak axis, while others buckled out of the tower plane.

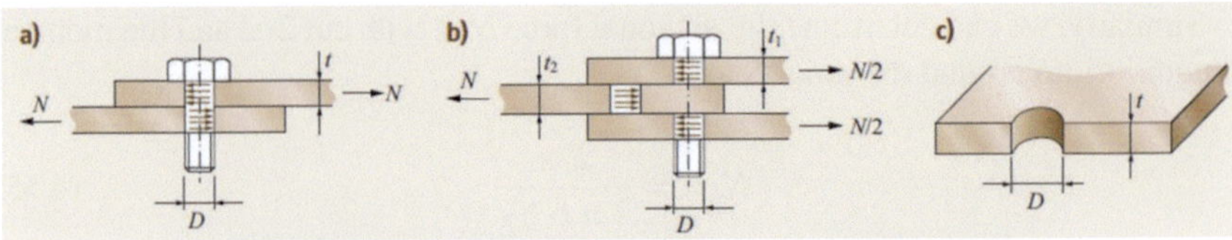

Fig. 4.113 Schematic of single (**a**) or double shear (**b**) and bearing area on bolted connections (**c**)

Fig. 4.114 Stability failure of the diagonals of a steel lattice tower under compressive stress (*Source* Omexom)

Fig. 4.115 Bearing failure around the bolt; the bolt itself is still intact (*Source* Omexom)

4.6.6 Tubular Towers

This tower variant, also called *pole*, is preferred for alternative tower designs, (Sect. 4.4.5 *Applications in compact lines*). Due to their narrow body, the use of insulating cross-arms, predominantly as rotating crossarms at suspension towers, is preferred.

Steel Poles

In general, round and polygonal cross-sections are used for tubular poles. Due to transportation and manufacturing constraints, section lengths may be limited, and depending on the voltage level and pole size, segments are connected either by overlapping or bolted joints. The profiles are formed using hydraulic presses, and the individual sections are welded together. Crossarms—sometimes in varying designs—are manufactured using the same principles. A variety of configurations are in use, including tubular poles with steel lattice crossarms, as well as crossarms made from solid or hollow composite elements (Fig. 4.63).

Tubular poles (Fig. 4.116) are also used in regions with difficult access. Iceland is a notable example. Overhead lines there are exposed to extreme environmental conditions influenced by volcanic activity, including glacial meltwater floods, earthquakes, coastal erosion, high winds, ice, saltwater pollution, and avalanches in the eastern and northwestern parts of the country. Wet snow accretion can reach up to 20 cm in diameter. The first overhead lines were constructed in Iceland in 1906. Today, dozens of ice monitoring stations are distributed across the country. In addition, test facilities have been recording sagging behavior due to icing, temperature, and humidity since 1972. As a result, transmission operators maintain extensive climate and icing databases, giving line designers access to highly reliable environmental data.

Fig. 4.116 Iceland, 400 kV steel poles crossing an avalanche area (CIGRE Colloquium Reykjavik 2011)

Avalanches are a major design consideration and played a significant role in the planning of a recent 400 kV line project in Iceland. The new line replaces an aging 220 kV line and crosses an avalanche-prone area with 85 towers. To minimize the visual impact, tubular steel poles were selected for the main structures, eliminating the need for additional avalanche deflectors. Predominantly suspension towers were used, with load profiles tailored to tower heights above ground. The design loads were extreme: wind speeds during snowfall ranging from 5 to 50 m/s, snow density of 300 kg/m^3, and additional dynamic loads from debris carried by avalanches. In total, the line includes 70 suspension towers, 13 angle tension towers, and 2 terminal towers, with tower heights ranging from 24 to 40 m, weights between 27 and 65 tons, and base diameters from 1.4 to 2.7 m. The poles are flanged constructions and support a single circuit in a horizontal conductor configuration—necessitating two parallel lines to carry both circuits. The structures are duplex-coated for corrosion protection and equipped with avalanche monitoring sensors at various heights.

Concrete Poles

Historically, on-site cast concrete poles and spun concrete poles were used for overhead line structures (Fig. 4.68). Today, concrete poles in the high voltage range are typically manufactured using reinforced, high-strength, compacted spun concrete (Fig. 4.64). In this production method, the pole is formed in a horizontal centrifuge, where centrifugal forces compact the concrete to increase its density and strength.

Single lengths exceeding 30 m can be produced. For longer structures, segments are connected either by an internal steel tube (slip joint) or by flanged joints with bolted flange plates. Internal bolted flange connections require sufficiently large tube diameters; otherwise, external bolting systems are used. Not all concrete poles are made entirely of concrete—in taller pole designs, the lower portion may consist of concrete, with steel tubes attached or flanged to the top section. This hybrid approach enables pole heights of up to 100 m.

Concrete poles are commonly equipped with concrete crossarms, although crossarms made of steel lattice or steel tubular elements are also in use. While environmental color coatings for concrete towers do exist, they remain relatively uncommon.

Lattice Towers with Steel Tubes

Particularly tall or structurally demanding towers often require specialized designs. In such cases, tubular profiles are frequently used instead of the traditional angle profiles commonly found in steel lattice towers. This approach is typically applied in transmission lines operating at the highest voltage levels, such as 1000 kV and above. Figure 4.117 shows the construction site of a 1000 kV AC transmission line in Japan, featuring towers constructed with tubular steel profiles.

Fig. 4.117 Assembly of a steel lattice tower with tubular profiles for a 1000 kV AC overhead line in Japan

Tubular profiles can also be an effective solution for 380 kV towers, especially in unconventional designs. The tubular crossarms are fitted with walkways and handrails to enable safe maintenance access. Figure 4.118 shows such a tower prior to conductor installation.

In Chile, the 2487 m long crossing of the waterway "Chile Chacao Channel Crossing" between the mainland and an island, a double-circuit AC 500 kV overhead line with triple AACSR bundle conductors is planned. The height of the shipping passage is 60 m, the conductor sag is 242m. For the design, a return period of the climatic loads of 500 years was considered, which is the highest value mentioned in IEC 80826. Also dynamic loads from earthquakes are considered, a constant wind speed of 250 km/h is overlaid with downbursts from thunderstorms—so-called downdrafts—resulting in calculated windspeeds of 360 km/h (including a partial safety factor for loads of 1.4, see Table 4.1). To be able to withstand these loads, the towers of the crossing are built with tubular profiles. The tubular bracings have a diameter of 1.1 m with a wall thickness of 30 mm [134].

Fig. 4.118 Triangular 380 kV tension tower with main legs and bracings made of tubular profiles in the Netherlands near Wateringen

4.6.7 Composite Towers

Such towers are made of *fiber-reinforced polymer composites* (FRP). This material has been replacing traditional materials for many years in aviation, ships and vehicles, in construction, sports equipment, in military areas, and—in this context—in composite insulators (Sect. 4.4.5). Fibers made of glass, carbon, aramid (Kevlar) and basalt are used. Examples of composite towers are also covered in Chap. 7 while details about their production and installation are extensively described in [126]. Composite towers are used as line supports and their crossarms—mainly for lower voltage levels—and also in outdoor substations. For medium voltage lines, towers with steel bodies and composite crossarms are also used.

Overhead lines are long-lasting structures that stand in public spaces for many decades—often 80 to 100 years or more. They are exposed to a range of environmental influences, such as wind and ice loads, temperature fluctuations, UV radiation, and must reliably maintain their properties throughout this time. As a result, utilities are understandably cautious about adopting new materials. For this reason, it was necessary to thoroughly investigate and clarify the following issues for composite towers:

- Reasons for using FRP
- Advantages and disadvantages
- Experiences, challenges
- Applicable for which voltage levels
- Temperature range for application
- Lifespan
- Design parameters (climatic loads, conductors, etc.)
- Material properties, manufacturing methods

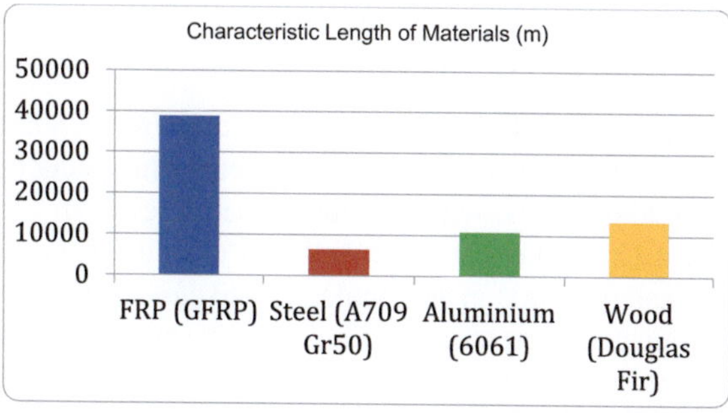

Fig. 4.119 Comparison of the characteristic length in m (ordinate) of different materials (abscissa) used in tower construction

- Foundations, grounding systems
- Coloring (warning coatings red-white-red, camouflage paint in green), maintenance coatings
- Durability, UV resistance, resistance to electric and magnetic fields
- Operation, maintenance, and safety systems for linemen
- Vandalism, damage, repair methods
- Dismantling and recycling, environmental relevance, life cycle assessment
- Cost analysis depending on the voltage, number of systems, conductor types, width of line corridor (ROW)
- Availability of product standards
- Comparison with other materials (wood, steel, concrete).

The decisions have been made in favor of composite poles in many cases. The light and easy transport of the parts of the poles is a major advantage of these line components, especially in difficult or inaccessible areas. Figure 4.119 shows the comparison of the strength (in MPa) to the density (in kg/m^3), the so-called *characteristic length* (in m), of GRP with other materials used for line supports.

Unlike wooden poles, they do not require impregnation against moisture, weathering, infestation by insect (termites) and the pole is not a food supplier for beetle-seeking animals (woodpecker). In some regions of the world, wooden poles are being cut, stolen and used as firewood. This is also eliminated with composite poles and is also an argument in favor of concrete or steel poles (Fig. 4.120).

Fig. 4.120 *Left*: the woodpecker, a nice bird, but dangerous for wooden poles—no problem with composite poles; *right*: easy transport of a part of a medium voltage composite pole [126]

4.6.8 Tower Tests

Overhead line towers, especially with new designs, are often validated by so-called *prototype tests* [135]. The tower to be tested is set up in specially equipped test stations and loaded with pullers, which simulate the forces from conductors acting on the tower (Fig. 4.121). The test is considered "passed" if the maximum loads (100%) are maintained for 5 min in all simulated load cases without any failure. It is helpful to attach strain gauges at critical points of the tower to record the progression and magnitude of the member stresses. It is also common that the tower is further loaded until failure (Fig. 4.122). IEC Standard 60652 forms the basis for carrying out such tests [136].

Fig. 4.121 Steel lattice tower at the test site with pulleys simulating conductor loads. The detailed view shows the measurement setup for recording the displacement of the main leg

Fig. 4.122 Failure of a suspension tower on a 220 kV double-circuit line during testing

4.6.9 Compact Lines

As already mentioned, in recent years considerable public opposition has formed against both, new and existing overhead lines, and underground cables have been often proposed as an alternative. Concerns frequently relate to the alleged health effects of electric and magnetic fields (EMC: *Electromagnetic Compatibility*) in the vicinity of the lines (Sect. 6.1.8), as well as the visual intrusion of overhead lines in the landscape. It is therefore not surprising that utilities have addressed these issues intensively, leading to the development of more visually appealing compact line designs [137]. These designs typically require narrower *rights-of-way* (ROW), mostly can carry higher levels of power, and reduce magnetic field exposure at ground level due to the closer spacing of the phase conductors compared to traditional tower configurations. Figure 4.123 shows two examples of how to reduce the visual impact of overhead lines [138, 139]. The growing acceptance of composite insulators has been a key factor in this development, as explained in Sect. 4.4.4.

Fig. 4.123 Examples of compact lines

Fig. 4.124 Visual impact of a conventional 380 kV line and a compact line in Dubai (*left*), and the "Wintrack" line in the Netherlands carrying two 380 kV and two 132 kV circuits (*right*), the "kissing" configuration of the two 380 kV circuits is used to reduce magnetic fields at ground level (*Source* PFISTERER)

Essential planning parameters in the design of compact lines include, among others, the objectives illustrated in Fig. 4.123:

- *Left*, Visual Impact: A compact 500 kV AC line using tubular poles and insulating crossarms [66]
- *Middle*, Right-of-Way: A compact DC angle suspension tower with V-string configuration, implemented in Finland
- *Right*, Internal Clearances: Reduced conductor spacing enabled by using interphase spacers [66] (Fig. 4.124).

The double-circuit 380 kV line Redipuglia–Udine Ovest operated by Terna in Italy also uses tubular poles. Figure 4.125 shows these poles in comparison with a single-circuit 380 kV line using a horizontal "Lyra" tower configuration. Both lines are equipped with triple conductor bundles per phase.

Fig. 4.125 Single-circuit 380 kV line with steel lattice towers (*left*) and double-circuit 380 kV line with tubular poles (*right*), located in northern Italy near Udine

4.6.10 Lightning Protection

Lightning strikes are the primary cause of most unplanned outages. International statistics indicate that approximately 65% of such outages are due to lightning strikes on overhead lines [140]. To protect the conductors, high-voltage overhead lines are generally equipped with ground wires, which shield them from direct lightning strikes that can cause significant damage. These ground wires are typically positioned with a shielding angle (θ) ranging from approximately 10° to a maximum of 35°, as illustrated in Fig. 4.126 [141].

Fig. 4.126 Earth wire with protected area on a "Danube" tower [7]

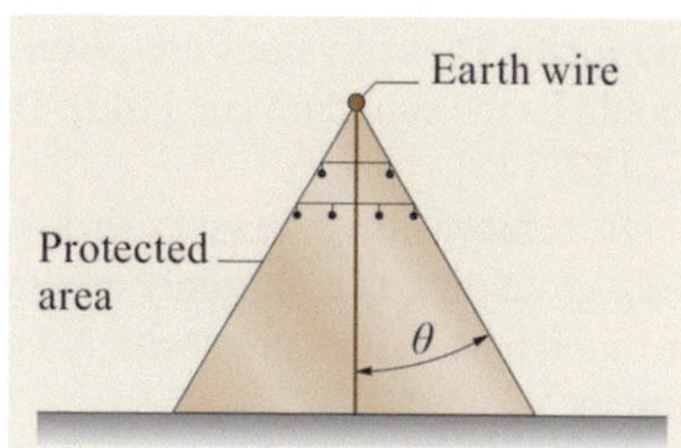

4.6.11 Earthing

Effective lightning protection depends on the reliability of the grounding systems for overhead line structures. These earthing systems must fulfill the following requirements:

- Ensure the safety of people and animals (e.g., grazing livestock) exposed to voltages that occur during earth faults.
- Prevent damage to third-party property and facilities.
- Maintain the reliability and availability of the line.
- Be adequately designed for the highest expected earth-fault currents.
- Possess sufficient mechanical strength and corrosion resistance (e.g., in the grounding grid).

The first requirement—ensuring the safety of people and animals—is of primary importance. The measures required to protect people and animals near earthed towers from electric shock caused by power frequency faults are summarized in [142]. These measures also apply to faults caused by lightning strikes and the associated overvoltages.

For overhead lines, it is common to connect the ground wires at each tower as well as the tower itself to the ground. Structures made of conductive materials, such as steel, are connected to the ground through the tower foundations. However, additional, sometimes complex grounding measures are often necessary to achieve the required properties of the earthing, e.g. for towers at locations with low soil conductivity, like in rocky, mountainous areas.

When a lightning strike is discharged to the ground via the tower, an overvoltage develops across the tower due to its surge impedance, which is influenced by the tower's ground resistance. If this overvoltage, combined with the power-frequency voltage of the conductors, exceeds the insulation strength of the insulators, a *back-flashover* from the tower to the conductors can occur.

The *ground resistance* of the tower is therefore an important parameter and is normally limited to values of max. 10 Ω. However, this strongly depends on the *soil resistivity* at the prevailing soil conditions. Cohesive soils (clay, marl, loam and silt) have a significantly lower soil resistivity than non-cohesive soils (sand, gravel, stones and rock). Typical values are given in [2]. For more accurate results, soil resistivity should be measured on site [142]. Details on grounding calculations can be found in [2, 142, 143].

To achieve the necessary ground resistance in high-resistivity soil conditions, deep grounding is commonly applied. This approach primarily uses copper profiles or round bars, which are securely connected to the main tower legs and driven deep into the ground.

Many transmission line operators commonly use band grounding systems made of copper or galvanized steel, also wires or strips, or copper-clad steel conductors. These are typically installed radially or in a ring configuration in trenches approximately

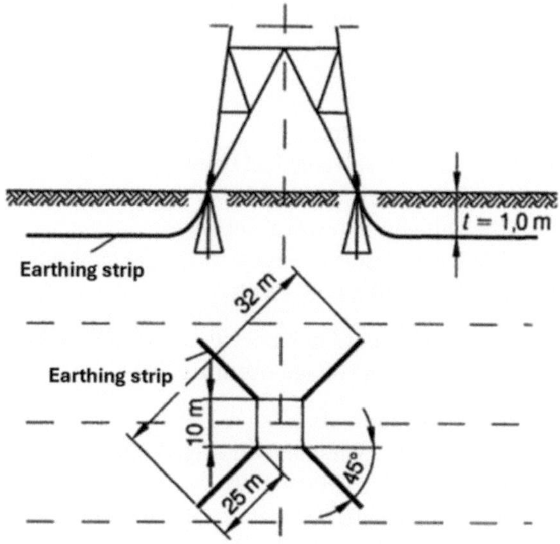

Fig. 4.127 Tower grounding with earthing strips [7]

1 m deep and are connected directly to the tower (Fig. 4.127). If this method alone does not achieve the required low ground resistance, it can be supplemented with deep grounding techniques, combining both approaches to improve overall performance.

4.6.12 Foundations

Overhead line tower foundations are typically made of concrete and come in various forms. Their primary structural function is to support the tower and its associated loads—including the self-weight of the tower, the weight of conductors, and environmental loads such as wind and ice—and to safely transfer these forces into the ground without causing settlement. As such, foundations are critical to the overall reliability and safety of the transmission line.

Foundation design poses significant challenges, largely due to its strong dependence on local soil conditions, which can be highly variable and are often difficult to characterize. For this reason, detailed geotechnical investigations at each planned tower location are essential for ensuring sound structural performance [144, 145].

For foundation design, soils are generally classified as either cohesive or non-cohesive. Cohesive soils—such as clay, marl, loam, and silt—have low load-bearing capacity and can become soft or even liquid when exposed to water. Non-cohesive soils—such as sand, gravel, stones, and rock—do not lose strength when wet and generally provide a solid, load-bearing ground.

According to [146], the subsoil is divided into the following *soil classes* (SC in the following):

- SC 1: Topsoil
- SC 2: Fluid soil types
- SC 3: Easily resolvable soil types
- SC 4: Moderately resolvable soil types
- SC 5: Difficult to resolve soil types
- SC 6: Easily resolvable rock and comparable soil types
- SC 7: Rock difficult to resolve.

These significantly influence the type of foundation. The following foundations are typical for high voltage overhead lines:

- Block foundation
- Pad and chimney foundation (Fig. 4.128)
- Shaft foundation (Fig. 4.129)
- Drilled pile (shaft) foundation, also micro-piles (Fig. 4.130)
- Rammed pile foundation (Fig. 4.131)
- Raft (Plate) foundation (Fig. 4.132)
- Anchor foundation, also micro-anchors (Fig. 4.133)
- Grillage foundation.

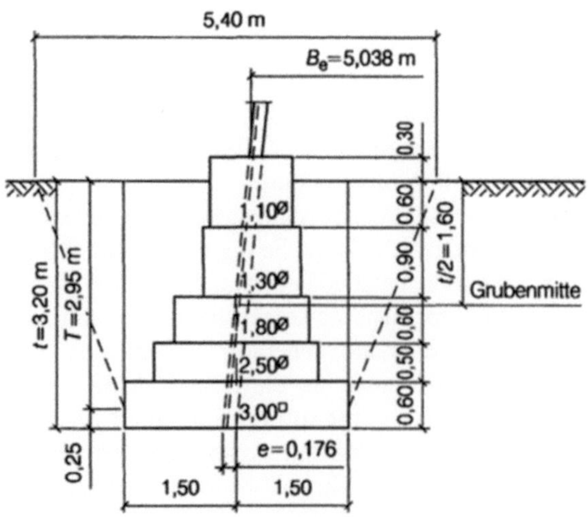

Fig. 4.128 Pad and chimney foundation with typical dimensions [7]

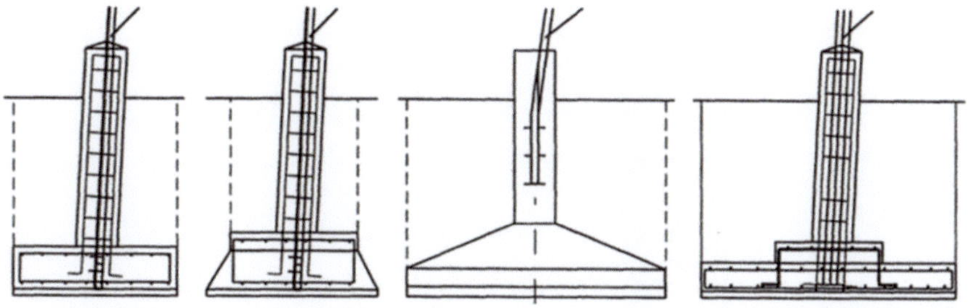

Fig. 4.129 Various designs of shaft foundations [7]

Fig. 4.130 Drilled pile foundation (*Sources* [7], Omexom)

Table 4.12 Forces acting on the legs of steel lattice towers [147]

Tower type	Voltage (kV)	Uplift (kN)	Compression (kN)
Suspension	110	10–25	12–30
	220	20–40	25–50
	380	30–150	40–180
Tension	110	30–60	35–80
	220	40–100	80–120
	380	80–200	100–350
Special (e.g., river crossing)		≤400	≤450

Table 4.12 provides an indication of the magnitude of the forces per main leg on the foundation for various voltage levels and tower types [147].

These relative high forces must be safely transferred from the main legs of the tower into the foundation—and from there into the surrounding soil. This is done with additional angle cleats, which increase the cohesion between steel and concrete (Fig. 4.134).

Fig. 4.131 Installation of a rammed pile foundation; the rammed piles are lying on the meadow (*Source* Omexom)

Fig. 4.132 Raft foundation during concrete pouring (*left*), after completion (*right*) (*Source* Omexom)

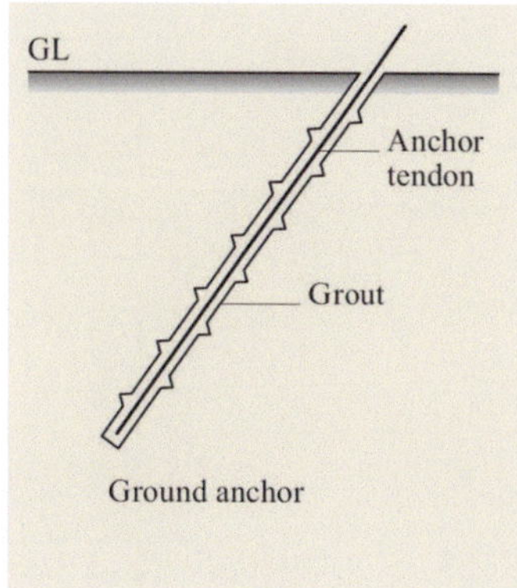

Fig. 4.133 Rock anchor [144]

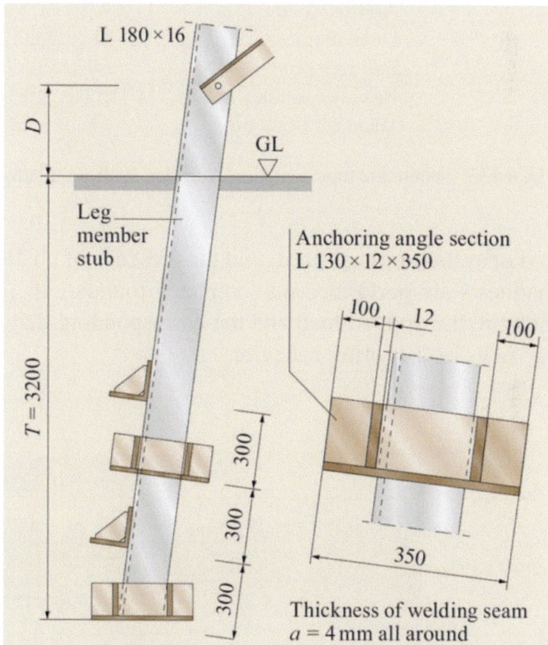

Fig. 4.134 Anchoring of the main legs with cleats in the concrete foundation (*Source* [7])

The design and calculation of tower foundations will only be briefly outlined here. In Germany, the semi-empirical method defined in DIN EN 50341-1 (VDE 0210-1):2013-11, Annex M.3, is applied. For tensile load verification, the static earth load method with an earth load angle—depending on the soil—is used. In the case of shaft foundations, the tensile force of the tower leg (V_u) is resisted by the weight of the supporting earth cone (P_1), while in the case of drilled foundations, it is resisted primarily by the frictional resistance (T) along the shaft (Fig. 4.135). In both cases, the self-weight (P) of the foundation can be considered. For compressive load verification, the permissible soil pressure at the foundation base must be observed.

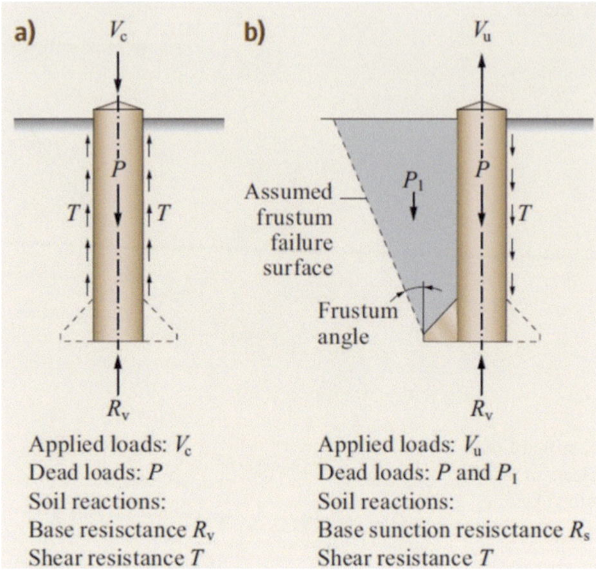

Fig. 4.135 Schematic representation of drilled shaft foundations: **a** compression, **b** uplift

For system-critical lines or in the presence of challenging soil conditions, on-site load tests are performed on completed foundations (Fig. 4.136) [144]. These tests measure the applied load and the corresponding displacement of the foundation to verify its load-bearing behavior.

Fig. 4.136 On-site testing of a rock anchor foundation (*Source* [144])

4.7 Line Construction

The construction of a high-voltage overhead line is a highly complex endeavor, as the site usually stretches across many kilometers. This demands advanced logistical planning as well as significant geotechnical expertise, since soil conditions—and thus the foundation requirements—can vary greatly along the route. In addition, the process of conductor installation, or "stringing," involves their careful handling, as they can be easily damaged, making it a highly challenging technical task.

High standards of quality must be met throughout, especially since network operators design overhead lines for service lives of over 50 years—and many lines have remained operational for nearly a century. Further complications arise from difficult access to tower sites and the fact that work is performed simultaneously across multiple, widely dispersed locations, often by different assembly teams.

4.7.1 Surveying and Tower Spotting

On-site work begins by transferring the selected route from the planning documents to the actual terrain. This is followed by the survey of the longitudinal profile, whereby the terrain and all objects within the *right of way* (ROW), such as roads, waterways, buildings and trees, are recorded. The output of the survey is a route plan and a longitudinal profile of the line, containing all relevant information. The decisions for the tower locations, the so-called *tower spotting,* once done manually using so-called *sag templates* to calculate conductor sags in each span, is now almost entirely handled by specialized software. The surveying of diagonal profiles at tower locations in inclined terrain concludes the on-site surveying work.

Surveying methods have also greatly benefited from digitization and automation. While in the past surveys were carried out with classic topographic methods, i.e. with tachymeters, the process is nowadays accelerated, simplified and its accuracy has improved. This is done either by automatically recording theodolites or with the help of real-time GPS systems (GPS: *Global Positioning System*). Both methods enable measurement, correction and immediate evaluation directly on site with a mean accuracy of the profile points of less than 10 mm over distances of up to 3 km [7].

The recently developed LIDAR method (*Light Imaging, Detection and Ranging*) uses rotating laser scanners mounted on aircraft, helicopters, or drones to capture detailed data of the earth's surface and existing infrastructure—including overhead lines.

4.7.2 Preparation of the Construction Site

Line construction is essentially a logistical problem, as heavy material and equipment must be transported through sometimes difficult terrain and correctly delivered to widely separated locations. This task is associated with the construction of temporary access roads and, if necessary, with the clearing of the route from vegetation, as growth can later cause serious operational problems. The scale of the effort involved in these preparatory works is illustrated in Fig. 4.137.

Fig. 4.137 Intermediate storage on the construction site (**a**) and temporary construction roads (**b**)

Fig. 4.138 Excavation of the foundation pit for a large tower

4.7.3 *Foundations*

The publication [148] provides a comprehensive overview of the various foundation options and includes an installation guide, an overview of safety aspects as well as an assessment of the environmental impacts of foundations as well as various remedial measures. On the other hand [149] deals with the installation of micropiles and ground anchors, which are often used in poor soil conditions or terrain difficult to access. Another reliable source about foundation installation is [150].

Excavation

The process of foundation construction begins with the excavation (Fig. 4.138). Hydraulic excavators are often used for excavation in normal soils like clay. In rocks, hydraulic hammers or explosives may be required. During the excavation, water should be prevented from entering the pit. If necessary, a suitable drainage system is used. During the foundation works, extensive and sometimes costly measures are necessary to minimize the effects on the environment and the surroundings.

Fig. 4.139 Installation of tower footings for a large tower

Concrete and Reinforcement Work

In the next step, the so-called *tower footing* is positioned in the foundation pit (Fig. 4.139), before the steel reinforcement of the foundations is installed. Its correct alignment is supported using *templates*.

The concrete used for the foundation can either be mixed directly on the construction site or, as is common today, be delivered from a concrete mixing plant to the construction site, e.g. with the grade C20/25—for on-site-mixed concrete—and C30/37—for ready-mixed concrete—according to [151] being used. When bringing in the concrete into the pit, the free height of fall of the concrete mix should be kept as low as possible, i.e., not exceed 100 cm. The usual method for concrete compaction after pouring is vibration, as correct curing of the concrete is essential to achieve the desired strength. Concrete requires, especially in summer, adequate humidity, as water evaporation can lead to an undesired shrinkage of the concrete. Therefore, it must be protected from water losses due to wind and temperature by covering it with foil or moistening with water.

Backfilling

The backfilling (Fig. 4.140) is of crucial importance to achieve the desired stability of the foundation. The main goal of proper backfilling is to achieve soil properties that are close to those of the natural soil. However, directly at the foundations, compaction of the filling material is necessary. For this purpose, a non-cohesive backfill should be introduced in layers of 20 to 25 cm and compacted with vibrating plates or vibration hammers. The compaction can be improved by moistening the filling with water.

Fig. 4.140 Backfilled foundation pit

Another method to increase the load-bearing capacity is stabilization with cement. Cohesive soils should be compacted in layers of 20 to 25 cm e.g. with tamping rollers or vibrating plates.

4.7.4 Tower Assembly

Preliminary Work

Good logistics planning is essential for the efficient assembly of towers. Even the delivery of components to the central construction storage area—known in industry jargon as the "cemetery", (Fig. 4.141) requires considerable coordination:

- Precise access planning
- Construction sequence planning
- Orderly storage of the tower components according to the assembling sequence
- Delivery is hardly done tower by tower anymore (utilization of the load capacity of the truck)
- Control of the tower parts, quantity check, quality check, recording of transport damages
- Use of polypropylene round slings for loading and unloading work (protection of the zinc layer and coating of the members of steel lattice towers)

Fig. 4.141 Construction storage ("cemetery") for a major construction site

This is followed by the *tower supply to the site* (Fig. 4.142). The following must be observed:

- All tower components are to be placed on squared timbers (prevention of bending, twisting, tensioning and damage to the coating and zinc layers)
- Correct placement of tower components at the tower site, e.g. corner legs and bracing sections in their designated positions.
- Access routes, road construction, assembly and crane set-up areas to be determined

Pre-assembly work follows, explained here in the context of steel lattice towers:

- All workshop drawings are available on the construction site in weatherproof foil
- Workshop drawings contain all information to erect the towers
  - Position numbers and location of the main legs, diagonals, plates, climbing bolts, cover plates, etc.
  - Type and design of but joints
  - Installation of the horizontal and wind braces
  - Number and size of the bolts and nuts
  - Diameter of the bore holes
  - Drawing errors must be recorded and faulty drawings revised

Fig. 4.142 Delivery and storage of the tower components at the site for a steel lattice tower, picture detail bottom right: bundled steel profiles stored gently

Before the tower assembly, the following needs to be examined and clarified:

- Comparison of local site conditions with the tower design (e.g. terrain, access, and environmental constraints)
- Identification of required areas for assembly and crane set-up
- Adjustment for height differences, potentially through terrain adaption (e.g. for outriggers of cranes)
- Selection of a suitable assembly method, including the choice of climbing derrick based on tower part size and weight
- Determination of appropriate crane size, considering total lifting weight and site-specific requirements
- Use of suitable lifting and assembly technology
- Crane selection criteria, including:
  - Load capacity
  - Tower height
  - Distance from crane pivot point to tower center
  - Required outreach
- Preparation and engineering of detailed lifting plans

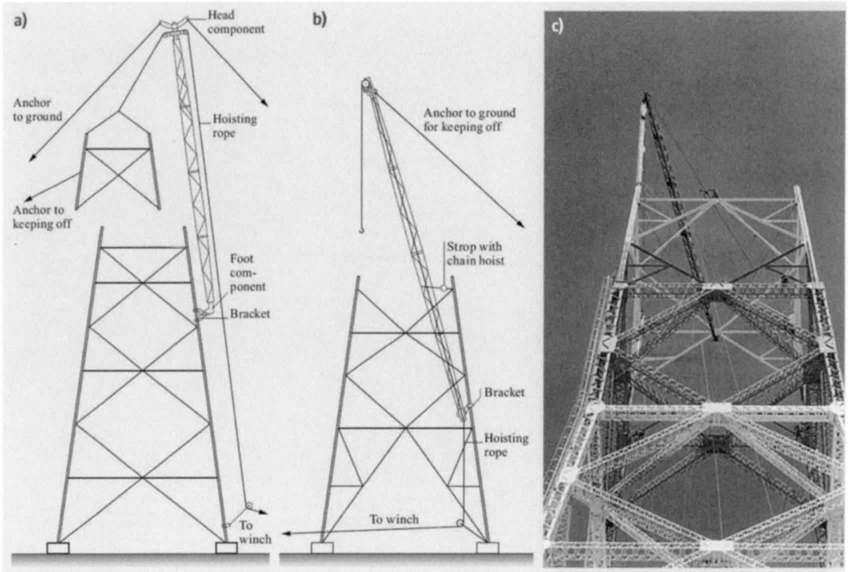

Fig. 4.143 Assembly with a gin pole outside (**a**), inside (**b**), and at the centre (**c**) of the tower body (*Source* [7])

Assembly of Steel Lattice Towers

For this demanding task, the following methods have become established:

- Gin pole
- Crane: corner-section, wall-section, or final-section lifting
- Helicopter

Assembly with the Gin Pole

The assembly using the *gin pole* involves face-wise handling of tower panels suspended inside the tower structure. The gin pole is guyed to the main legs for stability. In most cases, it consists of an aluminum framework beam approximately 10 m long. (Fig. 4.143).

Assembly with the Crane

With this method, the whole tower or a part of it (face- or panel-wise) is pre-assembled on the ground and then lifted with a mobile crane (Fig. 4.144). This is the most productive method for tower assembly, especially in flat terrain with good access conditions.

Fig. 4.144 Assembly of the tower top with the crane

Assembly with Helicopter

The use of helicopters can be advantageous when towers need to be erected in a short time, or when lines are being rebuilt, or access is difficult, e.g., in mountainous terrain. However, the use of helicopters is expensive, associated with additional stress for the personnel, and highly dependent on the weather. To achieve an economic advantage, the tower panels are pre-assembled on the ground according to the capacity of the helicopter used (Fig. 4.145).

Installation of Bolts

After completion of the tower assembly, the bolts and nuts that were installed during the process but left partially tightened to allow structural flexibility must now be fully secured. The following points must be observed:

- Knowledge of the tightening torques of the bolted connections according to customer specifications/standards
- Correct length of bolts: commonly nut completely, plus one thread
- Securing bolts according to the client's requirements (spring washer, punching, coating, tightening torque)
- Use of the correct strength class
- The prescribed torques are controlled with the torque wrench (Table 4.13).

Fig. 4.145 Tower assembly with the helicopter (*Source* Eduard Steiner AG/CH)

Table 4.13 Torque list for mounting of bolts

Bolt type	M12	M16	M20	M22	M24	M27
Bolt diameter (mm)	12	16	20	22	24	27
Torque (N m)	40–60	80–120	130–180	190–230	300–340	475–610

4.7.5 Stringing of Conductors

The *stringing* is the final discipline and the most demanding activity in the construction phase of a power line. There are usually specialized teams that deal exclusively with the stringing. This is necessary since there are many causes of errors during stringing and the conductor installation. They may not be recognizable during the acceptance tests of the line and can later lead to errors and line failures. Problems can start with abrasions on the conductors due to slipping suspension clamps or incorrectly pulling off the conductors from the drums, or due to driving with the line car along the conductors, or from the clashing of the sub-conductors before the installations of spacers, incorrectly set and not greased clamps, broken wires in the clamps, and unfortunately many more.

In general, stringing without touching the ground or any obstacles is required, as damage on the conductor's surface affects the mechanical and electrical properties. In addition, the conductors could corrode due to chemical components of the soil. Minimum requirements are compiled in [152]. The same principles apply analogously to conventional ground wires as well as to optical ground wires (OPGW).

The equipment arrangement during stringing operations is illustrated in Fig. 4.146, while Fig. 4.147 provides a bird's-eye view of the drum site, highlighting its considerable dimensions.

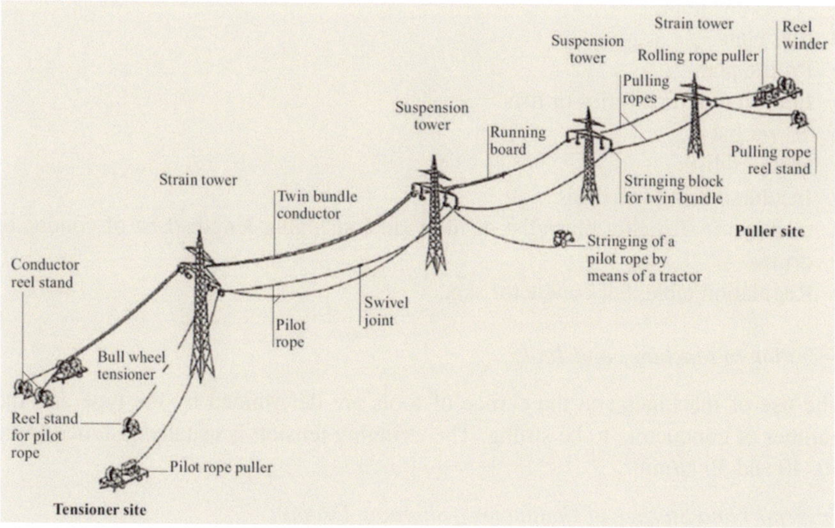

Fig. 4.146 Stringing of a line section

Fig. 4.147 Conductor drum site in bird's view

Workflow

The workflow for the stringing process is outlined below in key steps:

Required Construction Plans

- Overview plan
- Site plan
- Profile plan
- Intersection directories or lists
- Tower list
- Material lists
- Insulator string drawings
- Conductor stringing plans/list of individual stringing lengths/List of conductor drums
- Regulation tables for conductor sag.

Selection of Machines and Tools

The use of machines and the choice of tools are determined by the type and the number of conductors to be strung. The stringing tension is usually assumed to be 30, 40 and 50 N/mm^2.

Transport and Storage of Conductor—Shipping Drums

Conductor drums must always be stored and transported in an upright position. The rolling direction is clearly marked on the drum; the pull-out direction is opposite to this. To prevent damage to the conductor, the formwork must not be removed until the drum has been properly jacked up.

Shipping drums are typically made of wood and steel. Each drum is labeled with key information, including delivery length, weight, conductor type, drum number, manufacturer's name, and product designation.

Prior to stringing, the inner end of the conductor, which is routed through the drum opening, must be released. The manufacturer's installation guidelines must also be observed without exception.

Protective Scaffolding

At intersections, e.g. with roads, highways, railways, etc., protective scaffolding must be installed (Sect. 2.7).

Temporary tower anchoring using installation anchors

Temporary anchors are fixed points used as auxiliary means to transfer anchoring forces to the ground or to tower components during assembly. These anchors are either secured on structures or embedded in the ground. Suitable ground anchors include so called *dead men* (buried wooden beams) or anchoring frames weighted with concrete blocks. Screw anchors, which are hydraulically driven into the ground, are also commonly used.

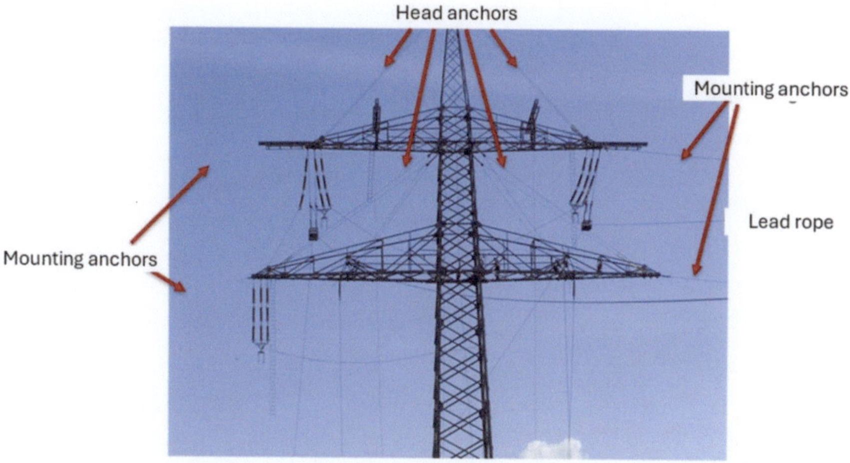

Fig. 4.148 Placement of assembly and head anchors

The purpose of tower anchoring is to prevent overloading during conductor stringing operations (i.e., to limit assembly loads on the tower and its crossarms). The crossarms of tension towers located at the drum and winch site are anchored. To reduce the vertical load on the crossarms, the anchor ropes should be tensioned as flat as possible—typically at an angle between 30° and 45° to the horizontal.

It is advisable to develop a site-specific anchoring structural design (anchor statics). The anchor locations must be determined on site, and it must be ensured that the selected anchor points do not interfere with buried infrastructure.

Steel ropes may be used as anchors, provided that the safety distances from live parts are observed in accordance with regulations, e.g. for Germany the *Implementation Guidelines for DGUV Regulation 3* (German Social Accident Insurance).

Anchoring to railway tracks, transportation infrastructure, machinery, vehicles, or equipment, as well as to live transmission line towers, are not permitted.

Additionally, head anchors may be installed from the tips of the crossarms upward to the tower peak or shaft. These absorb the increased loads acting on the crossarms during conductor stringing operations (Fig. 4.148).

Mounting of Insulator Strings and Stringing Blocks

This important work includes:

- Transport of fittings and insulators to the suspension- or tension towers
- Installation of strings
- Hoisting and installation on the crossarms (Fig. 4.149)
- When assembling the insulators, their length tolerances must be observed (especially with ceramic insulators)
- The stringing blocks are mounted on the lower string fittings (Fig. 4.150)

Fig. 4.149 380 kV Danube type tower with stringing blocks

Fig. 4.150 *Pulleys*, also called *stringing blocks*, for single and bundle conductors; these are coated with rubber for sensitive conductors, such as high-temperature conductors (middle)

- A transfer rope is used to guide the conductor from the running block to the insulator string for final attachment
- Minimum diameter of the wheels and possibly rubber inserts to be observed: conductor Ø 200 mm, OPGW Ø 300 mm).

Stringing the Pilot Line

To string the conductors, a pilot rope made of steel or synthetic materials is first pulled through the stringing blocks on the towers between the winch and drum sites. The type of pilot rope is selected based on the total cross-section of the conductor to be installed and the required stringing tension. These pilot ropes are typically connected to auxiliary pilot ropes (which have been installed before, following the same procedure). They are often made of steel and, as a rule, have a diameter of up to 10 mm.

Depending on the terrain and environmental conditions, the pilot rope (including the auxiliary rope) is deployed using appropriate methods—either manually, mechanically, or increasingly, with the use of drones (Fig. 4.151) or helicopters. All stringing equipment and the operator's sites must be effectively grounded to avoid induction currents.

Stringing of the Conductor

At the drum site, the pilot rope is connected to the conductor or conductor bundle. A *pulling sleeve* is mounted onto the conductor (Fig. 4.152a). A *swivel* (Fig. 4.152b) is installed between the pilot rope and the pulling sleeve to prevent the transfer of torque from the pilot rope to the conductor, which is firmly connected to the pulling sleeve.

The pilot rope is then hauled in with the puller, thereby pulling out the conductor. During this process, the conductor is braked and pulled out with as little dragging as possible.

Fig. 4.151 Stringing of a pilot rope with a drone (*Source* Hexapilots, LTB)

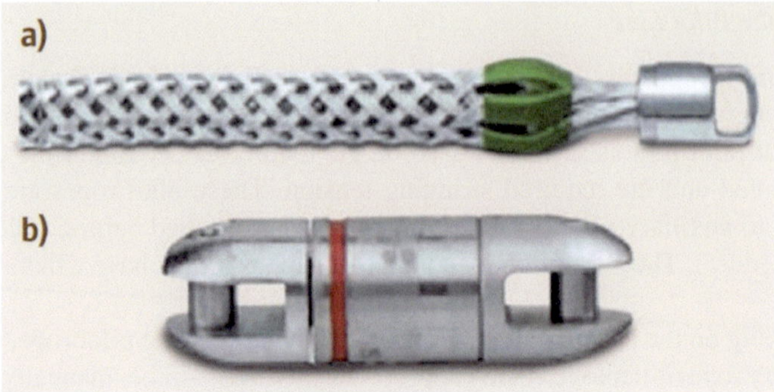

Fig. 4.152 Pulling sleeve (**a**) and swivel (**b**)

When stringing a triple or quad bundle—using a so-called *pulling carpet* (Fig. 4.153)—care must be taken to ensure that the lower sub-conductors are pulled with slightly less tension than the upper ones.

The compensation of the tension is carried out at the puller (Fig. 4.154), where the distance between the machine and the tower should be approximately three times the height from the ground to the stringing blocks.

During the stringing process, there is constant radio contact between the puller and the drum site. A lineman equipped with a radio communication device observes the connection point between the pilot rope and the conductor. In case of disturbances, the observer has to stop the stringing process immediately. When the conductor has been pulled out beyond the tension tower by the length of the jumper, it is stopped.

Fig. 4.153 Bundle stringing with pulling carpet

Fig. 4.154 Puller with a pulling/braking force of 400 kN

Now the final assembling work starts. By lowering the stringing blocks, the hanging tension string can be connected to the conductor. Finally, the *jumpers*—for current carrying connection between the two sides of a tension tower—are mounted (Fig. 4.155).

Conductor Stringing with Helicopter

The use of helicopters in difficult terrain and with poor access to the tower locations significantly shortens the construction time. The work process differs slightly from the conventional conductor stringing just described, in that the linemen are first

Fig. 4.155 Assembly of the jumpers on a 380 kV quad-bundle

brought to the individual tower locations by helicopter, where they climb the towers and mount the stringing blocks on the suspension strings. Then the *auxiliary pilot rope* made of synthetic materials—for less weight—is hooked onto the helicopter and pulled through the stringing blocks. The *pilot rope* is made of steel or synthetic materials then is attached to the auxiliary pilot rope and pulled through the stringing blocks in the opposite direction with the helicopter. Afterwards, the conductor is connected and pulled over the wheels with the *puller*. Finally, the conductor is hung in the suspension clamps and the stringing blocks are replaced by the suspension strings at each suspension point.

Sag Adjustment

Conductor stringing is completed with the adjustment and tensioning of the conductors [153]. This is done with the help of *regulation tables* to set the sag of the conductor at the temperature at the time of regulation so that the loads underlying the conductor design and the maximum sag at the permissible conductor temperature are maintained (Sect. 4.3.4). The creep of the conductors (Sect. 4.3.1) must also be considered, which is why after a first *rough adjustment* immediately after stringing, a second, so-called *fine adjustment*, takes place after one to two weeks, so that the conductor can stabilize.

Installation of Conductor Fittings

During the installation of conductor fittings, such as spacers, dampers or warning spheres as well as bird protection markers or similar devices in the span, so-called *line cars*, as shown in Fig. 4.156, are used [154]. This is particularly the case when the use of lifting platforms, cranes or helicopters is not possible. Line cars are work platforms equipped with wheels that can safely travel along single and bundle conductors as well as ground wires.

Fig. 4.156 Installation of spacers on a horizontal twin bundle using a line car

4.8 Inspection and Maintenance of Overhead Lines

4.8.1 Maintenance Activities and Strategies

Overhead lines are long-lasting technical installations that, if properly designed and manufactured, can remain in operation for 80 years or more. Achieving such longevity requires a reliable maintenance concept, which allows for regular inspection and assessment of their current condition.

Throughout their service life, some components will likely need to be replaced. For example, conductors may require renewal depending on the mechanical stress they have endured from wind, ice, and snow loads as well as thermal stress. The tower coatings also need to be renewed; if corrosion protection was originally applied using the so-called factory-applied DUPLEX process, the coating has a lifespan of about 80 years, typically requiring renewal once after 40 years.

Additionally, the foundation chimneys may need renovation, along with the grounding system. Fittings and insulators should be carefully inspected to evaluate their remaining service life.

Due to the challenges involved in constructing new lines and reinforcing existing ones and since many lines have already reached or exceeded their planned service life, implementing a robust maintenance strategy is essential to ensure their integrity and continuing reliable operation.

Defects and failures of overhead lines occur for the following reasons among others:

- External causes include lightning strikes, earth faults in the network, extreme weather conditions, and—though very rare—sabotage, as well as mechanical impacts such as vehicle collisions with structures.
- Internal causes such as wear and tear, aging, deformation, corrosion and poor design or sub-standard materials, poor assembling and installation
- Operational aspects such as electrical overload, lightning and switching overvoltages, malfunction of protective devices.

The maintenance work of an overhead line can be divided into three basic categories:

- Regular, normal maintenance
- Preventive maintenance including planned repair work
- Emergency renovation or emergency repair.

The regular, *normal maintenance* consists of activities like:

- Regular visual assessment of the line by walking or aerial survey
- Vegetation management
- Control and improvement of corrosion protection
- Cleaning of insulators.

Fig. 4.157 *Left*: mobile landing platform on a truck for line inspection and repairs; *right*: cleaning of insulators by helicopter, Korea, DC Research Center of KEPRI

The appropriate intervals between these activities can be determined based on a risk and cost assessment [155, 156].

Preventive maintenance or repair of components includes activities like:

- Replacement of damaged insulators
- Repair or replacement of conductor and ground wires
- Replacement of damaged fittings
- Replacement of spacers, dampers, warning systems
- Repair of towers or foundations
- Adjustment of reduced safety clearances, e.g. due to creep of the conductors.

Finally, the *emergency restoration and repair* are usually the result of the failure of a component and thus of a line failure, for example because of major storm damage, from sabotage or neglected maintenance [157] (Fig. 4.157).

4.8.2 Defects

The following section outlines typical defects in different line components and the inspection methods currently employed for their in-situ detection.

Conductors and Ground Wires

Corrosion

Corrosion of the steel core of ACSR conductors is primarily caused by industrial and marine environmental pollution in the presence of moisture. Corrosion of the aluminum wires is caused by sulfur-containing pollutants that deposit on the outer surface of the conductor which can lead to pitting on the wire surface. Normally, this type of corrosion is evenly distributed on the surface of the conductor and does

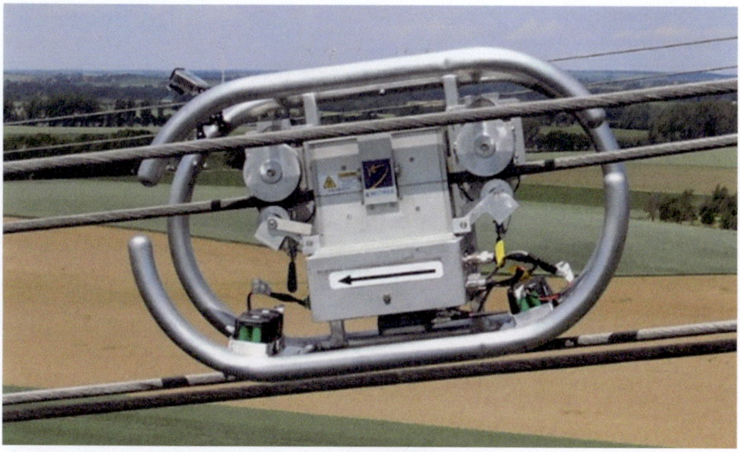

Fig. 4.158 LineVue™ in use for conductor inspection [158]

not reduce its strength to the same extent as internal corrosion; it can be detected by visual inspection of the affected areas or by discoloration of the aluminum wires.

Inspection Methods

Inspection methods include visual inspection from the ground or from a helicopter, with the help of a video camera by drones, as well as corona measurements. The latter helps to detect broken wires, as their sharp edges lead to increased field strength that causes high-frequency corona discharges. Meanwhile, devices are now available that allow non-destructive measurement of the intact cross-section of conductors, as well as the detection of broken aluminum strands and corrosion of the steel core—both in live and de-energized conditions (Fig. 4.158) [158].

Conductor Joints

Most conductor joints used with ACSR conductors are compression joints. Insufficient compression, e.g. due to the use of incorrect tools, can lead to the formation of aluminum oxide within the connection, and thus to an increase in resistance and excessive current flow in the steel core of the conductor, which can then overheat and subsequently lose its mechanical strength. Consequently, the electrical resistance of a connection is the most important factor for assessing its condition. Measurement of the joint temperature from a distance (ground, helicopter, drone) using *infrared thermography* (Fig. 4.159) and comparison with the simultaneously measured temperature of the conductor is a time-saving method to obtain the condition of the conductor joints along a line section [159]. A recommendation for the evaluation of such measurements is proposed in [160].

Fig. 4.159 Joint inspection from the ground with the IR (infra-red) camera [159]

For jumpers, on the other hand, parallel groove clamps (Fig. 4.84,) are almost always used. When assembling these clamps, it is important to ensure that the conductors to be connected are thoroughly cleaned within the clamping area using a wire brush to carefully remove any contamination or oxide layers on the aluminum surface. The bolts must be tightened alternately and in accordance with the manufacturer's specifications using a torque wrench.

Damage Due to Wind-Induced Vibrations

As explained in Sect. 4.3.11, overhead lines are often affected by wind-induced vibrations, which can cause the following damage:

- Falling, displaced or faulty vibration dampers (Fig. 4.160)
- Missing nuts from suspension clamps
- Missing cotter pins from cap-and-pin insulators
- Displaced or bent protective fittings
- Wire breakages (Fig. 4.37)
- Damaged or fallen warning spheres (Fig. 4.161)
- Loose or broken tower parts
- Heavily worn U-bolts (Fig. 4.162).

Fig. 4.160 Damaged Stockbridge dampers on the lower conductors (*Source* U. Cosmai)

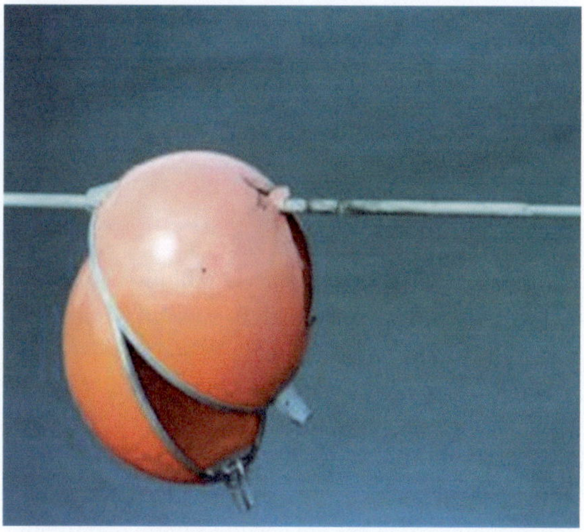

Fig. 4.161 Damaged warning ball due to wind-induced vibrations (*Source* U. Cosmai)

Fittings

As many lines, especially in countries with long-established power infrastructure, have been in operation for over 50 years, it is of utmost importance to have or to develop reliable test procedures for the evaluation of aged fittings. This allows us to decide whether fittings, which have been in operation for 30 or more years, can continue to be used safely or need to be replaced. For this [161] provides detailed information.

Fig. 4.162 U-bolts worn by wind-induced vibrations (*Source* D. Havard)

Insulators

Porcelain Insulators

Shattered insulator sheds can occur due to vandalism or errors during the manufacturing process. The main cause of failure of porcelain cap-and-pin insulators is corrosion of the ball, as the expanding corrosion products generate high circumferential tension, which can lead to radial cracks in the porcelain.

Publication [162] contains a guide for checking the condition of ceramic insulators and determining the appropriate time for their replacement.

Glass Insulators

Electrical discharges, e.g. due to pollution layer flashovers, can damage the glass surface. Since it is toughened glass, even relatively minor surface damage can lead to an imbalance of internal mechanical stress and cause the glass insulator to shatter.

Composite Insulators

Composite insulators, which have now established themselves as an alternative to porcelain and glass, can have degradations due to premature aging or manufacturing issues, which often are not easily visually identifiable. For aged or failed composite insulators, specific laboratory tests can be used to determine their remaining life expectancy [163].

Towers

Most overhead line structures are steel lattice towers, which are constructed from galvanized steel profiles. The usual maintenance practice is to provide them with protective coating at regular time intervals, usually every 10 to 25 years, depending on the climatic conditions at the location and the degree of pollution. Using the in-factory DUPLEX method (Sect. 4.6.5), this interval can be extended to about 40 to 50 years—depending on the environmental conditions. In addition, bolts can loosen

Fig. 4.163 380 kV Wintrack poles of Tennet in the Netherlands. *Left picture:* Entry hatch into the tubular pole; *right picture*: Interior of the poles with climbing rungs

if they are not properly secured, or individual members can bend due to exceptional loads or movements of the soil. Excessive wind loads can also lead to loosened bolts and to the failure of individual tower members.

Steel tubular poles also require corrosion protection, including on the inside of the structure. The interior of large tubular poles can be accessed through a hatch and climbed on a ladder (Fig. 4.163). In addition, corrosion protection as well as the connections of the individual joints of the pole's body, are regularly controlled (Fig. 4.164).

Fig. 4.164 380 kV tubular pole of Terna in Italy. The individual joints are bolted together with external flanges. The climbing steps are also clearly visible as climbing aids for maintenance work

Fig. 4.165 Climbing aids from *left*: climbing rungs and safety rope, safety rail Highstep^{TM}, safety climbing bolt "pig tail"

Tubular poles made of spun concrete are highly durable and, unlike steel poles, are not susceptible to corrosion. During inspections, particular attention is given to identifying cracks in the concrete and checking the connections at the crossarms.

For tubular poles, external inspection is typically carried out using a mobile crane or climbing aids, various such aids are employed. The most common system consists of steel rungs (bolts) affixed to the main legs, combined with a fall arrest system that uses either a vertically tensioned steel wire (Fig. 4.165, left) or so-called "pig tails," into which the safety rope is progressively inserted during ascent (Fig. 4.165, right).

A more modern climbing aid consists of a rail installed along the tubular pole or along the main leg of a steel lattice tower. This rail guides a manually operated or battery-powered climbing platform upward (Fig. 4.165, middle; Fig. 4.166).

Fig. 4.166 Battery-powered lift as a climbing aid and for material transport (*Source* Highstep Systems AG)

The following inspection techniques can be used to assess the condition of poles:

- Visual inspection during line inspection from the ground (binoculars, telescope, camera)
- Visual inspection when climbing the poles
- Drones
- Galvanizing condition for steel poles:
  - Electronic measurement of color and galvanizing thickness
  - Cross-hatching-cut test.

The latter is the method by which the adhesion forces for the paint on the members can be determined as specified in [164, 165].

Foundations

Although most foundations do not present significant problems when properly put in place, some network operators face problems with subsoil corrosion. Cracks in the upper part of the foundation, the so-called *chimney*, can also occasionally occur, as well as elevations of the structure, if the main legs are inadequately fixed in the concrete block of the foundation or due to corrosion of the main legs within the foundation concrete (in this example for separate footings). This can be detected with the help of suitable inspection techniques, ranging from simple visual methods to advanced procedures such as surface-penetrating radar and acoustic impulse echo. It is also necessary to regularly check the grounding conditions, as these can deteriorate, e.g., due to corrosion of the grounding system or poor contact of the grounding strips or damages by fieldwork (plowing). A related report is being prepared by the CIGRE Working Group WG B2.82 "Overhead line foundations for difficult soil and geological conditions".

Damage Due to Environmental Influences

Overhead lines interact closely with their surrounding environment, and this interaction must be carefully considered during line planning and design. Applicable standards define minimum requirements, which are often significantly tightened in areas with challenging terrain or severe climatic conditions. Despite these precautions, certain events can still cause damage to overhead lines.

Examples include avalanches and landslides in mountainous regions (Fig. 4.167, left), which are difficult to predict, as well as storms and hurricanes. Additionally, highly uneven ice loads between adjacent spans can result in structural damage to towers (Fig. 4.167, right).

Fig. 4.167 Damage to a 220 kV tower in the high mountains due to an avalanche (*left*); damage due to uneven icing on the 110 kV conductors (*right*), historical pictures from the 1950s

Powder avalanches are fast-moving mixtures of fine snow particles and air that exert particularly high dynamic pressure on towers and conductors. As they descend through valleys, they are redirected by the surrounding terrain, which can result in rapid acceleration and velocities reaching up to 100 m/s. When such an avalanche strikes overhead line components, it can generate forces of up to 5 kN/m^2 on the structures.

If routing a transmission line through avalanche-prone areas cannot be avoided—such as in the case of transporting electricity from alpine storage power plants—appropriate mitigation measures must be considered. These include positioning towers as far as possible outside the expected avalanche paths and by increasing the distances between the conductors in the span to mitigate the consequences on the structures.

Figure 4.168 shows a simulation of the deflection and subsequent return motion ("swinging out and back") of conductors impacted by a powder avalanche, using a 220 kV transmission line in the Austrian Alps as an example [166].

Bushfires present a significant hazard to overhead lines in many parts of the world. Line operators are typically alerted by meteorological services and can implement precautionary measures, such as switching off lines at risk. One preventive strategy is the controlled burning of vegetation within the line corridor, which helps reduce fire

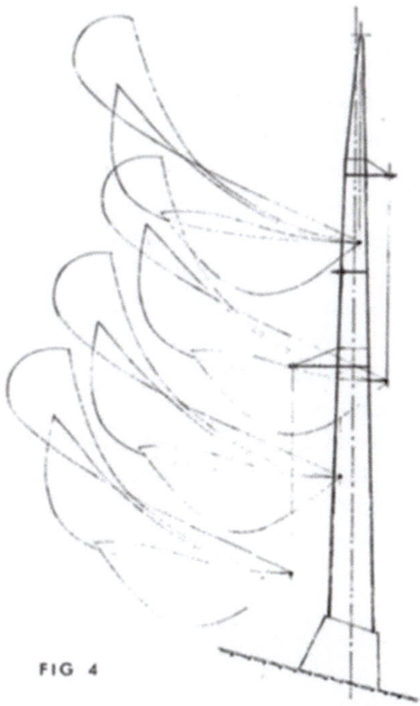

Fig. 4.168 Simulation of conductor deflection and return motion on a 220 kV transmission line in the Austrian Alps during a powder avalanche

load and associated risks (Fig. 4.169, left). Wooden poles used in medium-voltage lines are especially vulnerable to fire damage (Fig. 4.169, right).

Fig. 4.169 South Africa: preventive reduction of the fire load by controlled burning of vegetation in the line corridor (*left*); fire damage to a wooden medium-voltage line (*right*) [167]

Disturbances and Damage Caused by Animals

Power line towers often serve as attractive nesting sites for certain bird species. Elevated above the ground and inaccessible to most predators, these structures provide a safe environment for breeding. While this is a welcome development from a conservation perspective, it can pose operational challenges for line operators. Nesting materials—such as branches, twigs, and even wires—can come dangerously close to live conductors, potentially causing flashovers that endanger both the birds and the reliability of the power line. In addition, bird droppings can contaminate insulators, reducing their dielectric strength and increasing the risk of outages.

To mitigate these risks, many utilities install purpose-built nesting platforms on the towers at locations that are safe for both the birds and the electrical infrastructure (Fig. 4.170, left). A successful example of this approach is a collaborative conservation project by Austrian Power Grid, BirdLife Austria, and the Research Institute for Wildlife Science (FIWI) at the University of Veterinary Medicine Vienna. Through the targeted installation of nesting aids, the endangered Saker Falcon population was not only protected but also significantly increased (Fig. 4.170, right).

Fig. 4.170 *Left*: stork nest on a 380 kV overhead line, *right*: falcons like to use power line towers as a perch

Fig. 4.171 Live-line working: *left*, installation of a device for determining the conductor quality on the live conductor; *right*, working from a helicopter

4.8.3 Live-Line Working

Live-line working has been practiced safely around the world for decades and is recognized as a reliable method for carrying out maintenance activities. Many tasks traditionally performed with the line de-energized can also be carried out under voltage, provided that proper procedures are followed, and specific line conditions are carefully assessed. Not all transmission lines are suitable for live-line working. In some cases, this was already considered during the planning and design phase, while in others the need for live-line maintenance only became apparent after several decades—typically when the possibility of switching off the line was no longer feasible due to increased network load or operational constraints.

Key factors for enabling live-line working include the tower configuration and site accessibility. Access methods may involve insulated aerial work platforms, helicopters, insulating sticks, ladders, or rope access techniques (Fig. 4.171). Prior to any live-line operation, the minimum approach distances must be evaluated in accordance with the system's potential overvoltage conditions. Additionally, protective relay settings—particularly the disabling of automatic reclosing (momentary interruption)—must be addressed as part of the safety planning.

Responsibility and role allocation are especially critical in live-line work. Specific responsibilities must be clearly defined, and linemen must undergo dedicated training programs, with regular refresher courses to maintain certification. Specialized equipment and tools are required, along with routine testing to ensure their functionality and safety compliance.

Live-line work demands a detailed procedural framework that reflects the complexity of the task and adheres to applicable technical standards related to equipment, methodology, safety, and quality assurance. These factors lead to both technical and economic considerations—namely, whether the cost of performing live-line work is justified in comparison to the potential costs of a supply interruption or restriction.

Live-line working is particularly common in countries with extensive transmission networks, such as the United States, Canada, Russia, and several South American nations, as well as in parts of Europe including France [159, 168–170].

4.8.4 Robotics for Overhead Lines

The anticipated global increase in live-line working has spurred the development and deployment of robotic systems for the inspection and maintenance of overhead lines. These systems are capable of operating on energized conductors in environments that are difficult to access or unsafe for human workers. Drones, in particular, offer visual inspection capabilities that once required helicopters and now provide high-resolution imagery for detailed assessment.

The future of robotics in overhead line maintenance is highly promising. When integrated with artificial intelligence, these technologies enable the development of smarter, more efficient maintenance strategies—not only for overhead transmission lines but also for other areas of critical infrastructure.

Robots are primarily used in the following scenarios:

- When human access is impossible or poses significant safety risks,
- In areas with limited or difficult terrain accessibility,
- For long-range inspections using drones,
- For direct work on conductors, including minor repairs.

Drones

The development of drones (UAVs—Unmanned Aerial Vehicles) has advanced rapidly in recent years, with a growing number of models and manufacturers entering the market. In many countries, the use of UAVs is subject to legal regulations, covering aspects such as minimum and maximum flight altitudes, authorized operating zones, maximum distances from the operator, certification requirements, and more.

However, transmission line operators often receive exemptions, which significantly broaden the scope of drone applications. Fully autonomous drones can navigate along transmission lines using GPS support, carry out inspections as needed, and automatically return to their launch point—even at night (Fig. 4.172).

The areas of application for overhead line inspections are outlined in [171]. Drones support these tasks by capturing high-resolution images, videos, and thermal recordings, which are used to identify damage to conductors, clamps, and insulators. While the evaluation of this data is largely automated by use of artificial intelligence, the

Fig. 4.172 Drone for long-distance use (*left*) (*Source* Statnett/Schiebel); Picture of a defect vibration damper (*right*) (*Source* CIGRE Symposium Auckland 2013, PS3/Q3.1, F. Lirios)

Fig. 4.173 Drone "Multicopter" with camera for visual inspections of overhead lines and switchgear in substations (*Source* [171])

interpretation of results—such as determining the severity of defects and deciding whether maintenance, repair, or no action is necessary—remains the responsibility of qualified personnel. Artificial intelligence is taking over more and more of these evaluations. A typical drone used for tower and substation inspections is shown on Fig. 4.173.

Robots on Conductors

Robots on conductors are "extended eyes and arms" of the lineman, their main task is visual inspection. Worldwide, several institutes, universities, and the industry are working on the further development of robots for the testing of conductors and for minor repairs on these.

Other Types of Robots

Robots for climbing towers can climb smooth and vertical surfaces (Fig. 4.174 left) and are well suited for the inspection of tubular poles made of steel, concrete and composites, where otherwise aerial work platforms or mobile cranes would have to be used. Robots are also used for the inspection and cleaning of insulators (Fig. 4.174 right).

Fig. 4.174 Robot climbing a tubular pole (*left*); Robot inspecting a tension string made of porcelain cap-and-pin insulators (*right*) (*Source* [171])

Fig. 4.175 Prefabricated concrete weights for anchoring temporary structures

4.8.5 Emergency Restoration, Provisional and Temporary Structures

In the event of a catastrophic line failure resulting in an outage, service can often be restored quickly—well before full repairs are completed—using emergency restoration structures. These structures are also employed as temporary solutions during new line construction, for example, to facilitate rerouting of existing lines.

Emergency restoration structures (ERS) are designed based on a modular principle, using components that are as small, lightweight, and standardized as possible. This modularity ensures rapid and efficient transportation to the site—by truck, trailer, or, when vehicular access is not feasible, by helicopter. The use of interchangeable elements simplifies assembly and enables flexible installation in various terrain and emergency scenarios—ensuring that damaged transmission lines can—be restored quickly and reliably.

The same principle of simplicity and speed applies to the related foundations. These may be implemented as gravity-based (weight) foundations or anchored using guy ropes. The guy ropes themselves are secured by various means, including weighted anchors (Fig. 4.175), screw anchors (which are driven into the ground like large corkscrews), or buried beams—commonly known as "*dead men*".

Temporary structures may consist of portal-type lattice frameworks or single-pole "needle" structures (Figs. 4.176 and 4.177). The materials commonly used include steel and aluminum, while wood is occasionally employed for lower voltage levels. Looking ahead, it is expected that composite materials will play an increasingly important role in the construction of such structures, due to their favorable strength-to-weight ratio and corrosion resistance.

Fig. 4.176 Single system 220 kV temporary line with guyed portal structures

Fig. 4.177 Single-circuit temporary line with guyed towers, Lindsay USA [172]

Table 3.2 — Return period for temporary lines

Duration	Return periods (years)
≤ 3 days	2
≤ 3 months (but > 3 days)	5
≤ 1 year (but > 3 months)	10

Fig. 4.178 Table 3.2 of EN50341:2012

To ensure the rapid erection of a temporary overhead line, it is essential that transportation, assembly, and installation procedures are practiced regularly. Such exercises are especially valuable when a temporary section is integrated into an existing transmission line and energized under live operating conditions. The necessary components—including structures, foundations, conductors, fittings, insulators, and assembly tools—are typically stored in strategically located depots within the supply region. To ensure rapid deployment in the event of an emergency, these components must be regularly inspected, maintained in operational condition, and kept readily available for immediate use.

Temporary lines can be applied across a range of voltage levels, from medium to high voltage. Depending on requirements such as transmission capacity, span length, and available components, they may be configured as single- or multi-circuit lines. Similarly, conductor arrangements can range from single conductors to bundled configurations, depending on the structural design and electrical load demand.

In some countries, national and international standards allow for reduced structural requirements for temporary overhead lines compared to permanent installations. One example is the European Standard EN 50341:2012, which is applicable across all 34 European CENELEC member countries. According to this standard, the return period for climatic loads (e.g., wind, ice) may be adjusted based on the expected duration of the temporary installation (Fig. 4.178).

Another exemption applies to temporary lines erected during seasons when icing is not expected—in such cases, ice accretion does not need to be considered in the design. Examples of temporary line routes can be found in Sect. 2.7.

4.9 Uprating and Upgrading of Overhead Lines

As already mentioned, (Chap. 3), since permits for the construction of new overhead lines are not always easy to obtain, various concepts for uprating overhead lines have been developed in recent years [173]. These options play a key role in the NOVA philosophy—"Network Optimization before Network Reinforcement before Network Expansion" (Sect. 3.4.7)—which is actively applied in Germany.

4.9.1 Uprating

Uprating refers to the process of increasing the transmission capacity of an overhead line. The fundamental concept is to transmit more electrical power using the existing line routes. This can be accomplished by increasing the current or the voltage of the line, or both—though, in practice, increasing the current is more commonly implemented than increasing the voltage.

Table 4.14 provides an overview of the most common methods and technologies used for voltage and current uprating. Additional technical details can be found in [173].

Any line uprating initiative requires that the corresponding substation equipment either already supports the enhanced capacity or is suitably reinforced to meet the new operational demands.

The following factors should be considered before the transmission capacity of a line can be increased:

Costs of Losses

Increasing the current of a line without simultaneously increasing the conductor cross-section or reducing its specific resistance inevitably leads to increased Ohmic losses, which also cause higher costs.

Condition of the Current-Carrying Conductor Fittings

These, and especially conductor joints and tension compression clamps, must be able to carry the increased current, as they may otherwise mechanically fail. They should therefore be checked, preferably by resistance tests.

Table 4.14 Options for increasing the transmission capacity of an overhead line

Increasing	Method	Tool
Current rating	Increasing temperature	Increasing height attachment point
	Reconductoring	Retensioning
		Compact/smooth conductor
		High-temperature conductor
	Special engineering methods	Statistical approach
Voltage level	Reinsulation	Real-time approach
		Adding/substituting insulators, adjusting gaps
		Cross-arm modification
	Ground clearance	Increasing height attachment point
		Retensioning
	Phase-phase distance	Conversion of 2 to 1 circuit
		New tower top

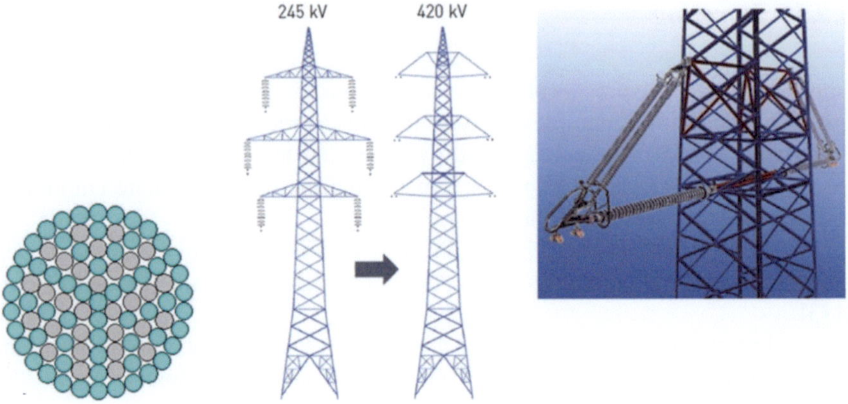

Fig. 4.179 Conversion of a 220(245) kV AC line into a 380(420) kV AC line, *left* the newly developed LWC conductor, *right* the converted line [174]

Condition of the Conductors

The conductors could have been damaged by lightning strikes, vibrations, poor attachment to the clamps or other incidents, even by shotguns or bullet impacts, and therefore need to be thoroughly checked along their whole length. Notched or damaged wires must be repaired with repair joints or repair helical fittings.

Example for Uprating

In [174] a study from Austria was presented, in which a double upgrade of the line was carried out. On the one hand, the original ACSR conductors were replaced by newly developed so-called Low-Weight-Conductors (LWC), i.e., conductors with the same proportion of aluminum as the original conductors, but with a larger diameter of 39 mm (Fig. 4.179 left). This increases their current carrying capacity due to higher convection and radiation cooling and reduces their corona noise. On the other hand, the metallic crossarms of the steel lattice towers were replaced by insulating crossarms with composite insulators (Fig. 4.179 right), which allowed the operating nominal voltage to be increased from 220 to 380 kV while maintaining the required safety clearances.

4.9.2 Upgrading

The main reason for the *upgrading* of an overhead line is to improve its availability respective reliability. This is becoming increasingly important in view of today's increasingly noticeable climate change, which can cause higher meteorological stresses than those for which the line was originally designed.

A second important point is the adaptation of the return periods based on the new climatic conditions. Furthermore, large parts of the existing networks worldwide are more than 50 years old and were therefore mostly designed according to old standards. This often means that the loading assumptions are no longer accurate, and the design methods have been further developed in the meantime. And finally, the desired increase in transmission capacity often requires the use of conductors with larger cross-sections, which in turn results in higher mechanical loads and thus related reinforcements of existing structures. The determination of the reinforcement needs and the relevant calculations are carried out according to standards, e.g. DIN EN 50341-2-4 in Germany. When using conductors with higher conductivity and HTLS conductors, however, the load on towers remains approximately the same. In this case, to give an example, the German *Committee Federal-State* "Electric Power Industry" (*in German: "Bund-Länder-Ausschuss Elektrizitätswirtschaft* ") has recommended the development of a new VDE application rule VDE AR-N 4210-4, in which reliability requirements for existing structures of overhead lines are defined and the related design procedures are described. This allows the network operator to provide proof that the safety of existing overhead lines is ensured in conformity with §49, Energy Industry Act [175].

Since the reliability of a line results from the combination of external loads and the material strength of the line components, it can be increased by either reducing the effects from climatic loads or increasing the characteristic strengths of the components. The methods and tools for the strengthening of overhead lines are summarized in Table 4.15 from [173], where further details can be found.

Aspects of upgrading (like social, environmental, economical, inspections, calculations of loads and strength, tests, approach) of all kinds of transmission towers can be found in detail in [108]

Table 4.15 Options for the strengthening of overhead lines

Method	Tool
Reducing impact of climatic loads	Compact/smooth conductor
	Decreasing ice accretion
	Decreasing number of sub-conductors
	Decreasing number of circuits
Increasing characteristic strength	Support: leg members
	Support: bracings
	Foundation: uplift
	Foundation: compression

Fig. 4.180 Additional main leg; the load is introduced via specially designed lap joints (*Source* Cteam)

Strengthening of Steel Lattice Towers

There are various possibilities:

- Exchange of bolts and tower member profiles
  - Exchange with higher material quality
  - Exchange with larger cross-section
- Additional profiles (members)
  - Sheet metal flanges
  - Double diagonals
  - Additional main legs (Fig. 4.180)
  - Installation of additional members/braces
- Replacement of complete tower parts
  - Ground wire peak
  - Crossarms
- Reinforcement/replacement of gusset plates (Fig. 4.181).

Fig. 4.181 Reinforcement at the intersection in the lower part of the tower by two C-shaped plates (*Source* Cteam)

Strengthening of Foundations

The reason for strengthening foundations is to prevent the exceeding of their stability by:

- Overloading on pulling out/tilting: Back-anchoring in deeper soil layers or increase of the superimposed load
- Overloading on bottom pressure/ground failure: Increase of the load transfer area
- Reinforcement of cleats for better load transfer of the tower leg into the foundation (Fig. 4.182 right).

Since a simple replacement is not possible in this case, the typical solution involves either reinforcing the existing foundation or constructing a new foundation around the partially dismantled original one.

This can be done depending on the subsoil, existing foundation and the loads acting on the existing foundation, through:

- Micropyles (Fig. 4.182 top left)
- Slab (Fig. 4.182 top right)
- Block (Fig. 4.182 bottom left).

Before construction begins, the feasibility of the strengthening measures must be thoroughly assessed—particularly with respect to implementation and ensuring structural stability during the execution phase. Temporary safeguards, such as anchors, should be installed prior to applying any additional loads. It is also essential to follow the correct assembly sequence to maintain safety and structural integrity throughout the process.

Fig. 4.182 Foundation reinforcement by micropyles (top *left*), slab (top *right*), block (bottom *left*) and reinforcement of the cleats (*right*) (*Source* Cteam)

4.10 Direct Current Overhead Lines

4.10.1 General

This chapter deals with the design of direct current (DC) overhead lines. The basics of direct current technology and the differences to alternating current technology are explained in Chaps. 1, 5 and 6. A comparison of basic properties of three-phase and direct current systems is given in Chap. 6. In summary, DC technology has the following advantages over AC:

- Suitable for transport of large amounts of power over long distances with overhead lines or underground cables
- Optimal for point-to-point connections
- Coupling two asynchronous networks
- The transmitted power can be controlled independently and is not affected by the surrounding three-phase networks
- Because of the ability to adjust the transmitted power, stability issues are avoided
- Only active power is transmitted, no line compensation is required, which is particularly beneficial for long underground cable routes
- For large overhead line lengths (depending on the power and configuration) the relative costs of the connection become lower despite the high costs for the converter stations than for AC lines, also because fewer phase conductors are required.

Disadvantages of DC transmission technology compared to alternating current technology are:

- Tie-offs in the DC line require the construction of separate converter stations
- Transforming the operating voltage (increasing or reducing) by transformers is not possible

- High construction costs for short line lengths due to the expensive converter stations
- Conventional types of circuit breakers cannot be used. The efforts to develop HVDC circuit breakers are enormous but still in the trial period.

Basically, all DC lines are connections between converter stations at their endpoints. Differences in the tower configuration result from the type of operation. From this, the following three basic forms of DC overhead lines can be distinguished.

The *monopolar HVDC transmission without return conductor* only requires one conductor per system. Return currents are discharged into the surrounding earth at the converter stations (ground return, Fig. 4.183).

The *monopolar HVDC transmission with return conductor* requires one conductor and one return conductor per system (*metallic return*), which can have a smaller cross-section than the phase conductor and does not need to have the same insulation level as the phase conductor. Such return conductors are usually installed at the top of the tower and can serve at the same time as a ground wire for lightning protection (Sect. 4.6.10). For shorter distances, underground cables are also an alternative for the metallic return (Figs. 4.184, 4.185 and 4.186).

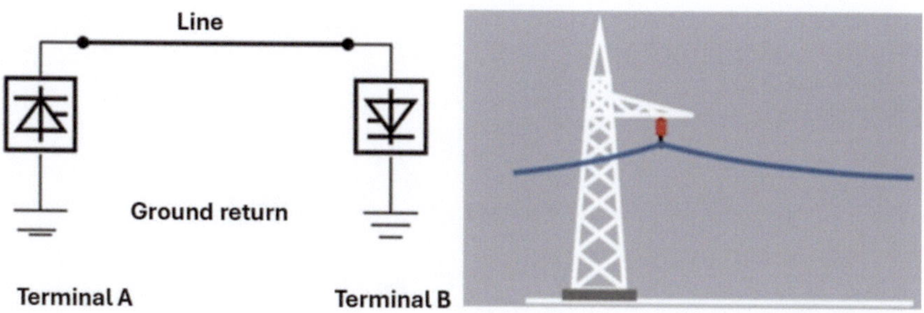

Fig. 4.183 Monopolar HVDC transmission *without* return conductor, *left*: single-line drawing, *right*: schematic

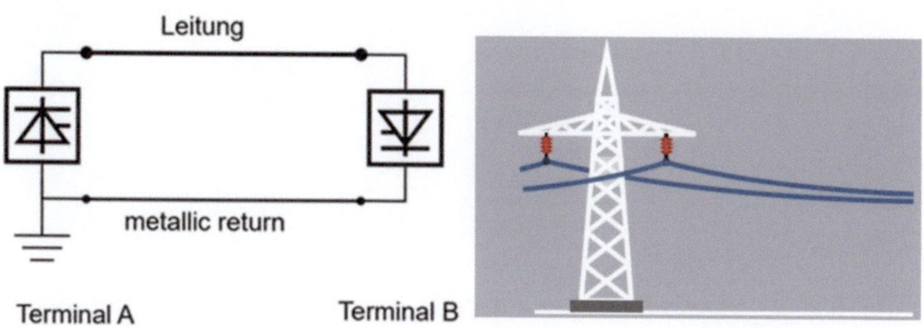

Fig. 4.184 Monopolar HVDC transmission *with* return conductor, *left*: single-line drawing, *right*: model drawing

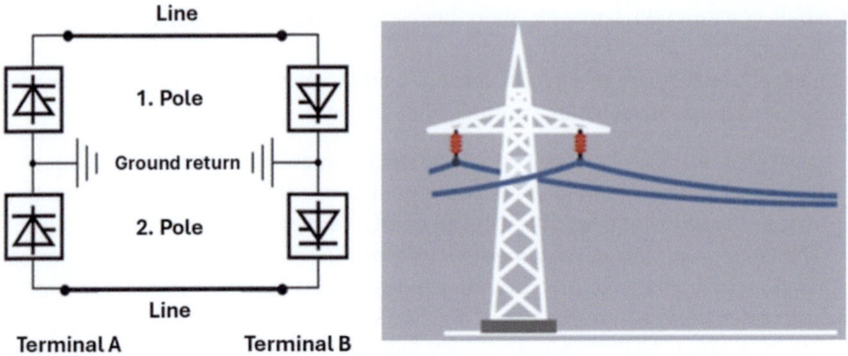

Fig. 4.185 Bipolar HVDC transmission with two conductors, *left*: single-line drawing, *right*: schematic

Fig. 4.186 Monopolar 400 kV DC line with return conductor for the submarine cable interconnection between Greece and Italy (*Source* PFISTERER)

The *bipolar HVDC transmission* requires two conductors per system. Return currents can be discharged into the ground at the converter stations (*ground return*), or a return conductor is installed on the towers (*metallic return*), which can have a

Table 4.16 "Creepage requirements for OHL insulators, recommendations on basis of existing IEC 60,815" *Source* Pfisterer

Creepage requirements		
Recommendations on Basis of existing IEC 60815		
Severity	Specific for AC in mm/kVp-p	Required for DC in mm/kV
Light (1.2)	16	33
Medium (1.35)	20	47
Heavy (1.5)	25	65
Very heavy	31	78…94

lower insulation level. Such return conductors can be installed at the tower peaks and provide simultaneously the function of a ground wire with lightning protection effect. For shorter distances, they can be replaced by an underground cable (Fig. 4.185).

Since there is no voltage zero crossing in DC as in AC, the insulation conditions are different and usually require longer creepage distances and thus longer insulators than with AC lines. Table 4.16 shows a comparison.

Long rod insulators have advantages over cap insulators here due to the shorter construction length. Glass cap insulators require special glass to avoid ion migration.

HVDC lines are especially suitable for long distances. Examples of this can be found in India with the 1800 km long ±800 kV HVDC line from Raigarh in the north to Pugalur in the South which can transmit 6000 MW. From Pugalur, a ±320 kV HVDC line continues over 180 km to Trichur with an approximately 30 km long XLPE underground cable section. This 320 kV line is the first "Multi Pole Line" in India with 2 × 2 multipoles, this means four phases on one tower. It transmits with twin bundles up to 2000 MW. It was planned with the aim of minimizing environmental impact. By installing all four poles on only one structure and with narrow tower bodies, a right of way of 44 m was achieved, compared to 52m to 64 m necessary with other tower configurations or higher operating voltages. Calculations and tests resulted in a pole arrangement according to Fig. 4.187 to minimize the electric field and to achieve a minimum ground clearance of 8.5 m [176].

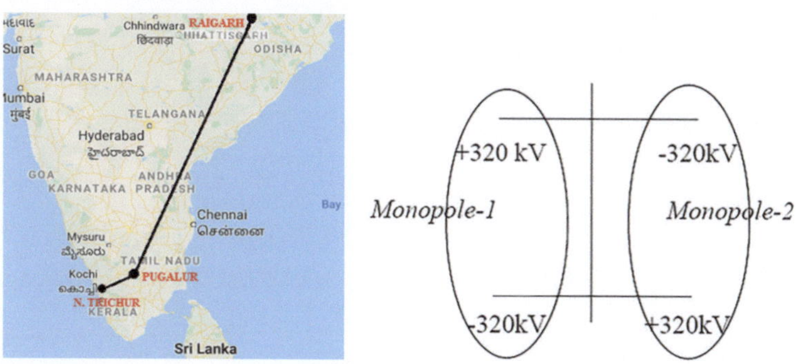

Fig. 4.187 *Left*: ±320 kV multipole HVDC line in India from Pugalur to Trichur; *right*: arrangement of the four poles [176]

The energy production from the “Three Gorges Project” in China, the largest hydroelectric power plant in the world so far, with 22 GW production, is transmitted via six bipolar ±500 kV HVDC overhead lines, each over hundreds of kilometers. Each bipole transmits a nominal power of 3000 MW [177].

4.10.2 AC/DC Hybrid Lines

Hybrid lines have AC and DC circuits on the same structure. Currently, in a pilot project in Germany, one of the two circuits of an existing 380 kV AC line between the regions North Rhine-Westphalia and Baden-Württemberg is being converted in a section into a 400 kV direct current bipole (Fig. 4.188). The third conductor is being used as metallic return (Fig. 4.189). In this case alternating current and direct current are transmitted together on one line, each circuit can carry up to 2000 MW. The essential features and differences of AC and DC transmission, as described in Sect. 4.10.1, also apply to a hybrid line. Particularly, in the case of a hybrid line:

AC circuit:

- Cost-effective transformation to the different voltage levels of distribution and transmission networks by transformers.
- Easy integration into a meshed network.
- Audible noise can occur in humid weather, less so in dry weather.

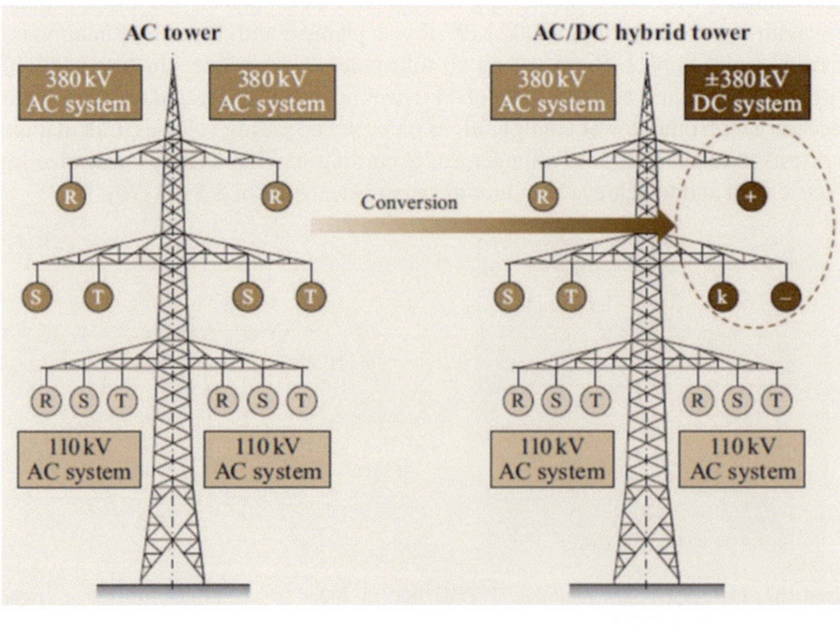

Fig. 4.188 Concept of a hybrid line (*Source* Amprion)

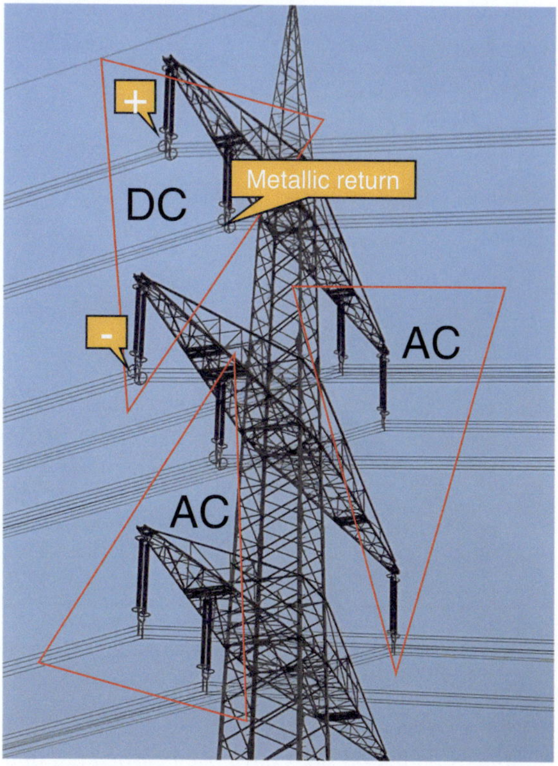

Fig. 4.189 Hybrid line in nature: *left* side AC; *right* side DC—upper crossarm inside: metallic return, outside: + pole, middle crossarm II outside: − pole); small picture on the right bottom DC string with composite insulators and lateral field control fittings (*Source* Amprion)

DC circuit:

- Audible noise can occur in dry weather, less so in humid weather.
- Required insulator length under pollution conditions can be an obstacle in adapting an existing AC structure to a hybrid line, as well as for ground clearances.

The design of the AC and DC circuits on the common tower must be coordinated to minimize undesirable influences. Different phase arrangements on the tower are available for this purpose.

An important precondition for the feasibility of this project was the use of the existing conductors and towers. This was only possible by replacing the installed long rod porcelain insulators with composite insulators, as these have significantly better behavior under pollution, especially with direct current (Sect. 4.4.1) [72] and thus could be accommodated in the existing tower top geometry. Further details on hybrid lines can be found in [178] and about this project in [179].

The different circuits AC and DC interact with each other, e.g. inductive, capacitive and ion current coupling, saturation phenomena in transformers, optimization of the conductor bundle arrangement, transient DC currents in the AC circuit especially in case of failure, adaptations to the line protection, requirement for transposition of the AC system, switching overvoltages. These effects need to be and are investigated.

Fig. 4.190 Parallel lines: AC line 230 kV and HVDC line ±450 kV, Canada/USA (*Source* G. Kendler, Technik-Dialog Hybridleitung, Lecture 2017)

Studies on hybrid lines have been carried out in the USA and Canada since the 1980s. In the line Québec–New England, separate parallel AC and DC overhead lines were erected in sections. The line is therefore referred to as a "hybrid overhead line" and can be considered as forerunner to a hybrid line on a single tower as discussed above. The length of the ±450 kV DC line is 1100 km (Fig. 4.190).

In Switzerland, the conversion of an AC 380 kV double circuit line to a hybrid line with one DC system was investigated in the 2010s for various DC voltages: ±350, ±400, and ±450 kV, (Fig. 4.191). Environmental impacts were considered with focus on audible noise and electric field at ground level from the AC and DC conductors and yielded encouraging results [180].

From South Korea, a study from the 2010s on a 154 kV AC/80 kV DC—hybrid line is known. The project receives special importance today because of the high solar energy feed-ins from private producers, so-called *prosumers*. A massive upgrade of medium voltage lines is becoming necessary. It was confirmed that with AC-DC hybrid lines many effects known from DC lines dominate. In comparison, the effects from the AC circuit are minor. Environmental impacts can also be kept low at medium voltage by an optimized arrangement of the DC phases [181].

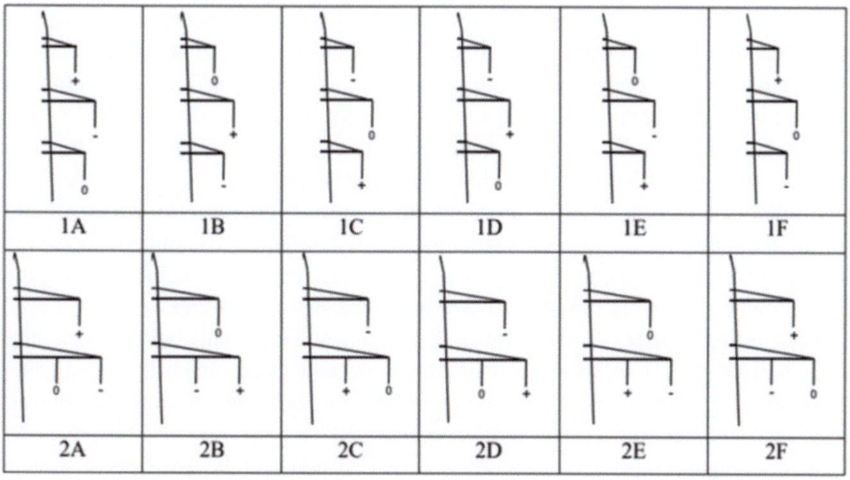

Fig. 4.191 Various DC phase arrangements with reference to environmental aspects [180]

4.10.3 *Converting an Existing AC Overhead Line to DC Operation*

Basically, the findings from Sect. 4.10.1 apply, but here the complete conversion of all AC circuits of an existing AC line to DC is discussed. Such a conversion is one of several methods to increase the performance of a line, as shown in Fig. 4.192. As expected, the lowest costs for the conversion bring the smallest increase of power, and the highest costs the greatest increase. This presentation is of a fundamental nature; in practice each project has to be assessed separately.

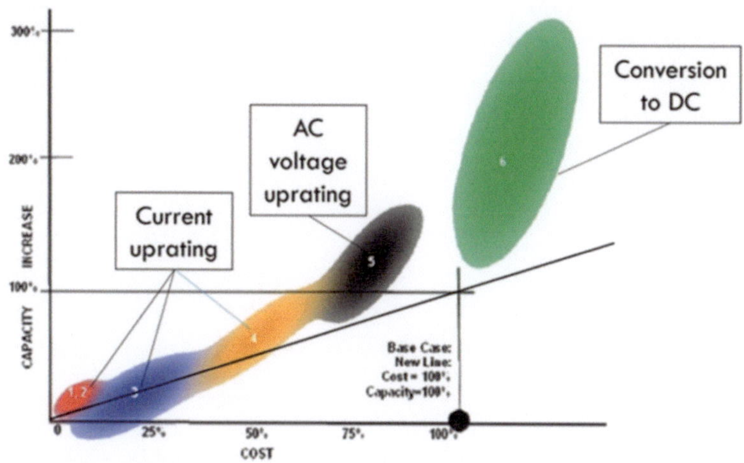

Fig. 4.192 Relative cost and ability of technologies to increase transfer capacity [182]

An AC to DC conversion changes also the impact on the environment. For example, corona noises from an DC line can occur in dry weather, they are a decisive characteristic for a conversion and must be examined in detail. This also applies to changes in the magnetic and the electric field, with the limit values for DC usually being higher than for AC. The magnetic DC field at ground level under the line is in the order of the earth's magnetic field and thus not acritical.

Insulators typically need to be replaced to ensure reliable performance under pollution conditions. Composite insulators are often selected to enable the operating voltage at which the expected increase in transmission capacity is achieved. Live line working when changing the insulators, can reduce shutdown times.

Minimum clearances—on the tower, in the span, and to ground, buildings, or other infrastructure—must be reassessed based on the new overvoltage levels expected from lightning strikes and switching operations. The behavior of converter stations must also be taken into account in this process.

A fundamental design consideration in DC systems is the absence of voltage zero-crossings, which has implications for insulation coordination and arc extinction. Additionally, the positive and negative poles may exhibit different behaviors under certain conditions, such as during pollution events or under transient overvoltages, which further influences insulator and clearance design.

An example from Italy shows how by switching to DC and increasing the operating voltage from AC 220 kV to DC 400 kV can be achieved. It was necessary to change the metallic crossarms against insulating crossarms, and to change the conductors with those allowing a higher operating temperature. With these measures, the transmission power can be doubled and even exceeded when using high-temperature conductors (Fig. 4.193). The use of ground wire as return conductor would also be possible and would bring the same power gain [183, 184].

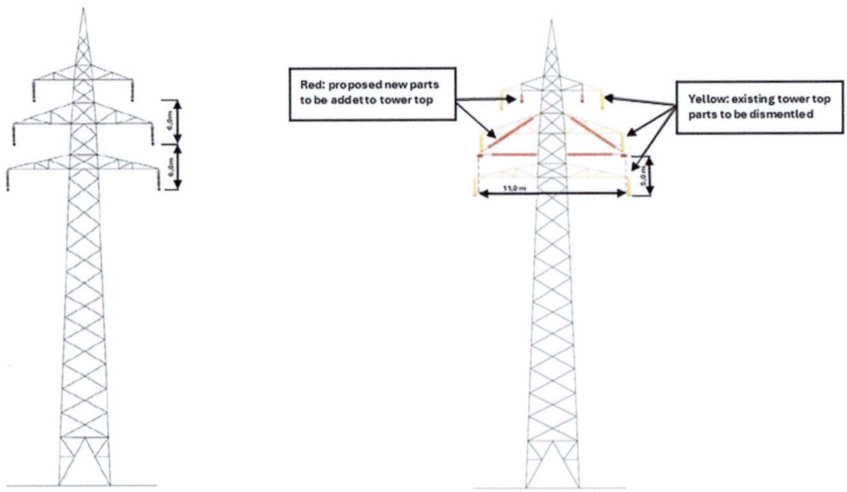

Fig. 4.193 Possibility of the conversion of an AC 220 kV line to a bipolar DC 400 kV-line with a metallic return conductor, suspension tower, *left*: existing AC tower; *right*: conversion to DC [183, 184]

Table 4.17 Possibility of converting an AC 220 kV line to a bipolar DC 400 kV line with a metallic return conductor, comparison of voltages and conductors [183, 184]

	I [A]	AC power		DC power		Standard values [mm/kV]	Design [mm/kV]	String length [mm]	Creepage/ string length
		Vac [kV]	[MVA]	Vdc [kV]	[MW]				
ACSR a 55°	420	220	320	400	672	14,400	14,810	3595	4.12
ACSR a 75°	810	220	617	400	1.296	14,400	14,810	3595	4.12
Hi-STACIR 480A 130 °C	1400	220	1.067	400	2.240	14,400	14,810	3595	4.12

It should also be considered whether any existing transmission reserves may have decreased after such a conversion and whether the line is now operating closer to its thermal limit under normal conditions (Table 4.17).

4.11 Recycling of Overhead Lines [90, 185]

Every technical structure has a limited life expectancy. This can lead to a replacement because of insufficient transmission capacity, a change in the route due to modified land use, or because the effort for maintenance becomes disproportionately large and expensive. The route can often be reused for a new line. When dismantling an overhead line at the end of its life, which can be 80 to 120 years, provided appropriate design and maintenance, almost all materials can be reused. In some countries, the extent of dismantling is mandated by law or regulatory decrees, which may define specific triggering criteria.

Steel towers—lattice and tubular structures—are cut into pieces and melted down. New tower steel always contains a certain proportion of old material. Whether some elements of the old towers can be stored as spare parts, or even reused for emergency structures, depends on the design. In the case of galvanized or coated towers, the protective materials are separated during the recycling process.

The dismantling of concrete poles and concrete foundations is primarily a mechanical task. The components are usually cut in pieces on site and brought to a recycling plant, where they are further crushed. Foundations are usually removed only to a certain depth. Shredded concrete is used, for example, in road construction, in civil works, and as backfill material. The recycling rate of this material is very high.

The general approach is not to burn dismantled wooden poles. Instead, they can be handed over to residents close-by for other use such as fences, or timber structures. However, this only applies to those poles that are sufficiently free of wood preservatives. The residents usually must declare in writing that they will use those wooden towers only as specified by the line operator.

Fig. 4.194 Line dismantling: cutting off a crossarm (*left*), demolition of a concrete foundations (middle), storage of the dismantled tower parts on steel plates to protect the soil (*right*)

Grounding systems are made of steel or copper. They can reach several meters in the ground far from the foundation and are pulled out for dismantling, depending on the laying depth. Both materials can be easily and efficiently reused.

During dismantling, conductors and ground wires are pulled off and wound onto drums. They are then sent to recycling facilities, where they are cut into short lengths and separated into their different materials using standard processes. Conventional conductor and ground wires consist largely of aluminum and its alloys, steel, zinc (from the galvanizing of the steel wires), and protective grease against corrosion. HTLS conductors (High Temperature Low Sag) with cores of non-metallic elements (e.g. Kevlar, carbon fibers) as well as ground wires with glass fibers for telecommunications (OPGW) are treated similarly, the composite cores or glass fibers require special treatment. Materials obtained from conductors are reused for everyday products.

Insulators made of glass and porcelain can be reused for new insulators or otherwise in everyday processes. For composite insulators, the recycling concepts depend on the materials, with reference also being made to related industries (e.g. the automotive industry).

Some situations during the dismantling of overhead lines are shown in Fig. 4.194.

References

1. Papailiou KO (ed) (2021) Springer handbook of power systems. Springer, Singapore
2. Papailiou KO (ed) (2016) CIGRE green book “overhead lines.” Springer, Cham
3. Bückner W (1961) Wirtschaftliche Gestaltung von Hochspannungs-Freileitungen SIEMENS Mag 10
4. CIGRE TB 388 (2009) Impacts of HVDC lines on the economics of HVDC projects. CIGRE, Paris
5. EN 50341-1 (2012) Overhead electrical lines exceeding AC 1 kV – Part 1: General requirements – common specifications. CEN, Brüssel
6. DIN EN 50341-1 (2013) Freileitungen über AC 45 kV – Teil 1: Allgemeine Anforderungen – Gemeinsame Festlegungen; Deutsche Fassung. VDE, Berlin
7. Kiessling F et al (2001) Freileitungen. Springer, Berlin/Heidelberg
8. EN 50341-2-4 (2019) Freileitungen über AC 1 kV – Teil 2–4: Nationale Normative Festlegungen (NNA) für Deutschland. VDE, Berlin

9. CIGRE TB 105 (1996) The mechanical effects of short-circuit currents in open-air substations. CIGRE, Paris
10. CIGRE TB 48 (1995) Tower top geometry. CIGRE, Paris
11. CIGRE TB 348 (2008) Tower top geometry and mid span clearances. CIGRE, Paris
12. Rawlins CB (2005) Analytical elements of overhead conductor fabrication. Fultus, Palo Alto
13. CIGRE TB 345 (2008) Alternating current (AC) resistance of helically stranded conductors. CIGRE, Paris
14. The Aluminum Association (1989) Aluminum electrical conductor handbook, 3rd edn. Aluminum Association, Washington, D.C.
15. EN 50326 (2001) Conductors for overhead lines – characteristics of Greases. CEN, Brüssel
16. BICC (1967) Overhead conductor design. British Insulated Callender's Cables Limited, London
17. Frigerio M (2016) Quantifying the effect of time and temperature on the mechanical behaviour of Aldrey overhead line conductors. PhD thesis 23027, ETH Zürich, Zürich
18. Frigerio M et al (2016) Analysis of the tensile response of a stranded conductor using a 3D finite element model. Int J Mech Sci 106:176–183
19. EN 50182 (2001) Conductors for overhead lines – round wire concentric-lay stranded conductors. CEN, Brüssel
20. Papailiou KO (1997) On the bending stiffness of transmission line conductors. IEEE Trans Power Deliv 12(4):1576–1588
21. Douglass D et al (2017) Overhead lines. In: Papailiou KO (ed) CIGRE green book. Springer, Cham
22. CIGRE TB 324 (2007) Sag-tension calculation methods for overhead lines. CIGRE, Paris. Revision (2016)
23. CIGRE SC22 WG22-04 (1974) A practical method of conductor creep determination. Electra 24:105–137. CIGRE, Paris
24. IEC 61395 (1998) Overhead electrical conductors – creep test procedure for stranded conductors. IEC, Genf
25. CIGRE WG 22.04 (1979) Recommendations for the evaluation of the lifetime of transmission line conductors. Electra 63:103–145. CIGRE, Paris
26. EPRI Transmission Line Reference Book–Wind-Induced Conductor Motion: The "Orange Book", EPRI, Palo Alto (2008)
27. Ziebs J (1970) Über das mechanische Verhalten von Aluminium-Stahl-Freileitungsseilen als Beispiel für Verbundseile. BAM-Bericht Nr. 3, BAM, Berlin
28. Thayer ES (1924) Computing tensions in transmission lines. Electr World Mag 84(2):72–73
29. Healy ES, Wright AJ (1926) Unbalanced conductor tensions. Trans Am Inst Electr Eng
30. Douglass DA et al (2017) Calculation accuracy of high-temperature sag for ACSR in existing lines. CIGRE Sci Eng 7:39
31. IEC 1597 (1995) Overhead electrical conductors – calculation methods for stranded bare conductors. IEC, Genf
32. CIGRE TB 643 (2015) Guide to the operation of conventional conductor systems above 100 °C. CIGRE, Paris
33. Scheel J (2010) Transitionspunktberechnung bei Hochtemperaturseilen und dessen Berücksichtigung in Durchhangsberechnungen. Diplomarbeit, Fachhochschule Gießen-Frieberg
34. CIGRE TB 601 (2014) Guide for thermal rating calculations of overhead lines. CIGRE, Paris
35. Morgan VT (2003) Effects of alternating and direct current, power frequency, temperature and tension on the electrical parameters of ACSR conductors. IEEE Trans Power Delivery 18(3):859–866
36. Puffer R (2012) Freileitungen, Vorlesungsmanuskript. RWTH Aachen, Aachen
37. Farzaneh M et al (2013) Electrical design of overhead power transmission lines. McGraw-Hill, New York
38. CIGRE TB 020 (1974) Interferences produced by corona effect of electric systems. CIGRE, Paris

39. CIGRE TB 838 (2021) Coatings for protecting overhead power networks against icing, corona noise, corrosion and reducing their visual impact. Paris
40. Straumann U, Weber HJ (2010) Potential reduction of audible noise from new and aged overhead transmission line conductors by increasing their hydrophilicity. CIGRE session 2010, paper B2-113. Paris
41. Grigsby LL (ed) (2012) The electric power engineering handbook. CRC Press, Boca Raton
42. Tenbohlen S (2014) Elektrische Energienetze I. Universität Stuttgart, Stuttgart, Vorlesungsmanuskript
43. CIGRE TB 746 (2018) Design, deployment and maintenance of optical cables associated to overhead HV transmission lines. CIGRE, Paris
44. CIGRE WG22-11 (1998) Guide to fittings for optical cables on transmission lines. Part 1: Selection and use. Electra 176. CIGRE, Paris
45. CIGRE WG22-11 (1998) Guide to fittings for optical cables on transmission lines. Part 2A: Testing procedures for optical ground wire fittings and optical phase conductor fittings. Electra 176. CIGRE, Paris
46. CIGRE WG22-11 (1998) Guide to fittings for optical cables on transmission lines. Part 2B: Testing procedures, all-dielectric self-supporting cable fittings and optical attached cable fittings. Electra 176. CIGRE, Paris
47. IEC 60889 (1987) Hard-drawn aluminum wire for overhead line conductors. IEC, Genf
48. CIGRE TB 244 (2004) Conductors for the uprating of overhead lines. CIGRE, Paris
49. CIGRE TB 695 (2017) Experience with the mechanical performance of non-conventional conductors. CIGRE, Paris
50. Lugschitz H (2024) Die moderne Freileitung—Geschichte, Status, Zukunft einer erfolgreichen Übertragungstechnik. Hochspannungssymposium Stuttgart
51. Stephen RG et al (2008) AC to DC Conversion and other techniques to uprate transmission lines based on input from planners. Paper B4-104, CIGRE Session 2008, CIGRE, Paris
52. CIGRE TB 969 (2025) Forecasting dynamic thermal line ratings. Paris
53. CIGRE TB 498 (2012) Guide for application of direct real-time monitoring systems. CIGRE, Paris
54. Sturm RG (1936) Vibration of cables and dampers: Part I, Part II. Trans Am Inst Electr Eng 55(6)
55. Helms R (1964) Zur Sicherheit der Hochspannungsfreileitungen bei hoher mechanischer Beanspruchung, VDI Forschungsbericht 506. VDI, Düsseldorf
56. Claren R, Diana G (1969) Mathematical analysis of transmission line vibration. IEEE Trans Power Appar Syst PAS-88(12)
57. CIGRE SC22 WG01 (1989) Report on Aeolian vibration. Electra 124:40–77. CIGRE, Paris
58. CIGRE TB 273 ((2005)) Overhead conductors safe design tension with respect to Aeolian vibrations. CIGRE, Paris
59. Shillai K (2019) Enhanced fretting fatigue resistance of conductors for high voltage overhead lines. Dissertation, vol 25661. ETH Zürich, Zürich
60. Möcks L, Metha PR, Swart RL (1969) Die statische Biegespannung im Leiterseil als möglicher Einflußfaktor auf die Seillebensdauer, ALUMINIUM, Heft 4, pp 210–220. Aluminium-Verlag GmbH, Düsseldorf
61. Papailiou KO (1987) Schwingungsmessungen an Freileitungsseilen. SEV/VSE Bulletin 23. Electrosuisse, Fehraltorf
62. CIGRE WG B2.11 (1995) Guide to vibration measurements on overhead lines. Electra 163:125–137. CIGRE, Paris
63. IEEE Standard 1368 (2006) Guide for Aeolian vibration field measurements of overhead conductors. IEEE, Piscataway
64. Poffenberger JC, Swart RL (1965) Differential displacement and dynamic conductor strain. IEEE Trans Power Appar Syst PAS-84(4):281–289, Discussion pp 508–513. Errata p 732
65. Cosmai U et al (2017) Conductor motions. In: Papailiou KO (ed) Overhead lines. CIGRE green books. Springer, Cham

66. Papailiou KO (1997) On the bending stiffness of transmission line conductors. IEEE Trans Power Deliv 12(4):1576–1588, Discussion and closure. IEEE Trans Power Deliv 15(4):1329–1331 (2000)
67. Zetterholm OD (1960) Bare conductors and mechanical calculation of overhead conductors. CIGRE Report No. 223, CIGRE, Paris
68. Hearnshaw D (1974) Spacer damper performance – a function of in-span positioning. IEEE Trans Power Appar Syst PAS-93(5):1298–1309
69. Farzaneh M, Chisholm WA (2022) Techniques for protecting overhead lines in winter conditions. CIGRE green book – compact studies. Springer Nature Switzerland AG, Cham
70. IEC TS 60815 (2008) Selection and dimensioning of high-voltage insulators intended for use in polluted conditions. Part 1–4, IEC, Genf
71. DIN EN IEC 60383 (2021) Isolatoren für Freileitungen mit einer Nennspannung über 1 000 V – Teil 1: Keramik- oder Glas-Isolatoren für Wechselspannungssysteme. DIN EN IEC 60383-1; VDE 0674-106-1:2021-02
72. CIGRE WG 22.03 (1995) Service performance of composite insulators used on HVDC lines. CIGRE, Paris
73. IEC 60383-1 (1993) Insulators for overhead lines with a nominal voltage above 1000 V – Part 1: Ceramic or glass insulator units for A.C. systems – definitions, test methods and acceptance criteria. IEC, Genf
74. IEC 60383-2 (1993) Insulators for overhead lines with a nominal voltage above 1000 V – Part 2: Insulator strings and insulator sets for A.C. systems – definitions, test methods and acceptance criteria. IEC, Genf
75. IEC 61365 (1995) Insulators for overhead lines with a nominal voltage above 1000 V – Ceramic or glass insulator units for D.C. systems – definitions, test methods and acceptance criteria. IEC, Genf
76. Bauer E, Brandt E et al (1982) Dynamic processes during load transposition in multiple sets with long rod-type insulators, session paper 22-03. CIGRE, Paris
77. IEC 60433 (1998) Insulators for overhead lines with a nominal voltage above 1000 V – ceramic insulators for A.C. systems – characteristics of insulator units of the long rod type. IEC, Genf
78. IEC 60672-1 (1995) Ceramic and glass insulating materials – Part 1: Definitions and classification. IEC, Genf
79. IEC 60672-3 (1997) Ceramic and glass insulating materials – Part 3: Specifications for individual materials. IEC, Genf
80. IEC 60120 (1984) Dimensions of ball and socket couplings of string insulator units. IEC, Genf
81. IEC 60471 (1977) Dimensions of clevis and tongue couplings of string insulator units. IEC, Genf
82. Baumgartner C (2018) Insulator manufacturing processes. In: ININMR power engineers handbook, Ch. 13, INMR, Quebec
83. Bauer EA, Miller EH (1976) Plastic composite insulators to the system Rodurflex. In: Presentation to the IEEE nonceramic. Composite Insulator Working Group, New York
84. Schmuck F et al (2012) Assessment of the condition of overhead line composite insulators. Paper B2-214. CIGRE, Paris
85. Papailiou KO, Schmuck F (2013) Silicon composite insulators, materials, design, applications. Springer
86. CIGRE WG 22.03 (2000) Worldwide service experience with HV composite insulators. Electra 191. CIGRE, Paris
87. EPRI (2004) Survey of utility experiences with composite/polymer components in transmission class (69–765 kV class) substations. EPRI, Palo Alto
88. Papailiou KO (1999) Composite insulators are gaining ground – 25 years of Swiss experience. IEEE Transm Distrib Conf New Orleans 2:827–833
89. Papailiou KO (2019) Innovative tower solutions & line uprating. In: INMR power engineers handbook. INMR, Quebec

90. Papailiou KO, Schmuck F (2022) Silikon-Verbundisolatoren – Werkstoffe, Dimensionierung, Anwendungen, 2. Springer-Verlag GmbH Deutschland, Auflage
91. Prenleloup A (2008) Analyse de l'état de contrainte et de l'endommagement d'assemblages sertis en matériau mixte métal-composite sollicités en traction ou en flexion. Doctoral thesis No. 4005, EPFL, Lausanne
92. Bisnath S, Pillay T (2005) The planning, design and construction of overhead power lines. Eskom Power Series, Crown, Johannesburg
93. Ansorge S (2015) Improvement of silicone rubberfor high voltage applications by addition of fillers. Doctoral Thesis ETH Zürich. https://doi.org/10.3929/ethz-a-010577999
94. Meyer LH, Cherney EA, Jayaram SH (2004) The role of inorganic fillers in silicone rubber for outdoor insulation – alumina tri-hydrate or silica. IEEE Electr Insul Mag 20(4):13–21
95. Lumb C, Papailiou KO (1997) Unterschätzte Dauerkraft von Verbundisolatoren. SEV/VSE Bull 88(7). Electrosuisse
96. IEC 61109 Ed. 2 (2008) Composite suspension and tension insulators for A.C. overhead lines with a nominal voltage greater than 1000 V-definitions, test methods and acceptance criteria. IEC, Genf
97. IEC 61952. Insulators for overhead lines – composite line post insulators for AC systems with a nominal voltage greater than 1000 V – definitions, test methods and acceptance criteria
98. Prenleloup A et al (2007) Acoustic emission inspection and analysis of crimped metal-composite joints subjected to bending. In: Proceedings of the 4th international conferene on NDT, Chania (2007), S 11–14
99. Stephen R, Iglesias J (2024) Compact overhead line design – AC and DC lines. Springer, Cham
100. Ammann M, Papailiou KO (1998) A new 400 kV line with compact towers and composite insulated cross-arms. CIGRE Sess., Paris, Paper 22/33/36-06
101. George J, Iqbal M, Papailiou KO, Huiber W, Schmuck F (2007) Dewa's new 420 kV compact line as an example of maximizing capacity of overhead lines in an urban environment. In: GCC power conference, Dubai, UAE
102. Goffinet JF (2015) Experience at Elia, the Belgian TSO: designing, testing, building & maintaining compact towers with insulated cross-arms. INMR World Congress, Munich
103. CIGRE SC22 WG01 (1989) Guide on the use of bolted suspension clamps. Electra 123. CIGRE, Paris
104. Ramey, Townsend JS (1981) Effects of clamps on fatigue of ACSR conductors. J Energy Div 107(1):103–110
105. CIGRE TB 653 (2016) Safe design tensions for single conductors fitted with elastomer cushioned suspension units. CIGRE, Paris
106. Maier HA, Schlotz H et al (1989) Ein neues Berechnungsverfahren von Schwingungsdämpfern für Freileitungen. Elektrizitätswirtschaft 88(23):1616–1623
107. Möcks L (1969) Das Verhalten des Leiterseiles unter dem Einfluss der Armaturen bei Freileitungen. Elektrizitätswirtschaft 68:336–341
108. CIGRE TB 964 (2025) Increasing the strength capacity of existing overhead transmission line structures. Paris
109. Hildmann C (2016) Zum elektrischen Kontakt- und Langzeitverhalten von Pressverbindungen mit konventionellen und Hochtemperatur-Leiterseilen mit geringem Durchhang, Diss. TU Dresden, Dresden
110. Raupp D (2009) Wirkungsweise einer gewalzten Seilendverbindung unter Zug- und Zugschwellbeanspruchung. Dissertation, Institut für Fördertechnik und Logistik der Universität Stuttgart
111. CIGRE TF B2.11.03 (2006) State of the art survey on aircraft warning markers. Electra 224:24–29. CIGRE, Paris
112. Möcks L (1982) Lichtbogenschutzarmaturen für Isolierstrecken in Freileitungen und Schaltanlagen. ETZ-Report 16, VDE, Berlin, Offenbach
113. Phillips AJ et al (2008) Electric fields on AC composite transmission line insulators. IEEE Trans Power Deliv 23(2):823–830

114. CIGRE WG 22.03 (1992) Use of stress control rings on composite insulators. Electra 143. CIGRE, Paris
115. CIGRE TB 365 (2008) On the use of power arc protection devices for composite insulators on transmission lines. CIGRE, Paris
116. Stockbridge GH (1925) Overcoming vibration in transmission lines. Electr World 86:1304–1305
117. IEC 61897 (1998) Overhead lines – requirements and tests for Stockbridge type Aeolian vibration dampers. IEC, Genf
118. Diana G (ed) (2018) Modelling of vibrations of overhead line conductors. CIGRE Green Books, Springer, Cham
119. Whitehead JB (1910) System of electrical transmission. U.S. Patent No. 1078711
120. Markt G, Mengele B (1932) Drehstromübertragung mit Bündelleitern. Elektrotechn Maschinenbau 50:293–298
121. Markt G, Mengele B (1930) Elektrische Leitung mit Bündelleitern. Österreichisches Patent 121:704
122. CIGRE TB 277 (2005) State of the art survey on spacers and spacer dampers. CIGRE, Paris
123. CIGRE TB 214 (2002) The mechanical effects of short-circuit currents in open air substations. Companion Brochure to No. 105, CIGRE, Paris
124. Lilien JL, Papailiou KO (2000) Calculation of spacer compression for bundle lines under short-circuit. IEEE Trans Power Deliv 15(2):839–845
125. Manuzio C (1965) An investigation of the forces on bundle conductor spacers under fault conditions. IEEE Trans Power Appar Syst PAS-86(2):166–185
126. CIGRE TB 818 (2020) Transmission line structures with fiber reinforced polymer (FRP) composite. CIGRE, Paris
127. CIGRE TB 384 (2009) Comparison of general industry practices for lattice tower design and detailing. Electra 244. CIGRE, Paris
128. EN 1993-1-1 (2005) Eurocode 3: design of steel structure – Part 1: General rules for buildings. CENELEC, Brüssel
129. EN 10025 (1994) Hot-rolled products of non-alloy structural steels, technical delivery conditions. CENELEC, Brüssel
130. EN ISO 898-1 (1999) Mechanical properties of fasteners made of carbon steel and alloy steel, Part 1: Bolts, screws and studs. CENELEC, Brüssel
131. Bückner W, Rieger H (1956) Freileitungsmaste mit drei Eckstielen. SIEMENS Zeitschrift, Juni 1956, H. 5–7, S. 334 ff
132. CIGRE WG B2.38 (2015) Evaluation of high surge impedance loading (HSIL) solutions for increased natural capacity of 500 kV overhead lines. CIGRE Sci Eng 2:63. CIGRE, Paris
133. Silva JBGF et al (2017) Supports. In: Papailiou KO (ed) Overhead lines. CIGRE green book. Springer, Cham, pp 559–712
134. de Menezes RCR et al (2024) Dynamic analysis on the 2 × 500 kV overhead line crossing over the Chacao channel in Chile. CIGRE Colloquium Krakau PL
135. CIGRE TB 399 (2009) Improvement on the tower testing methodology. Electra 247. CIGRE, Paris
136. IEC 60652 (2002) Loading tests on overhead line structures. IEC, Genf
137. CIGRE TB 416/416A. Innovative solutions for overhead line supports
138. George J et al (2007) DEWA's new 420 kV compact line as an example of maximizing capacity of overhead lines in an urban environment. CIGRE/GCC Power Conf., Dubai
139. Papailiou KO, Schmuck F (2011) The past experience and future trends with compact lines to solve right-of-way issues. CIGRE international symposium, Bologna
140. Lewis WW (1965) The protection of transmission systems against lightning. Dover, New York
141. Dorsch H (1981) Überspannungen und Isolationsbemessung bei Drehstrom-Hochspannungsanlagen. Siemens, Erlangen
142. CIGRE TB 694 (2017) Ground potential rise at overhead transmission line structures during power frequency faults. CIGRE, Paris

143. Nolasco JF et al (2017) Electrical design. In: Papailiou KO (ed) Overhead lines. CIGRE green book. Springer, Cham, pp 47–190
144. Cuer NR (2017) Foundations. In: Papailiou KO (ed) Overhead lines. CIGRE green book. Springer, Cham, pp 933–1046
145. BS 5930 (2015) Code of practice for ground investigations. BSI, London
146. DIN 18300:2015-08 (2015) VOB Vergabe- und Vertragsordnung für Bauleistungen – Teil C: Allgemeine Technische Vertragsbedingungen für Bauleistungen (ATV) – Erdarbeiten. Beuth Publishing DIN
147. Rieger H, Fischer R (1975) Der Freileitungsbau. Springer, Berlin/Heidelberg
148. CIGRE TB 308 (2006) Foundation installation – an overview. CIGRE, Paris
149. CIGRE TB 281 (2005) Design and installation of micropiles and ground anchors for OHL support foundations. CIGRE, Paris
150. IEEE 977-2010 (2010) Guide to installation of foundations for transmission line structures. IEEE, New York
151. DIN V ENV 206:1990-10 (1990) Beton – Eigenschaften, Herstellung, Verarbeitung und Gütenachweis. Beuth Publishing DIN
152. DIN 48207:1978-07 (1978) Leitungsseile; Verlegen von Freileitungsseilen. Beuth Publishing DIN
153. IEEE 524 (2016) Guide for the installation of overhead transmission line conductors. IEEE, New York
154. CIGRE TB 471 (2011) Working safely while supported on aged overhead conductors. CIGRE, Paris
155. CIGRE TB 175 (2000) Management of existing overhead transmission lines. CIGRE, Paris
156. CIGRE TB 353 (2008) Guidelines for increased utilization of existing overhead transmission lines. CIGRE, Paris
157. CIGRE WG B2.13 (2005) Guidelines for emergency resource planning for overhead transmission line asset owners. Electra 222. CIGRE, Paris
158. Pouliot N et al (2016) Portable X-ray system for in situ detection of broken ACSR strands at suspension clamps – field results and introduction onto line scout roboting technology, session paper B2-106. CIGRE, Paris
159. Leblond A, Lindsey KE (2017) Maintenance. In: Papailiou KO (ed) Overhead lines. CIGRE green book. Springer, Cham, pp 1151–1208
160. CIGRE TB 216 (2002) Joints on transmission line conductors – field testing and replacement criteria. CIGRE, Paris
161. CIGRE TB 477 (2011) Evaluation of aged fittings. CIGRE, Paris
162. CIGRE TB 306 (2006) Guide for the assessment of old cap and pin and long rod transmission line insulators made of porcelain or glass: what to do and when to replace. CIGRE, Paris
163. CIGRE TB 481 (2011) Guide for the assessment of composite insulators in the laboratory after their removal from service. CIGRE, Paris
164. EN ISO 2409 (2013) Paints and varnishes. Cross-Cut Test, ISO, Genf
165. ASTM D3359 (2017) Standard test methods for rating adhesion by tape test. ASTM, West Conshohocken
166. Schauer H (1981) Effect of powdery avalanches on the design and behaviour of overhead transmission lines in Austria. CIGRE Symposium Stockholm, pp 22–81
167. CIGRE TB 767 (2019) Vegetation fire characteristics and the potential impacts on overhead line performance. Paris
168. CIGRE TB 561 (2013) Live work – a management perspective. CIGRE, Paris
169. Jordan Th (2014) Arbeiten unter Spannung (AuS) an Hochspannungsfreileitungen – Quo vadis Deutschland?, 9. ETP-Konferenz, Köln
170. CIGRE TB 865 (2022) Inspection and testing of tools, equipment and training for live-line work on overhead lines. Paris
171. CIGRE TB 731 (2018) The use of robotics in assessment and maintenance of overhead lines. CIGRE, Paris
172. N. N., Single-system temporary line with guyed single towers, Lindsay USA

173. CIGRE TB 294 (2006) How overhead lines are redesigned for uprating/upgrading – analysis of the replies to the questionnaire. CIGRE, Paris
174. Schichler U et al (2016) Test-run for uprating a 220 kV OHL to 380 kV using insulated cross-arms and coated conductors. Session paper B2-301. CIGRE, Paris
175. Just M (2024) Verstärkung bestehender Stützpunkte von Freileitungen, Fachseminar "Grundlagen und ausgewählte Themen des Hoch- und Höchstspannungsleitungsbaus (7). Tagungsgesellschaft MVV Regioplan GmbH, Karlsruhe
176. Kant Ch et al (2019) Design & construction of India's first multi-pole +/-320 kv HVDC transmission line. CIGRE Colloquium New Delhi
177. Dass R et al (2006) Operation experience from bulk power HVDC links from three gorges complex. B-208, CIGRE Paris
178. CIGRE TB 583 (2014) Guide to the conversion of existing AC lines to DC operation. CIGRE, Paris
179. Sander B et al (2014) Conversion of AC multi-circuit lines to AC-DC hybrid lines with respect to the environmental impact. Session paper B2-105. CIGRE, Paris
180. Sander B et al (2014) Conversion of AC multi-circuit lines to AC-DC hybrid lines with respect to the environmental impact. CIGRE 2014 B2-105
181. Lee DI et al (2012) A study for electric environment characteristics in 154kV AC / 80 kV DC hybrid transmission line D.I. LEE. CIGRE 2012-B2-109, CIGRE, Paris
182. CIGRE TB 425 (2010) Increasing capacity of overhead transmission lines—needs and solutions. Paris
183. CIGRE TB 583 (2014) Guide to the conversion of existing ac lines to dc operation. Paris
184. Colla L et al (2010) Analysis on the possible conversion of overhead lines from AC to DC. CIGRE Session Paris 2010, paper B2-201
185. CIGRE TB 265 (2004) Life cycle assessment (LCA) for overhead lines. Paris

Chapter 5
Underground Cables

Summary

Underground cables are used in place of overhead lines where the latter cannot be technically implemented—for example, when crossing bodies of water or nature reserves, or when supplying islands and offshore platforms. In addition, underground cables are indispensable in densely populated urban areas with high energy demands, particularly across much of the industrialized world.

This chapter introduces the fundamentals of cable construction, which—beyond conductor material and cross-sectional area—is largely defined by the type of insulation and the protective measures against contact and mechanical damage. It begins with key characteristics of the two most commonly used conductor materials, copper and aluminum, along with an explanation of the skin effect and proximity effect. Next, the concepts of *service life* and *operating capacity* are discussed, followed by detailed descriptions of the main high and extra-high voltage cable types: mass-impregnated cables, oil-filled and gas pressure cables, extruded insulation cables (including submarine cables), and gas-insulated transmission lines (GIL). Subsequent sections address the specific design differences between alternating current (AC) and direct current (DC) applications. Under the topic of cable accessories, joints and terminations are covered, including their construction as plug-in types and the standard grounding techniques—such as cross bonding. Various installation methods are then presented and compared in terms of their advantages and disadvantages, with particular attention given to cable transport and pulling procedures, including relevant calculation methods. The chapter also explores cable and soil heating, along with possible cooling techniques. Finally, the chapter concludes with an overview of commissioning tests, monitoring systems, and fault detection methods.

K. O. Papailiou et al., *High-Voltage Overhead Lines and Underground Cables*, Power Systems, https://doi.org/10.1007/978-3-032-07498-0_5

5.1 Cables in Energy Transmission

Cables can be laid in, and on the ground, in cable trenches, in tunnels, in water and indoors. They must be protected against moisture ingress and chemical decomposition, often also against UV radiation, and must be highly resistant to electrical and mechanical stress in order to achieve reliability values similar to those of overhead lines.

To meet a wide range of technical requirements, cables have been developed with varying numbers and shapes of conductors, as well as a broad spectrum of conductor cross-sections. In addition, a variety of insulation, shielding, and outer jacket combinations are employed to ensure long service life and minimize the need for maintenance.

Because symmetrical load conditions are typically present from the medium-voltage level upward, the use of a fourth conductor—commonly referred to as the neutral conductor—is generally unnecessary. Therefore, this chapter focuses exclusively on three-core and single-core cable designs.

The electrical behavior of a cable is defined by its fundamental electrical parameters. In alternating current (AC) operation—unlike in direct current (DC)—the cable is characterized not only by its ohmic resistance R and shunt admittance G (the reciprocal of the insulation resistance), but also by its inductance L, which describes the magnetic field, and its capacitance C, which represents the electric field.

In contrast to overhead lines, underground cables are significantly affected by the skin effect (current displacement within the conductor) and the proximity effect (influence of nearby metallic conductors). These phenomena alter current distribution and increase effective resistance, especially at higher frequencies. Their effects are described in more detail below.

Chapter 6 further explores the role of these electrical parameters and explains how they are applied in network calculations.

5.2 Cable Construction

Copper and aluminum are the primary materials used for power cable conductors. At 20 °C, the following values apply for the specific conductivity of Copper and Aluminum: $\kappa_{Cu} \approx 56\,\mathrm{m}/(\Omega\,\mathrm{mm}^2)$ and $\kappa_{Al} \approx 35\,\mathrm{m}/(\Omega\,\mathrm{mm}^2)$. For example, a copper conductor of 1 km length and 300 mm^2 cross-section, has a DC resistance of 59.52 mΩ and an aluminum conductor of the same size, 95.24 mΩ. However, cable manufacturers do not specify the geometric conductor cross-section, but the electrically effective nominal cross-section, which is determined from a DC resistance measurement at room temperature (20 °C).

As noted in Chap. 1, conductor resistance can increase by several percent during AC operation due to effects such as the skin effect. In addition, for both copper and aluminum, resistance increases by approximately 4% for every 10 K rise in temperature. Aluminum is significantly lighter than copper. The specific densities are approximately $\alpha_{Cu} \approx 8.92\,\text{kg/dm}^3$ and $\alpha_{Al} \approx 2.7\,\text{kg/dm}^3$. To achieve the same electrical resistance over a given conductor length, the cross-sectional area of an aluminum conductor must be about 60% greater than that of a copper conductor. Despite this increase in volume, the aluminum conductor weighs only about half as much as its copper counterpart. However, copper offers superior corrosion resistance compared to aluminum.

The conductor, as the core component of a cable, can be constructed as either a single-strand or a multi-strand element. For conductor cross-sections of 25 mm^2 and above, a multi-strand design is typically used to facilitate handling, transport, and installation.

The conductor shape is optimized to make the most efficient use of the cable sheath cross-section—achieving a high fill factor. Common designs include round conductors, uncompacted and compacted sector-shaped conductors, as well as segmental conductors divided into multiple parts.

In certain cable types, such as external gas pressure cables, single conductors with an oval cross-section are often used. These allow for more efficient compression of the surrounding sheath, supporting the mechanical and sealing requirements of the system. Segmented conductors are referred to as *Milliken conductors*, where the current is divided among many individual wires (Fig. 5.1). For this purpose, these are insulated from each other during production by a thin layer of lacquer and twisted, which results in a very complex connection technology.

Milliken conductors are designed to reduce the *skin effect*, which causes non-uniform current distribution in solid conductors under alternating current (AC) conditions. Unlike direct current (DC), where the current density is evenly distributed across the entire conductor cross-section, AC induces *eddy currents* that lead to a displacement of current toward the outer regions of the conductor. This phenomenon occurs because the alternating current generates alternating magnetic fields within the conductor (Fig. 5.2, green). These internal magnetic fields weaken the current density

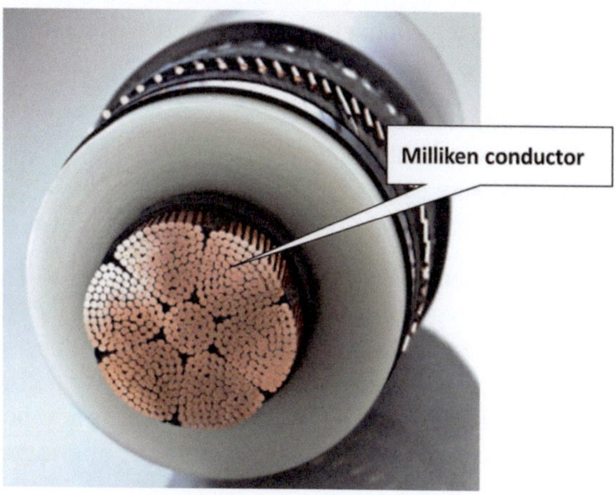

Fig. 5.1 Single-conductor cable with Milliken layout

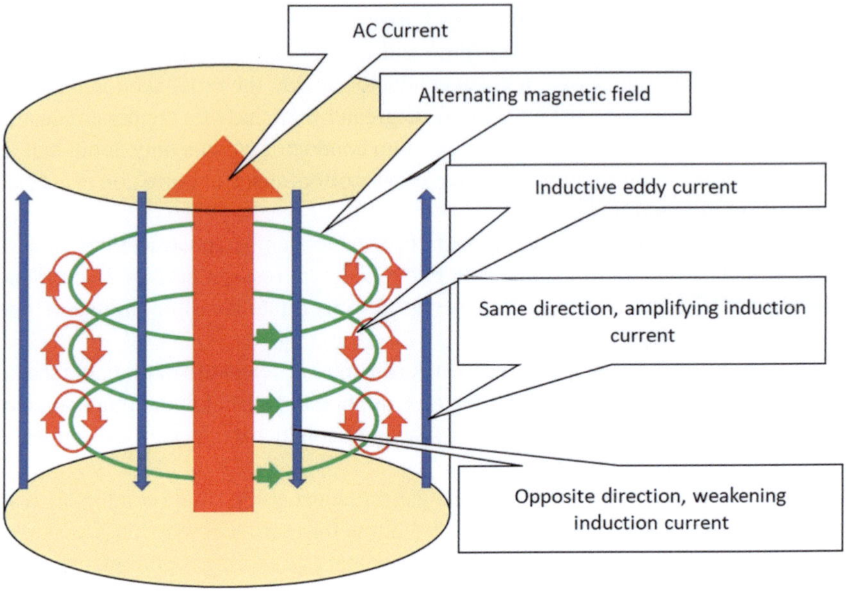

Fig. 5.2 Skin effect caused by alternating current in a conductor

in the core and strengthen it near the surface (Fig. 5.2, blue), effectively reducing the usable cross-sectional area of the conductor. To counter this effect, Milliken conductors are constructed by subdividing the conductor into multiple smaller, mutually insulated strands. This division significantly reduces the skin effect at standard power frequencies (50–60 Hz), resulting in conductor losses that are nearly equivalent to those experienced under direct current conditions.

Another phenomenon, the so-called *proximity effect*, is based on the principle of induction and can be considered an extended skin effect. The variable magnetic field generated by the alternating current causes current displacement in adjacent current-carrying conductors, which increases their effective resistance [1].

Cables are classified according to their *insulation* into:

- Mass impregnated cables
- Oil pressure cables
- Gas pressure cables
- Extruded cables.

In a broader sense, gas-insulated lines (GIL) can also be considered, as they combine the advantages of pressurized insulating gas with the compact spatial footprint typical of cable systems. Since the discovery of high-temperature superconductors (HTSC) in 1986, oxide-ceramic conductor materials have become available that can be cooled using liquid nitrogen. This development opens new possibilities for the transmission of high electrical power via cable systems, offering extremely low losses and high current densities. Further details on superconducting cables and their applications can be found in Chap. 7.

As described in Chap. 1, the earliest power cable insulation systems used *gutta-percha*—a rubber-like natural latex—which was soon replaced by paper windings embedded in a low-viscosity, resin-like mass. Cables constructed in this way are known as mass-impregnated cables, or simply mass cables. These cables are no longer used for high and extra-high voltage applications in new installations. One of the critical challenges with mass-impregnated insulation systems is their sensitivity to thermal expansion. Variations in electrical load cause temperature fluctuations, which can lead to expansion of the insulating material and the formation of cavities. These cavities contain air, which has significantly poorer dielectric properties than solid insulation. Under high electric stress, such voids can lead to localized electrical breakdowns that, over time, cause thermal and chemical degradation of the insulating material. The root cause lies in the uneven distribution of the electric field. Even minor defects in the dielectric can create localized field intensifications that are difficult to control and which contribute to long-term insulation failure. Figure 5.3 illustrates the effect of an air inclusion within the dielectric. The concentration or "squeezing" of electric field lines in the lower-permittivity region (air) leads to a local increase in field strength, triggering partial discharges and progressive material breakdown. As will be discussed later, similar electric field distortions can also occur when the conductor is made of multiple strands, or when the three phase conductors in a cable are placed within a common outer sheath.

To prevent the formation of cavities within the insulation, cable designs were developed in which the paper insulation is kept under pressure by filling the dielectric with low-viscosity oil or nitrogen. This pressure prevents void formation caused by thermal expansion under varying electrical loads, thereby avoiding partial discharges and the associated degradation of the insulation. Cables using this principle are referred to as *thermally stabilized cables*. However, these designs have largely been replaced by modern polymer-insulated cables, most notably those using polyvinyl

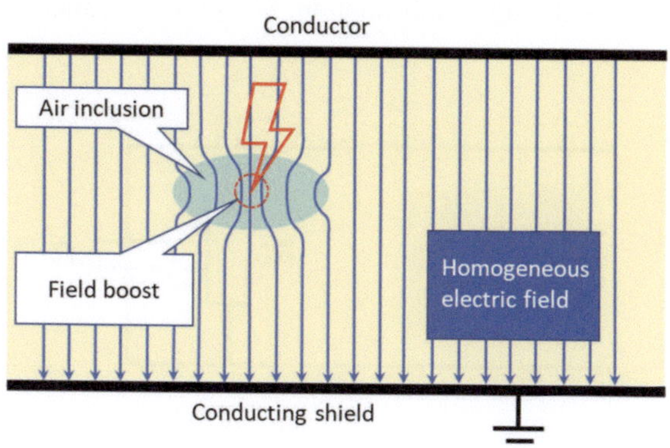

Fig. 5.3 Electrical field distribution in a dielectric with inclusion of air

chloride (PVC) and polyethylene (PE). In contrast to oil-filled systems, polymer-insulated cables eliminate the environmental risks associated with oil leakage in the event of mechanical damage.

Unfortunately, good electrical insulation properties go hand in hand with a good thermal insulation, which adversely affects the dissipation of heat from losses generated in the conductor through the outer sheath of the cable. Thus, cables for high voltages can only be operated with relatively low current density, as they would otherwise heat up too much.

The transferable power of a cable depends on several factors: the current-carrying capacity (determined by the conductor material and cross-sectional area), the operating voltage (determined by the insulation strength and quality), and the type of installation, which affects heat dissipation. As with overhead lines, the conductor temperature in steady-state operation is primarily influenced by the current and surrounding environmental conditions (Sect. 4.3.5). However, unlike overhead lines—where the conductor is typically the limiting factor—in cables, it is the insulation that defines the thermal limit. For example, in cables insulated with cross-linked polyethylene (XLPE), the maximum permissible operating temperature is 90 °C. Exceeding this temperature accelerates aging of the insulation material, leading to degradation and eventual failure of the cable.

The insulation in high-voltage cables must exhibit extremely low residual conductivity. However, once an AC voltage is applied, measurable dielectric losses occur—even under no-load conditions. As a general guideline, an XLPE-insulated cable operating at the 380 kV level typically exhibits dielectric losses of approximately 3–4 kW per kilometer and per phase. These losses are primarily due to *polarization* of dipoles within the insulation material, the dielectric. As a result, the phase angle between current and voltage deviates slightly from the ideal 90° of a perfect capacitor. This deviation is quantified by the *loss angle* or *dissipation angle* δ as illustrated in Fig. 5.4.

The purely capacitive current I_c, which leads the applied voltage by 90°, combines with the active current I_{loss}, representing dielectric losses, to form the total current flowing through the insulation. The following relationship applies:

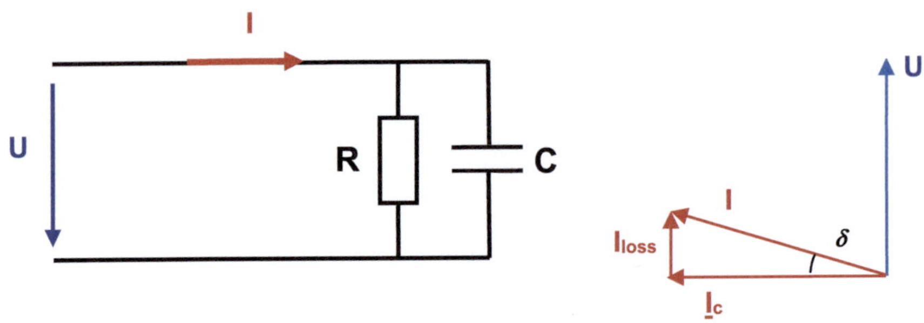

Fig. 5.4 Equivalent circuit of a dielectric with losses and vector diagram of voltage and current during operation with alternating voltage

$$I_{loss} = I_c \cdot \tan\delta \tag{5.1}$$

A corresponding relationship also applies to the power quantities. That is, the *power loss* P_{loss} is related to the *reactive power* Qc of the cable as follows:

$$P_{loss} = Q_c \cdot \tan\delta \tag{5.2}$$

wherein is:

I_{loss} Loss current (active current) (A)
I_c Capacitive current (A)
$tan\ \delta$ Dissipation factor (Loss factor) (–)
P_{loss} Power loss (W)
Q_c Capacitive reactive power (Var).

The expression $tan\ \delta$ is referred to as the *dissipation factor* or *loss factor* and reflects an important material property of the dielectric. It should be significantly less than one percent in high-voltage cables by using suitable materials.

The cable manufacturers provide a *rated current* I_r for the individual types and cross-sections, which refer to the "*standard load*" described below (Fig. 5.5) under certain soil conditions. If the environmental conditions or load profiles deviate from the values mentioned above, an empirically determined relationship between *rated current* I_r and *maximum permissible operating current* I_{max} is given using conversion factors. For example, $I_{\max} = 0.85 \cdot I_r$ applies if a continuous load is to be assumed

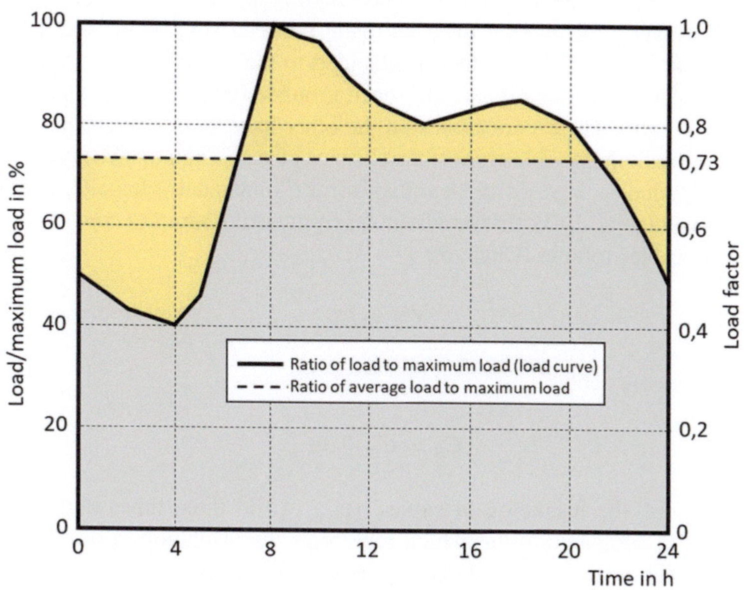

Fig. 5.5 Standard load and associated load factor

instead of a standard load. In the event of a short circuit, the cable is subjected to a significantly higher fault current until the fault is switched off. In the context of required short-circuit calculations, cable manufacturers typically specify a so-called *rated short-time current density*.

The influencing factors of the environment of cables include the soil material, its temperature and humidity, the nature of the ground surface, the laying depth and arrangement, and possible parallel laid (or occasionally also crossing) cables and, for example, district heating pipes. From the limiting temperature, the maximum permissible *operating current* I_{max} is finally determined for each cable type and cross-section.

Underground cables have a higher thermal time constant than overhead lines, meaning that their temperature reacts more slowly to changes in load. As a result, the temporal *load profile* plays a significant role in determining the thermal behavior and permissible loading of the cable. To characterize this profile, the *load factor m* is defined.

The load factor is determined by converting a 24-h load curve—normalized to its maximum value—into a rectangle of equal width (time) and equal area. The height of this area-equivalent rectangle corresponds to the load factor m. A load profile with $m = 1$ represents a continuous (constant) load, whereas a value of $m \approx 0.7$ is typical in practical applications and referred to as the *standard load*. The following illustration depicts such a load profile along with the corresponding load factor (here $m = 0.73$) (Fig. 5.5).

The inductance and capacitance of cables are strongly influenced by their geometric configuration. Basically, large distances between conductors of the same circuit, or conductor-shield distances, cause high inductance and small capacitance values. However, a distinction must be made between single-core and three-core cables. In the first, the electric field is shielded to the outside by a thin metal or a conductive paper layer or by a shield (each grounded), so that only a capacitance C_E exists to earth. Three-core cables on the other hand also have capacitances C_L between the individual conductors in the absence of such a conductive layer. Considering that C_E and C_L have different voltages (for C_L the rated voltage U_r and for C_E the phase voltage $U_r/\sqrt{3}$), this results in an equivalent *operating capacity* C_B. For single-core cables follows (Chap. 6):

$$C_B = C_E \tag{5.3}$$

and for three-core cables

$$C_B = C_E + 3\ C_L \tag{5.4}$$

Furthermore, the insulation of cables has a two to three times higher *relative permittivity* ε_r compared to air, which represents the insulation in overhead lines. Because of this and due to their geometry, cables have much higher capacities, while their inductances are in the same order of magnitude as those of overhead lines. These differences play a role wherever cables and overhead lines coexist in a power supply grid, whether they are in series or parallel connection (Fig. 5.6).

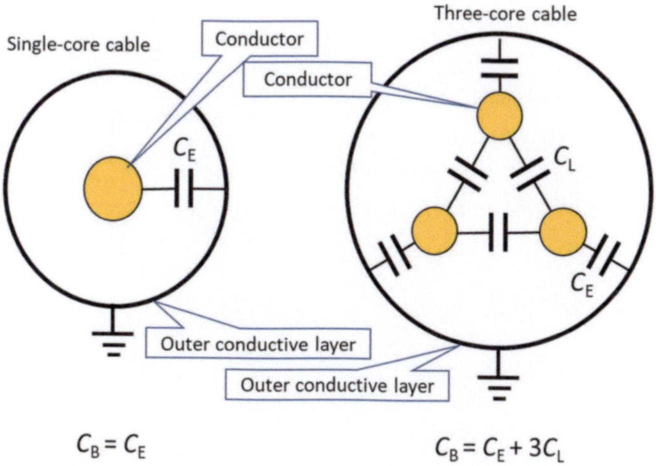

Fig. 5.6 Capacitance in a single-core cable (*left*) and in a three-core cable (*right*)

Furthermore, the insulation materials used in cables have a relative permittivity which is typically two to three times higher than that of air—the insulating medium in overhead lines. As a result, and due to their compact geometry, cables exhibit significantly higher capacitance values. In contrast, their inductance remains in the same order of magnitude as that of overhead lines.

For network calculations, the primary interests are the type and length of the line, including the conductor material and cross-section, as well as the transferable power depending on the type of installation. Other electrical characteristics are provided by the manufacturer as *relative quantities* R' (ohmic resistance), G' (shunt conductance), L' (conductor inductance) and C_B' (operating capacity) per unit length. The significance of these characteristics for network operation is explained in more detail in Chap. 6. There, links to daily practice is established using various examples.

Figure 5.7 illustrates the fundamental design of a single-core cable commonly used in electrical power systems.

Single-core cables with a solid conductor have a radially symmetric electric field, with the highest field strength occurring at the surface of the conductor, as the field line density is highest there. For larger cross-sections, several individual wires are used (except for the special case of the Milliken conductor), to avoid too large bending radii. To achieve homogeneity of the field, the individual wires are surrounded by an *inner semiconductive layer*. This layer consists of metallized paper or a wrapping made from weakly conductive plastic and is not related to semiconductors used in electronic applications. In the following, it will be referred to as the inner conductive layer. Its primary function is to homogenize the electric field in the region of highest field strength, as illustrated in Fig. 5.8 (right).

At the outer edge of the dielectric, a conductive and grounded layer is also applied—known as the outer conductive layer. Together, these layers ensure that

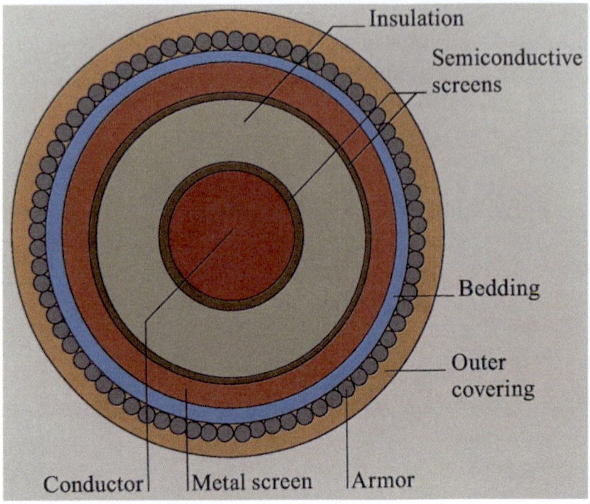

Fig. 5.7 Basic structure of a single core cable [2]

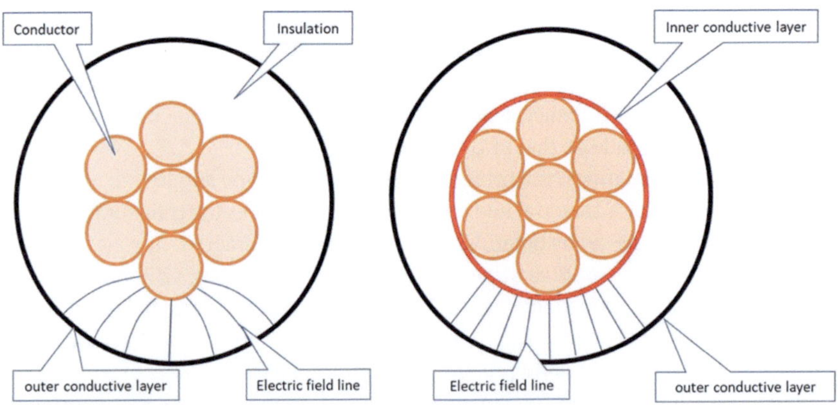

Fig. 5.8 Distribution of the electric field without (*left*) and with inner conductive layer (*right*)

the electric field is entirely confined within the dielectric, preventing field distortion and enhancing the operational reliability of the cable.

When each conductor in a three-core cable is enclosed by a separate, grounded outer conductive layer, electric field formation between the individual conductors is effectively prevented. Cables with such field control are referred to as *radial field cables*. A three-core cable of this type, developed at the beginning of the twentieth century, is known under the name *Höchstädter cable*.

In three-core cables, the magnetic fields generated by the individual phase conductors—due to their varying instantaneous currents—largely cancel each other out externally. This mutual cancellation significantly reduces external magnetic effects. However, in the case of three single-core cables laid in parallel, this cancellation does not occur in the same way. The magnetic fields from each core penetrate the surrounding conductive layers without obstruction and cannot be neutralized solely by the arrangement of the conductors. As a result, to minimize the resulting eddy current losses in such systems, non-magnetic materials should be used for the outer metallic sheath or armor. Alternatively, special techniques such as cross-bonding or single-point grounding of the cable shielding should be implemented, as discussed in Sect. 5.6.

Compared to overhead lines, cables have various advantages as well as disadvantages: They are generally much more expensive; on the other hand, their transmission capacity is independent of the weather, unlike overhead lines.

Once installed, underground cables are invisible except for any existing jointing structures and require no maintenance. At the same time, they are largely protected against atmospheric and weather influences. By using significantly larger conductor cross-sections, they tend to transmit higher currents than overhead lines.

The *ampacity* of a cable system can only be expanded by completely replacing it with cables of larger cross-section or higher voltage. Alternatively, at a later stage, a second cable system can be laid in parallel.

In the event of a fault—such as a lightning strike—the air as an insulating medium in overhead lines is almost always self-healing. This means that an arc extinguishes if the line is temporarily disconnected by the network protection. In contrast, damaged cable insulation no longer works, and the affected cable must be repaired by time-consuming replacement of the damaged section.

The high capacity values compared to overhead lines require compensation and monitoring devices for longer cable routes in AC grids. This means higher construction and operating costs and requires additional structures along a cable route.

5.3 Structure of High Voltage Power Cables

All cable types used for high-voltage energy transmission feature a relatively complex, multi-layered structure. Their primary function is to safely reduce the high voltage potential of the conductor to ground potential within a tightly confined space.

The electric field in a high-voltage power cable is restricted to the region between the inner and outer conductive layers of the insulation. Depending on the cable design and intended application, additional layers—such as *bedding tape* and a *metallic shield*—are typically applied outside the insulation system. Today, the metallic shield is commonly composed of copper or aluminum wires embedded in conductive swelling tapes. These tapes serve as *longitudinal water protection*: in the event of damage to the outer sheath that allows moisture ingress, the tapes swell upon contact with water, effectively blocking further penetration along the length of the cable.

The grounded metallic shield fulfils several critical functions. It conducts the charging current of the cable under normal operation and, in certain fault conditions, carries the earth fault current—both of which are explained in detail in Chap. 6. It also serves as a physical protection against accidental contact with live conductors.

To ensure mechanical robustness, a protective *armor layer* is typically applied, consisting of steel tape or flat steel wires. This reinforcement protects the cable against physical damage and enhances its tensile strength during handling and installation.

To protect against moisture ingress from the outside, a *metallic sheath*—commonly made of lead or aluminium—is applied. When using lead, the material must be a specially formulated alloy (known as *cable lead*) with a low copper content, as pure lead would be too brittle.

Over the metallic sheath, a *polymer outer sheath* is applied, most commonly made from High-Density Polyethylene (HDPE). This sheath serves as both electrical touch protection and corrosion protection for the underlying metal layer. In applications with stringent fire safety requirements, such as long tunnels, the outer sheath is made from flame-retardant polymer compounds.

For DC voltage sheath testing—carried out during commissioning and routine inspections (Sect. 5.11)—a very thin conductive layer is applied over the polymer sheath to serve as a test electrode.

Modern XLPE-insulated cables often have a simplified construction. In many cases, they are manufactured without armor and consist only of a metallic shield, possibly a layer for longitudinal water protection, and an outer sheath made of polyethylene.

5.4 Cables for High Voltage Systems

The following section presents the cable types most commonly used at the high-voltage level. In German technical literature, cable type designations follow the system outlined in [3], which uses a concatenation of short type symbols. This designation begins with the letter *N*, indicating standardized cables, followed by abbreviated symbols representing the cable components from the conductor outward.

Designations for copper as the conductor material, impregnated paper insulation, and in the case of extruded cables—the inner and outer conductive layers are typically omitted from the type code. Following the main type symbols, additional short codes are used to describe the *conductor structure*, offering information about its shape and configuration.

For design and calculation purposes, the most important parameter is the *conductor cross-section*, as it directly determines the current-carrying capacity (*ampacity*) of the cable.

Electrical characteristics such as DC resistance, shunt admittance, capacitance, and inductance per unit length may vary slightly between manufacturers. Resistance values are usually specified for direct current at a reference temperature of 20 °C. For practical applications at operating temperatures between 60 °C and 90 °C, these values must be adjusted. A practical rule of thumb is that the resistance increases by approximately 4% for every 10 °C rise in temperature.

The resulting rated current I_r and rated apparent power S typically refer to flat-laying configurations of three single-core cables in soil, under standard load conditions. Because the reactive power component contributes to the total apparent power, transmission capacity is generally specified in terms of apparent power. When no reactive power is transmitted, the apparent power (expressed in MVA) is equal to the active power (in MW).

Chapter 6 provides examples of typical cable calculations and practical applications, including key electrical characteristics for several standard cable types.

5.4.1 AC Cables

As previously discussed, the transmission capacity of a cable system depends on various factors, including the voltage level, conductor cross-section and material, load profile, and installation conditions. As such, generalized figures offer only limited insight. However, as a rough reference: in German high-voltage distribution networks, transmission capacities in the range of 80 to 100 MVA are typically achievable using copper cables with a conductor cross-section of approximately 300 mm^2 at the 110 kV level.

With the introduction of higher voltage levels in the early twentieth century, the demands on cable insulation materials increased significantly. The resin-like substances originally used to impregnate paper insulation were gradually replaced by low-viscosity oil or gas fillings under pressure, which effectively prevented the formation of cavities in the dielectric. Starting in the 1970s, these technologies were largely superseded by cables with extruded plastic insulation, which offer several advantages—most notably, very low dielectric losses.

Due to the required conductor sizes and insulation thicknesses, such cables are generally manufactured as *single-core cables*, each with its own complete sheath. This design allows for longer delivery lengths per drum, facilitates installation and repair, and improves heat dissipation into the surrounding soil. However, in submarine cable systems, this may not always be advantageous, as the number of joints becomes a critical factor in overall system reliability.

Mass-Impregnated Cables

This type of cable derives its name from the insulation layer, which consists of high-grade paper impregnated with a viscous insulating fluid—referred to as the "mass." Developed at the end of the nineteenth century, these *mass-impregnated cables* were initially used in medium-voltage networks and were later adapted for high and extra-high voltage applications. To this day, they remain relevant in certain DC transmission systems [4, 5], as the likelihood of partial discharges is lower under DC voltage than with AC. However, compared to modern *polymeric insulations*, this type of insulation quickly loses its dielectric strength when exposed to moisture. To protect the insulation, mass-impregnated cables are enclosed in metallic sheaths made of lead or aluminum, which also serve as electric screens. Due to the susceptibility of these metal sheaths to corrosion, an additional protective layer—typically a bitumen-saturated jute wrapping—is applied.

In installations with elevation differences, long-term migration of the impregnating compound due to gravity can occur, posing reliability concerns. Furthermore, if the cable is mechanically damaged (e.g. during laying), leakage of the insulating mass can cause environmental contamination. For these reasons, mass-impregnated underground cables are no longer considered state of the art in modern power systems.

Oil Pressure Cables

This cable type—commonly referred to as an *oil-filled cable* or more precisely as a *Self-Contained Fluid-Filled (SCFF) cable*—uses oil-impregnated paper as its insulation system. The conductor is constructed as a hollow core and filled with low viscosity insulating oil under pressure (typically 2–6 bar). This category is known as *low-pressure oil-filled cables.* To maintain consistent internal pressure despite fluctuations in electrical load and cable temperature, compensation tanks are installed at regular intervals along the route. This ensures thermal stability of the insulation and classifies these cables as *temperature-stable.* However, maintaining the oil pressure can be complex—particularly in routes with elevation differences. In such cases, *stop joints* (pressure-sealing joints) may be required at section transitions to stabilize pressure gradients. If a cable is damaged, the insulating oil can leak, potentially contaminating the surrounding soil. Oil must then be replenished, often requiring significant effort and cost. Due to these challenges and environmental concerns, oil-filled cables are generally no longer used in new installations.

A typical example of such a cable is the German-designated type *NÖKUDEY* [3]. Its construction includes: a copper conductor, an inner conductive layer, oil-impregnated paper insulation (which prevents the formation of air voids), an outer conductive layer, a lead sheath, padding, a pressure-protection bandage, and finally a PVC outer protective jacket (Fig. 5.9).

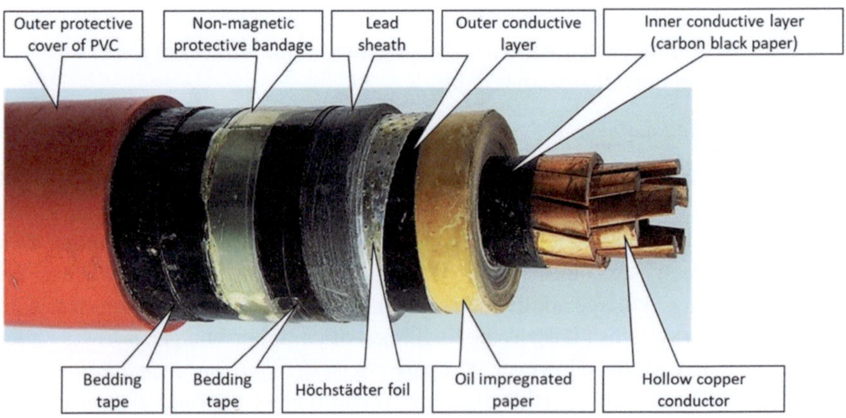

Fig. 5.9 Structure of a 110 kV oil pressure cable of the German "NÖKUDEY" type

Internal Gas Pressure Cables

In *internal gas pressure cables*, nitrogen is used to maintain pressure and prevent cavity formation within the dielectric. This type of cable is usually designed as a three-core cable. The paper insulation around the stranded conductors is soaked with high-viscosity mass. At pressures of about 16 bar, nitrogen has similar dielectric properties to the impregnating mass of oil cables and becomes part of the insulation through diffusion. If the cable is designed with three conductors, it requires less space than a system consisting of three individual cables with comparable transmission properties. The steel tube also provides effective protection against mechanical damage, such as that caused by construction work in confined urban spaces. Additionally, it eliminates inductive interference with nearby communication or control lines (Fig. 5.10).

External Gas Pressure Cables

In *external gas pressure cables*, cavity formation is prevented by a different approach. The conductor, insulation, and gas-tight lead or aluminum sheath are designed with an oval cross-section, making them more readily compressible. The cable assembly is housed within a steel tube pressurized with nitrogen at approximately 15 bar. This external gas pressure acts on the lead sheaths and the mass-impregnated paper insulation of the individual conductors, effectively preventing cavity formation during thermal expansion and contraction. Unlike internal gas pressure cables, the nitrogen here serves a purely mechanical purpose—it does not contribute to the insulation

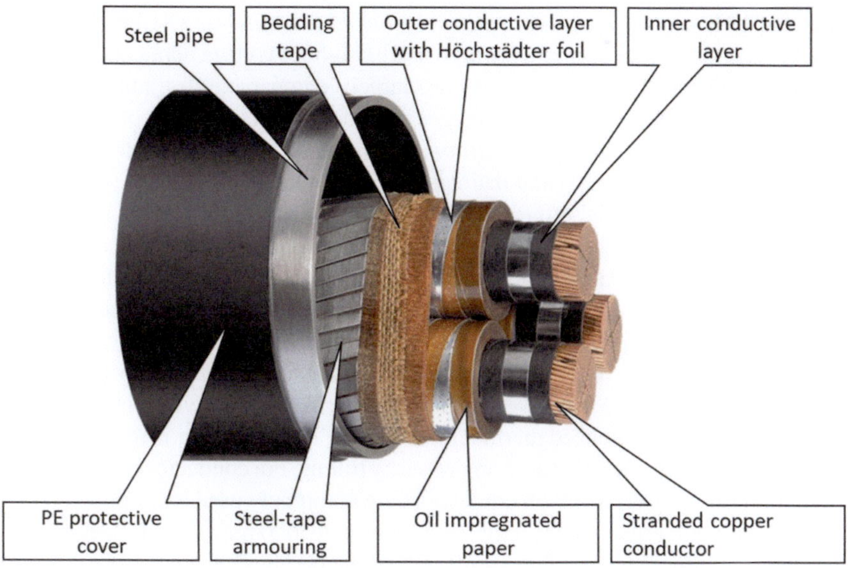

Fig. 5.10 Structure of a 110 kV internal gas pressure cable (*Source* NKT)

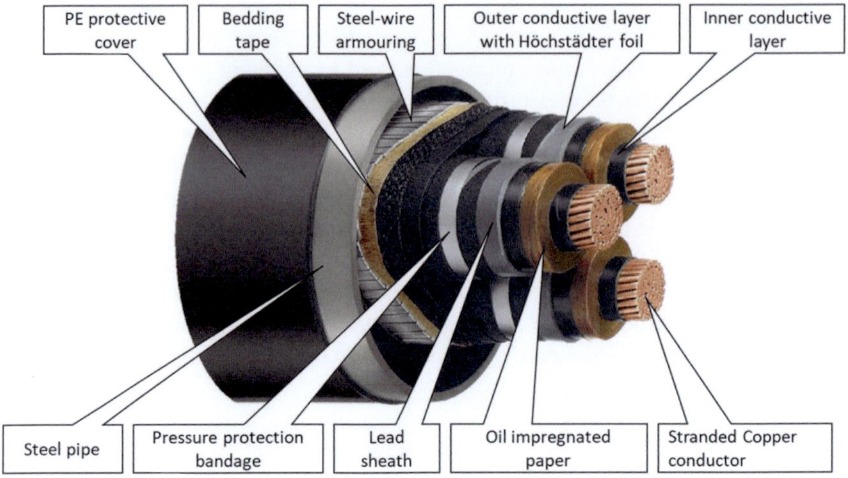

Fig. 5.11 Structure of a 110 kV external gas pressure cable (*Source* NKT)

system. The previously mentioned benefits of steel-tube installation, such as mechanical protection and shielding from external influences, also apply in this configuration (Fig. 5.11).

Extruded Cables

Since the 1970s, *extruded cables with cross-linked polyethylene* (XLPE) insulation have become the global standard for high and extra-high voltage applications. Structurally, apart from the dielectric, they are largely comparable to traditional mass-impregnated cables. Their key advantages include low weight, ease of installation due to small bending radii, straightforward assembly of joints and terminations, environmental compatibility, and maintenance-free operation.

This insulating material offers very high electrical strength, which is maintained over extended periods under permissible operating conditions, as confirmed by extensive long-term testing. Another notable benefit is its very low dissipation factor, with $\tan\delta$ values in the range of 0.3 to 0.5×10^{-3}—significantly lower than those of oil-impregnated paper, which typically range from 1.5 to 3×10^{-3}.

With XLPE cables, oil or gas leakage is no longer a concern. However, other measures are required to effectively prevent the formation of cavities in the insulation layer due to load-induced thermal fluctuations. The conductor is made of round copper or aluminum wires that are compacted by rolling. For conductor cross-sections of 800 mm^2 and above, Milliken conductors are commonly used to minimize losses caused by the skin effect. Today, cross-sections up to 3600 mm^2 (copper) are technically feasible; however, such large sizes reduce the permissible delivery lengths per cable drum due to transport constraints.

In high and extra-high voltage applications, the dielectric of extruded cables is manufactured together with the inner and outer conductive layers in a single, carefully controlled process known as *triple extrusion*, which is briefly described below.

Local electric field intensifications—such as those caused by minute cavities or the formation of gaps between the dielectric and the conductive layers—must be particularly avoided near the boundaries of the insulation. To achieve this, the inner and outer conductive layers must possess material properties closely matching those of the dielectric, ensuring that all three components expand and contract uniformly in response to temperature variations resulting from electrical load fluctuations. This uniform behavior is essential to maintain a permanently tight bond between the layers and thereby ensure high long-term electrical stability.

The dielectric material is polyethylene (PE), while the conductive layers are produced by incorporating a small amount of a specific type of carbon black into polyethylene. To enable subsequent *cross-linking*—i.e., the transformation from PE to XLPE—a peroxide is typically added to the granulate, which becomes active only at elevated temperatures.

In the triple extrusion process, the dielectric and the two conductive layers are separately extruded as heated granulates through a triple extrusion head onto the preheated conductor. Prior to extrusion, the granulate is subjected to a high-speed air stream ("wind sifter") and passed through a magnetic separator to remove fine dust and any magnetizable particles. The entire extrusion takes place under clean-room conditions to prevent contamination of the critical interfaces between layers (Fig. 5.12).

To minimize transport distances and avoid mechanical conveying, the granulate is fed vertically from above and flows into the extruder by gravity alone. The cross-linking process follows immediately after extrusion in a dedicated heating zone at around 200 °C. The continuously produced cable is then slowly cooled to ambient temperature in a downstream cooling stage before being wound onto a cable drum [4, 6, 7].

Following the outer conductive layer, the cable includes a metallic shield, typically made of copper or aluminum wires or tapes. For all extruded cables from this voltage level on, the shield area is also designed to be longitudinally watertight. This is achieved by incorporating textile or fleece tapes impregnated with swelling powder, which expands upon contact with moisture to block water ingress along the cable's length.

To ensure protection against radial (cross-sectional) water penetration, a coated aluminum foil—commonly referred to as a "layer sheath"—is applied. This foil is often bonded directly to the outer protective sheath, which is usually made of abrasion-resistant polyethylene. A widely used example of this cable design is shown in Fig. 5.13.

For a long time, mass-impregnated cables were the only viable option for power transmission at the extra-high voltage level. Over time, however, XLPE cable types have been developed that offer comparable—or even superior—performance characteristics. Today, conductor cross-sections exceeding 3600 mm^2 are technically feasible.

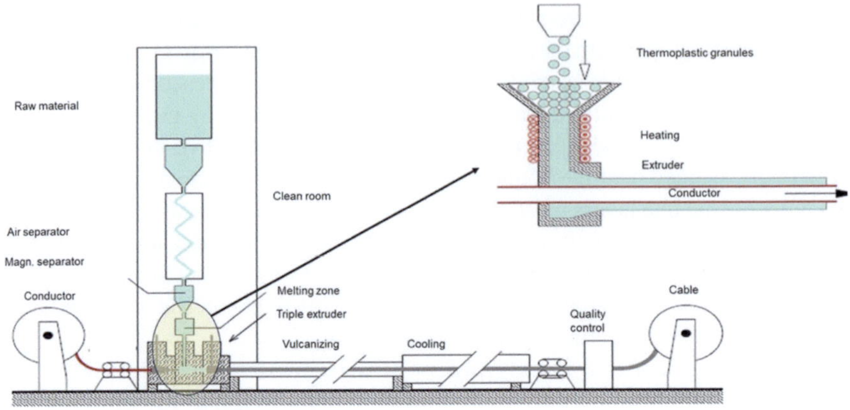

Fig. 5.12 Process of triple extrusion, schematic illustration [8]

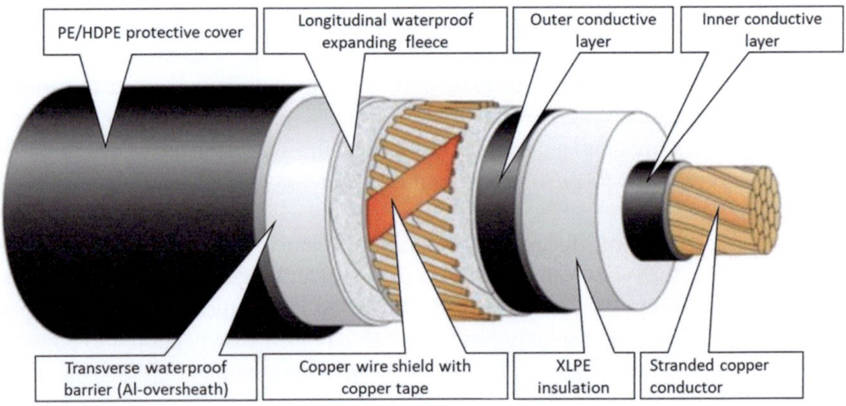

Fig. 5.13 Structure of a 110kV-XLPE cable (*Source* TFKABLES)

As in high-voltage applications, the conductor is made of round copper or aluminum wires compacted by rolling. For cross-sections from approximately 800 to 1200 mm^2 and above, segmented Milliken conductors are used to minimize AC losses caused by the skin-and proximity effects.

The insulation is produced using the triple extrusion process, performed under clean-room conditions and subject to stringent quality control. This ensures that the electric field within the dielectric follows an ideal, radially symmetrical distribution—analogous to the field in a cylindrical capacitor formed by the inner conductive layer, the dielectric, and the outer conductive layer. This precise field control enables the use of insulation layers less than 30 mm thick for 380 kV systems.

Following the outer conductive layer, a metallic shield made of copper or aluminum wires or tapes is applied. Its cross-section is dimensioned according to the anticipated fault currents it must carry. To ensure longitudinal water tightness, the shield area incorporates textile or nonwoven tapes embedded with swelling powder

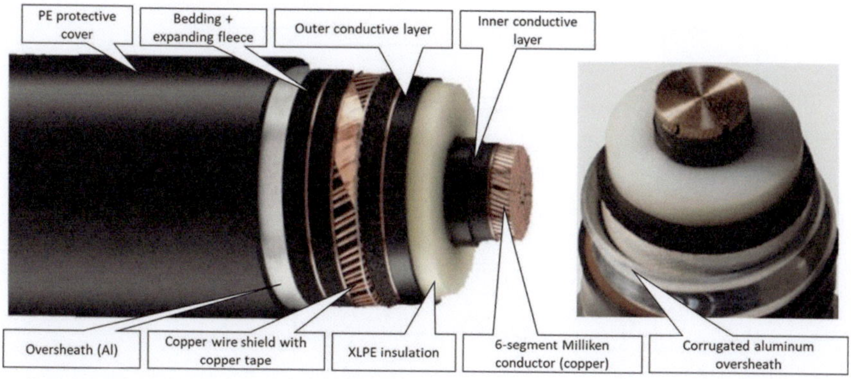

Fig. 5.14 380 kV XLPE Cable with insulation as multilayered jacket, *left* part of the picture (*Source* Südkabel), and as corrugated jacket cable, *right* part of the picture (*Sources* Nexans, 50 Hertz)

that expands upon contact with moisture, thereby preventing water propagation along the cable.

For monitoring purposes, optical fibers are often integrated into the shield wires—for instance, to detect and localize hotspots along the cable route. Transverse water tightness is typically achieved using a coated aluminum foil. Cables employing this layered configuration are referred to as *multilayered jacket cables*. Optional enhancements may include flame-retardant outer sheaths and, where required, additional lead jackets for environmental or shielding purposes.

An example of an XLPE-insulated 380 kV cable is shown in the left section of Fig. 5.14. With a conductor cross-section of 2500 mm^2, an outer diameter of 140 mm and a mass of 36 kg/m per single conductor, this cable has a transmission capacity of 900 MVA. As an alternative, transverse water tightness can also be achieved using a corrugated or smooth aluminum sheath. Cables designed in this way are known as *corrugated jacket cables*, as illustrated in the right section of Fig. 5.14.

5.4.2 DC Cables

DC transmission technology offers key advantages, including reduced conductor material requirements and the elimination of the need for reactive power compensation. The reason for this, namely the high cable capacity, does not play a role in the case of DC transmission systems.

Following the development of high-performance thyristors in the 1970s, DC transmission was initially adopted in countries requiring the transfer of large amounts of power over long distances typically via overhead lines. Soon thereafter, DC technology was also applied to submarine cable systems, such as those used to connect islands to mainland grids.

Converter systems for AC–DC and DC–AC transformation based on thyristor technology are known as *Line Commutated Converters* (LCC) and have proven highly

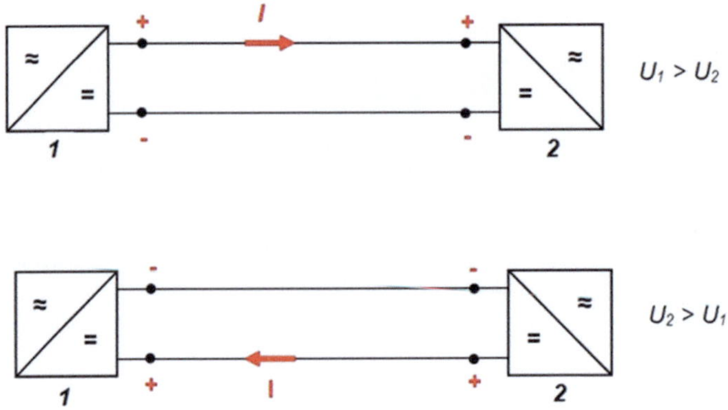

Fig. 5.15 Reversal of power direction in LCC transmission facilities

reliable in HVDC transmission for over 50 years [9, 10]. Unlike conventional AC systems, LCC-based DC transmission does not exhibit the skin effect, allowing for the use of solid (non-segmented) conductors in cables. Additionally, since inductive phenomena are absent under DC, the magnetic properties of cable armoring are irrelevant, and AC-specific cable design measures—such as cross-bonding or single-point grounding of the metallic shield—are not required (Sect. 5.7).

In LCC-based systems, power reversal, as required for example in pumped storage power plants, is accomplished by reversing the voltage polarity, while the current direction through the semiconductor valves remains unchanged (Fig. 5.15). This presents no operational issues when using overhead lines or mass-impregnated underground cables, but is does with XLPE-insulated types.

The reason for this lies in the space charges, which form when the dielectric is heated and can lead to failure of the insulation in the event of a sudden reversal of the electric field. Polyethylene, as an insulating material, has some unfavorable properties when it is exposed to a field with constant polarity instead of an alternating field. Its otherwise extremely low conductivity increases with rising temperature; it can change by several orders of magnitude between ambient and operating temperature. After switching on a cold XLPE cable, the field strength, which runs radially according to the field image of a cylindrical capacitor, is the highest near the conductor. The electric heat losses in the conductor result in a temperature gradient from inside to outside and *space charges* form, which lead to a change in the field course. Thus, also at the shield, the field strength can increase. The space charges only slowly equalize after the voltage is switched off. If the polarity at the cable must now be swapped to reverse the power flow, these space charges do not dissipate quickly enough, which can lead to an electric breakdown and destruction of the cable. Therefore, XLPE cables are connected with conventional LCC converter technology to this day, and mass-impregnated cables are still used there. It should be noted, that polypropylene laminated paper (PPLP) cables do not have this limitation. Currently, XLPE cables can only be operated in the case of direct current at a maximum conductor temperature of 70 °C. However, it is expected that the industry will soon develop cables that can guarantee a limiting temperature of 90 °C as with alternating current.

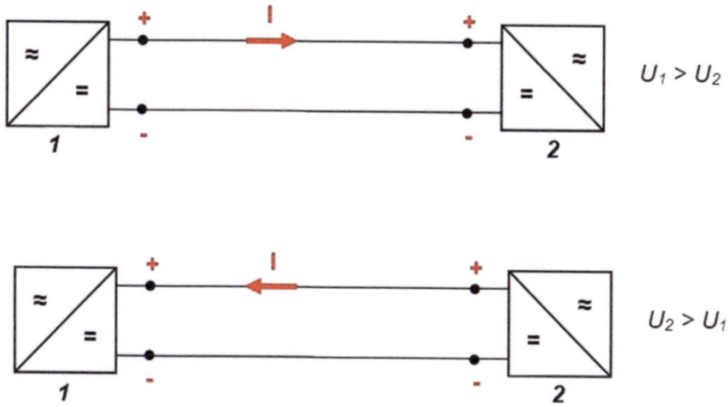

Fig. 5.16 Reversal of power direction in VSC transmission facilities

Another converter technology with the name *Voltage Source Converter (VSC)* uses power transistors, which, unlike thyristors, can be switched on and off at any time. This allows not only a power flow in both directions, but also the generation or consumption of reactive power. Chapter 6 covers this topic in greater detail.

VSC systems do not require a change in voltage polarity for power reversal and can therefore be well combined with XLPE cables (Fig. 5.16).

To distinguish, LCC technology is often referred to as *classic*, while VSC technology is described as *modern*.

Figure 5.17 shows a high-voltage direct current (HVDC) cable with XLPE insulation rated for ±525 kV. In HVDC systems, the voltage is typically specified as the conductor-to-earth voltage; therefore, the corresponding system voltage is 1050 kV. With a copper conductor cross-section of 3000 mm^2, the cable has a transmission capacity of up to 3 GW, according to the manufacturer's specifications.

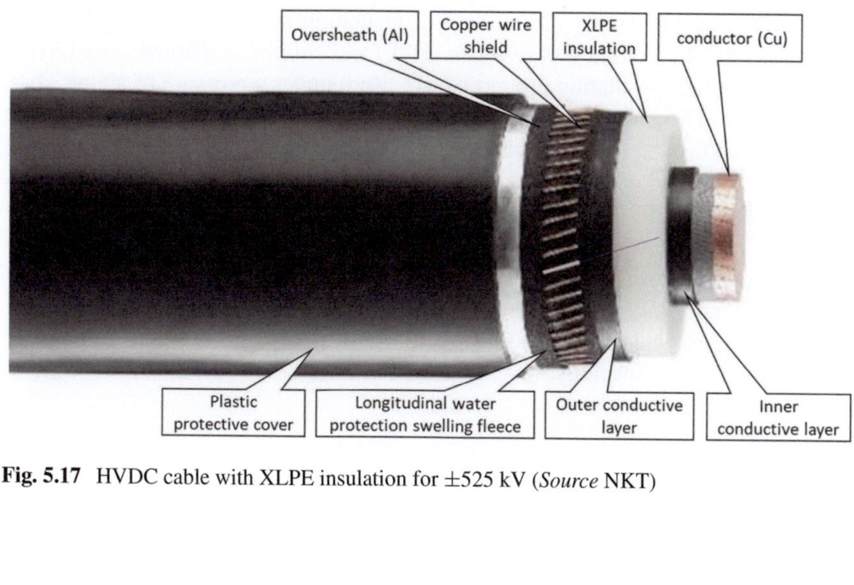

Fig. 5.17 HVDC cable with XLPE insulation for ±525 kV (*Source* NKT)

Table 5.1 Typical characteristics of AC and DC cables

Cable features	AC	DC
Utilization cross section of conductor	Not completely due to skin-effect	Completely
Length limitation	Due to (capacitive) charging current	No limitation
Dielectric materials	No restriction	Restriction due to space charging effects and (still) voltage level
Operation temperature	No restriction with respect to insolation material	Restriction due to insolation material
Dielectric losses caused by	Polarization, shunt conductance currents	Shunt conductance currents
Shield losses caused by	Induction	No losses
Partial discharges in cavities	Possible due to polarity changing	No discharges
Pattern of electric fields lines in layered dielectric materials (in cable terminals and joints) according to	Capacitive voltage divider, not depending on temperature	Resistive voltage divider, depending on temperature

Finally, Table 5.1 compares important features of AC and DC cables.

5.4.3 *Gas-Insulated Lines (GIL)*

A special form of both AC and DC energy transmission at very high voltages are *gas-insulated lines (GIL)*. These are metal pipes in which the current-carrying conductor is concentrically mounted with the help of insulating supports. In today's second generation of GIL, a mixture of *nitrogen* (80%) and *sulfur hexafluoride* SF_6 (20%), both with excellent insulating properties, are used, under a pressure of about 7 bar, as a dielectric. The geometry of the arrangement and the use of a gaseous dielectric mean that gas-insulated lines have much lower capacities than comparable cables and can therefore be used for greater distances than cables without compensation. Gas-insulated lines (GIL) exhibit negligible dielectric losses. Unlike conventional cables, their non-flammable insulating medium is—according to manufacturers—self-healing in the event of a fault, allowing the system to maintain operation without catastrophic failure. In this regard, GILs are designed to behave similarly to overhead lines. A particularly noteworthy feature is that they do not constitute a fire load, making them especially advantageous for installation in tunnels or confined underground spaces.

Gas-Insulated Line systems combine the structural advantages of cables with the electrical characteristics of overhead lines. Their cylindrical form categorizes them mechanically as cables, while electrically, their gas-based insulation resembles that of overhead lines. A key benefit is the seamless transition to *Gas-Insulated Switchgear* (GIS), as both components are constructed using compatible design principles. However, the use of sulfur hexafluoride (SF_6)—a potent greenhouse gas with a global warming potential approximately 20000 times that of CO_2—is increasingly subject to regulatory restrictions. In many countries, its use is limited to specific applications approved by environmental authorities. As a result, intensive research is underway to develop environmentally friendly alternatives, especially as a complete ban on SF_6 is under discussion [11].

Gas-Insulated Lines are engineered to handle high power ratings with large conductor cross-sections and exceptional electromagnetic shielding. The inner conductor typically has a cross-section of up to 6300 m^2, with diameters ranging from 150 to 200 mm. The outer tube, also made of aluminum, has an outer diameter of 500 to 550 mm and a cross-section comparable to the inner conductor. According to manufacturers, GIL systems can transmit power levels of up to 3 GVA at 380 kV. Unlike cables, GIL systems are not constrained by transport weight, which allows for virtually unrestricted conductor sizes. The low current density in the conductor results in moderate thermal losses, which are dissipated via the large surface area of the outer tube—either to the surrounding air or the soil. The electric field is fully contained within the outer tube, which also carries a high induced current. However, the large cross-sectional area keeps ohmic losses low. Moreover, the concentric geometry ensures that the opposing magnetic fields of the inner and outer conductors cancel each other, resulting in no measurable external magnetic field. To ensure the purity of the insulating gas, particle traps are integrated along the outer tube. These are specially shaped to attract and capture dust or metal particles via the electric field, guiding them into field-free zones. During operation, the maximum temperatures reached are typically 105 °C on the conductor and 60 °C on the outer tube. The outer aluminum tube, coated for corrosion resistance, provides excellent mechanical robustness and environmental protection (Fig. 5.18).

For logistics and on-site installation, GIL sections are prefabricated in 20 m lengths and welded together using orbital welding. Special insulating spacers segment the system into gas-tight sections of up to 1000 m, limiting the potential release of SF_6 in the event of leakage. Each section is equipped with gas sensors as well as monitoring systems for partial discharges and arc faults. Thermal expansion—up to approximately 1 m per kilometer—is absorbed through axial compensators and plug-type conductor connectors. Due to the high material usage and associated costs for manufacturing, transport, and assembly, GIL systems have thus far been limited to short distances and extra-high voltage applications - such as vertical connections in cavern power stations or urban transmission corridors [12–15]. Further discussion of the evolution of insulating gases and environmental considerations can be found in Chap. 7.

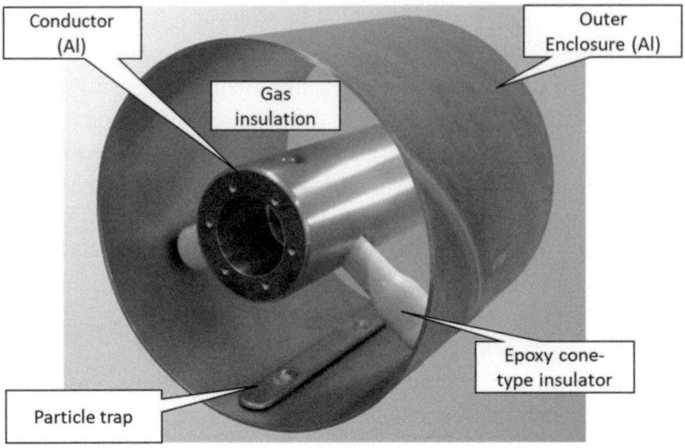

Fig. 5.18 Gas-insulated line (GIL) for the extra-high voltage level (*Source* Siemens)

5.4.4 Submarine Cables

Submarine cables have been in use since the 1850s, initially for the transmission of telegrams and later telephone signals between continents. Their basic structural principles were subsequently adapted for power transmission across bodies of water. Unlike early telecommunications cables, however, modern submarine power cables are not simply laid on the seabed, but are typically buried at depths of around 1.5 m. In sensitive areas—such as the Wadden Sea or near shorelines—burial depths of 2 to 3 m are often required to protect the cables from submarine landslides, seabed erosion, tidal forces, ship anchors, fishing activity, and eve sabotage.

To minimize the number of joints and enhance reliability, submarine cables are manufactured in large continuous lengths and laid using specialized cable-laying vessels.

Such vessels can carry cable loads of up to 10000 tons, enabling individual cable lengths of up to 150 km. Once laid on the seabed, the cable is mechanically embedded into the soil to the intended depth using jetting, plowing, or trenching systems (Fig. 5.19).

One of the main technical challenges in deep-water installation is the high self-weight of the cable, which leads to significant tensile forces during deployment. This necessitates reinforced armor layers, which further increase the overall weight. A design dilemma arises when choosing between single-core and three-core cables:

- Single-core cables are lighter and easier to handle, but require three separate laying operations, increasing installation time and cost.
- Three-core cables streamline installation but can weigh over 100 kg per meter, limiting the feasible drum length and complicating logistics.
- In addition, jointing operations (e.g., for repair) are more complex and time-consuming with three-core cables

Fig. 5.19 Cable laying ship (*Source* Nexans)

In the field of *high-voltage direct current* (HVDC) transmission, both mass-impregnated (MI) and extruded cable technologies have proven technically equivalent for submarine applications. As with AC systems, the question of whether to implement single-core or two-core designs arises and must be assessed based on mechanical, electromagnetic, and logistical factors. A notable consideration in HVDC systems is the magnetic field generated by the conductors. If the two poles of a bipolar system are laid with too much lateral separation, the resulting magnetic field may interfere with ship navigation by causing compass deviations—a known concern in maritime safety [9, 16].

For HVDC or AC submarine cables that are connected to floating offshore platforms, the riser section—where the cable transitions from the seabed to the floating structure—must be engineered with exceptional flexibility and mechanical resilience. This area is constantly subjected to dynamic forces from ocean currents, wave action, and tidal movement, both in vertical and horizontal directions. In such cases, lead sheathing is not suitable, as it is prone to mechanical damage and fatigue under cyclic loading, Fig. 5.20.

The design and production of dynamic submarine cables pose significant engineering challenges. Not only the cable's structural integrity and flexibility, but also its installation, monitoring, and long-term maintenance must be addressed. A notable example is presented in [17], which describes the design, testing, and validation of a 66 kV AC export cable insulated with Ethylene Propylene Rubber (EPR) and equipped with copper conductors. This dynamic cable, developed for the *Eolmed project* (also known as the *Gruissan project*), connects a 30 MW floating offshore wind farm located off the southern coast of France.

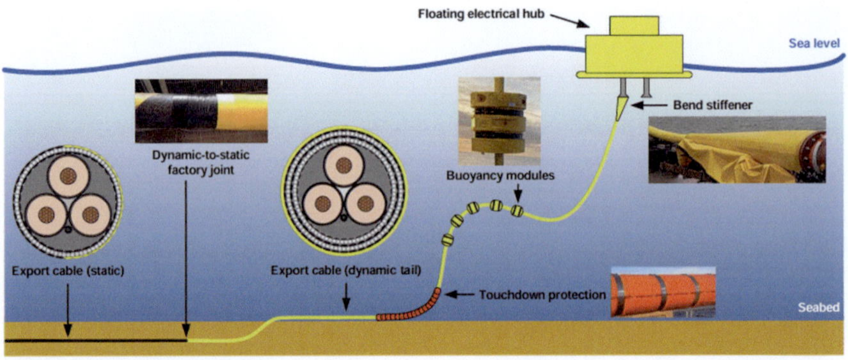

Fig. 5.20 Connection of a dynamic power export submarine cable to the swimming platform [17]

5.4.5 *Temporary Construction Cables*

During construction work in switchgear and substations, or for temporary rerouting of overhead lines or underground cables, it is often necessary to install temporary construction cables to avoid power interruptions or at least minimize their duration. Where space allows and the expected construction period is longer, such connections are frequently implemented using temporary overhead lines. These lines are supported by modular steel towers, which can be assembled from prefabricated components in a flexible, customized manner. Unlike permanent structures, these towers do not require traditional foundations; stability is instead achieved using anchor ropes and ballast weights, as discussed in Chaps. 2 and 4. However, in locations where the erection of temporary overhead lines is not feasible—due to space constraints, time limitations, or cost considerations—construction cables are used. These are preassembled, ready-to-deploy cable sections, typically ranging from 50 to 500 m in length. They are designed for quick installation, not only for planned construction activities but also as part of emergency response measures in the event of faults or outages. Transportation of these cables is facilitated using specialized steel cable drums, which are equipped with separate compartments to securely house and protect the pre-installed terminations. This enables rapid deployment and connection in the field, even under demanding conditions.

Unlike underground cables, construction cables are typically laid directly on the ground or mounted on flatbeds, which eliminates the need for earthworks but may require protective measures, such as temporary fencing, to ensure safety and prevent damage. These cables are now available for voltages of up to 300 kV, and are manufactured with fine-stranded copper or aluminum conductors, with cross-sections of up to 300 mm^2 and current-carrying capacities reaching 750 A. In high-voltage applications, they are implemented as single-core cables.

The insulation consists of high-quality XLPE, designed with reduced thickness compared to conventional power cables, which helps minimize weight and facilitates handling. For rapid and flexible installation, dry, open-air terminations are

Fig. 5.21 Flexible termination combined with a surge arrester (*Source* Technical Report 889 "Installation of underground HV cable systems" CIGRE, Paris 2023)

pre-installed at both ends. These allow for parallel operation—useful for increasing total transmission capacity—but not series connection.

As type-tested equipment, construction cables can be deployed reliably in all weather conditions. Standard bending radius limits for cables must be observed, though particular care should be taken to avoid bending the terminations. Similar to underground cable installations, minimum spacing requirements must be followed to allow adequate heat dissipation, and the cables must be fully unwound from the drum during use. Furthermore, to protect the insulation material, exposure to direct sunlight should be minimized wherever possible.

Because the terminations cannot withstand axial mechanical forces, they must be secured with clamps. In the event of a short circuit, it is essential that mechanical forces along the entire cable path are safely controlled. Additionally, cable shields must generally be grounded at both ends and, if needed, protected with surge arresters to ensure reliable operation under transient overvoltage conditions (Figs. 5.21 and 5.22).

Construction cables are designed for repeated use, provided they are handled with appropriate care. When not in operation, they can be stored on dedicated steel drums and held in reserve as stock. Prior to each deployment, however, the cables must undergo mandatory safety inspections and testing to ensure operational readiness. Thanks to their reduced insulation thickness, construction cables offer clear advantages in terms of transport and ease of installation. However, this same feature also results in increased sensitivity to mechanical stress. As such, both the cables and their terminations must be handled with greater care than conventional power cables. To avoid damage that could compromise performance or lead to failure, construction cables must be protected from mechanical impacts during storage, transport, and installation. Special attention should be given to shielding the insulation and terminations from crushing, bending, or abrasion, which could otherwise result in latent defects or immediate operational faults (Fig. 5.23).

Fig. 5.22 Construction cables installed directly on the ground (*left*) and suspended as a temporary road overpass (*right*). (*Source* Tennet)

Fig. 5.23 Failure on a construction cable due to damage from improper handling [18]

Temporary construction cables and associated components follow a defined operational cycle: they are stored, transported to the site, assembled, commissioned, and put into operation. After completion of the temporary installation, the process is reversed—the system is decommissioned, disassembled, returned to storage, and made ready for future use. This cycle may repeat multiple times throughout the cable's service life, assuming it is handled with the necessary care and precision.

Since this process involves various stakeholders—including logistics, engineering, commissioning, and maintenance teams—it requires strict coordination and adherence to quality standards to ensure safe and reliable operation.

According to internal company guidelines [19], the following procedures apply:

- Before initial reuse, the cable must undergo:
  - An AC voltage withstand test
  - A partial discharge (PD) measurement
  - A thermographic inspection using a thermal imaging camera

- For repeated deployments, routine requalification is limited to:
  - Partial discharge measurement
  - Thermal imaging check

These inspections help ensure that damage from previous use or handling has not compromised cable integrity and that the system is safe for reinstallation.

Application Example 1

Between Norway and the Netherlands, a direct current line was used to transmit 700 MW over 576 km (including the lines relocated on land) that went into operation in 2008 after several years of intensive planning. This NorNed project comprises a mass impregnated paper cable with a copper cross-section of 700 mm^2 per conductor (in the underwater section) and a voltage of ±450 kV. It was then the longest submarine cable connection in the world. According to the operator, it has approximately 4% transmission losses at nominal operation [16]. Following the model of Baltic Cable (Germany–Sweden) and Kontek (Germany–Denmark), it was initially planned to execute NorNed unipolarly with the earth as a return conductor. However, there were concerns about corrosive influences of the electrodes at the two converter stations on nearby infrastructure such as gas pipelines. These fears, along with other strict environmental regulations, led to a redesign with the result of building the line with two poles, i.e., with a forward and a return conductor. As already mentioned, direct current cables laid in the sea, where the two individual cables run separately, can cause compass deviations in ships due to their magnetic field. The requirement to maintain a maximum deviation of 5° could not be met in areas of shallow water depths. For this purpose, a double cable was developed, in which the two conductors are very close together and are surrounded by a common sheath. Thus, the magnetic fields formed by the forward and return conductors almost completely cancel each other out. This cable also offers advantages in terms of laying and embedding processes as well as the assembly of joints at sea. However, compared to individual cables, only shorter lengths can be transported on the laying ship for weight reasons. It was decided to lay a double cable in areas of shallow water depth under 50 m over a length of a total of 270 km, which has a weight of 85 kg/m with its reinforced sheath. For deeper waters (up to over 400 m depth at 306 km distance), single cables with a weight of 38 kg/m each have been implemented (Fig. 5.24).

Application Example 2

The United Kingdom is electrically linked to Norway, France, the Netherlands and Belgium. One of these connections is called the *Viking link* between Denmark and Great Britain [20]. It crosses Danish, German, Belgian and British stretches of water and is currently (2024) the worldwide longest cable connection. The HVDC project with a voltage of ±525 kV started in 2019, while its commercial operation began in 2024. The total length of the system is 761 km, of which 619 km is sea cable and it is designed to transport up to 1400 MW. The installation took 2.5 years.

The cables of the land section in Great Britain with 2×65 km and the subsea cable section with 2×619 km have copper conductors 1800 mm^2. The land-section

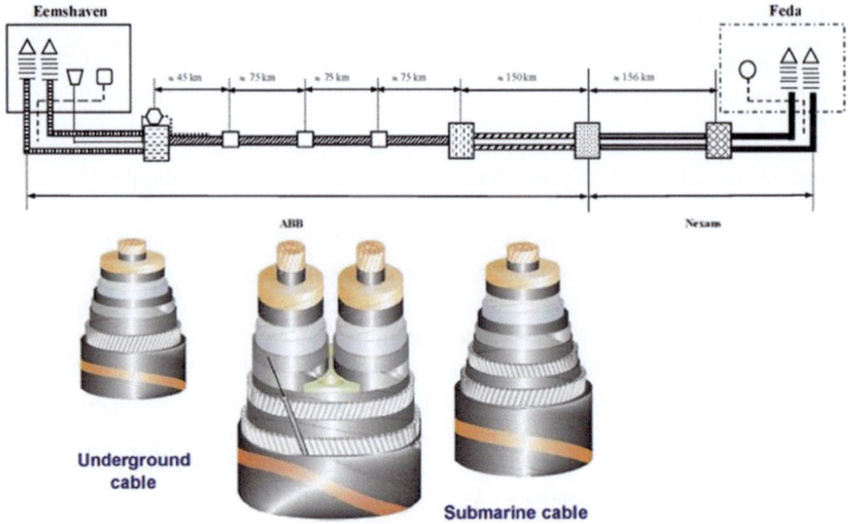

Fig. 5.24 Line scheme and cable types of the NorNed project [16]

in Denmark with 2 × 76 km uses conductors with 2110 mm^2. The section on land has 222 land joints, the subsea section has 53 joints, 40 of the latter are factory joints and 13 are offshore installation joints.

Cables are buried both on land and on the seabed, providing effective protection against mechanical damage and environmental influences. For further details regarding commissioning tests and fault location techniques—particularly those employing Time Domain Reflectometry (TDR)—refer to [21].

5.5 Cable Accessories for High and Extra-High Voltage Systems[1]

Cable accessories are used at the beginning and end of a cable route, at intermediate points where cable sections are joined, and at transitions to overhead lines or switchgear. These accessories are typically categorized into *terminations* and *joints*. In the high- and extra-high voltage range, both components must meet stringent technical requirements, as they involve the interface of three different materials—conductor, insulation, and air—each with distinct electrical properties, confined to a relatively small space.

[1] The explanations regarding terminations and plug-in connections are based in part on information kindly provided by Brugg Cables and Pfisterer, and are used here with their permission.

Modern solutions benefit from advanced numerical field simulations and the ability to manufacture weakly conductive plastic components in complex geometries. As a result, the market today offers a wide range of accessory systems that meet the high standards for electrical performance, mechanical robustness, and ease of installation—tailored to diverse applications and installation environments.

At the point where the cable insulation (dielectric) is terminated, the highly stressed electric field is interrupted. Without specific measures, the abrupt transition from the dielectric to air would cause extreme compression of the electric field lines, leading to a sharp increase in local field strength. This results in a high risk of surface discharges, which can degrade insulation and compromise system reliability. The critical region where the conductor, dielectric, and ambient air meet—known as the *triple point*—requires special attention. To ensure safe operation, the electric field must be carefully shaped and controlled in this area through techniques collectively known as *field control* [8].

The most commonly applied technique is *geometric field control*, achieved by installing a *deflector*, as schematically illustrated in Fig. 5.25. The curved edge of the electrode ensures that the field lines emerge uniformly and perpendicularly, thereby distributing the field evenly and reducing peak field intensities. This optimized field geometry is essential for preventing localized over-stressing and ensuring the long-term performance of high-voltage terminations.

In the insulation zones of cable accessories, multiple materials with differing dielectric properties are often employed. When these materials are arranged in layers, they form a series dielectric system, where the electric field strength is distributed inversely to the relative permittivity ε_r of each material. As a result, the material with the lower ε_r experiences a higher electric field, especially when the difference in permittivity between the materials is significant. This phenomenon is commonly referred to as *field displacement*. In systems operating with alternating voltage, this effect is utilized deliberately in what is known as *capacitive field control*, where the choice and arrangement of dielectric materials are used to shape and manage the electric field distribution within the accessory. Proper coordination of ε_r values is essential to avoid local overstressing of any single material layer and to ensure long-term electrical integrity.

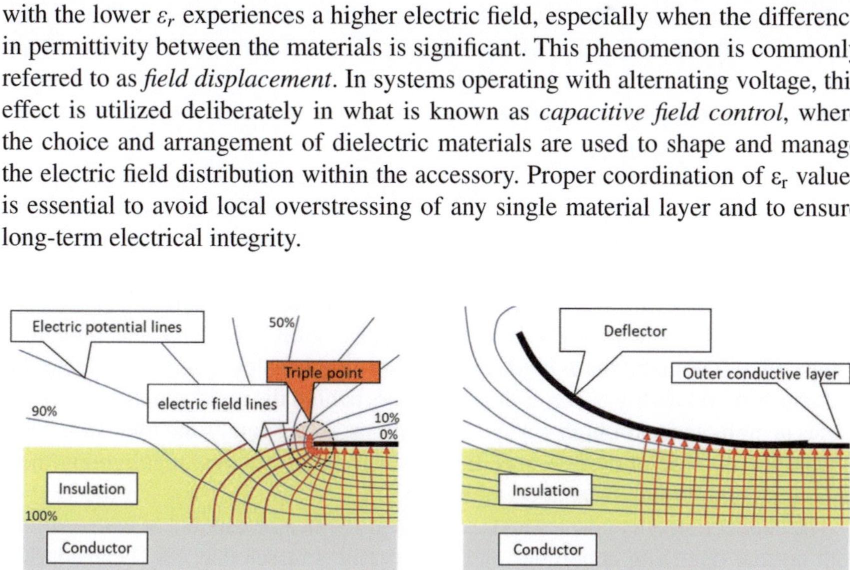

Fig. 5.25 Electric potential and field strength curve without (*left* part of the sketch) and with field control (*right* part of the sketch) at the point where the cable insulation is cut off

In DC systems, the dominant factor in field distribution is the differing residual conductivities of the insulation materials used. Unlike in AC systems where dielectric permittivity governs field behavior, in DC applications, charge carriers slowly migrate through the material, leading to a redistribution of the electric field over time. This phenomenon is known as *ohmic* or *resistive field control.*

Furthermore, the electrical conductivity of many insulating materials is highly temperature-dependent, meaning that both thermal conditions and the duration of electrical stress significantly influence the field distribution. These time-dependent processes—driven by conduction rather than capacitance—are collectively referred to as *field migration* [7].

For this reason, the long-term field behavior under operational conditions must be carefully considered during the design of insulation systems in DC cable accessories, especially for high-voltage and extra-high-voltage applications.

5.5.1 Cable Joints

High and extra-high voltage underground cables weigh up to about 40 kg/m and can therefore only be transported in limited lengths of up to about 1000 m (in exceptional cases 1500 m as special transport) on public roads. Where this is not possible, transport by ship is chosen as an alternative. Subsea-cables are produced at much longer individual pieces and thus the fabrication takes place close to the sea or near waterways navigable with large ships.

For the connection of the individual cable lengths, *joints* are needed. The number of joints in a cable line thus depends not only on the route length, but also on the possible delivery length on a drum. Higher voltage levels require larger and heavier cables and thus require more joints over comparable distances. Figure 5.26 shows this relationship exemplary.

Faults in cable networks occur most frequently in the cable accessories and only rarely in the cable sections themselves. As the length of a transmission route increases, the number of joints required also grows- naturally raising the risk of failures. In most cases, such failures are attributable to assembly errors. With XLPE-insulated cables, however, joints can now be prefabricated and electrically pre-tested, offering significant advantages in terms of installation simplicity and reliability compared to traditional mass-impregnated or oil-filled cables. This ease of handling and improved reliability is one of the key factors behind the rapid adoption of XLPE cables in high- and extra-high voltage applications. Accordingly, the following illustrations focus on joint designs specific to XLPE cable systems, highlighting their structural features and installation principles.

The connection of conductor ends within a joint must be capable of withstanding both normal operating conditions and the extreme stresses that may occur during short-circuit events. Since this connection represents the primary electrical and mechanical interface between adjacent cable sections, it must be designed to be particularly robust and durable.

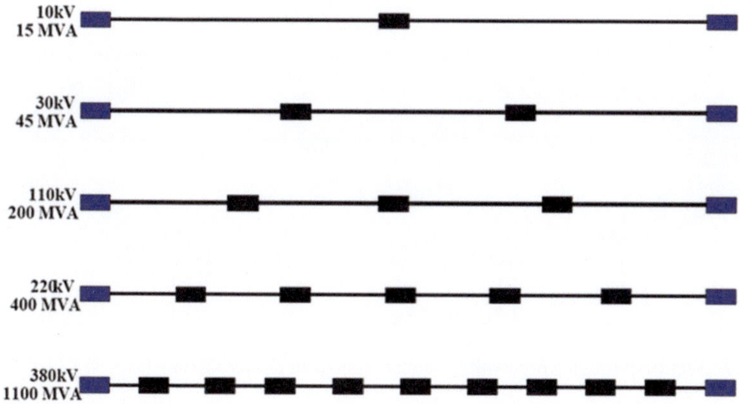

Fig. 5.26 Basic relationship between voltage level, delivery lengths and number of joints (*Source* APG)

Depending on the jointing technology used, the connector may maintain the same diameter as the conductor or increase in thickness. This design choice significantly influences the electric field distribution within the joint. In cases where the connector is thicker, a *shield electrode* at conductor potential—also known as a field-controlling electrode—must be applied to smooth the electric field and avoid local overstressing.

Several methods are available for forming the conductor connection, including:

- *Soldering*, where sleeves (matching or exceeding the conductor diameter) are used. Care must be taken to limit soldering temperatures, especially with XLPE or EPR insulation, to prevent thermal damage.
- *Welding*, commonly used for aluminum conductors, typically without sleeves. However, the high temperatures involved pose risks for heat-sensitive insulating materials, such as those used in modern plastic-insulated cables.
- *Crimping* (mechanical pressing), which is the most widely used method today, due to its lower thermal impact and high mechanical integrity. This method requires material compatibility between the sleeve and the conductor ends, and the pressing forces involved can reach the equivalent of dozens of tons to ensure a gas-tight and electrically stable connection.

The basic structure of a high voltage connection joint with shield electrode and deflectors on both sides is shown in Fig. 5.27 in sectional view.

The connector and the conductive deflectors are surrounded by the insulating joint body which is covered by a thin conductive layer on the outside. Like the outer semi-conductive layer of the cable, it acts as an earth electrode. This ensures that the vicinity of the joint remains field-free. Figure 5.28 shows the field distribution within the joint body as determined by simulation.

The further construction elements of a joint for high and extra-high voltage cables must include the mechanical protection of the joint body, a metallic protection against

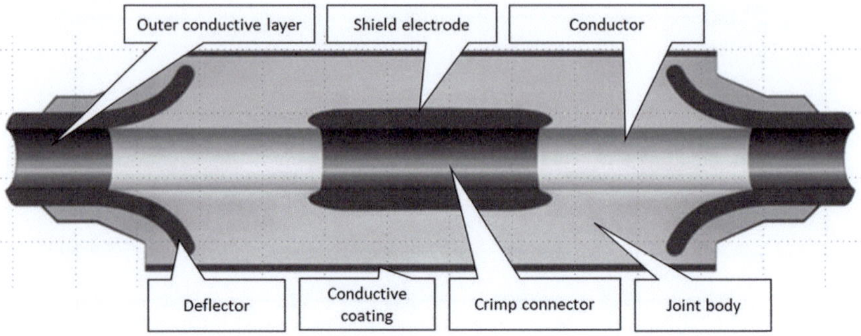

Fig. 5.27 Section through a high voltage connection joint (*Source* Brugg Cables)

Fig. 5.28 Field distribution inside a joint. Blue colored: areas of low field strength, red: areas of high field strength. (*Source* Brugg Cables)

penetrating water plus protection against corrosion. Depending on the manufacturer and intended use, there are a variety of different design forms.

When the *cross-bonding* technique is employed (see section below), the outer conductive layer of the cable must be interrupted at the joint and equipped with appropriate connection terminals on both sides. This allows the induced currents-which arise due to magnetic coupling in long cable systems—to be properly routed to the associated cross-bonding boxes. The location and design of the separation point must be carefully selected to preserve the predominantly radial symmetry of the electric field within the joint body. Any asymmetry introduced at this point could lead to local field distortions, increasing the risk of partial discharges or long-term degradation of the insulation system.

In practice, special joints are occasionally required—for example, when connecting cables of different types or from different manufacturers. This is especially relevant when older cable types are involved, such as during damage repairs or system retrofits. In such cases, custom joints must often be fabricated to match the specific materials and dimensions of the cables being connected. Once a joint is installed, only a partial discharge (PD) test can be performed during the commissioning phase of the cable route. Therefore, extremely careful assembly by specially trained personnel is essential. Joint installation is a time-intensive process, and the quality of workmanship is critical to long-term reliability. In this context, it is important to emphasize the role of quality assurance measures, which must be applied both in the manufacturing facilities and on the construction site. These inspections are intended to ensure that all components—cables and accessories—meet specifications and are free from defects before installation begins.

In the high- and extra-high voltage range, there are also special joint designs in which the cable shields of the three phases are cyclically transposed. This practice addresses the inductive coupling between the phase current and the surrounding metallic shield, which can otherwise cause unbalanced shield currents and elevated touch voltages. To mitigate this, both ends of longer cable sections are typically grounded. However, this can lead to circulating currents in the earth, as illustrated in Fig. 5.29, resulting from the potential difference between grounding points and the induced voltage drop along the cable shield.

If the grounding points of the cable shields are cyclically interchanged at regular intervals along the route, the induced voltages tend to cancel each other out. By dividing the cable route into equal-length sections and interconnecting the shield conductors as shown schematically in Fig. 5.30, the shield currents—due to the 120° phase shift between the three phase conductors—effectively sum to zero after three transpositions.

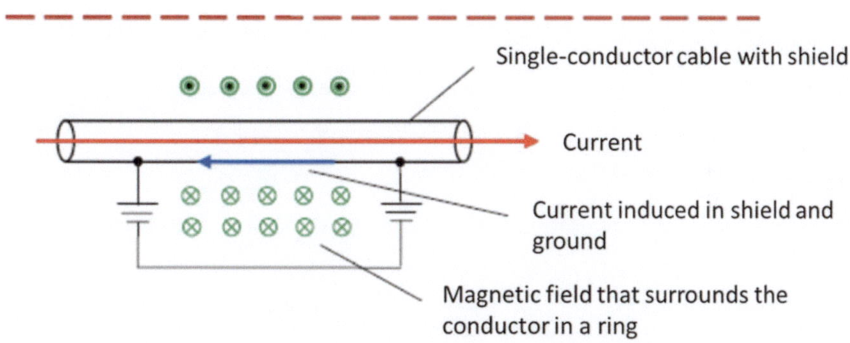

Fig. 5.29 Induction of circulating currents in the cable shield and surrounding soil due to magnetic coupling and multiple grounding points

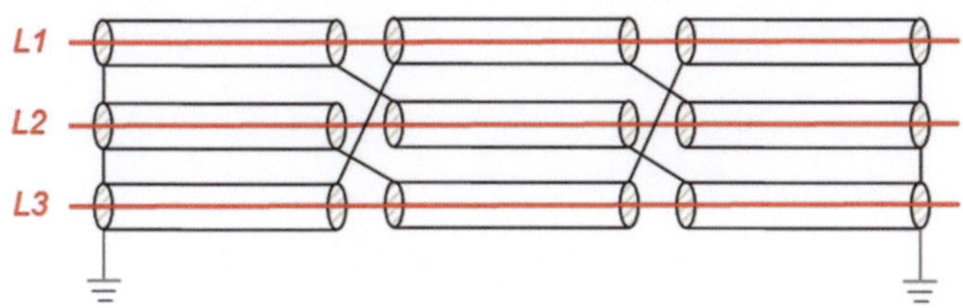

Fig. 5.30 Principle of cross-bonding of cable shields for three equal sections lengths

To protect against potential overvoltages at the grounding points, surge arresters are installed. This technique, known as *cross-bonding*, minimizes power losses caused by induction effects in both the shields and the surrounding soil, while at the same time limiting touch voltages to safe levels. It is widely used in high-voltage and extra-high-voltage underground cable systems to ensure both operational efficiency and personnel safety.

As cable sections expand slightly when heated under load, joints are often mechanically fixed to the ground to avoid longitudinal displacement. To accommodate this, accessible joint bays or structures—typically made of concrete—are sometimes constructed, especially in high-voltage systems. These structures can reach considerable dimensions, as the individual cable phases are spatially separated, and the joints are installed in a staggered arrangement to allow adequate space for assembly, inspection, and maintenance (see Fig. 5.31). Within such enclosures, one typically finds not only the joints themselves, but also cross-bonding facilities, surge arresters, and potentially even online monitoring systems for partial discharge or temperature. These installations play a critical role in ensuring long-term operational reliability, thermal stability, and ease of access for diagnostics and repairs.

Fig. 5.31 Staggered joints in an extra-high-voltage cable (*Source* Südkabel)

5.5.2 Cable Terminations

Cable termination is needed to introduce a cable into a switchyard—e.g., a gas-insulated switchgear (GIS)—or to enable the transition to an overhead line or a transformer. Depending on the type of component to be connected, terminations are referred to as indoor, outdoor, GIS or transformer terminations, which have different properties depending on the application.

The central component of a high-voltage cable termination is the *field control element*, typically made of an *elastomer* such as *silicone rubber* or *ethylene-propylene-diene rubber* (EPDM), and incorporating an integrated *deflector*. This component ensures a controlled and radially symmetric distribution of the electric field, similar to that achieved in high-voltage joints, and is particularly effective in systems using XLPE-insulated cables.

At the cable entry point, the metallic shield at earth potential is connected to the local grounding system, ensuring safe dissipation of shield and fault currents. The conductor is connected via a contact element (*connector*), which establishes the electrical interface with the switchgear or the transformer.

To protect the termination and connected cable from overvoltages caused by atmospheric discharges or switching operations, *surge arresters* are typically installed. These devices play a critical role in preventing destructive voltage transients from entering the cable system.

Figure 5.32 illustrates a schematic cross-section of an outdoor termination, featuring an oil-filled insulation system, electromagnetic shielding, and a metallic toroid for field grading. The figure also shows the plot of the equipotential lines between the toroid and the grounded supporting structure, highlighting the effectiveness of the field control design.

Indoor Terminations

This type of termination is protected from weather influences and strong temperature fluctuations. One distinguishes between dry terminations and those versions that

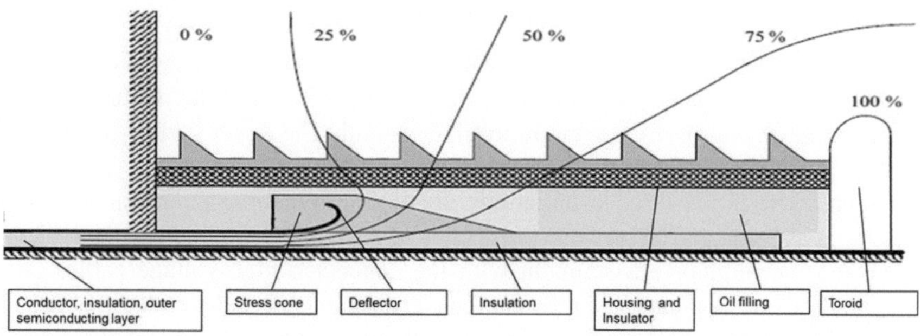

Fig. 5.32 Schematic cross-section of a high-voltage cable termination, showing field control elements, oil filling, shielding components, and toroid for outdoor use [8]

contain an insulating fluid inside. The latter is traditionally silicone fluoride and serves to avoid cavities. Today, a special compound is also used, which partially solidifies after filling. It has better thermal properties and cannot leak in case of leakages.

Outdoor Terminations

To ensure reliable operation under varying climatic conditions and environmental influences, high-voltage outdoor terminations are enclosed in hollow-core insulators with integrated sheds. These housings are made either of porcelain or composite materials. Composite insulators typically consist of a glass-fiber reinforced core tube, externally coated and equipped with sheds made from silicone elastomer. Like their indoor counterparts, outdoor terminations incorporate a field control system, including a deflector surrounded by a control cone. Owing to the elasticity of the field control elements, only three to four standard sizes are typically required to accommodate the full range of cable cross-sections within a given voltage level. A critical requirement is the moisture-tight sealing of the connector outlet to prevent ingress of humidity, which could lead to insulation degradation. Depending on the pollution level, installation environment, or proximity to coastal regions, outdoor terminations are available with different creepage distances, similar to overhead line insulators. In regions subject to large temperature fluctuations, the thermal expansion of the insulating fluid within liquid-filled terminations must be compensated. For this purpose, expansion vessels are used to absorb the volume increase of the fluid under elevated temperatures. These vessels may also be equipped with pressure sensors, enabling monitoring of the fluid level or pressure as a diagnostic measure. Expansion vessels are essential when the termination is installed horizontally or at a steep angle, rather than vertically.

Today, special environmentally friendly insulating compounds are often used, particularly in water protection areas. While insulating fluid-filled terminations still dominate in high-voltage applications, there has been a noticeable trend in recent years toward dry, plug-in type terminations, offering simplified handling and reduced environmental risk (Figs. 5.33 and 5.34).

Terminations for Gas-Insulated (GIS) Systems

Figure 5.35 shows two designs of high voltage terminations for entries of cables into gas-insulated switchgear (GIS) or transformers. The left part of the picture shows a traditional termination filled with an insulating fluid. The lower part of the picture shows a plug-in termination.

In contrast to liquid-filled terminations, dry-type terminations feature an insulator that serves as the interface between the insulating gas of the gas-insulated switchgear (GIS) and the outdoor environment with the connected high-voltage cable. This design allows for pre-installed terminations, enabling cable-independent testing of the GIS. The cable can later be connected without interrupting the gas insulation, which is considered a major advantage in terms of safety, modularity, and operational flexibility. Additional benefits of dry terminations include the absence of insulating fluid, which reduces maintenance requirements and eliminates the need for expansion

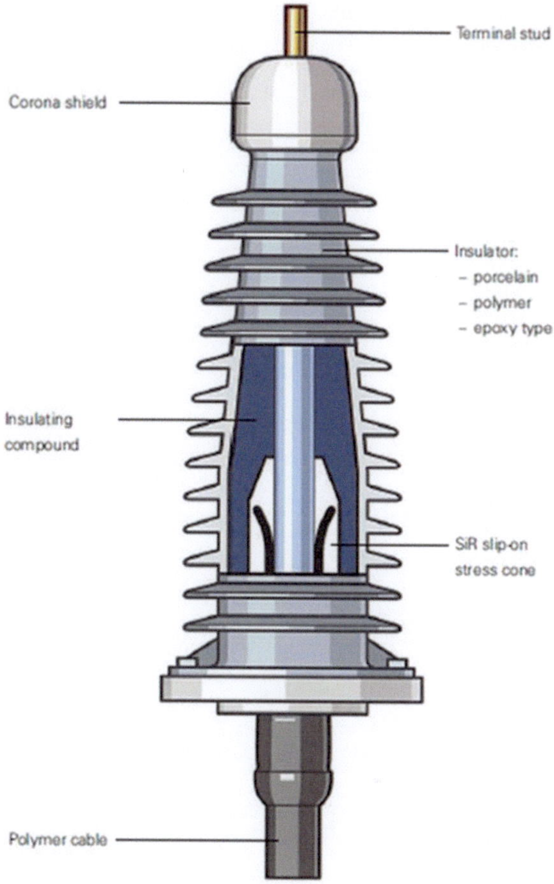

Fig. 5.33 Schematic cross-section through a vertical termination (*Source* Brugg Cables)

Fig. 5.34 Bushing with porcelain (*left*) and with hollow core composite insulator (*right*) (*Source* Brugg Cables)

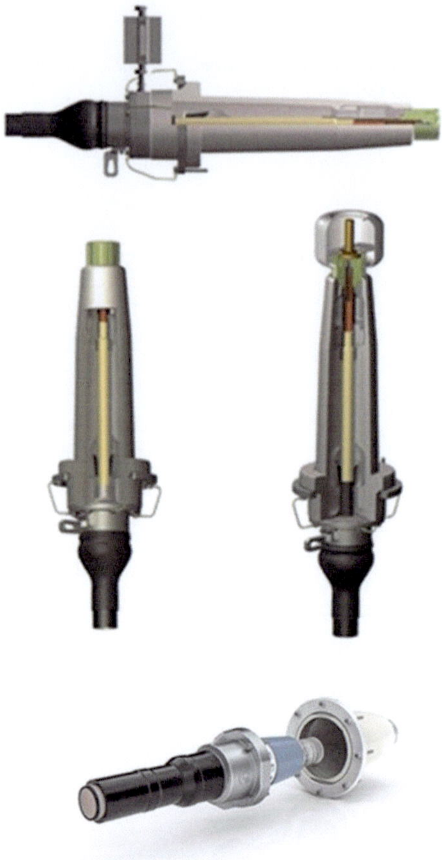

Fig. 5.35 Traditional GIS termination filled with an insulating fluid in horizontal and in vertical arrangement (top and middle) and dry plug-in GIS termination (bottom). (*Source* Pfisterer)

vessels. As a result, these terminations can be installed in any orientation, including horizontal or inverted positions, without added complexity. However, the demands on the field control system in dry-type GIS terminations are significantly higher. The field control element must ensure high dielectric strength not only at the interface to the cable insulation, but also at the transition to the epoxy GIS bushing, which requires precise design and high manufacturing quality to prevent partial discharges and long-term degradation.

Plug-In Connection Systems

In space-constrained environments—such as transformer terminations or switchgear bays—*plug-in connection systems* offer an efficient and flexible alternative. These systems are currently available for conductor cross-sections of up to 2500 mm^2 and rated currents of up to 2500 A [22].

The connection of high-voltage cables to oil-filled transformers or gas-insulated switchgear (GIS) requires exceptional care and strict adherence to safety and reliability standards. These connections must exhibit minimal contact resistance to accommodate high current levels, while also maintaining the required insulation level across the entire assembly. In conventional power systems, such connections were designed for long-term operation with a fixed and static topology.

However, the ongoing energy transition, characterized by the integration of decentralized renewable energy sources and the decommissioning of conventional power plants, demands new connection strategies. Frequently, existing interfaces must be modified or upgraded. At the same time, the evolution toward smart grid technologies imposes increased flexibility and adaptability on both transformers and switchgear.

In this context, cable connections can no longer be regarded as static terminal points but must be seen as dynamic system interfaces—integrated components within a flexible and future-oriented network architecture.

In response to these evolving requirements, plug-in connection systems, originally developed and proven in medium-voltage applications, have been progressively adapted and extended to high-voltage levels. While the primary objective has been to enhance operational flexibility, this technology also offers a range of additional benefits:

- Reduced installation space: The required footprint in the connection area is significantly minimized.
- Standardized interface: Bushings, surge arresters, and cables can all be connected via a common socket, simplifying configuration.
- Faster assembly: Installation times are shortened as there is no need to handle liquid insulating media—eliminating the need for oil or gas work within the substation.
- Versatile mounting options: Connections can be installed in horizontal, vertical, or inclined orientations, depending on site requirements.
- Simplified testing procedures: Routine and commissioning tests can be performed without temporary wiring or the need to open pressurized or fluid-filled compartments.
- Future-proofing: Provisions for future system expansions can be made by installing blind plugs in reserved connection points, enabling seamless upgrades later on.

These features make plug-in systems a valuable tool in modern grid planning and operation, especially in environments that demand compact, modular, and service-friendly solutions.

Some transformer manufacturers are now even offering plug-in connection technology as a standard feature. However, to enable a uniform and interoperable implementation across different manufacturers and system components, a clear definition of the interface standard is essential.

On the equipment side, plug-in cable connection systems typically comprise a cylindrical epoxy resin body that functions as a socket. Correspondingly, the cable-side component acts as the plug. To safely manage the high-voltage potential, the plug incorporates field control elements—analogous to those used in cable terminations and joints. This arrangement ensures a smooth field distribution and reliable insulation, as illustrated in Fig. 5.36.

Figure 5.37 illustrates a plug-in cable connection in use with a high-voltage transformer.

By integrating the socket directly into an electrical apparatus, such as a transformer or a switchgear bay, the plug-in connection system becomes part of it, rather than being classified solely as a cable accessory, as was traditionally the case. This shift in system boundaries has important implications for applicable standards and testing procedures. Responsibility for compliance now extends across multiple components—including the cable termination, the gas-insulated switchgear (GIS), and the transformer—all of which must be considered in a coordinated manner during design, testing, and certification.

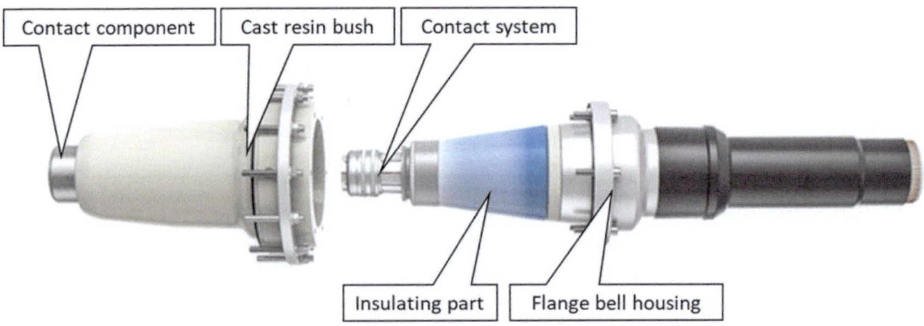

Fig. 5.36 Cross-section through a plug-in connection system (*Source* Pfisterer)

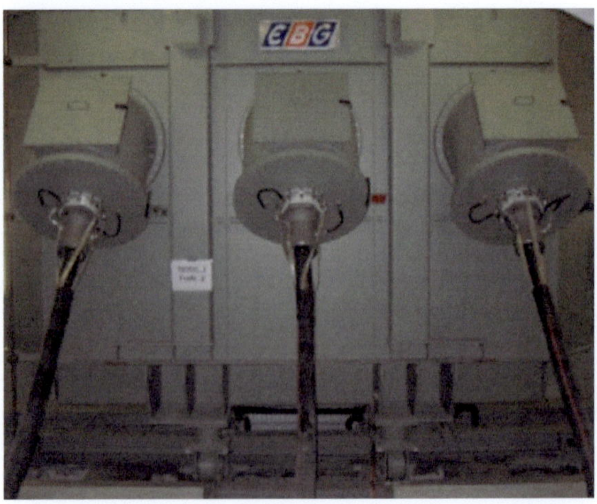

Fig. 5.37 High voltage transformer with plug-in cable connection (*Source* Pfisterer)

Accessories for High Voltage Direct Current Systems

As previously discussed, in AC systems, the distribution of electric field strength across two different insulating materials—such as insulation oil and porcelain—is governed by the ratio of their *relative permittivities*. In high-voltage cable accessories for polymer-insulated cables, this ratio is typically close to unity, resulting in a relatively uniform field distribution.

In contrast, under DC conditions, the electric field distribution is determined by the ratio of the *ohmic resistivities* of the respective insulating materials. In practice, this ratio can differ by several orders of magnitude—for instance, between the XLPE used in cable insulation and the EPDM or silicone commonly employed in cable accessories.

This poses a significant design challenge for DC cable accessories at high voltage levels. If the resistivity of the joint insulation is substantially higher than that of the cable insulation, excessive electric field stress builds up beneath the deflector in the joint body, potentially leading to electrical breakdown. Conversely, if the joint insulation's resistivity is significantly lower, the electric field becomes overly concentrated near the center electrode, creating a different failure risk.

In addition, the *thermal behavior* of the materials must be carefully considered. In radially symmetric designs, the highest temperatures—and thus the lowest resistivity—occur near the conductor. This causes the electric field to shift outward into areas that are otherwise less electrically stressed. Such redistribution is particularly critical when a *polarity reversal* occurs during grid operation, as required in some HVDC systems. As noted earlier, this effect is primarily driven by the formation of *space charges* within the insulation.

5.6 Grounding of High Voltage Cable Systems

For safety reasons, the shields of all high-voltage cables must be grounded. However, in AC systems, the current flowing through the conductor induces voltages in the cable shield. These induced voltages are proportional to both the conductor current and the length of the cable. If the shield were grounded at both ends, this would result in a significant circulating current through the shield. The associated ohmic losses would generate additional heat, thereby reducing the allowable transmission capacity of the cable.

To prevent such losses, several grounding methods are employed depending on the cable system's configuration and length.

Single-Sided Grounding

For short cable systems without joints, *single-sided grounding* is typically used. In this method, the shield is grounded at only one end of the cable. At the opposite end, a surge arrester is installed between the shield and ground. This arrangement ensures that no current flows in the shield during normal operation—even though induced voltages are present—because the arrester remains non-conductive under operating conditions (Fig. 5.38)

The surge arrester must be rated such that it withstands all power–frequency voltages, including those that could occur during short circuits in adjacent phases. It only becomes conductive in the presence of transient overvoltages—for example, from lightning strikes or switching operations—thereby protecting the cable sheath from damage.

Surge arresters typically have response (sparkover) voltages between 3 and 10 kV, depending on the maximum expected induced voltage. During normal operation, the voltage across the arrester remains well below this threshold.

Installation can be done directly above the base plate of a termination or within a *grounding box* designed for this purpose

Figure 5.38 illustrates the voltage profile along the shield in a single-sided grounding setup, while Fig. 5.39 shows a grounding box with an integrated surge arrester.

Grounding with Cross-Bonding Boxes at the Joints

For longer cable systems incorporating several joints, single-sided grounding is insufficient, as it would result in excessive induced voltages along the cable shield. In such cases, the voltage at the open end could exceed permissible limits under normal operating conditions. Moreover, the surge arrester would require a response voltage so high that it would no longer provide adequate protection for the cable sheath due to the correspondingly high residual voltage.

To address this, *cross-bonding* is applied: the cable shields are cyclically interchanged between the three phases over designated main sections. Owing to the 120° phase displacement in a three-phase system, the induced voltages in the shields effectively cancel out at the end of each cross-bonding section. This enables the direct grounding of all three shields without creating circulating currents. The resulting voltage profile along the cable system under cross-bonding is depicted in Fig. 5.40.

The principle of cross-bonding can be repeated for any length of a cable system. If there are an odd number of joints per conductor so that the induced voltages at the end of the cable route cannot add up to zero, cross-bonding can also be combined with single-sided grounding, as shown in Fig. 5.41.

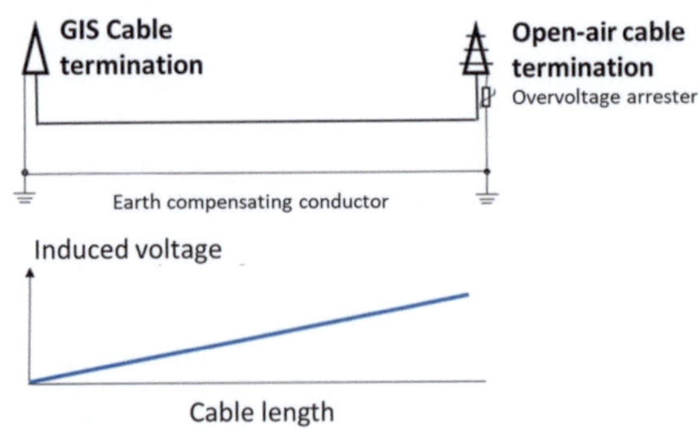

Fig. 5.38 Distribution of the induced voltage along the cable shield in a single-sided grounding configuration (*Source* Brugg Cables)

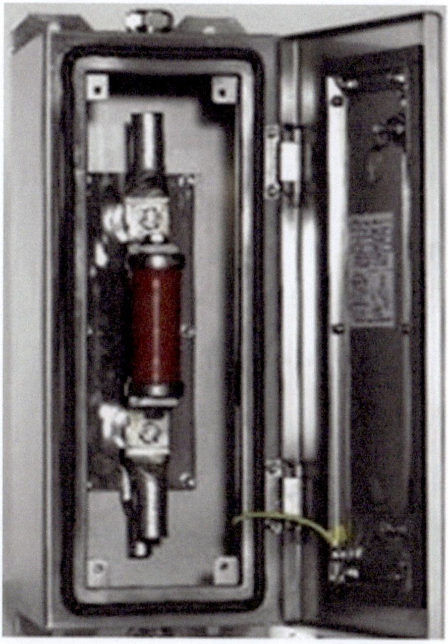

Fig. 5.39 Grounding box for a high-voltage cable shield, equipped with a surge arrester (*Source* Brugg Cables)

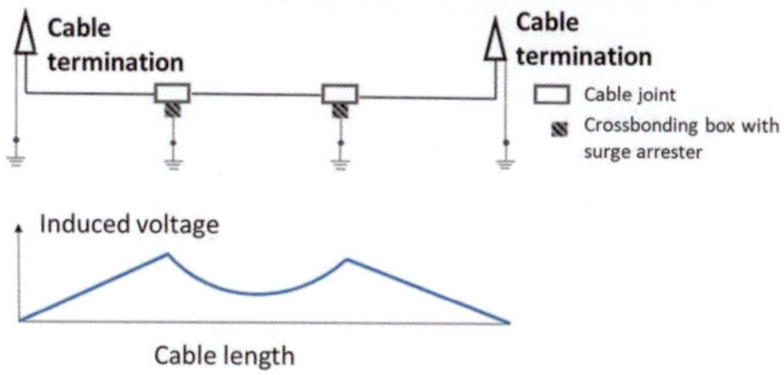

Fig. 5.40 Schematic illustration of the induced shield voltages in a cable system with cross-bonding (*Source* Brugg Cables)

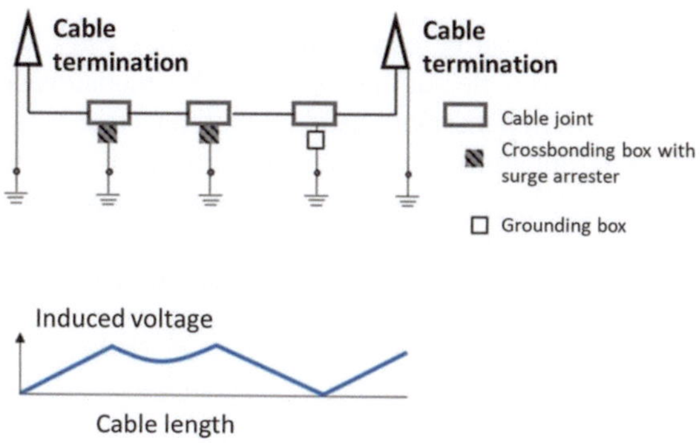

Fig. 5.41 Schematic illustration of the induced voltage in a cable system, where cross-bonding was combined with the concept of single-sided grounding (*Source* Brugg Cables)

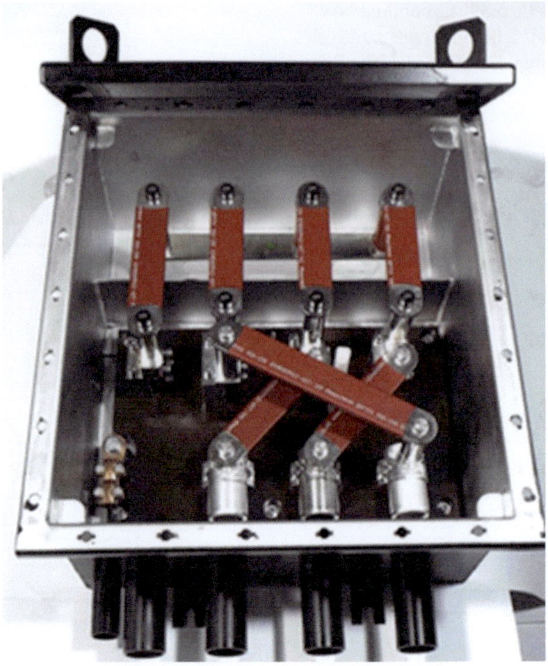

Fig. 5.42 Cross-bonding box (*Source* Brugg Cables)

The crossing of the cable shields generally takes place in cross-bonding boxes, which are installed close to the joints. These boxes usually also house surge arresters for the individual conductors. Figure 5.42 shows such a cross-bonding box.

5.7 Cable Laying

Cable laying can be carried out in various ways, as discussed earlier in Chap. 2 regarding possible routing options. On unobstructed routes, cables are typically installed in trenches with bedding materials that facilitate effective heat dissipation. In soil installations, cables must be laid in a manner that ensures adequate protection against mechanical damage—such as that caused by construction activities, vibrations, or ground movements like settlements. To allow unrestricted use of the land for agricultural purposes, the installation depth for high and extra-high voltage cables is generally around 1.50 m or more.

5.7.1 Cable Laying in Open Trenches

One of the most commonly employed methods for high- and extra-high-voltage cable installation is open trench laying, often referred to as the standard construction method. In some cases, plastic protective ducts are included in the trench. Their use is recommended when the exact cable type is not yet finalized at the time of civil works. Protective ducts may also be appropriate in situations involving:

- Anticipated future upgrades or cable replacements (e.g., at the end of service life),
- Construction zones with ongoing surface activity,
- Localized obstacles,
- The need for enhanced mechanical protection.

However, it should be noted that when cables are installed inside protective ducts, direct thermal contact with the surrounding soil is lost, resulting in reduced heat dissipation capabilities.

The design and execution of cable trenches must comply with applicable technical standards and construction guidelines. Additional specifications apply at crossings with railways, roads, bodies of water, pipelines (e.g., water, gas, chemicals), and other underground infrastructure, including telecommunication systems. Cable protection against mechanical damage and accessibility of joint bays for inspection and maintenance are also critical design considerations.

For single-core cables, two primary laying configurations are used:

- Flat formation: The three phase conductors are laid side by side in a horizontal plane.
- Trefoil (triangular) formation: The three conductors are arranged in an equilateral triangle.

Each arrangement has advantages. The trefoil formation requires less trench width and reduces the magnetic field at ground level. In contrast, flat formation allows superior thermal dissipation and is therefore commonly employed at the highest voltage levels.

For multiple cable systems (e.g., double circuits), the spacing between systems must account for the thermal conductivity of the soil. Where spacing is insufficient, thermal interference may occur, reducing the ampacity of the cables. This must be addressed during the thermal design phase of the installation (Figs. 5.43 and 5.44).

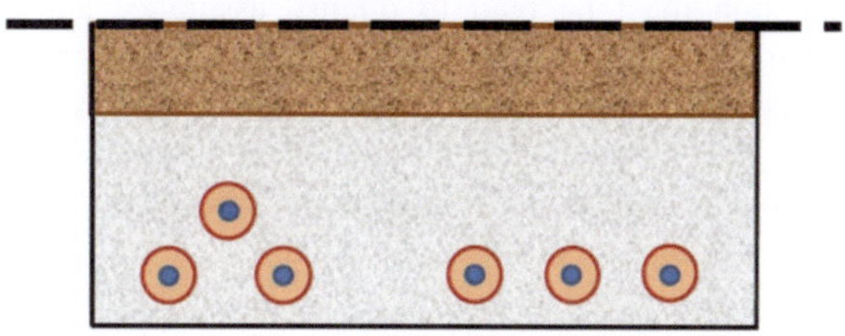

Fig. 5.43 Basic arrangement of the installation of high-voltage cables in triangular and in plane formation

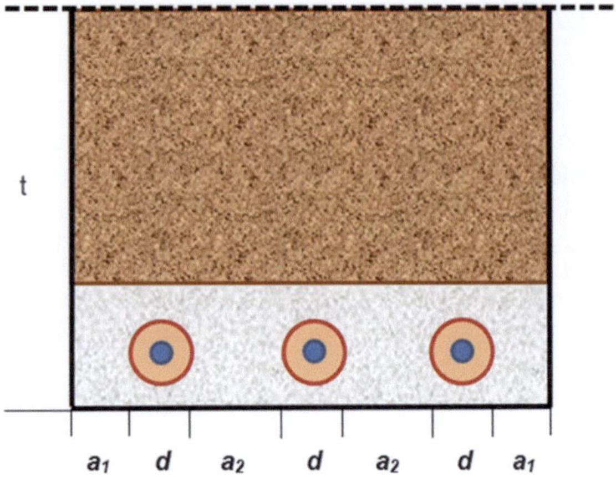

Fig. 5.44 Cable trench for a three-phase system with approximate dimensions, depending on the voltage level

Wherein is:

t Trench depth (1.50–2 m)
a_1 Distance cable—wall (20–30 cm)
a_2 Distance between the individual cables (25–50 cm)
d Cable diameter (5–15 cm).

Figure 5.45 illustrates the typical cross-sectional layout of a double-circuit 380 kV underground cable system designed for high transmission capacity. For thermal and electrical performance, each circuit consists of two parallel cable systems (twin configuration)—resulting in a total of twelve single-core cables (three phases per circuit, each doubled). The spacing between the two circuits is determined based on the thermal conductivity of the surrounding soil, in order to minimize mutual thermal influence and to maintain acceptable operating temperatures under full load. To improve heat dissipation, the cable trench is backfilled with a thermally favourable bedding material, shown in a lighter shade in Fig. 5.45. This material ensures homogeneous heat transfer and avoids thermal hotspots that could reduce the ampacity of the system. Additionally, for long-term reliability and to prevent root intrusion, a protective corridor with a width of approximately 2 m on each side of the trench is maintained. These corridors must remain free of deep-rooted vegetation, which could otherwise compromise the mechanical integrity of the cable installation or impede future access for maintenance.

During the time-intensive assembly of high-voltage cable joints, exceptional attention must be paid to quality assurance and environmental protection. To ensure optimal working conditions, the assembly area is temporarily enclosed using containers or tent-like structures, creating a clean, dry, and controlled environment. Once the joints are completed, they are embedded in specially formulated bedding compounds and then backfilled with either the native soil or engineered fill material, similar to the rest of the cable trench.

At the extra-high voltage level, dedicated underground joint structures—often referred to as *joint bunkers*—are commonly constructed along the cable route. These are typically reinforced concrete enclosures with dimensions of approximately

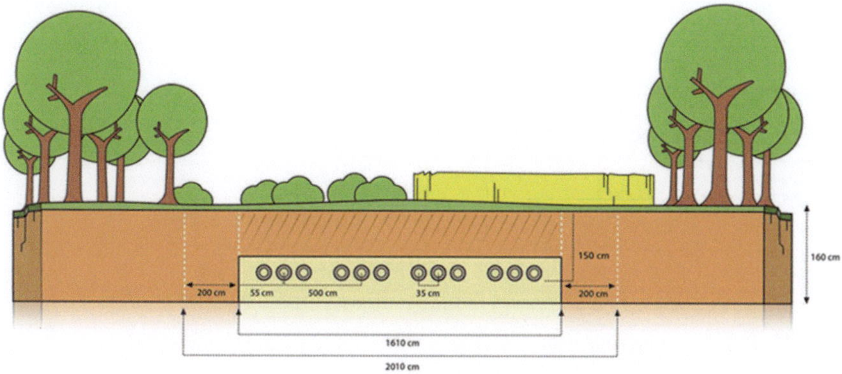

Fig. 5.45 Example of a 380 kV cable route, consisting of two circuits with twin cables each for a total power of 5 GVA (*Source* Joint Paper ENTSO-E, Europacable 2011)

12 m × 8 m (for two cable systems). While they are buried and no longer visible after installation, they must remain accessible for inspection via surface-level access shafts (see Fig. 5.46).

Where multiple cable systems are installed in parallel within the same trench, joint locations are staggered to optimize space and reduce electromagnetic interactions. To accommodate thermal expansion and contraction during operation, the trench is widened near joint zones to allow for compensation curves—gentle bends that relieve mechanical stress on the joints (Fig. 5.47).

In joint zones, the spacing between conductors increases, which can lead to localized magnetic field intensification. To address this, [24] proposes laying closed conductor loops above the joint sections. These loops carry induced currents due to the alternating magnetic field and, according to Lenz's law, produce counteracting

Fig. 5.46 Access shaft to joint structure with visibility aids [23]

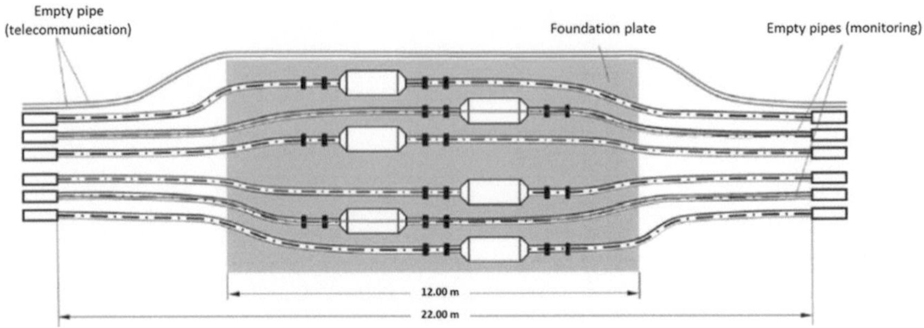

Fig. 5.47 Enlarged joint zone with workspace and thermal expansion buffers [23]

fields that partially cancel the original magnetic field. However, this approach introduces additional thermal losses due to heating of the compensation conductors and should be weighed against its magnetic shielding benefits.

Unlike overhead lines, expanding or modifying an underground cable system involves significant complexity and cost. When protective ducts are used, two main configurations exist: continuous ducting (for future upgrades) and localized crossings (e.g., under roads, railways, or waterways). Typically, new trenches must be excavated for expansions, entailing effort comparable to the initial installation.

To facilitate future extensions, empty plastic conduits (usually PE) can be installed during the initial build. These ducts, however, offer reduced heat dissipation compared to direct burial. Consequently, increased conductor spacing is required, which in turn slightly reduces the overall transmission capacity due to mutual thermal influence.

Recommended internal duct diameters (relative to cable diameter) are:

- 1.4–2.0 × cable diameter → one cable per duct
- 2.6–3.5 × cable diameter → three cables per duct.

Ducts are typically delivered in 5–20 m rod lengths, which must be securely joined at the site. When using thermoplastic materials (e.g., PVC, PP, PE), the high thermal expansion must be considered. A length change of 8–20 mm per 10 m for a 10 °C temperature fluctuation is typical. Improper accommodation of expansion may lead to gaps between segments, risking infiltration of backfill or water.

Prior to cable pulling, ducts must undergo:

- Cleaning
- Calibration (to verify internal diameter)
- Tightness checks.

These steps are crucial to prevent mechanical damage during cable installation and ensure long-term operational reliability.

5.7.2 Cable Transport and Installation

The transport and installation of high-voltage power cables, often referred as *cable pull*, require careful logistical and technical planning due to the significant weight and size of the cable drums. The load-bearing capacity of roads, bridges, and other infrastructure must be verified in advance to ensure they can support the transport vehicles and equipment.

Damage to cables during transport or installation is often caused by insufficient planning or inadequate project design. To avoid such issues, four essential principles must be strictly followed, regardless of the cable type:

- Respect the maximum permissible tensile force during pulling.
- Adhere to the allowable radial compression in curved sections.
- Maintain the minimum bending radius throughout handling.
- Observe the minimum permissible ambient temperature for installation.

These requirements are determined by the mechanical and thermal properties of the cable materials and are specified by the manufacturer.

Installation must always be performed by specially trained personnel, and the entire process must be carefully documented and monitored. Critical parameters such as:

- Cable type and configuration,
- Ambient temperature during installation, and
- Measured pulling forces, if applicable,

must be recorded and verified in accordance with the project specifications and applicable standards. Depending on the sheath material, installation is typically not permitted below −5 °C, as low temperatures can make the insulation and outer sheath brittle, increasing the risk of cracking.

In simple trench installations (e.g., for smaller cross-section cables or in medium-voltage applications), the cable may be pulled directly from a drum mounted on a *cable trailer* driving slowly along the trench. In such cases, special attention must be paid to the correct unspooling direction: cables must always be pulled against the indicated rolling direction on the drum to avoid twisting, internal stress, or damage to the insulation system (Figs. 5.48 and 5.49).

If manual installation is not feasible, the cable must be pulled mechanically with a *cable puller*, with particular attention to controlling the pulling forces to avoid damage (Fig. 5.50) Excessive pulling force can be mitigated by using a *cable power drive roller* (pushing device) positioned at the entry point of the section. For longer cable routes, additional power drive rollers may be required at intermediate points—ideally installed before curves to assist in force distribution and minimize cable stress.

Where feasible, reversing the pulling direction can further reduce stress—particularly when the section with the greatest number or sharpest curves is positioned at the drum end of the route. In cases involving elevation changes, the cable should always be pulled downhill to leverage gravitational assistance and minimize tension.

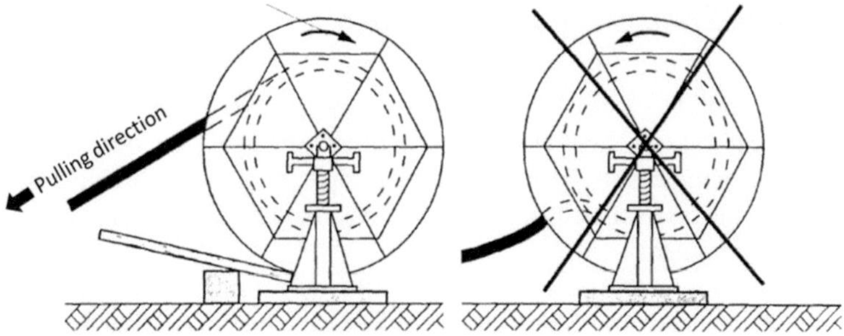

Fig. 5.48 Unwinding direction during cable pull [25]

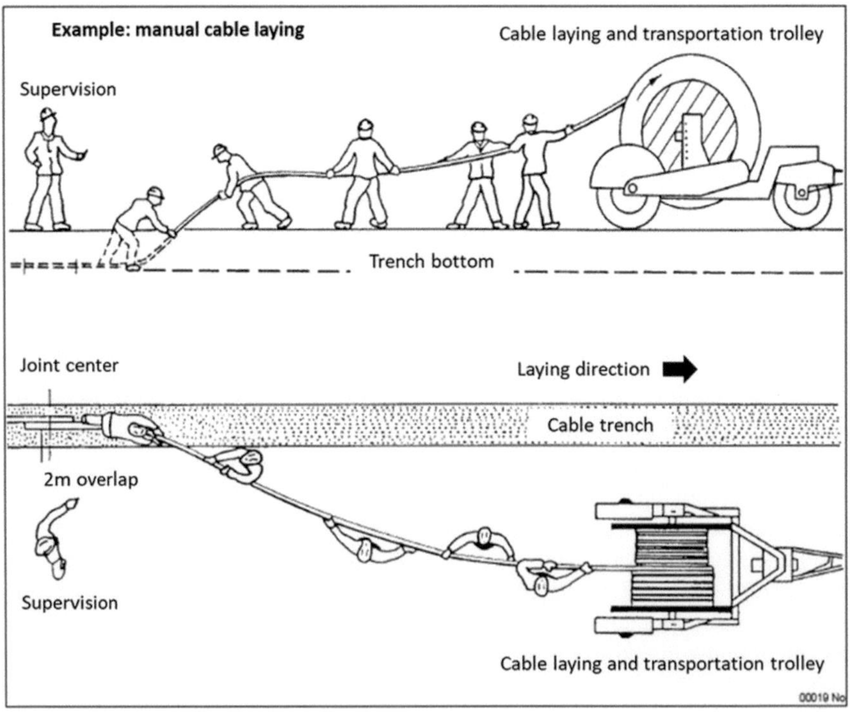

Fig. 5.49 Manual installation of lighter cable types [25]

To protect the cable sheath, direct dragging across the ground is strictly prohibited. Instead, the cable must be guided over properly secured cable rollers. These rollers must be immobilized to prevent shifting during both the pulling operation and any interruptions. At directional changes within the trench, corner rollers are required to guide the cable smoothly and maintain the bending radius.

To ensure that radial compression forces remain within permissible limits, the number and placement of rollers must follow manufacturer guidelines. A typical

Fig. 5.50 Cable puller with a pulling force of 75 kN (*Source* Zeck)

value is a maximum radial force of 10 kN per meter of curve. As a rule of thumb, the minimum bending radius should be at least 25 times the cable diameter D.

For single, localized bends—such as those at terminations—smaller bending radii may be permitted, provided the work is carried out by qualified personnel and appropriate bending templates are used.

A *braking device* must be installed on the cable drum to control the rotation speed and prevent the cable from uncoiling too quickly during pulling. At both ends of the installation section, the cable should be laid out with a surplus length beyond the designated connection point. This reserve ensures sufficient length for the electrical connection during jointing or termination work.

The pulling winch should be positioned at an adequate distance beyond the end of the trench to provide a straight pulling path and reduce mechanical stress. The connection between the cable and the pulling rope must be made using a pulling sock or pulling eye, and always include a compensating swivel ("*twist catcher*") to prevent cable torsion during installation.

In the case of open trench installations, cables must be immediately covered after laying to protect against mechanical damage caused by construction equipment, weather, or third-party interference.

5.7.3 Pulling Forces During Cable Installation

The maximum permissible *pulling force* F_{max} during cable installation is calculated using the formula:

$$F_{\max} = A \cdot \sigma_{\max} \tag{5.5}$$

wherein is:

F_{max} the maximum pulling force in N,
σ_{max} the permissible tensile stress in N/mm^2,
A the cross-sectional area of the cable component subjected to the pulling force (typically the conductor) in square millimeters (mm^2).

The permissible tensile stress $\sigma_{\max}$ depends on the type and construction of the cable and must be taken from the manufacturer's specifications. It is also influenced by the method of pulling—whether the force is applied directly to the conductor via a pulling eye or indirectly through a pulling sock (which transfers force via the outer sheath).

For direct pulling on the conductor, the following typical guide values for permissible tensile stress may be used:

- Copper conductors: $\sigma_{max} = 60$ N/mm^2
- Aluminum conductors: $\sigma_{max} = 30$ N/mm^2.

These values serve as general guidelines and should always be verified against the cable-specific documentation before installation.

For the pulling force calculation the *friction coefficient* μ is also needed. A distinction must be made between open installation (pulling over rollers) and closed installation (in the cable protection tube). For the latter, a lubricant is required, otherwise there could be an overheating of the cable sheath during pulling. The material in question also plays a certain role. The numerical values of μ, like other parameters, vary from manufacturer to manufacturer. In practice, when laying extruded cables, the following values are often used:

Laying rollers	$\mu = 0.2 - 0.3$
Concrete pipes with lubricant	$\mu = 0.4 - 0.6$
Plastic pipes with lubricant	$\mu = 0.1 - 0.2$.

This results in the following relationship for the pulling force F:

$$F = \mu \cdot m' \cdot g \cdot l \tag{5.6}$$

wherein is:

F Pulling force (N)
μ Friction coefficient (–)
m' Length-related mass (kg/m)
l Length of the pulling distance (m)
g Gravitational acceleration (9.81 m/s^2).

Calculation Example 1

For a 110 kV single conductor cable, the following data is available for calculating the pulling force: Length 500 m, cross-section 300 mm^2 Cu, mass 11 kg/m, open installation over rollers with a friction coefficient $\mu = 0.15$. According to the manufacturer's specifications, the maximum permissible tensile force is 15 kN when using a pulling eye.

Solution

$\mathrm{F} = \mu \cdot \mathrm{m} \cdot \mathrm{g} \cdot \mathrm{l} = 0.15 \cdot 11\,\mathrm{kg/m} \cdot 9.81\,\mathrm{m/s^2} \cdot 500\,\mathrm{m} = 8093\,\mathrm{N}$. The value is well below the permissible limit.

If the route has a *height difference*, this must be taken into account in the equation for the pulling force by the following extension:

$$F = m' \cdot g \cdot l \cdot (\mu \cdot \cos\alpha \pm \sin\alpha) \tag{5.7}$$

with α as the inclination angle of the route. The plus sign applies to upward pull and the minus sign to downward pull.

The increase in pulling force due to a *bend* in the route can be taken into account by a multiplication factor:

$$F_E = F_A \cdot e^{\mu\alpha} \tag{5.8}$$

with F_A, F_E, the pulling force before and after the bend and α the bend angle. This has to be calculated using the equation $\alpha_{rad} = \alpha_{deg} \cdot \frac{\pi}{180}$ to convert from degrees to radian (an exact solution to the problem is provided by the so-called *rope friction formula* by Euler and Eytelwein).

If, for example, at the end of the 500 m long route from the above example, a curve of 60° follows, then with $F_E = F_A \cdot \mathrm{e}^{\mu a}$ and the same friction coefficient μ, it follows $F_E = 8093\,\mathrm{N} \cdot \mathrm{e}^{0.1571} = 1.17 \cdot 8093\,\mathrm{N} = 9469\,\mathrm{N}$. The required tractive force increases because of the curve by 17%. The length of the curve hardly influences the calculation. With a radius of 2.50 m and an angle of 30°, it is about 1.30 m and can generally be neglected.

Calculation Example 2

The calculation is more complex for several cable sections. There is a single-core 110 kV cable with the following manufacturer specifications: related mass $m' = 11\,\mathrm{kg/m}$, maximum permissible tensile force 15 kN, minimum bending radius 2.2 m. As friction coefficient, the values 0.1 for installation in a plastic pipe and 0.2 for open installation are to be assumed. All curves have a radius of 2.50 m. Thus, the required minimum value for the bending radius is maintained (Fig. 5.51).

Section 0–1: 200 m in plastic pipe horizontally, then curve 30° horizontally. Section 1–2: 50 m horizontally like section 1, then curve 10° upwards. Section 2–3: 100 m with 10° incline, open laid, then curve 10° and continue horizontally.

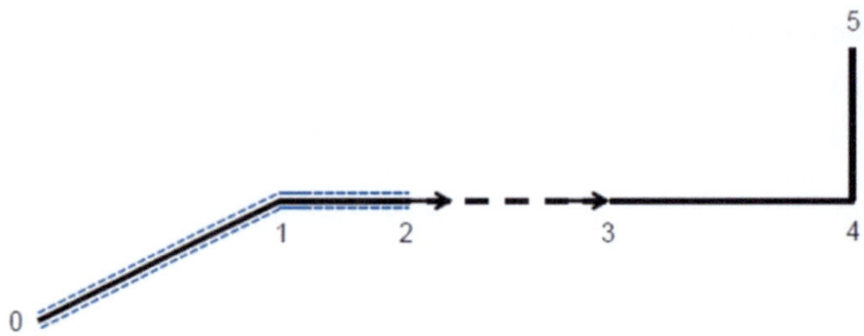

Fig. 5.51 Course of the cable route in the above example (view from above, not to scale)

Section 3–4: 130 m horizontally, open laid, then curve 90° horizontally. Section 4–5: 60 m horizontally, open laid.

The calculation of the pulling forces is done section by section whereby the intermediate results have been rounded to one decimal place. The following auxiliary parameters are needed.

Factor for curve 30°: 1.054; for curve 10°: 1.036; for curve 90°: 1.369; Bracket expression for $\mu = 0.2$ and $\alpha = 10°$: $(\mu \cdot \cos\alpha + \sin\alpha) = 0.371$.

Solution

Route 0–1: $F = \mu \cdot m' \cdot g \cdot l = 2158.2\,\text{N}$.

With a curve 30° at the end: $F_1 = 1.054 \cdot F = \underline{2274.7\text{ N}}$.

Route 0–2: first part section 1–2: $F = \mu \cdot m' \cdot g \cdot l = 539.6\,\text{N}$

$$F_2 = F_1 + F = 2814.3\,\text{N}$$

With a 10° curve at the end: $F_2 = \underline{2915.6\text{ N.}}$

Route 0–3: first part section 2–3: $F = m' \cdot g \cdot l(\mu \cdot \cos\alpha + \sin\alpha) = 4003.5\,N$

$$F_3 = F_2 + F = 6919.1\,\text{N}$$

With a 10° curve at the end: $F_3 = \underline{7168.2\text{ N.}}$

Route 0–4: first part section 3–4: $F = \mu \cdot m' \cdot g \cdot l = 2805.7\,N$

$$F_4 = F_3 + F = 9973.9\,\text{N}$$

With a 90° curve at the end: $F_4 = \underline{13654.3\text{ N.}}$

Route 0–5: first part section 4–5: $F = \mu \cdot m' \cdot g \cdot l = 1294.9\,N$

$$F_5 = F_4 + F = 14949.2\text{ N}$$

In this case, the theoretically expected pulling force lies just below the permissible value of 15 kN.

As an alternative, the pulling force for the opposite direction is determined. The same forces apply in the individual sections, except for Section 2–3 (reversal of the slope); only the curves change the overall result, which is around 8200 N. The significant difference to the above result has mainly two reasons: On the one hand, the 90° curve now only acts on a short section, and on the other hand, the pulling force in Section 2–3 downhill is much lower than when pulling uphill.

5.8 Cable Warming and Cooling Strategies

Cable warming and effective heat removal are central considerations in the operation of power cables, as they directly influence the transmission capacity of a cable system. For underground cables, heat dissipation is primarily governed by the thermal properties of the surrounding soil or bedding material. In contrast, cables installed in ducts or tunnels rely on specifically engineered ventilation systems to remove the generated heat.

In high-load applications, additional cooling methods—such as water-based systems—may be employed to manage thermal performance. When heat is dissipated solely through the surrounding ground or air without any mechanical assistance, the system is said to operate under *natural cooling* conditions. By contrast, *forced cooling* techniques—ranging from air or fluid-based systems to active conductor cooling—can significantly enhance the current-carrying capacity of the cable.

A further distinction is made between *direct cooling*, where the cooling medium is in contact with the cable, and *indirect cooling*, where the medium acts through an intermediate surface or structure.

5.8.1 Underground Cables with Natural Cooling

When backfilling cable trenches the thermal characteristics of the surrounding soil can vary both locally and over time. It is critical to ensure that the heat generated by the cable's thermal losses can still be effectively dissipated—even in areas where the soil exhibits poor thermal conductivity. If insufficient dissipation occurs, the temperature in the immediate vicinity of the cable rises, which in turn increases the conductor's ohmic resistance and accelerates heat generation. This feedback loop can lead to premature aging of the insulation, a reduction in the cable's service life, or in extreme cases, thermal failure.

Moreover, the heat emitted by the cable may cause partial drying of the surrounding soil. Since the thermal conductivity of soil decreases as moisture content declines, this drying effect can further impair heat dissipation, compounding the problem over time.

If no precise data about soil conditions are available, standard assumptions are applied for thermal calculations. For example, German standard DIN VDE 0276-620

Fig. 5.52 Placement of thermally stabilized bedding material (liquid soil) using weights to prevent buoyancy of the protective ducts during backfilling [26]

specifies a dry soil with a *thermal conductivity* $\lambda = 0.4$ W/(K·m) as a conservative basis for evaluating high thermal loads near the cable. Further away from the cable, moist soil with a higher conductivity of 1.0 W/(K·m) is assumed. This variation is modeled using the *two-layer thermal model*, which accounts for the reduced thermal conductivity of the cable bedding material in the immediate vicinity of the cable.

In general, the thermal conductivity of soil ranges from approximately 0.4 W/(K·m) for dry gravel to 5 W/(K·m) for fully water-saturated sand. For heavily loaded cable systems, specially engineered thermal backfills have proven effective. These include materials such as liquid soil (a flowable, self-compacting backfill) or cement-sand mixtures, which after curing form a compact layer with consistently high thermal conductivity (Fig. 5.52).

According to manufacturers, such *thermal backfills* can achieve values up to $\lambda = 3$ W/(K·m), ensuring efficient heat dissipation even under high electrical loading conditions [26].

The thermal and hydraulic interactions between underground cables and the surrounding soil are governed by a complex interplay of transport phenomena and environmental parameters. However, under the assumption of homogeneous bedding material and constant thermal properties, steady-state temperature distributions can be approximated using simplified analytical methods.

In the following section, a thermal equivalent circuit is introduced—drawing an analogy to electrical circuits—to represent heat transfer within the cable system and its immediate surroundings. By analyzing the individual heat sources and the thermal resistances of the involved materials, it becomes possible to calculate the permissible current load that keeps the cable temperature within its specified limit. To illustrate

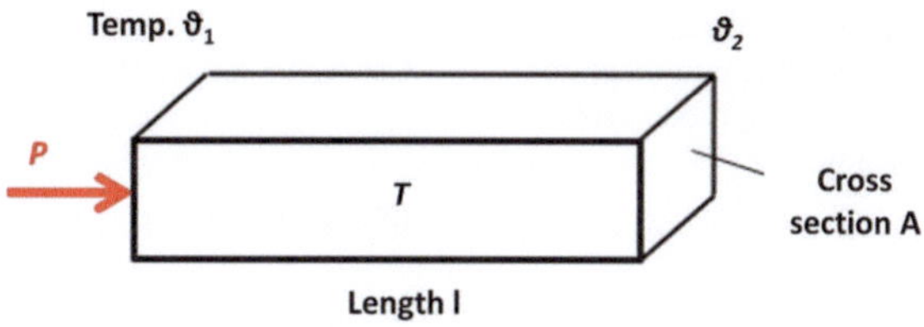

Fig. 5.53 Heat conduction in a rod

the basic principles, the thermal behavior of a simplified model—a rectangular rod cross-section—is first considered (Fig. 5.53).

Wherein is:

P Cable heat loss (W)
ρ Specific thermal resistance (K·m/W)
λ Specific thermal conductivity (reciprocal of ρ) (W/(K·m))
T Thermal resistance (K/W)
ϑ Temperature (in Kelvin) (K)
l Rod length (m)
A Rod cross-sectional area (m^2).

For steady-state heat conduction through the illustrated body, the heat transfer can be described analogously to Ohm's law (in the form $U = R \cdot I$) by the following relation:

$$\Delta\vartheta = \vartheta_1 - \vartheta_2 = T \cdot P \tag{5.9}$$

The thermal resistance T is calculated according to the following equation:

$$T = \rho \cdot \frac{l}{A} \tag{5.10}$$

and is analogous to the electrical resistance $\mathrm{R} = \rho_{\mathrm{el}} \cdot 1/\mathrm{A}$ in electric circuits.

In the case of a cable sheath, i.e. for a cylindrical geometry (hollow cylinder), the thermal resistance for radial heat flow from the inner surface to the outer surface is given by the following expression:

$$T = \frac{\rho \cdot \ln\left(\frac{r_2}{r_1}\right)}{2\pi \cdot l} \tag{5.11}$$

In this context, l represents the axial length of the cable, while r_1 and r_2 denote the inner and outer radius of the cylindrical cable sheath, respectively.

In the case of a cable laid in the ground, there are several heat sources contributing to the power loss P which must be dissipated via its surface to the surrounding soil, such as the conductor with ohmic losses P_1, the insulation with dielectric losses P_2, the metallic sheath and the—also metallic—armor with loss components P_3 and P_4.

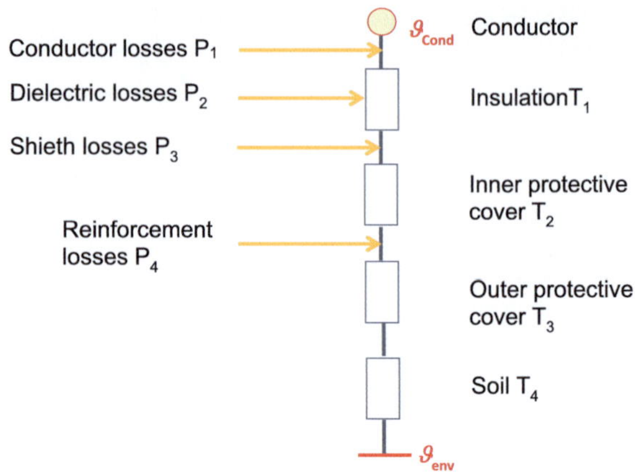

Fig. 5.54 Thermal equivalent circuit diagram for an underground cable (steady state)

On the other hand, there are the thermal resistances T_1 of the insulation, T_2 and T_3 of the inner and outer protective sheaths, and the thermal resistance T_4 of the surrounding soil.

The thermal resistance of the metallic parts of the cable can be neglected here.

Consequently, the following equivalent thermal circuit diagram for the stationary state applies to an underground cable.

From the equivalent thermal circuit (Fig. 5.54), the following equation can be derived for the temperature difference $\Delta\vartheta = \vartheta_{cond} - \vartheta_{env}$ between the conductor and the undisturbed environment. The designations for the individual losses and thermal resistances correspond to those on Fig. 5.54:

$$\begin{aligned}\Delta\vartheta =& P_1(T_1 + T_2 + T_3 + T_4) + P_2(0,5 \cdot T_1 + T_2 + T_3 + T_4)\\ &+ P_3(T_2 + T_3 + T_4) + P_4(T_3 + T_4)\end{aligned} \tag{5.12}$$

If the relevant parameters are known, the relationship between the cable current and the resulting temperature can be determined. However, the above equation is valid only under steady-state conditions - that is, for constant current and thus constant power losses. It is important to note that the loss components P_1, P_3 and P_4 (representing ohmic losses in the conductor, shield, and sheath, respectively) are proportional to the square of the current. In contrast, the dielectric losses P_2 depend primarily on the applied voltage and are largely independent of the load current.

Additional factors influencing the temperature distribution in the cable trench include its geometric dimensions, the number of parallel cable systems and their installation configuration (e.g., flat or triangular), the spacing between individual conductors or systems, and external thermal influences such as adjacent district heating pipelines. The thermal conductivity of protective conduits and the surrounding soil also plays a critical role. Seasonal variations in ambient temperature and surface conditions (e.g., grass cover versus asphalt) further affect the thermal regime. More detailed information is available in [27].

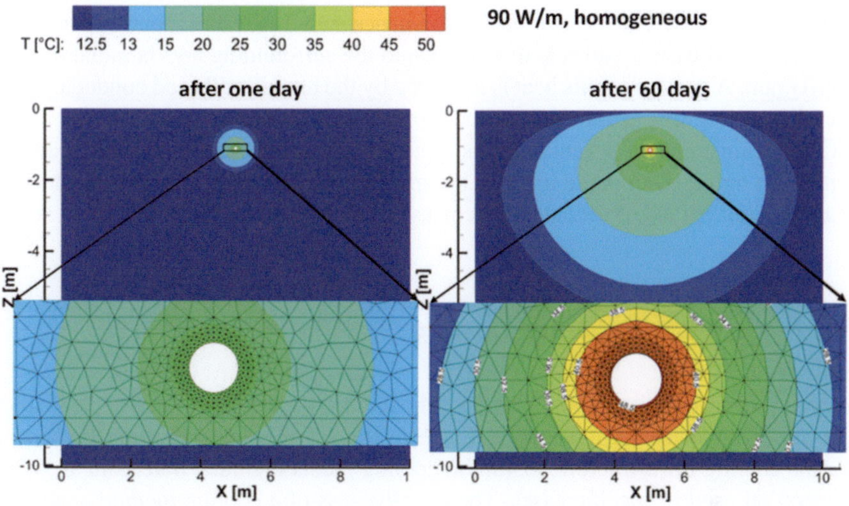

Fig. 5.55 Determination of temperature distribution in the neighborhood of the cable using a numerical process model, based on the assumption of continuous heat input of 90 W/m [28]

In cases involving heterogeneous soil layers, protective piping, or other inhomogeneities and particularly under transient conditions such as time-varying electrical loads or advective (flow-induced) heat transport—numerical methods are employed for accurate thermal analysis. These methods involve solving the governing differential equations both spatially and temporally across a discretized numerical grid. This allows for the assignment of spatially varying material parameters, enabling detailed modeling of thermal inhomogeneities in the cable environment.

An example of such a simulation, illustrating both the temporal evolution and spatial distribution of temperature, is shown in Fig. 5.55. These advanced modeling techniques provide essential input for selecting appropriate bedding materials and for reliably calculating the thermal load capacity of the cable system, thereby helping to prevent long-term degradation such as soil desiccation.

In cases of challenging soil conditions or proximity to other heat sources, it is essential to consult the applicable standards and thoroughly assess the thermal behavior of the cable route. This should be done using validated calculation tools or simulation software provided by cable manufacturers.

5.8.2 Cables in a Tunnel with Natural Cooling

As an alternative to trench installation, cable routing through tunnels is a viable option, particularly in urban environments where surface disruption must be avoided [29]. Tunnel installation can also accommodate other infrastructure systems, offering efficient use of subsurface space. A key requirement, however, is that the tunnel must remain accessible for inspection and maintenance activities. In this configuration, the heat generated by the cables is dissipated into the surrounding air via radiation and convection. A portion of this heat is absorbed by the tunnel walls and conducted into the surrounding soil, while the remainder part is removed through airflow and ventilation openings. Natural ventilation is typically driven by the temperature gradient within the tunnel. Compared to direct burial, tunnel installation generally results in a higher thermal rating of the cable system, as well as a reduced risk of soil desiccation.

5.8.3 Cables with Forced Cooling

Forced cooling can significantly enhance the transmission capacity of a cable system but comes at the cost of higher capital investment and ongoing operational expenses for the cooling infrastructure. Common cooling media includes water and air, with mineral oil used in specific cases. The effectiveness of a cooling method generally increases the closer the cooling medium is to the actual source of thermal losses. However, this proximity typically entails more complex and costly implementation.

The thermal performance of each coolant depends on its *specific heat capacity*, c, which quantifies the amount of heat energy required to raise the temperature of a unit mass of the substance by one degree Celsius (or one Kelvin). Higher specific heat capacities indicate a greater ability to absorb and transport thermal energy, making such materials more effective for dissipating heat in cable systems.

The following reference values apply to water with $c = 4190\,\mathrm{J/(K \cdot kg)}$ and mineral oil with $c = 1860\,\mathrm{J/(K \cdot kg)}$. Because of its expansion behavior, the value for air is temperature-dependent. For 40 °C, as an example, $c = 1000\,\mathrm{J/(K \cdot kg)}$ or $1200\,\mathrm{J/\left(K \cdot m^3\right)}$.

To determine the cooling performance P of a system, both the desired temperature difference $\Delta\vartheta$ (in Kelvin) and the flow characteristics of the cooling medium must be known. These include either the mass flow rate Q (e.g., for liquids like water or air in ducted systems) or, in the case of air cooling via natural or forced ventilation, the ventilation cross-sectional area A and the airflow velocity v. The corresponding equations are:

For liquids:

$$P = c \cdot \Delta\vartheta \cdot Q \tag{5.13}$$

For air:

$$P = c \cdot \Delta\vartheta \cdot A \cdot v \tag{5.14}$$

In the most advanced variant—*conductor cooling*—the coolant (typically deionized water or, in some cases, low-viscosity mineral oil) is circulated directly through a metallic cooling channel integrated into the center of the conductor. Since the cooling medium is at high voltage potential, it must be decoupled from the system using a specially designed insulated terminal that safely brings the water to ground potential. Only then can the medium be cooled and recirculated. According to manufacturers, this method can multiply the transmission capacity of the cable system compared to passive (naturally cooled) configurations [30].

In *direct cooling*, one or more cables are installed inside ducts or protective conduits through which water flows. While this method avoids potential separation issues, it presents other technical challenges:

- Heat must be transferred through the cable insulation to the coolant, which reduces overall cooling efficiency.
- The mechanical movement of cables under load fluctuations may require compensation measures.
- Cooling of cable accessories (e.g., joints, terminations) must be separately ensured.
- Implementation of cross-bonding becomes more complex due to reduced accessibility to the cable shields.

Nevertheless, direct cooling can increase transmission performance by a factor of up to 3, depending on design and operating conditions [30].

In *indirect cooling*, cooling pipes are laid parallel and above the cables, outside their outer sheaths. Heat must then pass through the insulation, the outer sheath, and the surrounding thermally optimized bedding material before reaching the cooling pipes. Although this method is less efficient than direct conductor cooling, it offers significant advantages:

- The cable design remains identical to standard systems.
- Cable accessories require no special modifications.
- The system can still double the transmission capacity over natural cooling, depending on thermal properties and layout [30].

While all forced cooling options entail higher capital and operational costs, they also offer clear benefits: increased transmission capacity, extended service life, and improved protection against hotspots. In certain applications, the residual heat in the coolant can be recovered and repurposed for industrial processes, further improving system efficiency.

An example of indirect cable cooling with water is shown in Fig. 5.56.

In the following, some examples of systems with forced cooling as well as examples of tunnel systems are presented.

Calculation Example 3

The transmission capacity of a 110 kV cable is to be increased from 120 to 220 MVA by means of water cooling. The XLPE cable $3 \times 300\ \text{mm}^2$ Cu ($R' = 60\ \text{m}\Omega/\text{km}$) with a length of 2 km has a nominal current of $I = 630$ A. This results in heat losses of $P_{loss} = R \cdot I^2 = 47.6$ kW per conductor; with a current value $I = 1.15$ kA (corresponding to 220 MVA) they rise to 159 kW. The difference is to be cooled away via a plastic pipe through which water flows, whereby the water may be heated by $\Delta\vartheta = 15$ K. A flow rate of $Q = 2$ l per second per individual cable is provided.

Fig. 5.56 380 kV underground cable system with a total of four cooling pipes (*Source* Wiener Netze)

Solution

P=159kW−47.6kW=112.4 kW per conductor must be cooled away. According Eq. (5.13), the cooling capacity is P=4190 Ws/(K · l) · 15 K · 2 l/s=125.7kW. It is therefore sufficient.

Calculation Example 4

A ventilation system for a cable system in the tunnel is to be dimensioned. The specific heat capacity of air is $c = 1000\,\mathrm{Ws/(K \cdot kg)} \approx 1000\,\mathrm{Ws/\left(K \cdot m^3\right)}$ and an air velocity of $v = 5$ m/s is provided for a tunnel cross-section of $A = 10$ m^2 and a temperature increase of $\Delta\vartheta = 10\,\mathrm{K}$.

Solution

For the assumptions made, Eq. (5.14) yields the following result:

$$\mathrm{P} = 1000\,\mathrm{Ws/\left(K \cdot m^3\right)} \cdot 10\,\mathrm{m}^2 \cdot 10\,\mathrm{K} \cdot 5\,\mathrm{m/s} = 500\,\mathrm{kW}.$$

Application Example 3

A comprehensive network of 380 kV AC cable tunnels has been constructed in London. The favorable soil characteristics of "London Clay" facilitate the use of 3 to 4 m diameter tunnel boring machines. National Grid, the system operator for the approximately 50 km long and up to 40 m deep tunnel network, identifies several advantages in this method of installation. Most notably, surface disruptions to road traffic, businesses, and residents are minimized. Above-ground work is limited to the insertion of tunnel boring machines, cable drums (via shafts up to 15 m in diameter), and the delivery and removal of materials and excavated soil. However, similar disturbances would occur with conventional trench-based installation. Once completed, the tunnel system enables maintenance and repair activities to be conducted with minimal impact during daytime hours. Additionally, future system expansions can be implemented by simply pulling in new cables within the existing tunnel infrastructure [31, 32].

Application Example 4

Comparable arguments also support the use of tunnel-based cable installations in other urban areas. This was the case in Berlin (Germany), where a section of the 380 kV AC "diagonal connection" between the substations Mitte and Marzahn was constructed using two ventilated cable tunnels measuring 6.3 and 5.2 km in length. Constructed between 1998 and 2000, the tunnels lie 25 to 30 m below ground and traverse under rivers (such as the Spree), canals, railway lines, and existing cable systems. Two plastic-insulated cable systems were installed in tunnels with an internal diameter of 3 m. Multiple tunnel boring machines were operated simultaneously to complete the excavation. Prior to the main project, comprehensive qualification tests were performed on cables and joints supplied by several manufacturers [33, 34] (Fig. 5.57).

Application Example 5

One of the world's longest cable tunnels is located in Tokyo, Japan: the Sinkeiyo–Toyosu Line, spanning a total length of 40 km. The urban layout of Tokyo—with its system of canals and railway corridors—enabled efficient delivery of cable drums weighing up to 90 tons directly to the installation sites. The use of such large drums allowed for extended delivery lengths and therefore significantly reduced the number of required joints. The tunnel houses a 500 kV AC cable system using 2500 mm^2 copper XLPE-insulated single-core cables, which was commissioned in 2000 [35].

Fig. 5.57 Cable tunnel in Berlin with two three-phase circuits

Application Example 6

Tunnel-based cable systems have also been implemented in other countries. In Madrid, Spain, such a solution was adopted in the area surrounding the newly built Barajas International Airport. Here, a 380 kV double-circuit AC overhead line had to be rerouted underground for an overland section outside the city. The underground system consists of 2,500 mm^2 XLPE single-core cables, with a total of 2×3 conductors. The tunnel, extending 12.8 km, was constructed primarily using the cut-and-cover method with prefabricated concrete segments. At road and river crossings, trenchless drilling was performed down to 15 m depth. Due to the shallow tunnel depth—typically only 2 m below ground level—an active cooling system was necessary. The design included five fan systems, each rated at 38 kW, to maintain safe operating temperatures [36].

5.9 Commissioning Tests of High Voltage Cables

Before a cable system is put into service for power transmission, a series of electrical tests—referred to as *commissioning tests*—must be performed to verify its integrity and readiness. The most critical of these are:

- An *alternating voltage test* on the high-voltage components (optionally combined with a *partial discharge test*), and
- A *direct voltage sheath test* to assess the condition of the outer cable jacket.

Alternating Voltage Test

In its simplest form, the alternating voltage test consists of applying the system's rated voltage under no-load conditions for an extended period, typically 24 h. This method is fast, inexpensive, and does not require any external voltage source. However, a major limitation is that the test voltage cannot exceed the normal operating voltage.

If testing at higher voltages is desired—for example, to detect latent defects that only manifest under electrical stress—a resonance test system must be used. This setup enables testing at voltages up to 1.5 times the rated voltage, typically applied for one hour. Such tests are often performed in combination with partial discharge measurements, which provide additional insight into insulation condition and aging.

DC Sheath Test

The sheath test is a direct voltage test performed to ensure that the cable's outer sheath was not damaged during transport or installation. Damage to the sheath could allow moisture ingress, leading to corrosion of the cable's metallic water barriers and subsequent failure.

For this test, a DC voltage of 10 kV is applied for one minute between the conductor and the cable shield. If the voltage is maintained without breakdown or leakage, the test is deemed successful. A passed test indicates that the sheath provides sufficient insulation and protection against moisture ingress.

5.10 Cable System Monitoring

Monitoring a cable system involves the continuous observation of key system parameters to ensure operational reliability and to detect anomalies early. Typical parameters monitored include:

- Temperature distribution along the cable,
- Moisture ingress or humidity within the cable environment, and
- Partial discharges, which may indicate insulation degradation.

The primary objective of monitoring is early fault detection, enabling proactive maintenance to prevent major failures or extended outages.

In recent years, *Real-Time Thermal Rating* (RTTR) has become an important development in the context of high-voltage cable monitoring. RTTR enables dynamic load management by continuously assessing the actual thermal state of the cable system. This allows the optimization of power flows—especially valuable in meshed network topologies—by safely utilizing thermal reserves under varying load and environmental conditions.

Distributed Temperature Sensing (DTS)

The temperature profile along a cable system is now almost exclusively measured using *optical fiber* technology, a method known as *Distributed Temperature Sensing* (DTS). In DTS systems, optical fibers function as both sensors and transmission media, measuring temperature at thousands of points along the fiber's length.

Installation options include:

- Parallel laying of optical fibers alongside the power cables, or
- Integration into the cable design, where the optical fibers are embedded directly within the cable shielding.

In the latter case, the fibers are housed in a stainless-steel capillary tube embedded in the cable's metallic sheath. This ensures long-term mechanical and thermal stability while preserving signal integrity for high-precision measurements (Fig. 5.58).

Optical Fiber Monitoring in High Voltage Cable Systems

The use of optical fibers for monitoring temperature along high-voltage cable systems offers significant advantages. Optical measurement is inherently immune to electrical and magnetic interference and is galvanically isolated from high-voltage components. In addition to detecting hotspots, the system can monitor environmental influences such as soil drying, and it enables controlled short-term overloading of the cable based on real-time thermal conditions and thermal preloading.

Fig. 5.58 High-voltage cable with integrated stainless steel tubes in the cable shield housing optical fibers for Distributed Temperature Sensing (DTS) monitoring (*Source* Brugg Cables)

The optical fiber itself functions as a distributed temperature sensor, enabling spatially and temporally resolved temperature measurements along the entire cable route. This is achieved through evaluation of backscattered light. Multiple stainless steel capillaries, each containing individual optical fibers, are typically integrated into the cable shield. Both *single-mode* (SM) and *multi-mode* (MM) fibers are employed:

- Single-mode fibers provide higher spatial resolution (< 0.5 m).
- Multi-mode fibers support longer measurement distances (up to 70 km), which is advantageous for series-connected fibers.

In both cases, the temperature resolution is better than 1 K.

The principle of temperature sensing is based on *Raman backscattering*, as illustrated in Fig. 5.59. A pulsed laser injects light into the fiber. A small portion of this light is scattered back:

- Most is backscattered at the same wavelength (*Rayleigh scattering*).
- A smaller portion is scattered at slightly shifted frequencies (*Stokes and anti-Stokes lines*) due to interaction with molecular vibrations in the glass.

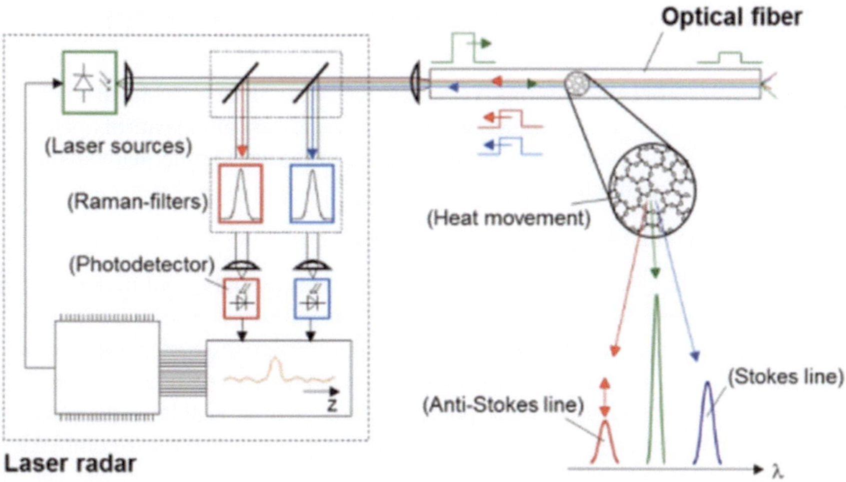

Fig. 5.59 Principle of temperature measurement using optical fibers in high-voltage cables via Raman scattering [37]

The ratio of the Stokes to anti-Stokes intensity is temperature-dependent. Measuring this ratio allows precise determination of the local temperature. The position of the measurement point is calculated from the transit time difference between the emitted and backscattered pulses.

In addition to temperature, *partial discharge* (PD) *monitoring* is a crucial diagnostic tool, especially for cable accessories, where assembly defects may cause early-stage discharges that precede insulation breakdown. PD sensors are commonly installed in grounding or cross-bonding boxes, eliminating the need to interfere with the high-voltage components. However, PD monitoring within the polymeric cable insulation itself is more challenging: the time between detectable discharge and breakdown is very short—often just a few seconds. Furthermore, external electromagnetic interference (e.g., from railways) complicates interpretation, requiring expert analysis.

Moisture monitoring can also be achieved via a *resistance wire* integrated into the cable shield. If moisture penetrates and causes contact between the resistance wire and the surrounding shield wires, the fault location can be precisely determined via resistance measurement.

In GIL, fiber-optic monitoring systems are used to measure:

- Gas density
- Temperature
- Partial discharges
- Arc detection during short circuits.

These monitoring methods are essential for maintaining system reliability and preventing failures in high-voltage transmission infrastructure.

5.11 Diagnostic Methods for High Voltage Cable Systems

Diagnostic testing of cable systems is essential for assessing their technical condition at a specific point in time. Such diagnostics may be necessary after years of operation, following high-stress events (e.g. short circuits), or during routine maintenance to ensure ongoing system reliability.

Partial Discharge (PD) Diagnostics

As with monitoring, *partial discharge measurement* is a central tool in diagnostics applied to both the cable and its accessories. The required test voltage can either be drawn directly from the system or generated externally using *resonance transformers.* Externally generated test voltage allows diagnosis at levels above nominal operating voltage, which helps identify weak points that may not be visible under standard conditions. The main disadvantage of this approach is the logistical complexity: transporting resonance equipment—including large coils—to the test site can be challenging.

For cable accessories, PD testing is particularly valuable, as manufacturing or assembly defects often manifest here before causing a breakdown. However, for older oil-filled cables, PD testing alone is of limited value due to the self-healing properties of their insulation and the occurrence of non-critical discharges.

Loss Factor (tan δ) Measurements for Ageing Assessment

To evaluate the aging of power cables, *dielectric loss factor (tan δ) measurements* are widely used in the medium-voltage level. These are typically performed at *very low frequencies* (VLF, e.g. 0.1 Hz) and at different voltage levels: half, full, and 1.5 times the rated voltage. Aging is inferred from the increase in tan δ with rising voltage. A significant increase in tan δ is indicative of insulation deterioration, often associated with *water treeing*—degradation process involving moisture-induced micro-cracks in the insulation that eventually lead to failure. In contrast, high-voltage cables, particularly those with metallic cross-water barriers, are less prone to water treeing, making tan δ testing more relevant for medium-voltage systems.

DC Sheath Testing

Another standard diagnostic technique is the *DC voltage test of the cable sheath*, similar to commissioning procedures, whereby a DC test voltage of around 5 kV is applied for approximately one minute. This identifies mechanical or moisture-related sheath damage (e.g. from excavation or ground movements) that may compromise the cross-water barrier or insulation integrity.

Dissolved Gas Analysis (DGA) of Liquid-Filled Terminations

For fluid-filled outdoor terminations, *gas-in-oil analysis* provides a fast and cost-effective diagnostic method, comparable to transformer oil testing. Thereby a small sample of insulating liquid is analyzed for trace gases generated by PD activity. *Hydrogen* and acetylene are key indicators of partial discharge, with *acetylene* being particularly significant even at low concentrations. *Methane*, while also present, is less conclusive, as it may originate from the diffusion of crosslinking by-products from XLPE into the oil over time.

Additional measurements can assess *moisture content* and *dielectric strength* of the fluid, providing early warning of potential termination faults before PD activity becomes critical.

Infrared thermography is increasingly used for condition assessments as it can detect abnormal temperature distributions in terminations or *hotspots* on exposed cable sections (e.g., in tunnels or open trenches). This non-invasive method is valuable for detecting local overloads, defective joints, or hot spots in forced-cooled systems.

5.12 Areas of Application of Cables

Historically, the high cost of power cables meant they were used primarily in locations where overhead lines were not feasible—such as densely populated urban areas, environmentally protected zones, or when crossing bodies of water. Subsea applications, including the supply of islands and the connection of offshore wind farms or oil and gas platforms, also necessitate the use of cables. The same applies within large industrial facilities and near airports, where space and safety constraints limit the use of overhead systems.

The global transition to renewable energy sources requires major adaptations to the power transmission infrastructure. While overhead lines remain the dominant method for long-distance transmission—particularly for bulk power at ultra-high voltage levels above 1000 kV—underground cables are increasingly being integrated into the grid.

Unlike overhead lines, underground cables must dissipate their thermal losses directly into the surrounding soil, without benefiting from cooling effects such as wind or ambient air temperature. To enhance their transmission capacity, several technical strategies are employed: increasing conductor cross-sections, using thermally optimized bedding materials, and implementing active cooling systems. In parallel, high-performance gas-insulated line systems have become a viable alternative in specific applications.

Another important development is the growing use of high-voltage direct current transmission via cables—an approach that Germany, among others, has adopted as part of its energy transition strategy. Long-distance HVDC cable transmission remains relatively uncharted territory and presents significant technical challenges. However, successful implementation could lay the groundwork for future overlay networks and enable large-scale integration of renewable generation, as discussed further in Chaps. 6 and 7.

References

1. Suchtanke R, Just H (2014) Numerische Untersuchung des Wechselstromwiderstandes von Energiekabeln, TU Berlin
2. Papailiou K (ed) (2021) Handbook power systems. Springer
3. DIN VDE 0292:2021-08, System für Typkurzzeichen an isolierten Leitungen, Beuth Verlag 2021
4. Peschke E, von Olshausen R (1998) Kabelanlagen für Hoch-und Höchstspannung, Publicis MCD Verlag
5. Gockenbach E, Frohne Ch (2021) Gleichstromkabel für die Hochspannungs-Gleichstrom-Übertragung. ETG J, Hannover
6. Cichowski R. Kabelhandbuch ew Medien und Kongresse
7. Küchler A (2024) HGÜ-Kabel und Kabelverlegung. Fachseminar Tagungsge-sellschaft/MVV Regioplan GmbH, Karlsruhe
8. Küchler A (2017) Hochspannungstechnik, Springer
9. Anon., High voltage direct transmission—proven technology for power exchange, Siemens Erlangen 2011
10. Fuchs K, Novitskij A, Berger F, Westermann D (2014) Hochspannungsgleichstromübertragung-Eigenschaften des Übertragungsmediums Freileitung, Ilmenau
11. Goll F (2020) Untersuchungen an neuen umweltfreundlichen Isoliergasen für kompakte gasisolierte Übertragungsleitungen, Siemens Gas and Power
12. Anon., Unterirdische Stromversorgung durch Gasisolierte Leitungen (GIL), Siemens Energiewendedialog Berlin 2013
13. Schöffer G, Kunze D, Smith I (2007) Gas insulated transmission lines—successful underground bulk power transmission for more than 30 years, Siemens
14. Sauer D (Hrsg.) Flexibilitätskonzepte für die Stromversorgung 2050
15. Oswald B (2007) 380-kV-Salzburgleitung. Auswirkungen der möglichen (Teil-) Verkabelung des Abschnitts Tauern-Salzach (neu), Hannover
16. Skog J, van Asten H, Worzyk T, Andersrod T (2010) NorNed—world's longest power cable. CIGRE Paris
17. Colla L et al (2024) Design and testing of the dynamic export cable of Gruissan offshore floating wind farm, CIGRE Paris
18. Pirker A, Machl A (2024) Development of an extended commissioning program for temporary 220-kV cable connections. CIGRE Paris
19. etn Energie-und Telecom Netze GmbH, Austria
20. Prysmian: Viking link—the world's longest interconnector
21. Hansen HR (2024) Challenge of TDR fingerprint on viking link. CIGRE Paris
22. Escobin A, Grund R (2018) Flexibilität im Höchstspannungsnetz—Steckbare Durchführungen als Schlüssel, Stuttgart
23. Tennet, Erläuterungsbericht 380-kV-Leitung Kreis Segeberg-Raum Lübeck, Bayreuth 2020
24. Brakelmann H, Jarass L (2019) Erdkabel für den Netzausbau. Norderstedt
25. E-ON, Tiefbau für unterirdische Versorgungsleitungen-Kabel-und Leerrohrlegung, 2020
26. Anon., Bettungsmaterialien für Höchstspannungskabel, Infra-Tech Essen 2014
27. Schlabbach, J. and Rofalski, K.H.: Power System Engineering, Wiley-VCH, 2014
28. Bauer S (2021) Durch Erdkabel induzierte thermische und hydraulische Prozesse im Untergrund. Seminar, Hochschule Karlsruhe
29. Woschitz R. Höchstspannungsübertragungsleitungen für die Verlegung in langen Tunneln, Graz 2008 (Buchprojekt S. 57)
30. Anon., Höchstspannungskabel zur Energieübertragung, Nexans Hannover 2001
31. Anon., London cable tunnels, Elstree—St. John's Wood 2005—West Ham—Hackney 2009
32. Weinlein A. 420-kV-VPE-Kabelsysteme im Projekt London Power Tunnels. EW 4/2019
33. Anon., Berlin im Stromverbund mit Europa, Fachreport 380-kV-Diagonalverbindung. 50Hertz Berlin

34. Fischer, Henningsen (2001) Long time operational experience with 400 kV XLPE cable systems in Berlin, Jicable
35. Anon., Construction of the world's first long-distance 500 kV XLPE cable line Japan
36. Anon., Spanish grid operator selects 400 kV cable system for airport expansion, RED Electra Espana 2010
37. Glombitza U (2021) Temperaturmonitoring bei Kabeln—Verfahren, Software, Kabelbetrieb. Tagungsgesellschaft Energie, Karlsruhe, 10.09.2021

Chapter 6
Comparison of Overhead Lines and Underground Cables

Summary

Overhead lines, underground cables, and gas-insulated lines (GIL) are collectively referred to as transmission lines, as they all serve the same fundamental purpose: to reliably and efficiently transfer electrical energy between network nodes.

This chapter begins by examining how the different technologies affect key electrical parameters such as current-carrying capacity and transmission losses. It then explores the behavior of the electric and magnetic fields in the immediate vicinity of each transmission type.

Statistical data on outages and damage provide insights into the frequency and duration of failures for each line technology. Based on these data, reliability parameters—essential for the planning and operation of electrical networks—can be derived using mathematical models, such as the Markov method. These and other operational considerations, such as neutral point treatment, are illustrated through practical examples.

Network calculation tools also rely on standardized line models. These models are introduced and applied to the main operating states relevant for planning: no-load, matched load, and short-circuit. Several step-by-step calculation examples are presented for different line types and lengths in the high and extra-high voltage range, with results visualized using phasor diagrams. This section also addresses reactive power and discusses compensation strategies.

Finally, the most important characteristics of overhead lines, underground cables, and gas-insulated lines are compared, both in alternating current (AC) and direct current (DC) systems. A dedicated section discusses the construction costs of the different line types, supported by indicative cost figures.

K. O. Papailiou et al., *High-Voltage Overhead Lines and Underground Cables*, Power Systems, https://doi.org/10.1007/978-3-032-07498-0_6

6.1 Electrical Behavior

As previously discussed, transmission lines differ fundamentally in their conductor materials and cross-sections, insulation media, and insulation distances. The design of insulation in three-phase networks above 1 kV is governed internationally by IEC Standard 60071-1: Insulation Coordination, which outlines the procedure for selecting standardized rated voltages for both conductor-to-conductor and conductor-to-earth insulation in high-voltage equipment.

In the high and extra-high voltage ranges, the primary stress factors are switching surges and lightning impulse voltages. These stresses, combined with the dielectric strengths of different insulation materials, determine the required clearance distances. Representative values are shown schematically in Fig. 6.1.

In overhead lines, conductors must withstand significant mechanical stress, which is why composite conductors made of aluminum and steel are commonly used (Chap. 4). In contrast, after installation, cable conductors are not subject to tensile stress, and conductor weight is less critical. Therefore, copper is typically preferred in cables for its superior conductivity, although aluminum is also used for cost reasons. As a result, cables generally exhibit lower ohmic resistance per unit length than overhead lines for the same transmission capacity.

Gas-insulated lines (GIL) currently use aluminum conductors exclusively. Thanks to their large cross-sectional areas, they achieve very low resistance values and minimal losses, even at high transmission capacities.

For example, at the 110 kV level, single-conductor overhead lines (e.g., 264-Al 1/34 ST1A, formerly Al/St 265/35) have a transmission capacity of approximately 130 MVA, comparable to that of extruded copper cables with $3 \times 1 \times 630\ \text{mm}^2$ conductors.

At the 380 kV level, a quadruple-bundle overhead line using the same conductor material can transmit around 1800 MVA, while a 2500 mm^2 copper XLPE cable

line types ▷ / Voltage level and type ▽	Overhead line 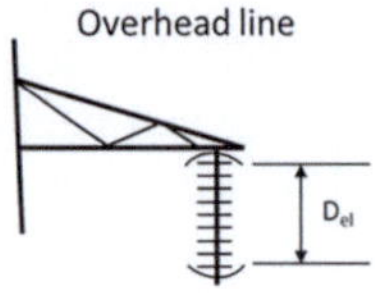	Gas-insulated line (GIL) 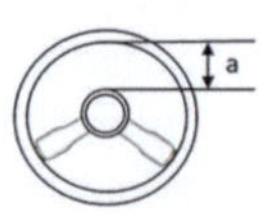	Cable with XLPE insulation 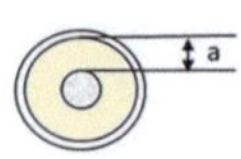
110 kV AC	1000	-	13
220 KV AC	1700	-	18
380 kV AC	2800	290	27
525 kV DC	-	-	26-28

Dimensions in mm

Fig. 6.1 Typical insulation distances for overhead lines, gas-insulated lines, and cables [1–3]

achieves around 1600 MVA, assuming good thermal conductivity in the surrounding soil. In addition, GIL systems, with a 6300 mm^2 conductor cross-section, can transmit up to 2000 MVA when buried (Chap. 5 and [3]). Even higher capacities can be achieved using larger cross-sections or high-temperature conductors (Chap. 4). Transition to a higher voltage level or adopting HVDC transmission is also an option.

In overhead lines, the ambient air serves as insulation, and the large conductor spacing prevents dielectric breakdown. Since air constantly renews itself, no aging occurs in the insulation medium. In the event of a fault-induced flashover, the arc often self-extinguishes after a short automatic outage (known as *automatic reclosure*) allowing immediate resumption of operation without repair.

In contrast, cable insulation is typically damaged during a fault, and the affected section must be physically replaced, often involving significant effort and cost.

Gas-insulated lines, used only at high voltage levels, resemble cables in structure but use compressed insulating gas instead of solid insulation. This allows for smaller insulation distances than overhead lines. Their rigid enclosures offer excellent protection from external damage; however, due to the limited number of installations to date, operational experience is still limited and there is hardly any information available on this.

During internal overvoltage events, arc breakdowns may occur within GIL systems. According to manufacturers, these are manageable through automatic reclosure, similar to overhead lines.

Because of the smaller spacing between conductors and ground and the higher relative permittivity of solid insulation materials compared to air, cables have significantly higher capacitance—up to 20 times greater than that of overhead lines.

By contrast, overhead lines typically have higher inductance. Simply put: large conductor-conductor or conductor-earth spacing leads to high inductance and low capacitance, while small spacing yields the opposite.

Bundled conductors in overhead lines (Chap. 4) allow for higher transmission capacity and reduce the local electric field strength. They also increase capacitance slightly and lower inductance, which in turn reduces *surge impedance* (Chap. 5).

GIL systems typically exhibit about one-quarter the capacitance of equivalent cables and one-fifth the inductance of overhead lines.

As will be shown later, the ratio of capacitance to inductance directly affects *surge impedance*, which in turn significantly influences system behaviour during operation, particularly when partial underground cabling is implemented within otherwise overhead networks.

Table 6.1 summarizes the key electrical characteristics of common high and extra-high voltage transmission systems. The significance of the line length-related parameters—R' (ohmic resistance at 20 °C), L' (inductance), and C_B' (operating capacitance)—as well as the length-independent values Z_s (surge impedance), *SIL* (surge impedance loading), and S_{th} (thermal limit power) will be explained in Sect. 6.7. All values provided are intended as indicative reference figures.

Table 6.1 Approximate characteristic values for overhead lines, cables, and GIL at high and extra-high voltage levels (*Sources* Manufacturer data and technical catalogs)

Line types	110-kV-overhead line AC	110-kV-cable AC	380-kV-overhead line AC	380-kV-cable AC	380-kV-gas-insulated line (GIL)	525-kV-cable DC
Number	1	2	3	4	5	6
Materials	Al/St	Cu	Al/St	Cu	Al/SI	Cu
Cross section in mm^2	264/34	630	$4 \times 264/34$	2500	6300/13600	3000
R' in $m\Omega/km$	110	28	27	10.8	5.5/4.2	6
L' in mH/km	0.94	0.52	0.81	0.6	0.22	
C_B in nF/km	8	186	14	246	55	
Z_s in Ω	343	53	240	50	63	
SIL in MW	35	230	600	2930	2300	
S in MVA (AC), MW (DC)	130	140	1800	1250	2000	2600

6.2 Transmission Capacity of Lines

The maximum power a transmission line can carry is fundamentally limited by two physical parameters:

- the current, which determines thermal loading of the conductor material, and
- the voltage, which dictates the stress on the insulation.

In the case of underground cables, heat losses must be dissipated through the surrounding soil, whose thermal conductivity remains relatively constant throughout the year. At typical burial depths of around 2 m, soil temperature and moisture exhibit minimal seasonal variation.

The conductor cross-section of a cable cannot be selected arbitrarily. While a larger cross-section reduces electrical losses, it simultaneously increases the amount of material required for insulation and joints, raising costs. An economic optimum is usually reached at *current densities* of around 1 A/mm^2.

Since cables are delivered to site wound on drums, there is a trade-off between cable length and cross-sectional area, due to transport weight limits. Longer cable sections reduce the number of joints required along the route (Chap. 5), but must remain within handling constraints.

For overhead lines, the conductor cross-section is similarly constrained–but for different reasons. The self-weight of the conductor and the resulting mechanical loads from wind and ice set practical limits. While insulation distances are not an issue in air-insulated systems, mechanical strength and tensile performance of the conductor material are critical, as they directly affect sag.

Heat losses in overhead conductors are dissipated to the ambient air. These losses may be higher under cool and windy conditions, and lower under hot, still weather. This principle is used in dynamic line rating systems, also known as overhead line monitoring, which optimize line loading based on real-time weather conditions (Chap. 4).

Underground cables do allow for short-term overloads, depending on their preload condition. The thermal time constant of the surrounding soil—typically several hours—is a key factor. For cyclic loads, a higher thermal limit may be permissible than for continuous operation. This thermal limit is expressed via the *load factor m* (Chap. 5).

However, care must be taken to avoid critical over temperatures, which may reduce insulation life or cause catastrophic failure. For this reason, continuous temperature monitoring along the cable route is recommended and is now standard practice in high voltage systems (Chap. 5).

In AC systems, a line's ampacity is not only governed by ohmic resistance, but also significantly influenced by its inductance and capacitance:

- Resistance (*R*) depends on conductor material and cross-section.
- Inductance (*L*) and capacitance (*C*), in contrast, can be actively compensated to enhance performance.

While resistances are to be regarded as fixed values, external series capacitors can be used to partially or fully compensate inductance in long overhead lines and shunt reactors (inductive components) to compensate capacitance in cable systems.

Such compensating equipment must be switched in steps, according to the voltage and load conditions, and require special switching and protection devices. Because of the associated complexity and cost, they are primarily used at the extra-high voltage level.

An alternative to mechanical switching is the use of power electronic systems, which offer fast, continuous control without the need for mechanical switches. These systems fall under the umbrella of FACTS (*Flexible AC Transmission Systems*).

Depending on their configuration, FACTS devices can be connected:

- in series with the line: to affect line impedance and current directly, or
- in parallel: to inject reactive power and regulate voltage at network nodes.

Examples of series-connected FACTS devices include:

- TCSC (Thyristor-Controlled Series Capacitor)
- TSSR (Thyristor-Switched Series Reactor)
- TSSC (Thyristor-Switched Series Capacitor)
- SSSC (Static Synchronous Series Compensator).

Examples of shunt-connected FACTS devices include:

- SVC (Static Var Compensator)
- TCR (Thyristor-Controlled Reactor)
- TSC (Thyristor-Switched Capacitor)
- STATCOM (Static Synchronous Compensator) [4].

Besides phase shifter transformers, these devices play a major role in voltage regulation, power flow control, and stability enhancement.

Lastly, the *stability* of an AC power line is determined less by the voltage magnitudes at the sending and receiving ends than by the *phase angle difference* between them. This angle reflects the electrical distance between the two nodes and limits how much power can be stably transmitted. Further details on this topic are provided in Sect. 6.7.6.

6.3 Electrical Losses

With any type of power transmission, there are losses in terms of the transported active power, which can add up to considerable values over the course of a year throughout the network. Through suitable conductor material and high-quality insulation in cables, the losses are below five percent in the high and extra-high voltage level.

Basically, a distinction must be made between current-dependent and voltage-dependent losses. Somewhat simplified, it can be said that the former only occur when a consumer is connected, while the latter are present even when a line is in open circuit mode. Accordingly, the two kinds of losses must be considered separately for calculation of the energy losses which occur during the observed period (usually one year).

6.3.1 *Current-Dependent Losses in Overhead Lines*

The dominant source of current-dependent losses in overhead lines is *ohmic resistance* in the conductors. The *active power loss* P_l can be calculated using the standard formula: $P_l = 3 \cdot R \cdot I^2$, where R is the resistance of one phase conductor, and I is the phase current (RMS value).

Most overhead line conductors are made of aluminum wires with a steel core (e.g., ACSR -Aluminum Conductor Steel Reinforced). The steel core, serving structural purposes, is not considered in the calculation of electrical resistance. Over time, the aluminum strands become surface-oxidized, resulting in electrical insulation between them, which mitigates the skin effect (Chap. 5). Therefore, DC resistance can be used to approximate the conductor resistance, with a correction for operating temperatures, which may reach up to 80 °C in standard conductors.

Because power loss depends quadratically on current, the energy losses over time must be determined by integrating or summing up the loss values at all relevant current levels throughout the observation period. Since exact load profiles are often unavailable, a useful approach is to use the *load factor m* (as introduced in Chap. 5), from which an *energy loss factor d* can be derived.

This energy loss factor allows estimation of annual energy losses based on nominal current and is approximated by the following empirical equation [1]:

$$d = 0.083 \cdot m + 1.036 \cdot m^2 - 0.12 \cdot m^3 \tag{6.1}$$

For a continuous load, where the load factor $m = 1$, the energy loss factor d also equals 1. In this case, the energy losses can be directly calculated from the *maximum power loss* $P_{l,max}$ and the *total operating time T* using the following equation:

$$W_l = P_{lmax} \cdot T \tag{6.2}$$

wherein is:

m	Load factor (–)
d	Energy loss factor (–)
W_l	Total energy loss over time (Wh or MWh)
P_{lmax}	Max. power loss (W or MW)
T	Observation period (h or a).

The above simplified relation applies only to cases of constant load and full utilization of the transmission capacity throughout the entire time period.

Calculation Example 1

For a 110 kV overhead line, operated at thermal power limit, the power losses as well as the expected annual energy losses shall be determined.

Data: R $= 4.3\,\Omega$ at constant operating temperature, thermal power limit $S_{th} =$ 130 MVA, operation time 8760 h and load factor m = 1.0.

Solution

The thermal limit current is $I_{th} = 682$ A. It follows a maximum power loss $P_{lmax} = 3 \cdot R \cdot I_{th}^2 = 6.0$ MW. Thus, with Eq. 6.1 we get for the work loss factor d a value of 1.0 and, according to Eq. 6.2 the expected energy loss follows with about $W_l = 53$ GWh.

6.3.2 *Voltage-Dependent Losses of Overhead Lines*

Voltage-dependent losses primarily consist of leakage currents along insulators and corona losses. These losses occur independently of the load on the line and are instead determined by the magnitude of the operating voltage. As such, they are also present when the line is under no-load conditions.

The leakage current is influenced by the type and condition of the insulators, for example, porcelain long-rod insulators, glass cap-and-pin types, or composite insulators. In practice, these currents are usually very small, especially due to the natural cleaning effect of rainwater. Consequently, leakage current losses can generally be considered negligible under normal operating conditions.

Corona losses, however, can become significant, particularly at very high voltages. They depend on the electric field strength at the conductor surface, which in turn is influenced by conductor diameter, surface condition, ambient humidity, and altitude (Chap. 4). These losses are highly variable and weather-dependent but typically lie in the range of a few kilowatts per kilometer at the 380 kV level and above [5, 6].

In technical terms, *leakage to earth* is often expressed using the conductance *G*, the reciprocal of resistance. Typically, length-related values *G'* amount to 40 nS/km for 110 kV overhead lines and 20 nS/km for 380 kV lines. The abbreviation S stands for the unit Siemens, commonly used in German-speaking regions, where $1\,S = 1\,\Omega^{-1}$. In Anglo-American literature, the unit "mho" (an inverted ohm) is sometimes still used.

Since the operating voltage of a line generally remains close to the rated value, the voltage-dependent energy losses over a year can be estimated using the annual operating hours, eventually up to 8,760 h (non-stop operation).

The charging current associated with the operating capacitance of overhead lines is comparatively small. As a result, the associated losses due to the line's longitudinal (ohmic) resistance are negligible and generally not considered in energy loss calculations for overhead transmission lines.

6.3.3 *Current-Dependent Losses of Cables and Gas-Insulated Lines*

In cables, the primary sources of current-dependent losses are the ohmic resistances of the conductors and the metallic shield. Both can be designed to be significantly lower than those of overhead lines.

For example, for a 110 kV cable with a thermal power limit of 140 MVA and a resistance of 1.5 Ω per conductor at operation temperature the ohmic losses are to be calculated. The result is 2.54 MW which amounts to approximately 1.8 % of the transmitted apparent power. The energy losses over a defined observation period can be calculated using Eqs. 6.1 and 6.2.

At alternating voltage, the skin effect causes a slight increase in conductor resistance. To limit this, Milliken conductors are typically used in cables with cross-sections exceeding 1000 mm^2 (Chap. 5). Conversely, for cross-sections up to 185 mm^2, the skin effect is negligible, making the range between 185 mm^2 and 1000 mm^2 the most relevant in practice.

Additional losses can occur in the metallic shield due to inductive coupling. These are minimized through design measures such as single-point shield grounding or cross-bonding with dual grounding at designated locations (Chap. 5).

Gas-insulated lines exhibit similar effects. However, due to their very large conductor and enclosure cross-sections and correspondingly low resistance values, such losses are significantly smaller than those found in cable systems.

6.3.4 *Voltage-Dependent Losses of Cables and Gas-Insulated Lines*

The high capacitance of cables causes a charging current to flow continuously, regardless of the actual load. This *charging current* can reach substantial values and, over a certain cable length, can fully utilize the current-carrying capacity of the system.

In regular operation—when transmitting a mixed resistive-inductive load—the charging current has a slightly compensating effect, reducing the amount of reactive power that must be delivered by the source. However, this capacitive current imposes limitations on the transmission performance of the cable and must be compensated for beyond certain lengths to ensure economic operation.

For this purpose, *shunt reactors* (compensation coils) are employed. These are connected between the three phases and ground and are typically installed at the beginning and/or end of the cable route. In high-voltage networks, multiple shunt reactors may be required at intervals of 20–30 km.

Shunt reactors require dedicated structures and generate ohmic losses in their windings, which must be accounted for in the overall loss assessment of a cable system. At the 110 kV level, for instance, ohmic losses of approximately 0.15% of the rated reactive power per three-phase reactor are typical. For a 50 MVar reactor, this corresponds to about 750 kW of losses.

In gas-insulated lines, the specific capacitance per unit length is significantly lower than in cables due to differences in geometry and insulating medium (as per manufacturer data). Consequently, compensation equipment is generally only required for GILs if the route length exceeds 100 km.

In addition, *polarization effects* occur in the cable dielectric, contributing to heating. The small but finite conductivity of standard insulating materials also results in dielectric losses. These are quantified using the *dissipation factor tan δ* (Chap. 5). The relationship between *tan δ* and the length-related shunt conductance G' is described by the following equation:

$$G' = \omega C'_B \cdot \tan \delta \tag{6.3}$$

wherein is:

G' Length-related shunt conductance (S/m)
ω Angular frequency (= 2π f) (s^{-1})
C'_B Length-related operating capacity (F/m)
$\tan \delta$ Dissipation factor (–).

Cable manufacturers typically specify a worst-case dissipation factor (tan δ) of 0.001 for standard polymeric insulation. In contrast, for gas-insulated lines, the loss factor is negligibly small.

Figure 6.2 summarizes the maximum transmission capacities and their corresponding current-dependent losses per kilometer for the various line types listed in Table 6.1. The numbering of the entries in the diagram corresponds directly to the table. For consistency, a uniform conductor temperature of 80 °C was assumed for AC systems and 70 °C for the DC cable.

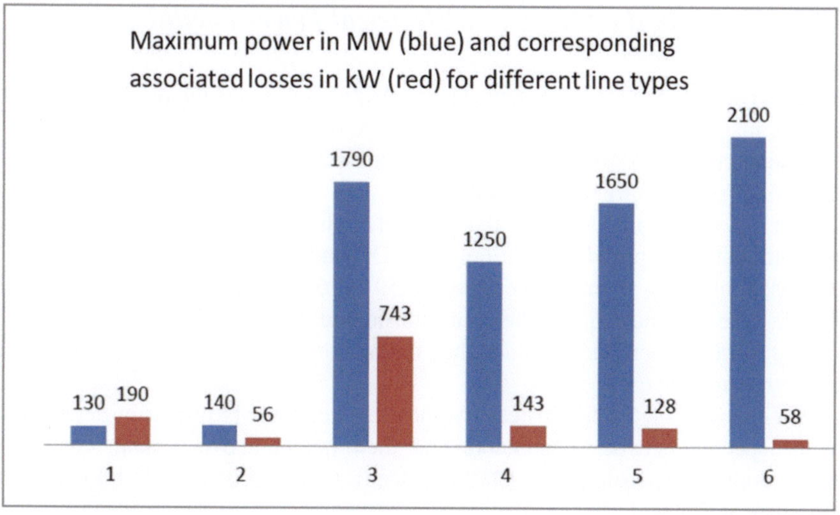

Fig. 6.2 Maximum transmissible power and associated current-dependent losses per kilometer for the example configurations presented in Table 6.1 (1 = Overhead line 110 kV, 2 = Cable 110 kV, 3 = Overhead line 380 kV, 4 = Cable 380 kV, 5 = GIL 380 kV and 6 = DC Cable 525 kV)

6.4 Electric and Magnetic Fields

Electric and magnetic fields are naturally present everywhere on Earth and in the atmosphere, independent of any man-made sources.

The *electric field strength E* is measured in volts per meter (V/m), while the *magnetic field strength H* is expressed in amperes per meter (A/m). However, because magnetic field strength is less directly measurable, the related quantity magnetic *flux density B*, measured in Tesla (T), is commonly used instead. In vacuum, air, and other non-magnetizable materials (such as the human body), the two quantities are related by the equation:

$$B = \mu_0 \cdot H$$

wherein is:

B Magnetic flux density (Vs/m^2 or T (Tesla))
H Magnetic field strength (A/m)
μ_0 Magnetic field constant $\left(0.4\pi \cdot 10^{-4}\,\mathrm{Vs/Am}\right)$ (permeability of free space).

For the natural electric field on earth, the values in fair weather are roughly between 0.1 and 0.5 kV/m. The magnetic flux density on earth ranges from 20 to 60 μT. In both cases, these are stationary fields, i.e., fields that maintain their intensity and orientation constant over time.

6.4.1 Fields in the Vicinity of Electrical Power Facilities

Voltages generated in power systems create electric fields, and the resulting currents produce magnetic fields. Depending on the type of supply—direct current (DC) or alternating current (AC)—these fields are either static (unchanging in direction) or alternating (periodically reversing direction).

Because of daily interaction with electrical appliances and infrastructure, people are continuously exposed to such fields. In DC and low-frequency AC systems—such as public power grids operating at 50 or 60 Hz, and railway systems at 16.7 Hz—electric and magnetic fields are considered separately. However, at frequencies of several kilohertz (kHz) and higher, the two become interlinked and are treated as *electromagnetic fields*.

The intensity of the fields near overhead power lines is primarily determined by the spacing between conductors and their height above ground. Generally closer conductor spacing and greater distance from the ground result in lower field intensities at ground level.

In underground cables, metallic shields are used to effectively confine electric fields within the cable structure. This shielding limits exposure and will be discussed in more detail later in this chapter.

To visualize electric and magnetic fields, the concept of *field lines* is used:

- Electric field lines originate from the energized conductor and terminate at ground potential.
- Magnetic field lines, in contrast, form closed loops encircling the conductor through which current flows.

Both field types are vector quantities, meaning they have both magnitude and direction, which may vary from point to point in space.

The illustration below schematically shows the characteristic patterns of electric and magnetic field lines in the vicinity of a long, current-carrying conductor:

In the case of the *electric field*, the field lines are formed by the applied voltage and are also present when no current is flowing (Fig. 6.3, shown in blue). In the case of an elongated single conductor, a radial field is formed. The electric field strength depends only on the voltage and the distance to the conductor. The following—simplified—distance law applies:

$$E \sim 1/r \tag{6.4}$$

wherein is:

E Electric field strength (V/m)
r Distance to the conductor (m).

In the case of a three-phase line, the tower configuration and -height as well as the number of circuits carried on the tower affect the electric field strength near ground. In the area of medium and high voltage, it is usually significantly below the limit values

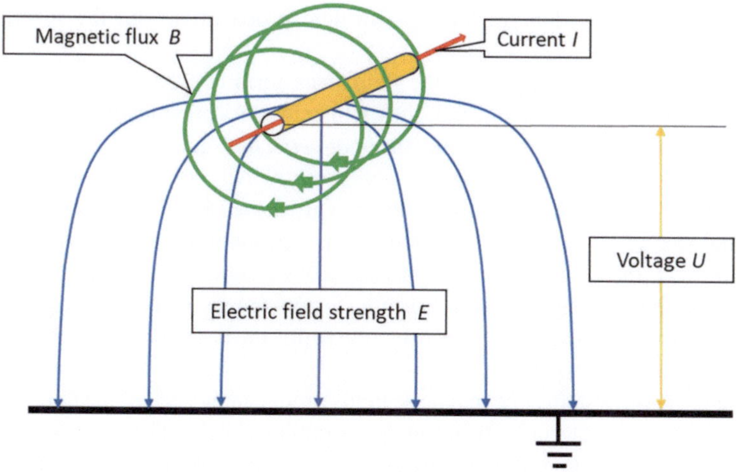

Fig. 6.3 Electric and magnetic fields in the vicinity of a current-carrying conductor

given in Sect. 6.4.2. For high-voltage lines this can be achieved by an optimized phase arrangement or by the construction as a *compact overhead line* (Chap. 4 and [7, 8]). However, compacting a line can lead to higher acoustic emissions and possible radio interferences, as this will increase the electric field at the conductor surface. This issue though can be addressed by using *bundle conductors* (Chap. 4).

Electric fields can be effectively shielded with simple measures. For example, the field lines end at the metal housings of electric devices, at the shields of cables, and at the casing tube of gas-insulated lines (Chap. 5). This requires grounding of these parts, which has anyway to be done for the sake of electrical safety. The environment of such arrangements is therefore field-free.

Electric fields generated by overhead lines are typically attenuated or redirected by surrounding structures such as vegetation and building exteriors. As a result, electric fields generally do not penetrate into interior spaces. Secondary effects may occur due to capacitive coupling (*electrical influence*), whereby voltages are induced in ungrounded metallic installations. Consequently, metallic structures such as handrails, fences, and railings located beneath high-voltage overhead lines or near substations are systematically grounded to prevent hazardous potentials.

In the case of direct current (DC) overhead lines, ionization of the surrounding air may lead to the formation of space charge clouds around the conductors. These ionized regions reinforce the local electric field and can result in significantly elevated field strengths near ground level, exceeding values predicted by electrostatic calculations. Under windy conditions, these charge clouds may drift laterally, causing field attenuation downwind from the line and spatial redistribution of the field intensity.

In contrast, underground cables and gas-insulated transmission lines do not emit external electric fields. In such systems, the electric field lines are fully contained within grounded enclosures—either the metallic cable sheath or the outer tube of a gas-insulated line—thus preventing any field propagation into the surrounding environment.

In the case of *magnetic fields*, the field lines form concentric circles around a current-carrying conductor (Fig. 6.3, shown in green). As previously noted, magnetic fields are typically described using the *magnetic flux density B*, which is more practical to measure than the *magnetic field strength H*.

Analogous to Eq. 6.4, the magnetic flux density decreases with increasing distance from the conductor, assuming a homogeneous surrounding medium such as air or soil. However, the distribution of field lines can be significantly altered by nearby materials with high magnetic permeability (ferromagnetic metals) which can guide and concentrate magnetic flux. This principle forms the basis for magnetic shielding, albeit with great effort [9].

In three-phase systems, the magnetic fields produced by each of the three conductors superimpose vectorially, resulting in a combined field. The magnitude of this combined effect is commonly expressed as the equivalent magnetic field strength. Due to the 120° phase displacement of the three currents, the resulting magnetic field does not remain static but forms a rotating elliptical field over time.

Figure 6.4 schematically illustrates the distribution of the equivalent magnetic field strength at a given vertical height h above an underground cable system, where the three conductors are arranged in a flat formation with spacing d between them.

If the cable is operated with equal current intensities I in the three phases, the resulting effective value B according to [1] becomes:

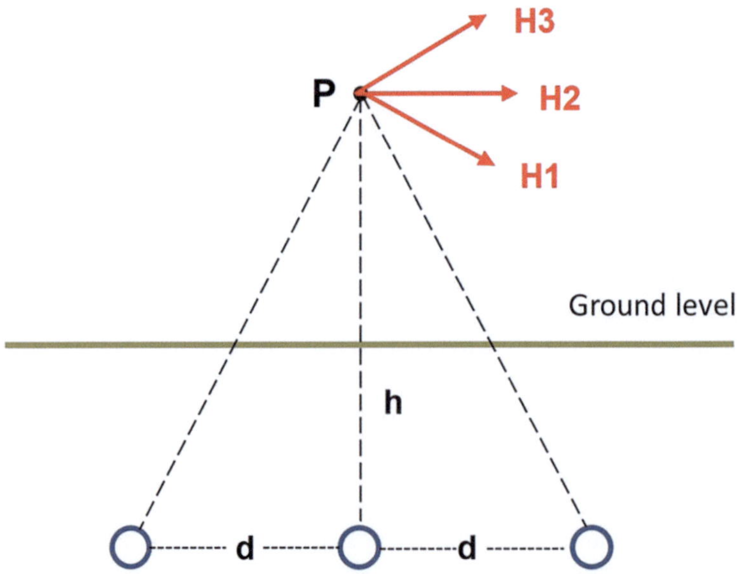

Fig. 6.4 Magnetic field vertically above a current-carrying three-phase underground cable with flat laying of the individual conductors

$$B = \frac{\mu_0 \cdot I \cdot d}{2\pi \cdot h(h^2 + d^2)} \sqrt{3h^2 + d^2} \tag{6.5}$$

wherein is:

B Magnetic flux density (T)
I Current (A)
h Height above the cable (m)
d Distance between the conductors (m).

Equation 6.5 applies also to the magnetic field perpendicular to an overhead line with horizontal conductor arrangement.

Calculation Example 2

An underground cable laid at a depth of 1.50 m in flat arrangement of the three conductors at 25 cm from each other is operated with a current of 420 A. What is the effective value of the resulting magnetic flux density perpendicular and centrally above the cable directly on the ground surface and at a height of 1 m?

Solution

Equation 6.5 gives the following flux density for the case of the cable at the ground level: $B = 15.8\,\mu\text{T}$. At a height of 1 m above the ground, a value of $B = 5.8\,\mu\text{T}$ is obtained.

Calculation Example 3

An overhead line in horizontal arrangement (distance between conductors 1.50 m, maximum sag in the middle of the span reaches 10 m above the ground) is considered. What is the effective value of the resulting flux density perpendicular and centrally under the overhead line directly at ground level and at a height of 1 m with a current of 420 A?

Solution

Applying Eq. 6.5 gives value $B = 2.1\,\mu\text{T}$ in the middle of the span at ground level and of and $B = 2.6\,\mu\text{T}$ at a height of 1 m above ground.

It is worth noting that, with regard to magnetic field effects, a triangular (compact) arrangement of the conductors is generally more favorable for both cables and overhead lines. However, specific minimum spacing requirements must be observed—due to the insulation properties of air in the case of overhead lines, and due to thermal dissipation constraints in the case of underground cables.

6.4.2 *Effects on Human Bodies*

In the 1970s, a topic came up regarding allegedly negative effects on human health caused electric and magnetic fields, especially in the vicinity of high and extra-high voltage overhead lines. In line with the mega-trend of the growing environmental awareness of the public, this and the aesthetics of overhead lines, has become an important concern in the industry [10]. The health issue in this area has been closely monitored for many years by a permanent special working group of CIGRE, Working Group C3.01. The group has published two basic reports on the latest status in the past decades [11, 12].

It all started with a study conducted in Denver (Colorado, USA) in 1979 [13]. The authors were looking for possible environmental factors within living houses where children who had leukemia were staying. The low-frequency magnetic field in these houses was estimated at about 0.2 T, and the authors suspected that this could be the reason for this association. Despite skepticism among experts, the study was taken very seriously by authorities, science, and industry and led to major research efforts on an international scale. However, all subsequent studies have shown that the development of childhood leukemia cannot be attributed to the magnetic fields generated by high-voltage lines [14]. Since electric fields are shielded or weakened by building envelopes and grounded objects, they play a subordinate role in the health discussion.

To exclude risks as safely as possible, legislators in nearly all countries in the world have set limits for the effective values of low-frequency electric and magnetic fields to protect against possible environmental impacts. In Germany for example, these are regulated in the current German Federal Immission Control Ordinance (Bundes-Immissionsschutz-Verordnung BImSchV) for the range of 50 Hz as follows:

"Lines must be constructed and operated in such a way that in their impact area in buildings or on properties that are intended—not only for temporary stay—for humanoccupancy, at maximum operational load, the *limit values of 5 kV/m for the electric field and 100* μT *for the magnetic flux density* are not exceeded" [10].

Both types of fields can either act directly on the human body or induce currents in it and cause stimulating effects on sensory-, nerve- and muscle-cells. The current densities are relevant, which are given in mA/m^2 (the human body has a cross-section of about 0.06 to 0.07 m^2).

In the range of interest here up to 50–60 Hz, current densities below 10 mA/m^2 show no observable scientifically confirmed biological effects. Acute dangers occur (especially regarding the heart function) only at local current densities in the body above 100 mA/m^2 on.

In addition, a magnetic flux density of a maximum of 500 μT was set for *direct current*. The listed limit values apply to "permanent occupied areas" in general. There is currently no legal regulation for areas for temporary stay. International bodies such as the ICNIRP (International Commission on Non-Ionizing Radiation Protection) recommend for areas in or at facilities for power supply with low frequency alternating current, in which specialist staff of network operators temporarily stay, for electric fields a maximum value of 10 kV/m and for the magnetic flux density such a value of 500 μT. For particularly short stays, even higher values are permissible [15, 16].

Due to its low electrical resistance, the human body behaves similarly to a metallic conductor when exposed to electric fields, thereby distorting the surrounding

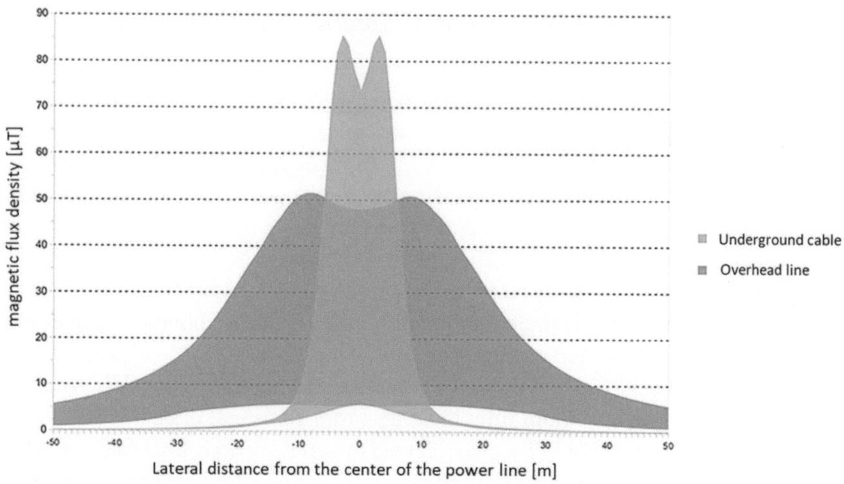

Fig. 6.5 Profile of the resulting magnetic flux density from a 380 kV underground cable and a 380 kV overhead line (each as a double-circuit system) at maximum current load, measured at 1 m above ground level [10]

field. Field lines, which would otherwise extend perpendicularly to the ground, are deflected to run nearly perpendicular to the surface of the body, resulting in internal *displacement currents*. The resulting current densities are proportional to the electric field strength. For low-frequency alternating electric fields, typical of power transmission systems, measurements show that these current densities are approximately 0.015 mA/m^2 per kV/m of the acting electric field [5, 6, 17, 18].

Similarly, low-frequency alternating magnetic fields induce *eddy currents* in the human body, which in turn generate internal electric fields and corresponding current densities. To produce a current density of 10 mA/m^2 within a homogeneous magnetic field—such as in the human head—external flux densities between 2.5 and 6 mT are required. For the chest and heart region, the necessary external field strength ranges from 1.4 to 6.7 mT.

In contrast, DC transmission lines do not induce such effects, as neither capacitive coupling nor induction occurs. Moreover, the legal exposure limit for magnetic flux density around DC lines is significantly higher—typically around 500 μT—than for AC systems.

Measurements and field calculations across a wide range of configurations and operating conditions consistently show that even under unfavorable scenarios, exposure levels remain well below established legal and safety thresholds.

To date, no conclusive scientific evidence has demonstrated a causal relationship between exposure to electric or magnetic fields from power transmission lines and specific health disorders.

Figure 6.5 illustrates, by way of example, how the magnetic flux density decreases asymptotically with distance from a 380 kV underground cable and a 380 kV overhead line, each designed as a double-circuit system.

6.5 Reliability and Lifespan of Transmission Networks

High- and extra-high voltage transmission networks are typically designed and operated in accordance with the (n − 1) principle, and increasingly also with the more stringent (n − 2) criterion (Chap. 1). For more detailed planning—particularly when comparing potential system expansion alternatives—specific methods for reliability assessment are applied. These approaches are based on the stochastic behavior of the system components, where repair or replacement is possible after failure.

In technical terms, *reliability* refers to the capability of a system component (or the entire system) to perform its intended function under specified boundary conditions over a defined time interval.

Consider a transmission line that is commissioned at time T = 0 and then experiences a fault after some random time interval. From statistical data, one can derive the *Mean Time to Failure* (MTTF), also referred to as the *average fault-free operating time*. Similarly, the average time required to repair the line after a failure is referred to as the *Mean Time to Repair* (MTTR). The sum of these two values defines the *Mean Time Between Failures* (MTBF). High system availability requires that the *downtime* T_A be significantly shorter than the *fault-free operating time* T_B.

The *failure rate* λ is defined as the probability that a component (e.g., a line) that is operational at time T will experience a failure in the interval $T + \Delta T$.

It applies:

$$\lambda = \frac{1}{T_B} \tag{6.6}$$

Accordingly, the *repair rate* μ is defined as the reciprocal of the average downtime, so

$$\mu = \frac{1}{T_A} \tag{6.7}$$

wherein is:

T_B Operating duration (a or h)
T_A Downtime (a or h)
λ Failure rate (a^{-1} or h^{-1})
μ Repair rate (a^{-1} or h^{-1}).

For many technical components, the failure rate over lifetime is not constant, but can be divided into three phases:

- *Early failures* are predominantly due to manufacturing defects, material or operating faults. The failure rate is reduced by "running in".
- *Random failures* or *normal failures* are characterized by an approximately constant failure rate.

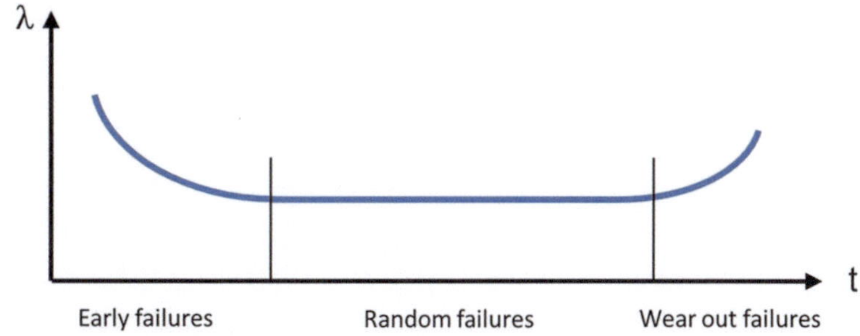

Fig. 6.6 Typical profile of the failure rate as a function of time

- *Late failures* or wear out failures occur due to wear or fatigue, where the failure rate increases after a certain operating duration. They can be countered by preventive maintenance measures to extend the life span.

The typical profile of a component's failure rate can thus be represented according to Fig. 6.6 and is often referred to as the "*bathtub curve*". The first phase is characterized by early failures. This is followed by a period of purely random failures, which finally changes over into the phase of age-related damage.

In the *Markov process*, which relies on certain simplifying boundary conditions, all possible states of a system and the associated transition probabilities of a single component are considered. This model restricts itself to two discrete states: "operation" and "failure". Intermediate states, such as reduced performance or standby operation, are not accounted for. Additionally, state transitions must be governed by linear differential equations.

To illustrate the Markov approach, consider a simple system: the supply of an electrical load via a single transmission line, assuming a fully reliable source. The corresponding *state diagram* defines the probabilities associated with the two system states, along with the relevant *transition rates* between them.

The *reliability* of a component—or an entire system—is defined as the probability that it is operational (or ready for operation) at a given point in time (Fig. 6.7).

The Markov method provides the probabilities P_1 and P_2 for the two states "On" and "Off" according to the following equations:

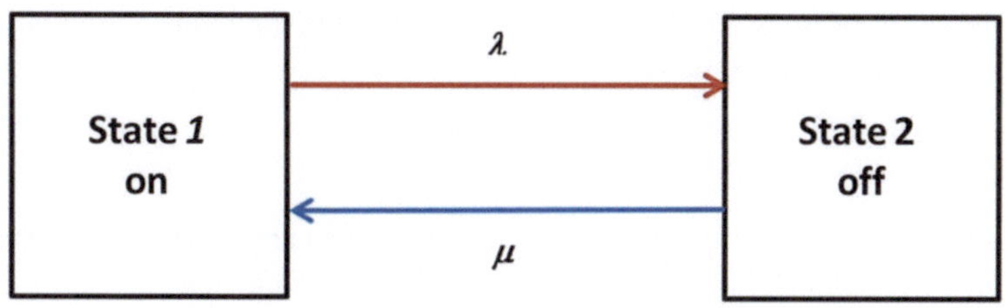

Fig. 6.7 State diagram for a single component

On:

$$P_1 = \frac{\mu}{\mu + \lambda} \tag{6.8}$$

Off:

$$P_2 = \frac{\lambda}{\mu + \lambda} \tag{6.9}$$

Calculation Example 4

For a component, it is known that it operates on average for 2.5 years without failure. After a failure, 17.52 h are needed for the repair. What is the probability that the component does not work?

Solution

From the information given, a failure rate $\lambda = 0.4\ \text{a}^{-1}$ is derived. This corresponds to 0.2% of the 8760 h of a year. Furthermore, a repair rate $\mu = 500\ \text{a}^{-1}$ is obtained. The probability of finding the component in the off-state (P_2) is calculated using Eq. 6.9 to be 0.0008, which corresponds to an off-time of about 7 h per year.

When analyzing the parallel connection of two identical components (e.g. lines), each capable of handling the full load alone, the probability for state 4 in Fig. 6.8, in which both components K1 and K2 have failed simultaneously, is of interest.

The Markov method provides the probability for state 4 (both "Off") with the following equation:

$$P_4 = \left(\frac{\lambda}{\mu + \lambda}\right)^2 \tag{6.10}$$

For the same transition rates as in the example above, Eq. 6.10 yields a probability $P_4 = 6.4 \cdot 10^{-7}$ corresponding to an off time of about 20 s per year.

Not considered in this analysis is the so-called *common-mode failure*, which refers to the simultaneous failure of both components due to a shared cause—for example, the collapse of a tower in a double-circuit overhead line. To account for such scenarios, the model must be appropriately extended [19].

Various literature references, e.g. [6, 19–21] provide further information on the failure behavior of lines. From this, the following rough guidelines, each based on 100 km line length, can be derived:

- Overhead lines: $\lambda \approx 0.35\ \text{a}^{-1}$; $\mu \approx 2900\ \text{a}^{-1}$,
- Cables: $\lambda \approx 0.7\ \text{a}^{-1}$; $\mu \approx 125\ \text{a}^{-1}$.

Detailed, voltage- and current-specific reliability data are provided, for example, in Germany by the VDN through its *Fault and Damage Statistics*.

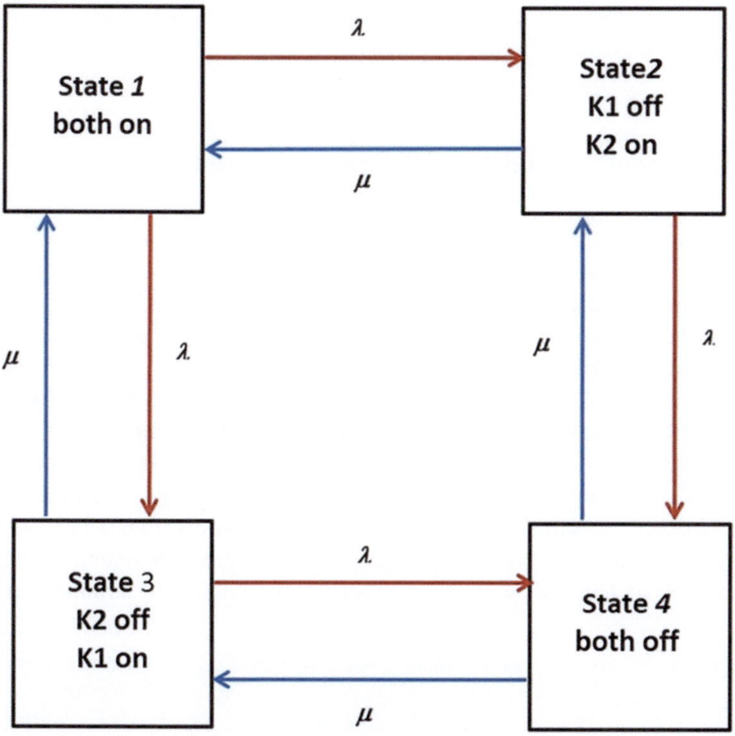

Fig. 6.8 State diagram for two components K1 and K2

The noticeably lower repair rates for cables compared to overhead lines can largely be attributed to the additional time required for fault location, excavation work, joint installation, and the subsequent high-voltage testing.

Statistical evaluations can also be used to estimate the expected service life of overhead lines, underground cables, and gas-insulated lines (GIL). Overhead lines, in particular, have a typical service life exceeding 80 years. This longevity is supported by the possibility of replacing conductors, fittings, or insulators as needed. In addition, regular maintenance of towers -including protective treatments based on environmental conditions—prevents premature corrosion.

For traditional oil- and gas-pressure cables, historical data indicate a typical service life of approximately 40 to 50 years. A key factor in this assessment is whether and to what extent the thermal limit capacity was exceeded during operation.

A comparable service life is generally assumed for XLPE-insulated cables, although long-term operational experience is still limited for this type of insulation.

The first gas-insulated lines were installed in the 1970s. These early systems consisted of flanged and bolted pipe sections, which over time have proven to be susceptible to gas leakage issues. Second-generation GIL, introduced in the 1990s, feature welded pipe connections and modified insulating gas mixtures. While comprehensive statistical data are not yet available, an expected lifespan of 40 to 50 years is also assumed for these systems.

6.6 Intermediate and Partial Conversion to Cable

The term *intermediate cabling* refers to the replacement of a section of an overhead line with an underground cable, typically for reasons of landscape preservation or to reduce visual and environmental impact in residential areas. In contrast, *partial cabling* usually describes the undergrounding of only the final one or two kilometers of a line leading into a substation.

Introducing a cable section into an overhead line significantly increases the overall cost of the transmission system. This is largely due to the need for complex transition structures at both ends of the cable section. At medium-voltage levels or occasionally at the 110 kV level, this may involve installing cable terminations and surge arresters on an additional crossarm of the existing termination towers (Fig. 6.9).

At higher voltage levels, however, a dedicated transition station is generally required. This includes its own support structures (portals), cable terminations, surge

Fig. 6.9 Cable transition compound of a single circuit 380 kV overhead line to two underground cable circuits

arresters, and potentially reactive power compensation equipment along with their associated switching devices (Chap. 2).

Key aspects influencing the layout and design of transition compounds include:

- Site selection: Environmental, societal, and safety considerations (both internal and external).
- Electromagnetic emissions: Assessment of electric and magnetic fields, corona discharge, and audible noise levels.
- Insulation coordination: Alignment of insulation levels between overhead lines and underground cables.
- Earthing and protection: Grounding arrangements and protective measures.
- Reactive power compensation: Requirements for compensation equipment (e.g., shunt reactors).
- Structural design: Mechanical layout and electrical configuration of the compound.
- Logistics and access: Availability of transport routes and space for the delivery of large components (e.g., cable drums).
- Operational requirements: Provisions for fault detection, operational security, and sabotage prevention.
- Maintenance strategy: Access for inspection and repair, availability of spare parts, and overall maintainability.

These factors influence not only the physical design of the compound but also its cost, operational reliability, and long-term maintainability.

The vast majority of transition compounds worldwide adhere to the classical substation-like design. However, alternative concepts for the layout and integration of such facilities have been proposed, as presented in [22] Fig. 6.10.

Another example for an alternative design is given in Application Example 2.

A special phenomenon can be observed at the transition between overhead lines and cables: Since the *surge impedances* of the two media differ considerably from each other (Sect. 6.7), traveling waves caused by lightning strikes into the overhead line or by switching operations in the network are refracted and reflected at the transition compounds. This can result in dangerous high voltage peaks. To protect the cables, surge arresters are used. However, partial cabling can also have positive

Fig. 6.10 Examples for alternative design of transition compounds [22]

effects on the operating behavior of the network, by protecting transformers in the substations from incoming traveling waves through short cable sections at the end of an overhead line, as will be shown below.

A sudden change in voltage with a steep slope (e.g. due to lightning) propagates on an overhead line from the point of impact in the form of two traveling waves running in opposite directions at approximately the speed of light. When hitting a cable, a part of the wave is refracted, and another part is reflected. The laws that apply to this change in surge impedance are similar to those in the field of optics.

With u = value of the overvoltage (peak value), Z_1 = surge impedance of the overhead line (1), Z_2 = surge impedance of the cable (2), Z_3 = surge impedance of the subsequent overhead line or the transformer (3), the subscripts f for forward and b for backward and the abbreviation b for refraction factor and r for reflection factor, the following applies:

$$u_{2f} = \frac{2 \cdot Z_2}{Z_1 + Z_2} \cdot u_{1f} = b \cdot u_{1f} \tag{6.11}$$

$$u_{1b} = \frac{Z_2 - Z_1}{Z_1 + Z_2} \cdot u_{1f} = r \cdot u_{1f} \tag{6.12}$$

wherein is:

u_f Forward voltage pulse (V or kV)
u_b Backward voltage pulse (Vor kV)
b Refraction factor (–)
r Reflection factor (–)
Z_1, Z_2 Surge Impedances (Ω).

Figure 6.11 shows the ratios for intermediate cabling and partial cabling:

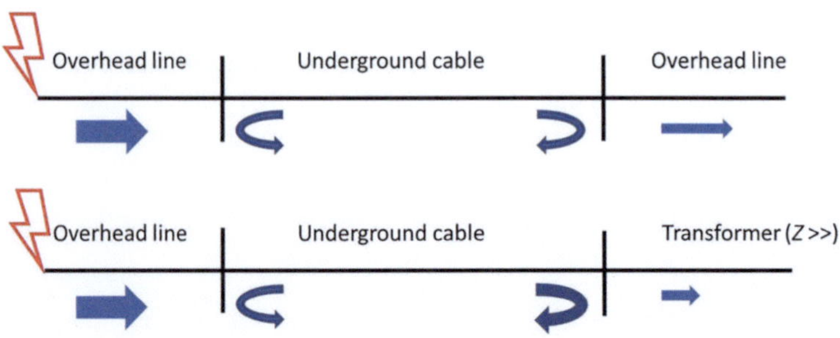

Fig. 6.11 Traveling waves schematically (top: intermediate cabling, bottom: partial cabling)

Calculation Example 5

Consider the case of a lightning strike with a voltage peak of 1000 kV into a partially cabled overhead line. For the overhead line $Z_1 = 250\ \Omega$ and for the cable $Z_2 = 50\ \Omega$. The partial lengths of the lines do not play a role in the following considerations.

Solution

From Eq. 6.11, the refraction coefficient is calculated as $b = 0.33$, and from Eq. 6.12, the reflection coefficient is $r = -0.66$. This means that the voltage travelling wave caused by the lightning surge enters the cable with one-third of its original magnitude, i.e., $u_{2v} = 333\,\text{kV}$. Two-thirds of the wave is reflected back into the overhead line—the negative sign indicating a reversal in direction.

In this case of partial cabling, another round of refraction and reflection takes place at the cable's far end, producing a reflected voltage wave of $u_{2r} = 222\,\text{kV}$. This process repeats within the cable, as forward and backward travelling waves interact, potentially building up to dangerously high voltage levels that can lead to dielectric breakdown.

Cables are particularly vulnerable to lightning strikes impacting the overhead lines ahead of them. To minimize this risk—especially in the vicinity of the cable transition—it is essential to take effective protective measures. The most effective method is the installation of two earth wires on the first one to two spans both before and after the cable section, in addition to the surge arresters that are required at the transition structures.

In cases of partial undergrounding (see Fig. 6.11, near the substation), nearly total reflection occurs at the transformer interface due to its significantly higher surge impedance compared to the cable ($Z_3 \gg Z_2$). This reflection protects the transformer, but it does so at the expense of the cable insulation, which is subjected to elevated stress.

In the following, several examples of intermediate cabling installations are presented.

Application Example 1

The 400 kV AC-connection Middlesbrough—York between Wales and Scotland with a length of 75 km has a 5.7 km long intermediate cable section. The network operator was against a continuous cable version because of the significantly higher costs. The government shared this view and has approved the overhead line except for the 5.7 km long cable section. The overhead line consists of two circuits (6 individual conductors), while the cable line is designed with four circuits, i.e., it contains twelve individual conductors. The connection went into operation in 2003. For the cable section, automatic reclosure in the event of a fault was intentionally omitted.

Application Example 2

This ± 250 kV direct current line Hokkaido–Honshu in Japan (operator J-Power) with a rated power of 600 MW connects the island of Hokkaido in the north of Japan with the main island of Honshu. The 167 km long line consists of three sections—starting from the converter station Hakodate (AC to DC) on Hokkaido—with 27 km of overhead line and 80 towers, followed by 43 km of submarine cable (crossing the "Tsugaru Straits" with a maximum laying depth of 300 m) and the 97 km long overhead line section on Honshu with 294 towers to the converter station Kamikita (DC to AC). The line complements old connections established with oil pressure cables since 1979. The cabling of the overhead line is carried out with the two live conductors with 810 mm^2 and two return conductors with 240 mm^2 (all ACSR). The two XLPE submarine cables with Cu-conductors each have a cross-section of 600 mm^2; one XLPE return cable has a cross-section of 500 mm^2 Cu. The cable terminations are housed in protective enclosures to ensure operational safety (Fig. 6.12).

Application Example 3

The 28 km long, single-system 380 kV line Turbigo–Rho in Italy was commissioned in 2007 as an overhead line featuring an 8.4 km section of intermediate cabling. This line enhances the network meshing and delivers an additional 1000 MW of transmission capacity from the Turbigo thermal power plant into the industrial region of Milan. The cable section runs through the densely populated suburbs of Rho and Ospiate in the Milan metropolitan area.

Fig. 6.12 DC connection with intermediate cabling in Japan. The image shows the enclosed cable termination and the corresponding terminal tower

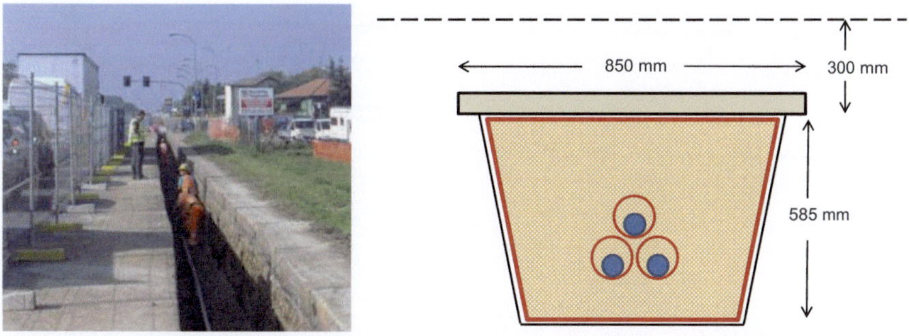

Fig. 6.13 Intermediate cabling of the 380 kV line Turbigo—Rho near Milan. *Left*: flat installation in the ground, *right*: installation to reduce the magnetic field strength in a triangle and in a ferromagnetic trough [23]

The overhead line section is equipped with triple-bundle conductors using 3 × 585 mm^2 ACSR. The extruded cable section comprises copper conductors with a cross-section of 2500 mm^2 and an aluminum corrugated sheath. Two cable systems were laid in parallel—predominantly in a flat configuration—directly in the ground.

In zones with sensitive land use, such as residential areas, Italy enforces a magnetic field limit of 3 μT. To meet this requirement, the cables in those areas were installed in a triangular configuration and enclosed within a ferromagnetic trough (Fig. 6.13).

Application Example 4

Intermediate AC undergrounding Riniken–Gäbihübel in Switzerland

In Riniken, a 380 kV overhead line project was submitted for approval. In 2011, the Federal Court mandated, based on expert reports, an investigation of an approximately 950-m-long section near Riniken to be implemented as an underground cable. The partial undergrounding project "Gäbihübel" was approved in 2016, and construction began in 2018. Four cable systems form the undergrounded section of the double-circuit overhead line.

Further key data of the cable section: the width of the route during construction was 25 m and during operation is 5 m (not including additional safety strips). Two transition compounds, each with an area of around 2000 m^2 and a height of 25 m, were constructed. The cabling was carried out in open trench construction, including the tunnelling of a railway line. The cables are installed in a concrete duct block, which provides good protection of the separately laid phases against external damage. In the event of a fault, the cable can only be pulled out and reinstalled along the entire length. Spare ducts for this purpose are available (Fig. 6.14).

The line went into operation in 2020. The operator, Swissgrid, has installed monitoring stations along the cable route and thus gains insights into cable operation, including the temperature of the soil near and farther from the cable area, questions of electromagnetic compatibility, and further relevant parameters.

Fig. 6.14 Intermediate cabling, cable basement (*Source*: Stadtwerke Tübingen)

6.7 Lines in Network Operation

The following considerations are limited to the steady-state operation of AC lines with equal loading on all three phases. This means that no fourth conductor is required in the representations below. However, a resistance-free return conductor would be needed in the case of single-phase equivalent circuits.

It has proven useful to describe voltage and current conditions including their magnitudes and phase angles using complex notation. *Complex quantities* are underlined in the following.

The electrical behavior of an overhead line, a cable, or a gas-insulated line (collectively referred to as a *line*) of length l can be described by four primary parameters: resistance $R = R' \cdot l$, inductance $L = L' \cdot l$, shunt conductance $G = G' \cdot l$, and operating capacitance $C_B = C_B' \cdot l$.

The dashed quantities represent specific values (per unit length) and are typically provided by manufacturers in units per kilometer. The term *operating capacitance* is explained in Chap. 5 and briefly summarized below.

From these primary parameters, three secondary characteristics can be derived, which facilitate the understanding of the subsequent *line equations*.

The *surge impedance* (also called *characteristic impedance*) $\underline{Z}_s$ describes the (constant) ratio of voltage to current for each point of the line. It is given by:

$$\underline{Z}_s = \sqrt{\frac{R' + j\omega L'}{G' + j\omega C'_B}} \tag{6.13}$$

The *propagation coefficient* $\underline{\gamma}$ is defined as:

$$\underline{\gamma} = \sqrt{(R' + j\omega L') \cdot (G' + j\omega C'_B)} \tag{6.14}$$

The third useful quantity is the *surge impedance loading* (SIL), often referred to as the natural load:

$$\underline{\text{SIL}} = \frac{U_r^2}{\underline{Z}_s *} \tag{6.15}$$

The asterisk denotes the conjugate complex value of $\underline{Z}_s$.
Wherein is:

$\underline{\text{SIL}}$ Surge impedance loading (complex) (VA or MVA)
$\underline{Z}_s$ Surge impedance (complex) ((Ω))
ω Angular frequency (= 2π f) ((s^{-1}))
R' Ohmic resistance per length unit (Ω/m)
G' Shunt conductance per unit length (Ω^{-1} /m or S/m)
L' Inductance per unit length (H/m)
C'_B Operating capacity per unit length (F/m)
U_r Rated voltage (V or kV).

The *surge impedance loading* (SIL) occurs when a line is operated with a load at its receiving end that equals its surge impedance. This operating condition is also referred to as *matched loading*.

Starting with a transmission line of infinitesimal length *dl*, characterized by the primary parameters mentioned above, the following representation is obtained (Fig. 6.15).

The following differential equations apply:

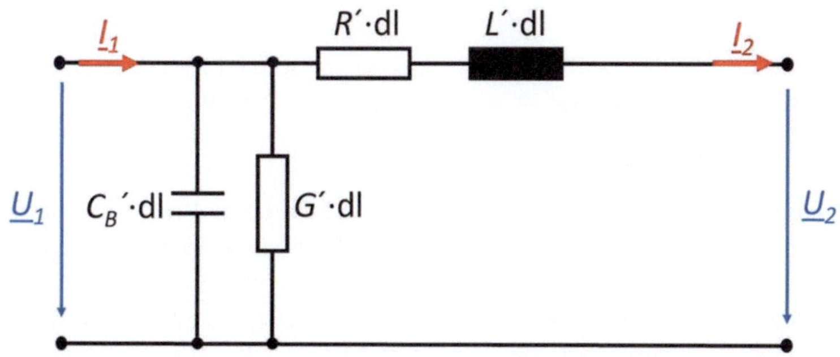

Fig. 6.15 Model of an infinitesimally short section of an AC line in single-phase representation

$$d\underline{U} = \underline{I}\cdot\left(R'dl + j\omega L'dl\right) \tag{6.16}$$

$$d\underline{I} = \underline{U}\cdot\left(G'dl + j\omega C_B'dl\right) \tag{6.17}$$

In network calculations, the voltage and current values at one end of a transmission line are typically of primary interest, while the corresponding values at the opposite end are generally known. If subsript 1 refers to the starting point and subscript 2 to the endpoint of a line of length l, the corresponding solutions to the above equations, after further differentiation for given values of voltage and current at the line's end, are as follows:

$$\underline{U}_1 = \cosh\left(\underline{\gamma}l\right)\cdot\underline{U}_2 + \underline{Z}_s\sinh\left(\underline{\gamma}l\right)\cdot\underline{I}_2 \tag{6.18}$$

$$\underline{I}_1 = \frac{1}{\underline{Z}_s}\sinh\left(\underline{\gamma}l\right)\cdot\underline{U}_2 + \cosh\left(\underline{\gamma}l\right)\cdot\underline{I}_2 \tag{6.19}$$

These two expressions represent the *line equations* and form the basis for calculations of load flow and short-circuit in power supply networks.

In matrix form, the line equations are:

$$\begin{pmatrix}\underline{U}_1\\ \underline{I}_1\end{pmatrix} = \begin{pmatrix}\underline{A} & \underline{B}\\ \underline{C} & \underline{D}\end{pmatrix}\cdot\begin{pmatrix}\underline{U}_2\\ \underline{I}_2\end{pmatrix} \tag{6.20}$$

For the elements of the above *transfer matrix* the following applies:

$$\underline{\mathrm{A}} = \underline{\mathrm{D}} = \cosh\left(\underline{\gamma}\mathrm{l}\right),\ \underline{\mathrm{B}} = \underline{\mathrm{Z}}_{\mathrm{s}}\cdot\sinh\left(\underline{\gamma}\mathrm{l}\right)\ \text{and}\ \underline{\mathrm{C}} = 1/\underline{\mathrm{Z}}_{\mathrm{s}}\cdot\sinh\left(\underline{\gamma}\mathrm{l}\right).$$

wherein is:

$\underline{U}$ Voltage phasor (V or kV).
$\underline{\mathrm{I}}$ Current phasor (A or kA).
Z_s Surge impedance (Ω)
l Line length (m or km).

Note: Hyperbolic sine and hyperbolic cosine are defined for complex arguments x as follows: $\sinh(x) = 1/2\left(\mathrm{e}^x - \mathrm{e}^{-x}\right)$ and $\cosh(x) = 1/2\left(\mathrm{e}^x + \mathrm{e}^{-x}\right)$.

Calculation Example 6

This example shows how the line equations are applied and illustrates the orders of magnitude of the propagation coefficient and the surge impedance. A 110 kV overhead line with a length of 40 km has the following specific values.

$R' = 126\,\mathrm{m\Omega/km}$; $G' = 40\,\mathrm{nS/km}$; $L' = 1.24\,\mathrm{mH/km}$ and $C_B' = 8.9\,\mathrm{nF/km}$. At its end, a mixed ohmic-inductive load of 85 MW and 36 Mvar at nominal voltage is applied. Voltage and current at the beginning of the line are to be determined.

Solution

The load indicated corresponds to a complex current $\underline{I}_2 = (0.446 - \text{j}\,0.189)$ kA.

In the case of a symmetric three-phase load, the question of whether it is connected in delta or star generally does not matter, as the current of interest here occurs in each of the three phases. Furthermore, in the *consumer reference arrow system* used here, the inductive component of the current has a negative sign. Details on this will be explained later using phasor diagrams.

To start with, the propagation coefficient and the surge impedance are determined according to the above equations to $\underline{\gamma} = (0.000174 + \text{j}\,0.001056)\ \text{km}^{-1}$ and $\underline{Z}_s = (377.994 - \text{j}\,59.610)\ \Omega$. Furthermore, $\cosh\left(\underline{\gamma} l\right) = (0.999133 + \text{j}\,0.000294)$ and $\sinh\left(\underline{\gamma} l\right) = (0.006963 + \text{j}\,0.042217)$.

Since the calculation refers to a single-phase equivalent, the load-side line-to-neutral voltage U_2 is $U_{2Y} = 110\,\text{kV}/\sqrt{3} = 63.508\,\text{kV}$. The imaginary part is set to zero, so the phasor $\underline{U}_{2Y}$ lies along the real axis and serves as the reference for the phase angles of all other phasors. Using Eq. 6.18, the voltage at the sending end of the line is calculated as $\underline{U}_{1Y} = 68.687 + \text{j}\,5.978\,\text{kV}$. This corresponds to a line-to-line voltage of $U_1 = 119.42$ kV, with a phase angle of approximately 5°. For the sending-end current, (6.19) yields $\underline{I}_1 = 0.446 - \text{j}\,0.182$kA. This corresponds to a magnitude of 0.48 kA and a phase angle of approximately –22°.

To illustrate single-phase voltages (conductor-to-earth voltages), the subscript "Y" was used in the above calculation. This notation also applies to the following calculation examples.

Showing the equations in matrix form offers the advantage that swapping the given and desired quantities becomes straightforward. All that is required is the *inverse of the matrix.*

In the case of the transmission line equations, the following relationship applies:

$$\begin{pmatrix} \underline{U}_2 \\ \underline{I}_2 \end{pmatrix} = \begin{pmatrix} \underline{EF} \\ \underline{GH} \end{pmatrix} \begin{pmatrix} \underline{U}_1 \\ \underline{I}_1 \end{pmatrix} \tag{6.21}$$

For the (2×2) matrix used here, the inversion of the original transmission matrix is particularly simple and yields the matrix elements $\underline{E} = \underline{H} = \cosh(\underline{\gamma} l)$, $F = -\underline{Z}_s \cdot \sinh(\underline{\gamma} l)$ and $\underline{G} = 1/\underline{Z}_s \cdot \sinh(\underline{\gamma} l)$.

Calculation Example 7

The case of the 110 kV overhead line calculated above is now to be analyzed in reverse, i.e., the values at the sending end of the line are now given, and the corresponding quantities at the receiving end are to be determined. The propagation coefficient and the surge impedance remain unchanged.

Solution

With the voltage $\underline{U}_{1Y} = (68.687 + \text{j}\,5.978)$ kV and the current $\underline{I}_1 = (0.446 - \text{j}\,0.182)$ kA, Eq. 6.21 yields the values $\underline{U}_{2Y} = (63.501 - \text{j}\,0.002)$ kV and thus $\underline{U}_2 = 109.99$ kV, $< 0°$ and $\underline{I}_2 = (0.446 - \text{j}\,0.189)$ kA. This corresponds exactly to the specification from the first example.

For given values of voltage and current at the beginning of the line, however, no calculations regarding required *power* at the end of the line can be carried out using the line equations, as both voltage and current and consequently also the transmission losses are initially unknown. The input variables would have to be varied until the required power at the end of the line is reached. This approach is not practical. The following will show how such calculations can be carried out more easily.

The line equations describe the behavior of a passive quadrupole without its own source for voltage or current. For practical reasons, the following considerations no longer use the T-equivalent circuit diagram shown above (Fig. 6.10). A symmetrical π-equivalent circuit is more suitable, in which the values in the shunt branch are halved and evenly distributed at the beginning and the end of the line. This avoids a node in the middle of the line element.

As long as an AC transmission line is symmetrically loaded, a single-phase representation is sufficient. This significantly simplifies the subsequent calculations without loss of generality for balanced three-phase systems.

When applying this method, the concept of *operating capacitance* C_B, as introduced in Chap. 5, becomes relevant. This parameter accounts for the fact that the conductor-to-conductor capacitances C_L and the conductor-to-earth capacitances C_E exist at different voltage levels. The effective or operating capacitance C_B, used as a substitute quantity in the equivalent circuit, is defined by:

$$C_B = C_E + 3C_L \tag{6.22}$$

Overhead lines inherently exhibit capacitances both between the individual conductors and between each conductor and ground (or the earth wire). In contrast, for cable systems, inter-conductor capacitances only arise if the conductor cores are not individually shielded and grounded. For single-core cables, typically used at voltage levels of 110 kV and above, this condition is generally fulfilled. Therefore, a common simplification hereby is $C_B = C_E$.

The hyperbolic functions appearing in the transmission line equations can be expressed as Taylor series. For small arguments, these series can be truncated after the first term without significant loss of accuracy. This simplification is valid because the propagation coefficient γ is typically small, and the line lengths l (particularly in Europe) are usually well below 500 km. Under these conditions, the arguments of the hyperbolic functions are sufficiently small, justifying the approximation. A detailed explanation of this simplification can be found in references [24, 25].

As a result, the steady-state (stationary) behavior of a transmission line is no longer modeled using a distributed parameter approach based on infinitely small elements. Instead, it is adequately described using lumped parameters: resistance R, conductance G, inductance L, and the operating capacitance C_B.

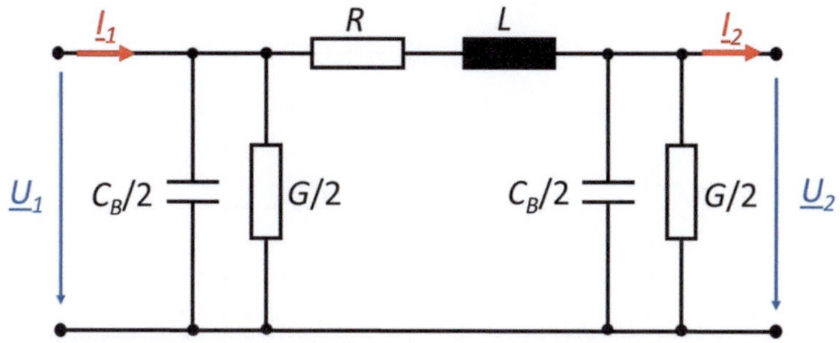

Fig. 6.16 Electrically short line with lumped elements, representation as a π-element, $\underline{U}_1$ and $\underline{U}_2$ indicate conductor-to-earth voltages

Combining all these considerations leads to the single-phase π-equivalent circuit for an *electrically short* transmission line with a return path that is to be assumed lossless (Fig. 6.16).

In the transmission of electrical energy, efficiencies well above 90% can be achieved when the line resistance and shunt conductance are minimized through the appropriate selection of materials. This condition is fulfilled when $R << \omega L$ and $G << \omega C_B$. Under these circumstances, the line is referred to as a *low-loss line.*

For basic or qualitative analyses, the line model can be further simplified by neglecting both resistance R and conductance G. However, if transmission losses are to be considered, the line resistance R must at minimum be retained in the model.

Through these simplifications, the *surge impedance* and *surge impedance loading* (SIL) become purely real quantities and are given by:

$$Z_s = \sqrt{\frac{L'}{C'_B}} \tag{6.23}$$

and

$$\text{SIL} = \frac{U_r^2}{Z_s} \tag{6.24}$$

wherein is:

Z_s Surge impedance (Ω)
SIL Surge impedance loading (VA or MVA)
U_r rated voltage (V or kV)
L' length-related inductance (H/m or H/km)
C'_B length-related Operating Capacity (F/m or F/km).

The corresponding equivalent circuit diagram is as follows (Fig. 6.17).

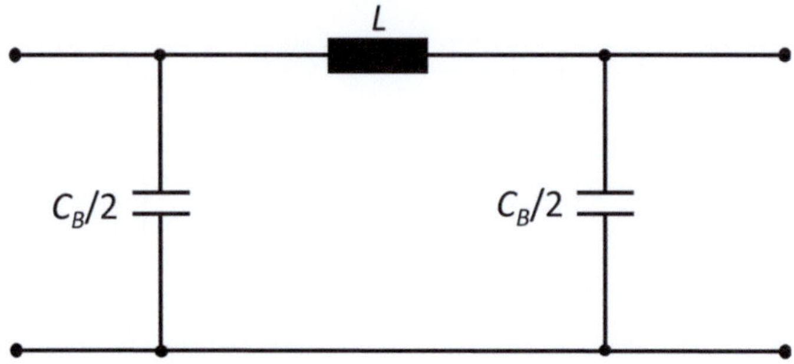

Fig. 6.17 Electrically short lossless line

6.7.1 No-Load, Matched Load, and Short Circuit

This section considers three fundamental operating conditions relevant in practical transmission systems: no-load, matched load, and short circuit. To clearly identify their characteristic behaviors, the analysis is initially based on the *lossless* line model, as illustrated in Fig. 6.18.

These conditions are first addressed in general terms and then examined through specific examples. As is customary and practical in power engineering, the starting point is the known voltage and current at the receiving end of the line. The corresponding sending-end quantities are then determined by applying Kirchhoff's voltage and current laws.

To aid visualization, the results are presented using combined voltage and current phasor diagrams. The reference phasor is defined as the voltage at the receiving end of the line and is placed along the positive real axis in the complex (Gaussian) number plane. In Fig. 6.19 and all subsequent figures, this axis is drawn vertically.

Note: A phasor can be interpreted as a vector rotating counterclockwise with an *angular frequency* ω. A phasor diagram provides a snapshot of the relative phase positions of voltages and currents at a given moment in time.

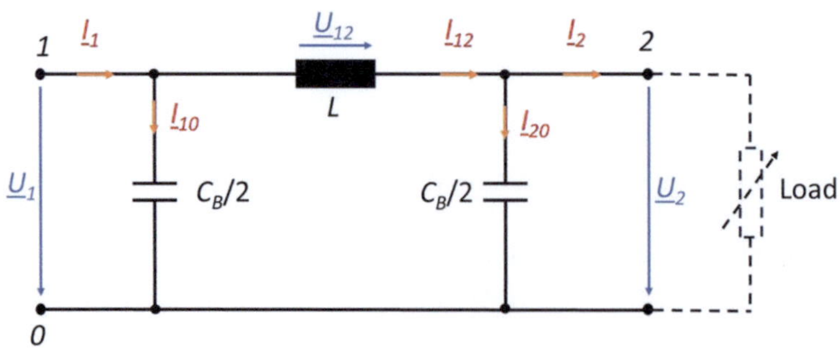

Fig. 6.18 Circuit diagram for analyzing various load conditions of a lossless transmission line

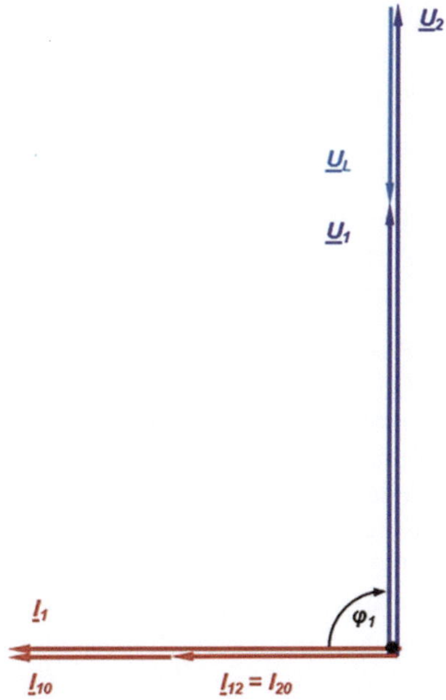

Fig. 6.19 Combined voltage- and current phasor diagram for a no-load, lossless power line

The *phase angle* φ is defined positively along the shortest path from current to voltage. In the consumer reference system used here loads with inductive components exhibit positive phase angles and loads with capacitive components exhibit negative phase angles.

It is also essential to distinguish between longitudinal quantities (voltages and currents along the line), and transverse or shunt quantities (i.e., towards the return conductor).

The phasor diagrams shown are qualitative (i.e., not to scale) and refer to rms (root mean square) values of phase (star) voltages and the associated currents.

For the following considerations, the line is assumed to be terminated with a variable resistive load that can take on any value between zero (short circuit) and infinity (open circuit). The circuit analysis and the derivation of the relevant equations are consistently performed starting from the receiving end of the line.

In the no-load case, the load resistance is considered to be infinitely large, meaning that no current flows into the load. As a result, the load current $\underline{I}_2$ is zero. However, a capacitive charging current $\underline{I}_{20} = \underline{U}_{2Y} \cdot (-\,\mathrm{j}\omega C_B/2)$ still flows through the operating capacitance at the end of the line. Since there is no load current, the longitudinal current through the line is equal to the charging current: $\underline{I}_{12} = \underline{I}_{20}$. This current flows through the series inductance of the line and causes a voltage rise in the direction opposite to the receiving-end voltage $\underline{U}_{2Y}$. The voltage drop across the series inductance is $\underline{U}_{12} = \underline{I}_{12} \cdot \mathrm{j}\omega L$. Therefore, the voltage at the sending end of the line becomes $\underline{\mathrm{U}}_{1\mathrm{Y}} = \underline{\mathrm{U}}_{2\mathrm{Y}} + \underline{\mathrm{U}}_{12}$. This voltage is higher than the receiving-end voltage due to the inductive voltage drop.

If the shunt current $\underline{I}_{10} = \underline{U}_{1Y} \cdot (-\,\mathrm{j}\omega C_B/2)$ is added to the longitudinal current $\underline{I}_{12}$, the total sending-end current $\underline{I}_1$ is obtained. The corresponding *phasor diagram* clearly illustrates these relationships. It shows how:

- The capacitive transverse currents at both ends of the line lead their respective voltages by 90°,
- The longitudinal current $\underline{I}_{12}$ contributes to an inductive voltage drop $\underline{\mathrm{U}}_{12}$,
- And the total sending-end current $\underline{I}_1$ results from the vector addition of these two components.

As illustrated in the diagram, under no-load conditions, the voltage at the receiving end of the line is higher than at the sending end. This phenomenon is known as the *Ferranti effect*. It becomes particularly pronounced in cable systems due to their significantly higher capacitive values and can lead to dangerously high voltages at the open end of the line, especially from a certain line length onward. Such conditions must be detected and mitigated by network protection systems.

The line capacitance, represented in the shunt branches of the equivalent circuit, generates reactive power that depends solely on the operating voltage. As a result, even under no-load conditions, this capacitive reactive power is present. In contrast, the series inductance of the line consumes only a minimal amount of reactive power in this case, since no active load current flows.

If one conceptually reduces the load resistance, the line current increases, and the reactive power balance begins to shift. The net capacitive effect diminishes as more current flows through the inductive series branch. To compensate for the increasing voltage drop along the line, the voltage at the sending end must be increased accordingly.

At a specific load resistance value, the capacitive and inductive reactive powers exactly cancel each other out. In this case, the line draws at its sending end precisely the active power that it delivers to the load. This theoretical condition—assuming a lossless line—is referred to as the *matched load* or line matching. Hereby, the load resistance equals the surge impedance of the line. The corresponding transmitted power is known as the *surge impedance loading* (SIL).

It should be noted that in cable systems, this matched condition is not practically achievable, since the resulting current would be extremely high and could only be maintained with forced cooling. In contrast, for overhead lines, both higher and lower load levels relative to SIL are technically feasible, as long as the thermal current limit of the line is not exceeded.

For completeness, Fig. 6.20 illustrates the general case of a resistive load and Fig. 6.21 shows the matched condition, where the load equals the surge impedance.

If the load resistance is finally reduced to zero, a *short circuit* occurs, which can damage not only the line itself but also other parts of the system. Therefore, it must be detected and disconnected by the network protection system in the shortest possible time.

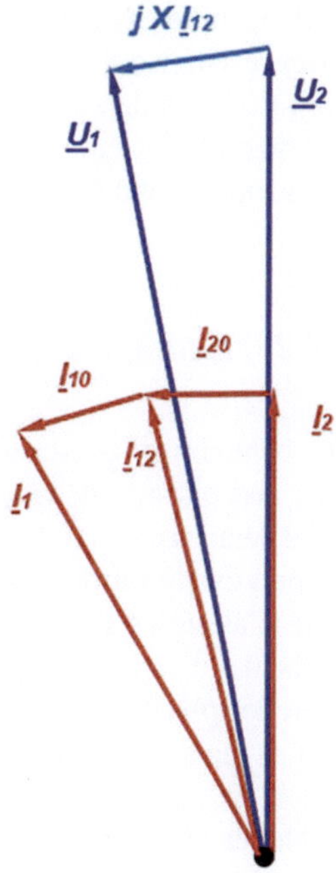

Fig. 6.20 Voltage- and current phasor diagram for a lossless power line with resistive load

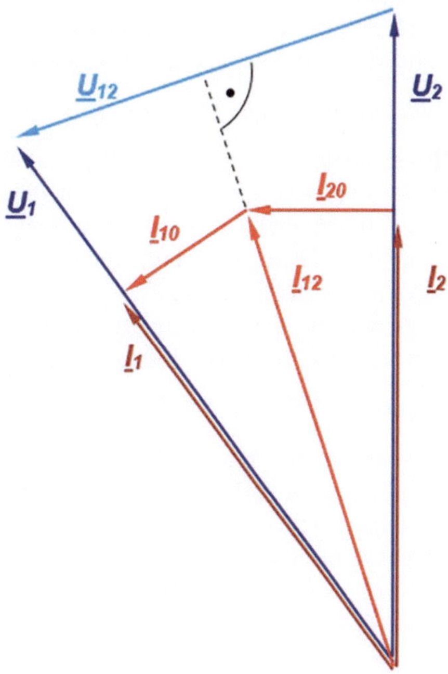

Fig. 6.21 Voltage- and current phasor diagram for a lossless power line in case of matching

If the three characteristic load cases are reconsidered while taking into account the series resistance R and the shunt conductance G, there are no significant differences from the considerations above, due to the relationships $R \ll \omega L$ and $G \ll \omega C_B$. For practical calculations, it is therefore sufficient to consider, in addition to the parameters L and C_B, only the resistance R (for determining line losses).

6.7.2 Examples of Different Load Cases

The behavior of overhead lines, cables, and gas-insulated lines at various levels of high and extra-high voltage under different load conditions will be examined in more detail using several practical examples. The calculations are carried out using the previously introduced equations, which apply to *lumped elements*.

It is assumed that in all examples the rated voltage with a phase angle of zero exists at the end of the line. This makes the voltage phasor at the end of the line the reference phasor for all calculated quantities. The required voltages and currents, as well as the active and reactive powers at the beginning of the line and the resulting losses, are determined.

The calculations are performed in single-phase equivalent (voltages are marked with the subscript Y), but results are presented as phase-to-phase voltages and three-phase power values. The same subscripts as in Fig. 6.18 are used throughout.

All examples are based on a frequency of 50 Hz, as this is a very common value worldwide. The shunt conductance is neglected in all cases.

Calculation Example 8

This example has already been treated as an application case for the line equations. A 110 kV overhead line with a length of 40 km has the following electrical parameters: $R' = 0.126\ \text{m}\Omega/\text{km}$ at operating temperature, $L' = 1.24\,\text{mH/km}$ and $C_B' = 8.9\ \text{nF/km}$. The mixed ohmic-inductive load at the end of the line consists of an active power of 85 MW and a reactive power of 36 Mvar at nominal voltage. What are the voltage, current, active and reactive power at the beginning of the line? How high are the active power losses of the line?

Solution

We start with the apparent power at the end of the line. In complex representation, it applies: $\underline{S}_2 = \sqrt{3} \cdot \underline{U}_2 \cdot \underline{I}_2^*$ (the asterisk means conjugate complex). This results in a current $\underline{I}_2^* = (0.446 + \text{j}\,0.189)$ kA or $\underline{I}_2 = (0.446 - \text{j}\,0.189)$ kA. The shunt current at the end of the line through half the operating capacity is $\underline{I}_{20} = \underline{U}_{2Y} \cdot \text{j}\omega C_B/2 = \text{j}\,0.004\,\text{kA}$. This results in the longitudinal current through the line to $\underline{I}_{12} = \underline{I}_2 + \underline{I}_{20} = (0.446 - \text{j}\,0.185)$ kA and the longitudinal voltage drop becomes $\underline{U}_{12} = (R + \text{j}\omega L) \cdot \underline{I}_{12} = (5.138 + \text{j}\,6.017)\,\text{kV}$. This results in the voltage at the beginning of the line: $\underline{U}_{1Y} = \underline{U}_{2Y} + \underline{U}_{12} = (68.646 + \text{j}\,6.017)$ kV or $\underline{U}_1 = 119.35$ kV, $< 5°$ (magnitude and phase).

Furthermore, the shunt current at the beginning of the line through half the operating capacity is needed. With $\underline{I}_{10} = \underline{U}_{1Y} \cdot \mathrm{j}\omega C_B/2 = \mathrm{j}\,0.003\,\mathrm{kA}$. We get a current value at the beginning of the line of $\underline{I}_1 = \underline{I}_{12} + \underline{I}_{10} = (0.446 - \mathrm{j}\,0.182\,\mathrm{kA})$ or 0.481 kA, <−22°.

For active and reactive power, the results are $P_1 = 88.53\,\mathrm{MW}$ and $Q_1 = 45.44\,\mathrm{Mvar}$. The associated apparent power $S_1 = 99.5$ MVA, lies below the thermal limit power $S_{th} = 125$ MVA.

The active power losses are generally determined for lines according to the equation $P_l = 3 \cdot R \cdot I^2$, In this case, however, it is easier to calculate them from the difference between the two active powers at the beginning and end of the line, i.e. $P_l = P_1 - P_2 = 3.53\,\mathrm{MW}$ (Fig. 6.22).

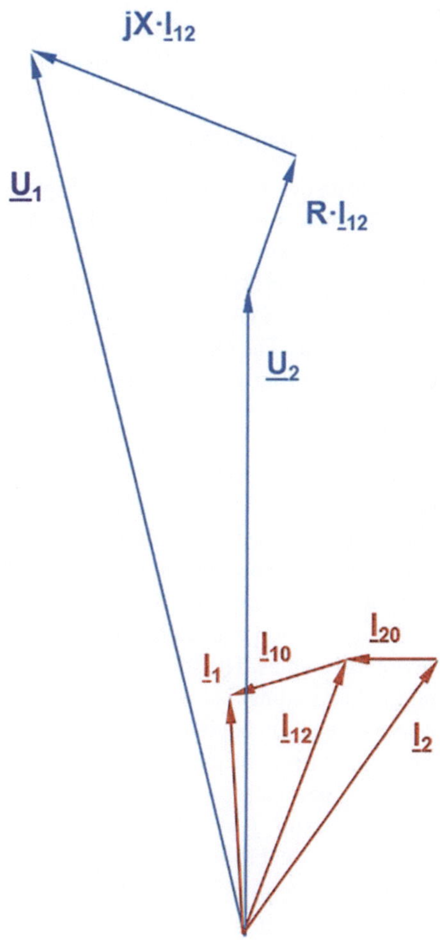

Fig. 6.22 Voltage- and current phasor diagram for a low power-loss line with ohmic-inductive load

Calculation Example 9

For comparison, the calculation is made for a XLPE-insulated 110 kV cable of the same length and load as in calculation example 8. The electrical data of the copper cable 3 × 300 mm² are $R' = 0.06\,\Omega/\text{km}$ at 20 °C, $L' = 0.61\,\text{mH/km}$ and $C'_B = 0.144\,\mu\text{F/km}$.

The calculation is to be performed assuming a conductor temperature of 70 °C.

Solution

The conversion of the specific conductor resistance results to $R'_{70°} = 1.20 \cdot R'_{20°} = 0.072\ \Omega/\text{km}$ (Chap. 5). Analogously to example 8, the following applies: $\underline{I}_2 = (0.446–\text{j}\,0.189)$ kA, $\underline{I}_{20} = \text{j}\,0.058$ kA, $\underline{I}_{12} = (0.446–\text{j}\,0.132)$ kA and $\underline{U}_{12} = (2.293 + \text{j}\,3.041)$ kV. Thus, it follows that $\underline{U}_1 = 114.09\,\text{kV}, < 2.6°$.

$\underline{I}_{10}$ has the value $(-0.003 + \text{j}\,0.060)$ kA and $\underline{I}_1 = 0.449\,\text{kA}, < -9.2°$. The power at the beginning of the line results in $P_I = 86.87$ MW and $Q_I = 18.25$ Mvar with transmission losses of 1.87 MW.

The example shows that the cable makes a significant contribution to the reactive power demand of the load. Due to its conductor material and cross-section, its losses are approximately half as high as those of a comparable overhead line.

Calculation Example 10

Parallel to an overhead line at the 110 kV level, a 110 kV cable is laid. Both are 40 km long. Parameters of the overhead line: $R' = 0.126\ \text{m}\Omega/\text{km}$, $L' = 1.24\,\text{mH/km}$, $C'_B = 8.9\,\text{nF/km}$. The thermal limit power of the line is 125 MVA. Data of the cable: $R' = 0.08\,\Omega/\text{km}$, $L' = 0.61\,\text{mH/km}$, $C'_B = 0.144\,\mu\text{F/km}$ and its thermal limit power 110 MVA. They jointly supply a load of 180 MW and100 Mvar inductive at rated voltage. The required voltage at the beginning of the line, as well as the load contributions of the overhead line and the cable, are to be determined.

Solution

From the apparent power $\underline{S}_2 = (180 + \text{j}\,100)$ MVA at the end of the line, the load current is $\underline{I}_2 = (0.9448–\text{j}\,0.5249)$ kA (to avoid significant rounding errors, this calculation is carried out using four decimal places).

The parallel transmission paths are first combined into an equivalent line with the following parameters: $R = 1.9573\ \Omega$, $X = 5.1353\ \Omega$ and $C_B = 6.116\ \mu\text{F}$. Further calculation with these equivalent values is carried out in the same way as the previous examples. The shunt current at the end is $\underline{I}_{20} = \underline{U}_{2Y} \cdot \text{j}\omega C_B/2 = \text{j}\,60.1$ A and the longitudinal current follows to $\underline{I}_{12} = (0.9448–\text{j}\,0.4639)$ kA. This results in a voltage at the beginning of the equivalent line of $\underline{U}_1 = 117.527$ kV, $< 3.3°$. The shunt current at the beginning is determined with $\underline{I}_{10} = \underline{U}_{1Y} \cdot \text{j}\omega C_B/2 = (-3.8 + \text{j}\,65.1\,\text{A})$.

The current $\underline{I}_{12}$ is to be divided between overhead line (subscript OHL) and cable (subscript CAB). For this, the two longitudinal impedances $\underline{Z}_{OHL} = (R_{OHL} + \text{j}\,X_{OHL}) = (5.04 + \text{j}\,15.657)\,\Omega$ and $\underline{Z}_{CAB} = (R_{CAB} + \text{j}X_{CAB}) = (3.2 + \text{j}\,7.661)\,\Omega)$ are required. It follows:

$\underline{I}_{12,OHL} = \underline{Z}_{CAB}/(\underline{Z}_{OHL} + \underline{Z}_{CAB}) \cdot \underline{I}_{12} =$ (0.3092–j 0.1735) kA and $\underline{I}_{12,CAB} = \underline{I}_{12} - \underline{I}_{12,OHL} =$ (0.6357–j 0.2905) kA. The shunt currents, which were determined for the equivalent line, must also be shared, in proportion to the capacitances. For the overhead line, this results in the beginning to $\underline{I}_{10,\mathrm{OHL}} =$ (− 0.22 + j 3.79) A and for the cable to $\underline{I}_{10,\mathrm{CAB}} =$ (− 3.57 + j 61.26) A. If the partial currents $\underline{\mathrm{I}}_{12,\mathrm{OHL}}$ and $\underline{\mathrm{I}}_{12,\mathrm{CAB}}$ are added, the incoming current of the overhead line has a value of $\underline{\mathrm{I}}_{1,\mathrm{OHL}} =$ 0.353 kA, < −28.8° and the one of the cable of $\underline{\mathrm{I}}_{1,\mathrm{CAB}} =$ 0.672 kA, < −19.9°.

The overhead line absorbs at its beginning an active power of 60.8 MW, a reactive power of 38.1 Mvar or an apparent power of 71.8 MVA. For the cable, the corresponding values are 125.5 MW and 54.1 Mvar or 136.7 MVA. The total losses of the transmission amount to 6.3 MW.

In this case the disadvantage of a cable, which is connected in parallel to an overhead line of the same length, becomes evident: While the cable is already overloaded, the overhead line is only operated at about 60% of its thermal limit power.

Calculation Example 11

The performance of a 380 kV overhead line of 240 km length is to be examined for the following load situations: No-load—Load lower than matched load—Matched load—Load higher than matched load—High mixed resistive-inductive load—Influence of parallel compensation with capacitors. At the end of the line, for all cases, the rated voltage is required. The line parameters are $R' = 27.3\,\mathrm{m\Omega/km}$, $L' = 0.81\,\mathrm{mH/km}$, $C'_B = 14.2\,\mathrm{nF/km}$ and $S_{th} = 1790$ MVA. In the respective load cases, only voltage and current along with the power at the beginning of the line as well as the power losses are of interest.

Solution

Case 1**:** In no-load condition, $P_2 = Q_2 = 0$. Thus, initially only the proportional charging current at the end of the line and the longitudinal voltage drop are of interest. Their values are: $\underline{I}_{20} = \underline{I}_{12} =$ j 0.117 kA and $\underline{U}_{12} =$ (−7.16 + j 0.77) kV. This leads to a voltage $U_1 = 367.6$ kV and a current $I_1 = 0.23$ kA. The associated powers are 0.3 MW and −147.1 Mvar (i.e. capacitive).

In this case, the *Ferranti effect* becomes clearly visible, resulting in a 3.4% voltage increase at the end of the line compared to the beginning.

Case 2: $P_2 = 500$ MW, $Q_2 = 0$ (lower than matched load). With the known calculation steps (see case 1) we get: $U_1 = 385$ kV, $\mathrm{I}_1 = 0.77$ kA, $P_1 = 511.6$ MW, $Q_1 = -48.4$ Mvar, $P_l = 11.6$ MW.

Case 3: For an exact calculation, the complex value for the surge impedance is required. According to Eq. 6.13, this is Zs = (239.17 + j 12.79) Ω. From this, the surge impedance load (SIL) is determined to be $\underline{\mathrm{S}}_2 =$ (602 − j 32.2) MVA. Further calculation yields: $U_1 = 385.6$ kV, $I_1 = 0.93$ kA, $P_1 = 619$ MW and $Q_1 = -30.76$Mvar. In this load case, the line itself practically neither generates nor consumes reactive power.

Case 4: $P_2 = 800$ MW, $Q_2 = 0$ (higher than matched load). The calculation results to: $U_1 = 402.9$ kV, $I_1 = 1.20$ kA, $P_1 = 829.3$ MW, $Q_1 = 109$ Mvar, $P_l = 29.3$ MW.

Case 5: $P_2 = 800$ MW, $Q_2 = 350$ Mvar (higher than matched load). The following values are obtained: $U_1 = 454.8$ kV (!), $I_1 = 1.21$ kA, $P_1 = 832.4$ MW, $Q_1 = 464.1$ Mvar and $P_l = 32.4$ MW. The necessary voltage at the beginning of the line is unrealistically high, while the apparent power flowing into the line of 953 MVA is far from fully loading it.

Case 6: $P_2 = 800$ MW, $Q_2 = 100$ Mvar (higher than matched load), where the high reactive power from the previous case was compensated to a residual value of 100 Mvar using a parallel capacitor. Now the respective values are: $U_1 = 417.6$ kV, $I_1 = 1.18$ kA, $P_1 = 829.1$ MW, $Q_1 = 200.2$ Mvar and $P_l = 29.1$ MW.

Calculation Example 12

The behavior of a 12 km long 380 kV gas-insulated line (GIL) with a cross-section of the inner pipe (the conductor) of 6 300 mm^2 and the parameters $R' = 5.8\,\mathrm{m\Omega/km}$ at operating temperature, $L' = 0.216\,\mathrm{mH/km}$, $C'_B = 0.054\,\mathrm{\mu F/km}$ and $S_{th} = 1645$ MVA, when buried [3]. The much smaller specific inductance and capacitance values allow the transmission of higher powers compared to overhead lines. According to example 11 (case 5), we start with the values $P_2 = 800$ MW and $S_2 = 350$ MVA (case 1). This is followed by a load case in which the thermal limit power is reached (case 2). We are only interested in the magnitudes of voltage and current at the beginning of the GIL and the power fed into it. In both cases the rated voltage is expected at the load.

Solution

Case 1: For $P_2 = 800$ MW and $Q_2 = 350$ Mvar, we get: $U_1 = 380.9$ kV, $I_1 = 1.31$ kA, $P_1 = 800.4$ MW, $Q_1 = 324.8$ Mvar and $P_l = 0.4$ MW follow. The GIL is operating with a load less than SIL.

Case 2: For $P_2 = 1551.5$ MW and $Q_2 = 600$ Mvar: $U_1 = 381.6$ kV, $I_1 = 2.51$ kA, $P_1 = 1551.5$ MW, $Q_1 = 586.0$ Mvar. This corresponds to an apparent power of $S_1 = 1{,}658$ MVA, which slightly exceeds the thermal limit power. The transmission losses are $P_l = 1.5$ MW.

Calculation Example 13

A 380 kV overhead line with high-temperature low-sag conductors (HTLS) of 30 km length is considered. It is equipped with quad bundles 4 × 249 mm^2 of *Super Thermal Resistant Aluminum Alloy* (Chap. 4) and has a specific resistance *R'* of 0.0275 Ω/km at 20 °C, 0.033 Ω/km at 70 °C and 0.0473 Ω/km at 200 °C, while *L'* and C_B' are identical to the data from example 11. However, it can transmit up to 90% more power compared to the line with conventional conductors of the same cross-section. The line is loaded at the end once with 1400 MW/500 Mvar (case 1) and with 2800 MW/1000 Mvar (case 2). In both cases, the rated voltage should be present at the end of the line. The resulting losses—separated into active and reactive power components—are to be determined, keeping in mind that the reactive power required by a line is commonly regarded as "reactive losses", as it must be supplied by the system despite not representing real energy consumption. For case 1, the line temperature is 70 °C and for case 2 it is 200 °C.

Solution

Case 1: According to the usual calculation method, we get a voltage $U_1 = 394.4$ kV and a current $I_1 = 2.25$ kA follows. The supplied powers are $P_1 = 1415.1$ MW and $Q_1 = 596.3$ Mvar. Thus, the transmission losses are approximately $P_l = 15$ MW (corresponding to 1.1% of the active load) and $Q_l = 96$ Mvar.

Case 2: Under these conditions, the following values are obtained: $U_1 = 413.7$ kV and $I_1 = 4.5$ kA. The supplied powers are $P_1 = 2887$ MW and$Q_1 = 1445$ Mvar. The active losses increase to about 87 MW (corresponding to 3% of the active load). Noteworthy in this operating case is the very high reactive power demand of the line. This effect occurs inevitably when transmitting high powers with high-temperature low-sag conductors and is caused by the line inductance.

Calculation Example 14

The transmission performance of an AC and a DC overhead line with the same consumption of conductor material (mass) are to be compared with each other. In both cases, an active power of 125 MW is to be transmitted over a distance of 100 km, which corresponds to a load current of 0.656 kA at the 110 kV three-phase voltage level. The conductor material for the case of AC has a specific resistance of 0.126 Ω/km. For easier comparison, it is assumed that the line is compensated using shunt reactors, meaning the effect of the line capacitance is cancelled. Additionally, a series compensation is applied to eliminate the effect of the line inductance. At the line end rated voltage is to be assumed.

Solution

For the AC case, the calculation is greatly simplified and leads to the following results: $U_1 = 124.3$ kV, $I_1 = I_2 = 0.66$ kA, $P_1 = 141.3$ MW and $Q_1 = 0$. The losses amount to: $P_l = 16.3$ MW. Considering the DC variant, i.e. when using two instead of three conductors, a specific conductor resistance of 0.084 Ω/km and a maximum current of 0.984 kA can be assumed. With the same conductor-to-conductor and conductor-to-earth distances as in the three-phase case, a rated voltage of ± 63.5 kV results (Chap. 1). This leads to $U_1 = \pm 71.77$ kV and losses of $P_l = 16.27$ MW. However, the losses in the associated AC/DC and DC/AC converter stations are not considered.

Calculation Example 15

Similar to example 14, the transmission behavior of a 380 kV AC cable will be compared to that of an equivalent DC cable, with 1250 MW to be transmitted over a length of 500 km in each case. Nominal voltage at the line end is required. In the case of AC, complete reactive power compensation is assumed, so that only active powers are to be considered. The AC cable has a copper cross-section of 2500 mm^2 with $R' = 10.8\,\mathrm{m}\Omega/\mathrm{km}$.

Solution

Here too, the calculation is greatly simplified and leads to the results $U_l = 397.8$ kV, $I_l = I_2 = 1.9$ kA, $P_l = 1308.4$ MW and $Q_l = 0$. The losses amount to $P_l = 58.4$ MW. For the DC variant, a specific conductor resistance of 7.2 Ω/km and a permissible current of 2.85 kA apply. Assuming the same insulation as in the three-phase case, the results are $U_l = \pm 231.3$ kV and $P_l = 58.5$ MW. Here too, the losses in the associated converter stations must be considered separately.

6.7.3 Reactive Power Behavior and Transmissible Power

The significantly different inductance and capacitance values of the various transmission media used in AC systems lead to marked differences in reactive power behavior during operation.

An open overhead line generates only a small charging power due to its relatively low capacitance. When the line is lightly loaded at its receiving end, its inductance consumes reactive power, thereby compensating for part of the capacitive charging power. This operating condition is referred to as operation below the surge impedance loading (SIL). As the load current increases, a point is reached where the capacitive and inductive reactive power components cancel each other out. This condition is known as matched loading or SIL operation. If the current continues to rise beyond this point, the line itself requires reactive power, leading to an operating condition referred to as above SIL. In the limiting case, if the load resistance approaches zero, a short circuit occurs.

In contrast, cables, due to their high capacitance, can only be operated in the below SIL range, where they generate more reactive power than they consume. To approach a matched condition, a very high current would be required—achievable only with external cooling measures.

Gas-insulated lines (GIL) behave similarly to cables in this respect. Across their entire transmission range, they operate in the capacitive region, but the reactive power levels are significantly lower than those of cables.

Figure 6.23 illustrates a comparison of the reactive power behaviour of the three transmission systems—overhead line, cable, and gas-insulated line—at the 380 kV level. The active power ($\cos \varphi = 1$) is assumed to be taken at the receiving end of a 50 km long line. The thermal limit power assumed for the calculation is 1700 MVA for the overhead line and the gas-insulated line, and 1200 MVA for the cable.

The reactive power behavior of cables limits their transmission capacity in contrast to overhead lines and gas-insulated lines. Together with the active current, the reactive current (geometrically added) results in the apparent current, which must not be exceeded (Chap. 1). This fact prevents the use of long and non-compensated AC cable connections (e.g. as submarine cables).

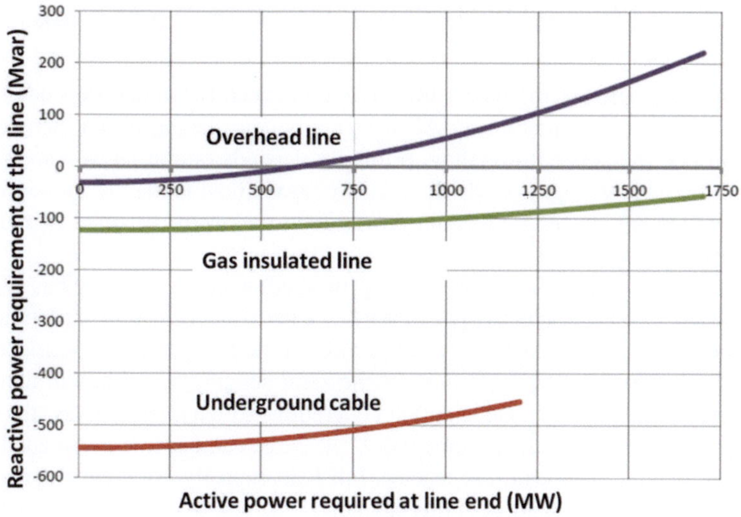

Fig. 6.23 Reactive power behavior of overhead line, cable and GIL each with 50 km length, depending on the transmitted active power

Cables in the higher voltage levels generate more reactive power, as its amount increases quadratically with the applied voltage: $Q_c \sim U^2 \cdot \omega C_B$.

The diagram in Fig. 6.24 shows the possible active power transport of an extruded cable for 380 kV as a function of its length. As can be easily seen, this effect very clearly limits the possible transmission distance of a cable, if not compensated.

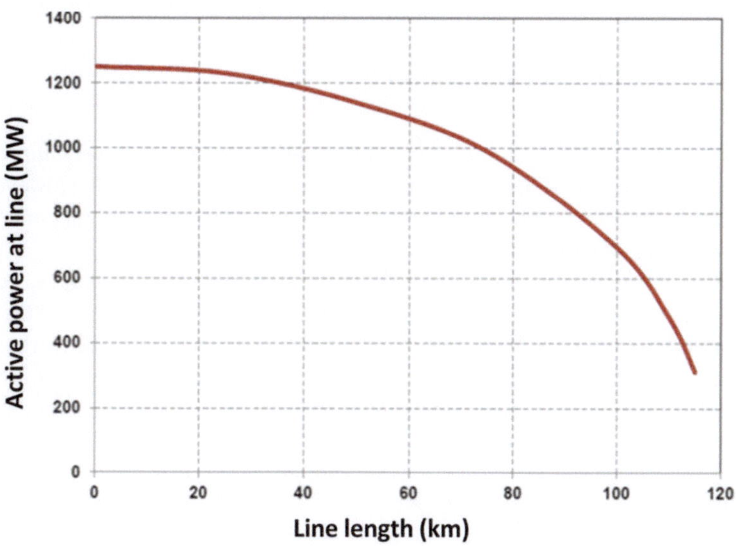

Fig. 6.24 Maximum transmissible active power of an uncompensated XLPE cable at the 380 kV voltage level as a function of line length

6.7.4 Load Flow and Short-Circuit Situations in a Network

While in a DC network the power transmitted between two nodes depends solely on their voltages, the situation in AC systems is considerably more complex. In AC networks, the active power flow primarily depends on the *angular difference* between the node voltages, while the reactive power flow is mainly influenced by the *magnitude difference.* As a result, power flows are always affected by multiple interdependent parameters.

The addition of a new line to an existing (meshed) network therefore has system-wide effects. This is particularly relevant when a network consisting predominantly of overhead lines is extended with a cable connection. Due to the lower impedance of cables compared to overhead lines of the same length, the cable will naturally carry a greater share of the load in parallel operation—often referred to as "the cable draws power onto itself." In extreme cases, the cable can become overloaded even before the parallel overhead line reaches full load capacity. Reference is made to Calculation Example 10 earlier in this chapter.

It should also be noted that reinforcing the network with additional lines and transformers increases the short-circuit current levels, which may require adjustments to the network protection settings. When planning such expansions, it is essential to evaluate whether circuit breakers, current transformers, or even busbars need to be upgraded to handle the increased fault levels.

As an alternative, it may be possible to limit rising short-circuit power by introducing network segmentation (i.e., "enmeshing" the grid). However, such a measure would likely come at the cost of reduced supply reliability and must be carefully weighed.

6.7.5 Neutral Point Connection

In undisturbed operation and in the event of *symmetrical* (three-phase) *faults* in AC power systems the type of neutral point connection is largely irrelevant. However, for *unsymmetrical faults*, which occur far more frequently, the neutral point connection becomes a critical factor.

Among unsymmetrical faults, the single-phase-to-earth fault—that is, contact between one phase conductor and earth or earthed components—is by far the most common and therefore receives special attention. In overhead line networks, such faults can result from mechanical damage, polluted insulators, or lightning strikes that trigger arc formation. In cable systems, they may occur due to excavation damage, thermal overload, or insulation failure.

For faults of this type, the response—either immediate disconnection or temporary continued operation—depends on several factors, including:

- the voltage level
- the extent and configuration of the network, and
- the type of neutral point grounding.

Under favourable conditions, the arc ignited by the fault may extinguish itself, avoiding the need for immediate shutdown.

For Germany, the following types of neutral point connection have proven effective in practice:

- *Extra-high voltage networks* (220 and 380 kV)
  - Solidly grounded (direct connection to earth)
- *High voltage networks* (60 to 110 kV)
  - Either solidly grounded or compensated grounding using earth fault compensation coils, depending on network size and structure
- *Medium voltage networks* (10 to 30 kV) in public supply:
  - Mostly compensated grounding, while larger or newer networks are often solidly grounded
  - Internal networks, such as those in power plants, are typically operated with an isolated (ungrounded or floating) neutral point.

The primary side of the feeding transformer (or generator) is not shown in the illustration above or in the ones that follow. For the purposes of further analysis, a greatly simplified network model is sufficient. This model includes only a single feed-in point, and the network lines are represented by a single equivalent impedance. Load currents can be neglected, as they are small compared to the short-circuit currents that occur during fault conditions (Fig. 6.25).

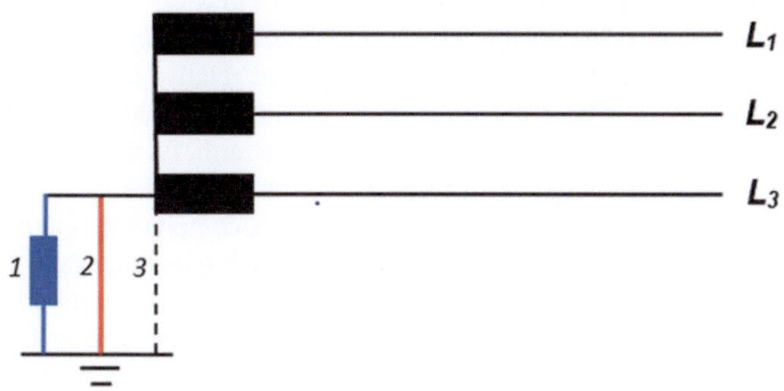

Fig. 6.25 Options for neutral point connection (1: compensated, 2: directly grounded, 3: isolated neutral point)

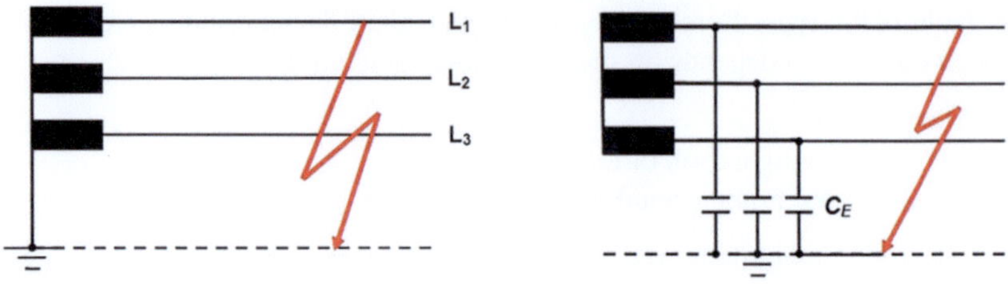

Fig. 6.26 Earth short-circuit (*left*) and earth fault (*right*) circuit in a three-phase network

In the case of low-impedance (direct) neutral grounding, a single-phase fault results in a high current in the affected conductor. This fault must be detected by the protection system, and the faulted line must be disconnected immediately. In pure overhead line networks, or networks with a low proportion of cables, the fault often clears on its own (e.g., when the arc extinguishes or the cause burns out). In such cases, *Automatic Reclosure* (AR) allows the line to be re-energized after a short delay—typically a few hundred milliseconds.

The return path for the fault current in this configuration runs through earthing systems, earth wires, cable shields, and the earth itself. Since these impedances are low, the resulting fault current is high, and is therefore referred to as the *earth short-circuit current*.

At the other extreme, in the case of an isolated (floating) neutral point, the fault current return does not flow through the low-impedance earth path described above, but instead via the earth capacitances of the healthy phases in the network. Due to the high impedance of this path, the resulting fault current is low and is referred to as the *earth fault current* (Fig. 6.26).

An isolated neutral point offers the important advantage that the network can continue operating in a faulted condition for a certain period of time (up to the hour range) until the damaged line is located and disconnected for repair. In Germany, for example, this method is used in specific cases such as the auxiliary power supply of power plants, where high supply reliability is critical. However, this setup introduces another issue, which is described by the term *displacement voltage* (Fig. 6.27).

If one considers the position of the neutral point relative to earth potential in both grounding cases, it becomes clear that a floating (isolated) neutral point can reach a displacement voltage of up to $U_r/\sqrt{3}$ against earth. As a result, the voltages of the two healthy conductors with respect to earth rise to the value of the rated voltage U_r. Consequently, the insulation of all network components is stressed to the full line-to-line voltage, increasing the risk that the fault may propagate.

If an inductance of suitable value is connected between the network's neutral point and earth, the capacitive earth fault current can be compensated by the inductive current from the coil. In this case, the fault current nearly disappears, with only a small *residual current* remaining, due to the ohmic resistance of the conductor material. Under these conditions, arc faults, which often occur in overhead line networks,

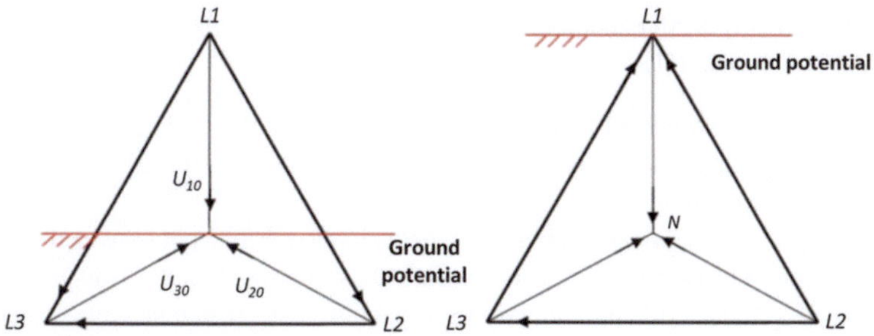

Fig. 6.27 Position of the neutral point in an earth short circuit (*left*) and in an earth fault (*right*), in both cases conductor *L1* is affected

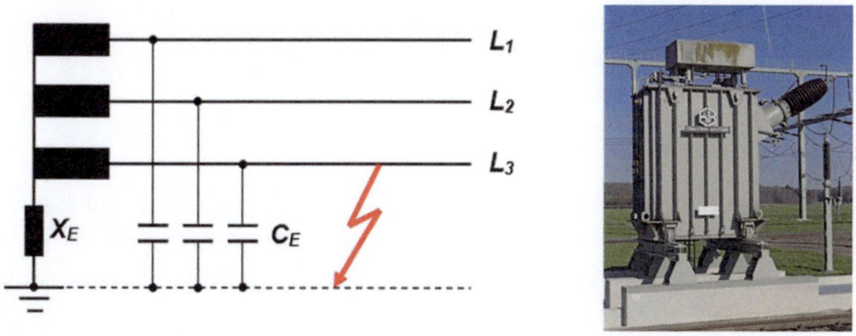

Fig. 6.28 Petersen Coil (*Source* right side: W. Dünkel, VDE Kassel)

typically extinguish themselves. This inductance is therefore referred to as an *Earth Fault Neutralizer*, or more commonly, a *Petersen Coil*, named after its inventor.

Since the total earth capacitance of a network can change depending on switching operations (i.e., lines being connected or disconnected), the inductance of the Petersen Coil must be adjustable. This is commonly achieved using plunger core coils or coils with taps (Fig. 6.28).

The equations required for the calculation are given in [1] as:

$$I_E = \sqrt{3} \cdot U_r \cdot \omega C_E \tag{6.25}$$

and

$$L = \frac{1}{3 \cdot \omega^2 C_E} \tag{6.26}$$

wherein is:

I_E Earth fault current (A or kA)
U_r Rated voltage (V or kV)
ω Angular frequency (= 2πf) (s^{-1})

C_E Conductor-to-earth capacitance (F)
l Coil inductance (H).

Equation 6.26 is derived from the requirement that the coil reactance must be equal to the reactance of the sum of the earth capacitances and is also referred to as the *resonance condition.*

Due to typically existing small asymmetries in the earth capacitances of the individual conductors in a three-phase network, it has proven advantageous in practical operation to set the coil inductance slightly below the resonant value calculated from the above equation. This deviation improves stability and performance under real-world conditions.

However, compensated network operation requires that the limit current for the self-extinction of freely burning arcs is not exceeded. For networks at the 110 kV level, this limit is typically around an empirical value of 130 A [26]. As a result, this type of neutral point treatment is most commonly used in medium-voltage networks and in high-voltage networks of limited extent.

Calculation Example 16

For a 110 kV network, the magnitude of the earth fault current and the required inductance of a Petersen coil for earth fault compensation are to be determined. For this purpose, a network is considered, which consists exclusively of overhead lines with a total system length of 270 km. The specific earth capacitance of the line conductor is assumed to be $C_E = 3.5$ nF/km.

Solution

According to the Eqs. (6.25) and (6.26) above, the following values are obtained: $I_E = 56.5$ A and $L = 3.6$ H.

6.7.6 *Stability in Power Transmission Systems*

Stability means, that a power system withstands disturbances so that no blackout can result as a consequence of a domino-effect. One has to distinguish between two kinds of stability [1, 5, 25]:

- *Steady-state stability* describes a power system's ability to maintain synchronism under small disturbances such as gradual load changes. These events can cause abrupt changes in the rotor angle of the associated synchronous generators. If the system can absorb the disturbance and the generator rotor can resynchronize within a few seconds, it is considered steady-state stable. If not, it results in system breakup.
- *Transient stability* is concerned with the possibilities to maintain synchronism following large, sudden disturbances, such as short circuits, line trips, or sudden loss of generation. During events such as automatic reclosing after a fault, the

voltage angle difference can momentarily exceed the steady-state limits. This underscores the importance of protective relays and system planning to avoid operation too close to the stability limit.

The first of the two types of stability can be easily explained with an example of a single-phase power line with the Impedance X, where resistance and shunt admittance are neglected for simplification.

If we assume $\underline{U}_1$ (phase angle 0°) and and $\underline{U}_2$ (phase angle or *line angle δ*) to be the complex phase-to-ground voltage at both ends, then follows for the current from node 1 to node 2:

$$\underline{I}_{12} = (\underline{U}_1 - \underline{U}_2)/\mathrm{j}\,X \tag{6.27}$$

Thus, for the corresponding power:

$$\underline{S}_{12} = 3 \cdot \underline{U}_2 \cdot \underline{I}_{12}^* \tag{6.28}$$

and for the active power:

$$P_{12} = 3 \cdot U_1 \cdot U_2 \cdot \sin\delta/X \tag{6.29}$$

The line angle δ is a key parameter. It is approximately 6° per 100 km under normal operating conditions in high and extra-high voltage overhead lines and about three times higher (17°) for underground cables. It can be calculated with the help of Eq. 6.14 (propagation coefficient) and becomes for a lossless line:

$$\delta = \omega\sqrt{(L' \cdot C_B')} \cdot \ell \tag{6.30}$$

Equation 6.29 shows that the transmittable active power is proportional to the sine of the angle δ between the sending and receiving voltages. Illustrating this relationship shows, that the graph reaches its maximum at $\delta = 90°$, beyond which power decreases even if δ increases (Fig. 6.29):

The peak of the curve represents the maximum power transfer limit. If the operating point is pushed beyond this peak due to a disturbance, the system might not recover, resulting in a loss of synchronism.

Expanding the considerations to a whole network, one has to take into account the behaviour of the feeding synchronous generators, too. When a generator is connected to a line, its load-depending pole wheel angle θ is to be included. Therefore, the transferable active power is limited by the expression sin (θ+δ).

To prevent transient instability, system operators enforce limits on the permissible angular deviation. In practice, the upper boundary for δ is around 30°, ensuring adequate margin to accommodate disturbances which can cause synchronous generator rotor angle to swing widely, potentially crossing the maximum stable angle of 90°. For this reason, the operating value of δ is kept safely away from the maximum

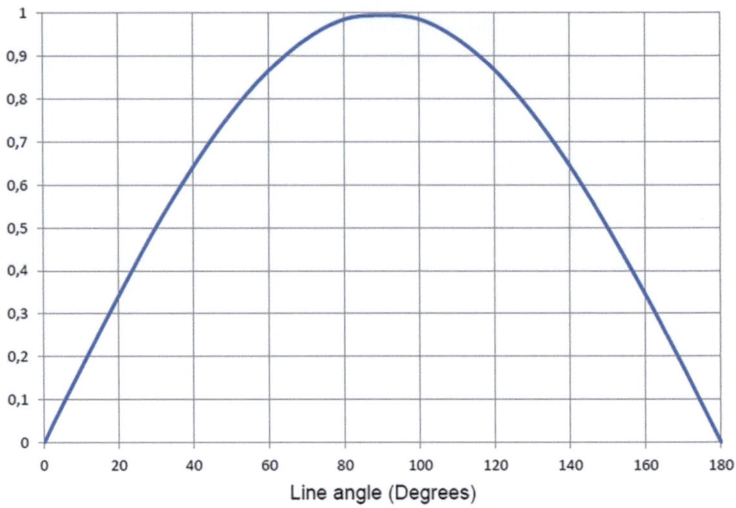

Fig. 6.29 Line angle δ and associated sine-value

angle. This angular restriction inherently limits the maximum transmittable power over long AC lines and in particular underground cables (transient stability).

A common technique to improve the angular stability and transmission capacity of long AC lines is the use of series compensation. By installing series capacitors, the effective inductive reactance X of the line is reduced. This reduction leads to a smaller line angle δ for the same transmitted power, effectively shifting the operating point farther from the critical 90° boundary and allowing higher power transfer.

Other countermeasures include:

- Power System Stabilizers (PSS) in generator excitation systems
- Flexible AC Transmission Systems (FACTS) devices like SVCs and STATCOMs
- Controlled switching of transmission elements.

These techniques enhance both steady-state and transient stability.

Calculation Example 17

Consider a (lossless) 380 kV overhead line with a typical length of 400 km ($R' = 0$, $L' = 0.81$ mH/km and $C_B' = 14$ nF/km (Table 6.1). Lies the line angle within a given limit of 30° and what can be done to reduce its value?

Solution

According to Eq. 6.30, the line angle is 0.1057 rad/100 km or 6.06°/100 km, thus it has a total value of 24.3° and it lies within the safe operating limit of 30°. However, regarding the transient stability, if the load increases or a disturbance occurs pushing the line angle beyond 45°, installation of a series capacitance is recommended.

6.8 Integration of DC Lines into Existing AC Networks

As already mentioned, the first power supply networks were built using direct current (DC) technology, but were later abandoned in favour of alternating current (AC) or three-phase systems. However, with today's advancements in power electronics, it is entirely feasible to develop DC systems that are functionally equivalent to the existing three-phase infrastructure.

Rectifiers based on semiconductor components can now convert an AC voltage of virtually any magnitude and frequency into a practically ripple-free DC voltage. Conversely, *inverters* can generate an almost perfectly sinusoidal AC voltage of any desired amplitude and frequency from a DC source.

In a similar way, *AC-AC converters* can produce an AC voltage with different amplitude and frequency from an existing AC supply. (The applications discussed for AC voltages also apply, of course, to three-phase systems.) Finally, *DC-DC converters* (also named controlers) make it possible to convert a DC voltage to a higher or lower level as required.

The various types of voltage and frequency conversion, along with the corresponding converter units, are summarized in Fig. 6.30.

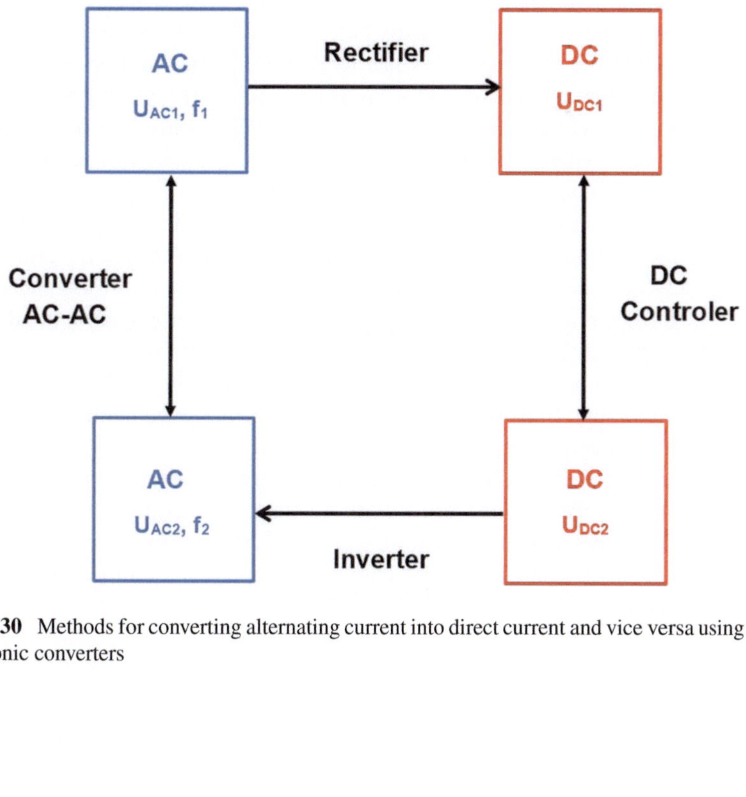

Fig. 6.30 Methods for converting alternating current into direct current and vice versa using power electronic converters

All the transformations mentioned above are now carried out with low losses, often supported by active filters to suppress harmonics as needed [27].

The main arguments that once justified the preference for three-phase and single-phase AC systems—such as their easy voltage transformation, the simple and robust construction of synchronous and asynchronous machines, and, not least, the ability to provide two voltage levels in low-voltage systems (via delta and star configurations)—are therefore of limited importance today.

However, since conventional AC systems still form the global backbone of power supply infrastructure, and nearly all electrical equipment is designed for them, a comprehensive and highly complex transition to DC networks appears unimaginable under current conditions.

However, within the existing alternating current infrastructure, a wide range of modern technologies—such as photovoltaic systems, battery storage units, charging stations for electric vehicles, fuel cells, and various components of the chemical industry (e.g., electrolysis equipment)—must be operated using converters to switch between direct and alternating current, depending on the application. The same applies to DC railways and trolleybuses.

Moreover, virtually all information and communication technology (ICT) equipment operates on direct current, and the necessary conversion from AC to DC continuously generates significant amounts of avoidable waste heat. As a result, data centres and large server facilities have already begun to implement internal direct current networks to improve energy efficiency and reduce conversion losses.

Deregulation in the 1990s significantly accelerated the trend toward cross-border power trading. On the one hand, this has facilitated mutual support with reserve power, but on the other hand, it has led to high utilization of the existing transmission networks (Chap. 7). A situation common to many countries illustrates the challenge: various parts of the network have already reached their capacity limits, particularly for the large-scale transmission of wind energy. If a line becomes overloaded or fails due to a fault, this can impact a larger section of the grid, often causing neighbouring lines to become overloaded as well. The typical response is to expand the grid, either by adding new lines or upgrading existing connections. However, if this expansion is done using conventional AC technology, faults can still propagate into adjacent network areas, potentially causing a domino effect of further failures.

This vulnerability is one of the key reasons for considering DC technology for new extra-high voltage transmission links, as it offers improved controllability and fault isolation compared to traditional AC systems.

In contrast, high-voltage direct current (HVDC) transmission is not affected by line angles, sudden load changes, or network frequency fluctuations (Chap. 1). This makes HVDC systems inherently more stable, as disturbances in one part of the grid do not propagate to other areas, thereby enhancing the overall quality and reliability of the power supply.

Furthermore, HVDC systems do not contribute to increased short-circuit power, since they do not transmit reactive power. As a result, when HVDC lines are integrated into existing grids, it is not necessarily required to upgrade or replace existing circuit breakers, busbars, or other substation components.

In addition to these advantages, HVDC systems offer further benefits, including savings on conductor and insulation material, lower transmission losses, and reduced corridor widths.

This presents significant opportunities for the use of direct current technology. The longer the transmission distance, the less dominant the cost of the *converter stations* becomes in the overall project. Today, transmission capacities of over 7000 MW per converter station are achievable over distances of several thousand kilometers, for example, in the transport of hydropower in China and North and South America.

Over the past fifty years, numerous HVDC transmission systems have been successfully integrated into existing three-phase AC networks around the world.

In many cases, this technology is also indispensable—for example, in the connection of remote offshore wind farms or the power supply of islands without local generation. Similarly, HVDC is essential for the interconnection of AC networks that operate at different frequencies or with different frequency control methods. Such configurations are referred to as *Back-to-Back* links.

Examples of Back-to-Back HVDC links can be found in the USA and its neighbouring countries, in Japan, between Brazil and Paraguay (linking 50 and 60 Hz systems), as well as in various regions of the Middle East.

The long-distance transmission of bulk power using direct current technology began its success story in the 1970s, driven by the development of high-performance semiconductor devices. In these systems—successfully in use since then—the conversion between alternating current (AC) and direct current (DC) in the converter stations is carried out using *thyristors.*

The commutation processes in the semiconductor valves—that is, the sequential takeover of current by the three AC phases—are triggered by the connected AC network; this technology is known as *Line Commutated Converter* (LCC).

To ensure proper operation, both connected AC networks must have a short-circuit power that is at least twice as high as the DC transmission power [28, 29].

LCC converters also draw reactive power from the AC network, which can be supplied by load-dependent, switchable capacitor banks. Additionally, to suppress harmonics caused by thyristor switching, complex harmonic filter circuits are required.

Since the 1990s, a new converter technology has been introduced that uses *Insulated Gate Bipolar Transistors* (IGBTs), the *Voltage Source Converter* (VSC) [28, 30, 31].

To differentiate between the two technologies, they are often described as "classic" (LCC) and "modern" (VSC). Although VSC technology does not yet support power ratings as high as those achievable with thyristor-based systems, its modular design allows for flexible scaling of both voltage level and transmitted power. This makes it easier to customize the converter design to meet specific project requirements.

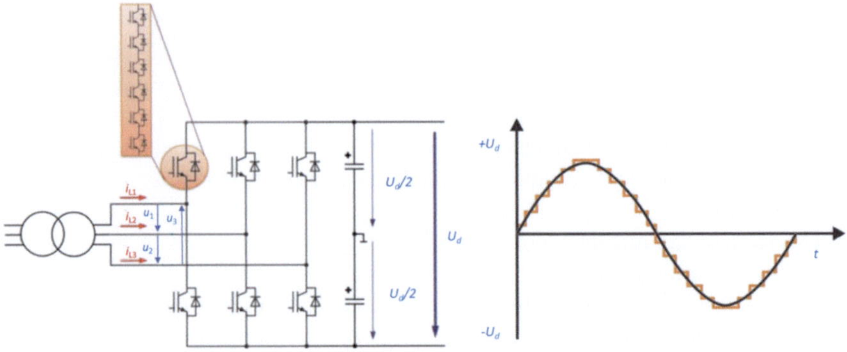

Fig. 6.31 Basic circuit diagram of a multilevel converter and the resulting sine wave on the AC side, generated through pulse-width modulation [31]

Unlike thyristors, IGBTs can be both turned on and turned off at will. One major advantage of this capability is that VSC converters can generate or absorb reactive power, regardless of the direction or magnitude of the active power flow. This key feature enables important functions such as the black start of a subnetwork—i.e., the restoration of power in an isolated area that has no local generation capability.

Today, *self-controlled multilevel converters* (MMC) are commonly employed in VSC technology, where a large number of identical submodules operate under independent control, enabling precise voltage shaping and improved scalability (Fig. 6.31).

At high switching frequencies in the range of a few kHz, the generated voltage achieves a very close approximation to the desired sinusoidal waveform, with little or no filtering effort required. During power reversal, the current direction remains unchanged, allowing the use of any cable types (Chap. 5). However, based on the current state of technology, converter losses in VSC systems are slightly higher compared to those in line-commutated (LCC) systems.

Converter stations include not only the actual converters, but also transformers, as well as switching and protection equipment. Depending on the system design, harmonic filters may also be integrated.

Due to the high voltages and the nature of direct current, it is not possible to interrupt DC currents using conventional circuit breakers. As a result, switching operations are typically carried out on the AC side, or by utilizing the control capabilities of the converter valves (Fig. 6.32).

In high-voltage direct current transmission, various systems have emerged, which are divided into *monopole* and *bipole* technologies. *Monopole* in this case means that each converter station contains only one converter unit with one plus and one

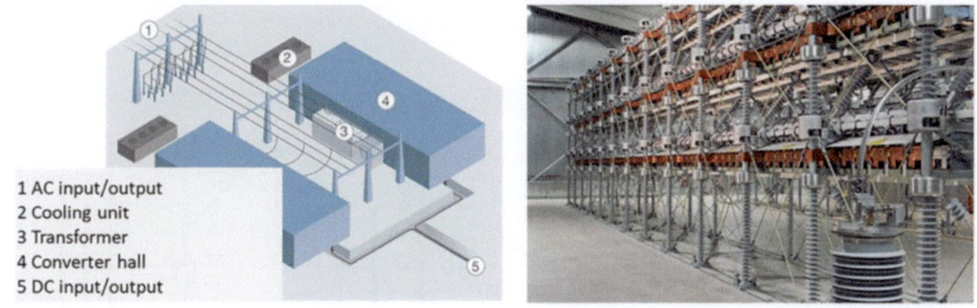

Fig. 6.32 Components of a converter station (*left*) and interior view of a valve hall (*right*) [31]

minus pole. Usually, the minus pole is grounded. The associated transmission line (cable or overhead line) contains two conductors, with the minus pole (as return conductor) grounded at both ends of the line. In older systems, this return conductor was sometimes omitted, with the earth taking over its function. However, the effort for the two earth electrodes is considerable, as otherwise dangerously high step voltages could develop.

In *bipole* systems, each converter is divided into two units in series. Here, the mid-point between the two units is grounded (Fig. 6.33). This design has the advantage that the system can continue to operate at half transmission power in the event of a malfunction or maintenance of one unit, if there is a line connection (possibly also via the ground) between the two centers. This option, which only requires an additional conductor with little insulation effort, is recommended for redundancy reasons.

Figure 6.33 schematically shows the structure of a commonly used DC transmission system today, with the possibility of using a third conductor as reserve.

In the AC networks of Western Europe, for example, it is unlikely that voltage levels above 380 kV will be introduced soon. Therefore, the current vision is to gradually build a meshed DC network that is superimposed on the existing extra-high voltage AC system.

With advances in direct current circuit breaker technology, it has now become feasible to implement *multi-terminal systems* instead of the previously common point-to-point connections. In such systems, three or more converters are interconnected via HVDC lines, allowing for the reliable, low-loss, and therefore economical exchange of bulk power over long distances between multiple partners.

This type of grid architecture—capable of operating at voltages of ±500 kV and above—is commonly referred to as an *overlay network*.

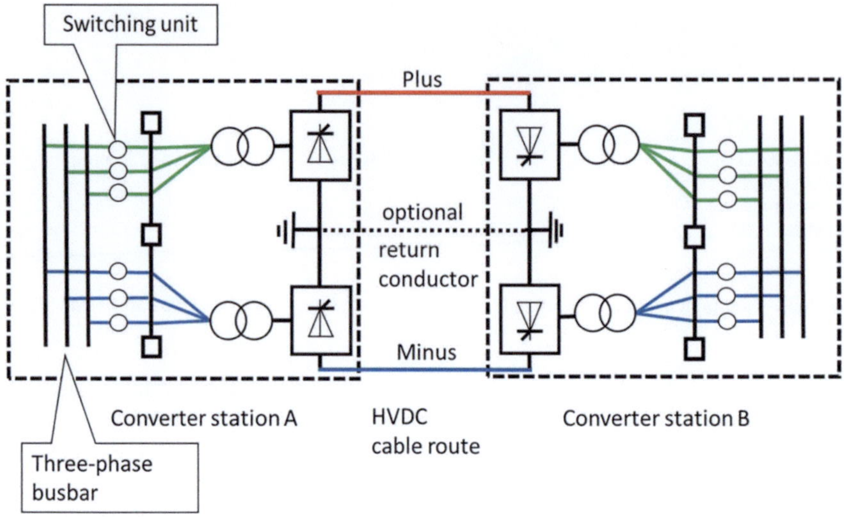

Fig. 6.33 Bipole direct current system including an associated line section, optionally with return conductor at ground potential

Application Example 5

Direct current connection through the mountain range "Pyrenees" between Spain and France. This DC cable line connects the substations Santa Llogaia in Spain and Baixas (near Perpignan) in France. It is 64.5 km long, the cables are mostly laid in trenches, and the Pyrenees are crossed with an 8.5 km long tunnel with a diameter of 3.5 m. The two affected transmission network operators jointly founded the company "Inelfe", which planned and carried out the project; it is an EU project named "France-Espagne-Portugal". The project was the first of its kind due to its length and the use of direct current technology.

The line has approximately doubled the transport capacity between Spain and France. It also strengthens the network of the Girona region and the operation of the high-speed train then under construction. The line is planned in the final expansion for two DC systems (4 XLPE insulated cables ±320 kV Cu, 500 mm^2 with aluminum sheath). The intended transmission power is 2 × 1000 MW. In the initial expansion, one system was installed. The single conductor cables were manufactured in individual lengths of 2000 m (Fig. 6.34).

Fig. 6.34 Direct current connection ±320 kV through the Pyrenees. *Left:* Tunnel drilling machine for the cables, *right:* Converter station (*Source* Inelfe)

6.9 Cable and Overhead Line Shares in Transmission Networks

The cable share in the high and extra-high voltage range of the European ENTSO-E network, for example, still plays a subordinate role to this day. This AC network comprises approximately 160000 km at the 380 kV le and around 115000 km at the 220 kV level. Currently, the cable portion accounts for about 1%, but it is expected to increase significantly, particularly in the form of submarine cables—due to the expansion of offshore wind power—and through the construction of high-capacity underground cable connections.

At the 110 kV AC level in Germany, the cable share is already around 8%. As explained elsewhere, many projects associated with the energy transition are being implemented using cable technology—and to a lesser extent, gas-insulated lines—largely due to their higher public acceptance.

In route sections where the conversion of existing overhead lines into hybrid lines (i.e., AC and DC circuits on a common structure) is feasible, HVDC systems can also be applied at the highest voltage levels (Chap. 4).

The different transmission technologies each involve different levels of environmental impact during the construction phase and exhibit distinct technical characteristics during later operation.

Table 6.2 summarizes the most important characteristics of the various extra-high voltage three-phase transmission technologies. Depending on the specific conditions of each project, deviations from the values presented may occur in practice.

Table 6.2 Key characteristics of AC overhead lines, cables, and gas-insulated lines in extra-high voltage networks. The numerical values provided are to be understood as reference values and may vary depending on specific project conditions

Aspect	Overhead line	Cable	Gas insulated line
Transmission capacity	1800 MVA/system	1 250 MVA/system	Up to 3000 MVA/ system
Availability	Failures due to atmospheric influences, but short repair times	Failures rare, but longer repair times	No reliable data available
Intervention into landscape	Limited in time for access roads and for rope pulling, earth works only in the area of towers	Limited in time for access roads, permanently for joint buildings and possibly for transition installations	Same as with cables, but only short lines verified up to now
Limitation of land use	Essentially in the tower area	No plants with deep roots, no dredging works. Space for compensation-, joint-, and transition units if needed	Same as with cables, but no compensation units
Security risk	Rope break, falling ice load	No risk in case of XLPE cables	Gas leak, but very low risk
Maintenance effort	Tower painting, rope or insulator changing	Check of joint area	No reliable data available
Lifetime	80 years or more	40–50 years	Expected same as with cables
Agreement of property owner	Necessary	Necessary	Necessary
Public consent	Necessary	Necessary	Necessary
Electric field	Limited to closer surrounding	No field outside of cable	No field outside of outer tube
Magnetic field	Limited to closer surrounding	Limited to closer surrounding	Negligible outside of outer tube
Noise generation	Corona, depending on weather conditions	No noises	No noises
Radio interference	Limited to closer surrounding	Not measurable	Not measurable

A comparison of basic properties of three-phase and direct current systems is given in Table 6.3.

Table 6.3 Comparison of basic characteristics of three-phase and direct current transmission systems

Alternative current system	Direct current system
Proven technology, standardized, different voltages easily to provide, generation and consumption units easy to integrate	Generation and consumption rather with low voltages or with AC, but converters required in this case. Transmission losses lower compared to AC systems
Larger distances problems with stability and easier with overhead lines, for cables compensation units needed	Any distances with overhead lines as well as with cables, no stability problems
Proven beaker technology for normal operation conditions as well as for short circuits	Breaker technology still well new
Meshed grids possible without problems	So far point-to-point transmission, multi-terminal operation possible (actually up to three converters)
Load flow according to impedance conditions. Otherwise use of phase shifter transformers or power electronics (FACTS)	Load flow control with the help of converter stations
Generation or consumption of reactive power depending on load situation	Generation of reactive power only possible together with VSC Technology or with the help of filters
Generally no filtering of harmonics needed	Filtering of harmonics at least needed together with LCC technology
Providing of short-circuit power for grid stabilizing in case of a fault	No grid stabilizing in case of a fault
No restriction concerning materials for overhead lines and cables, many years of experience	(Still) restrictions concerning space charges using plastic cables together with LCC converters, longer creepage distances due to deposits on insulators
Magnetizable materials close to lines can lead to additional losses (proximity effect)	No magnetic influences, higher limit values concerning flux density

6.10 Cost Comparison: Overhead Line Versus Underground Cable

The costs of overhead lines and underground cables differ significantly. Key cost components include not only investment costs, but also operating and maintenance expenses, as well as dismantling costs at the end of the system's service life.

As the share of underground cables in high and extra-high voltage networks increases, maintenance costs—which include statistically necessary and potentially lengthy repairs—must be considered in long-term planning.

The largest portion of total costs typically comes from line construction, which is heavily influenced by the foundation type and the feasibility of conductor stringing in the case of overhead lines. For example, foundations with separated footings on flat terrain and favorable soil conditions are much cheaper than those required in rocky or mountainous areas. In poor soil conditions, more complex foundations may be needed, such as rigid plate or block foundations, or driven/bored piles.

Likewise, significant cost differences arise depending on whether conventional tower erection and stringing methods can be used—where ample space allows—or whether helicopters, cranes, or ropeways must be employed in inaccessible terrain.

The cost differences are even more pronounced for underground cables. Here, construction costs largely depend on both the cable system and the installation method. Techniques range from relatively low-cost options like plowing, to more expensive solutions such as open trench installation, or trenchless pipe jacking through rock or tunnels.

The operating costs of overhead lines and cables differ mainly due to transmission losses. Cables, thanks to their larger conductor cross-sections and lower electrical resistance, generally exhibit lower losses, depending on the specific operating conditions.

Considering all these factors, it becomes clear that generalized cost statements are not meaningful. Instead, cost ranges must be specified, and these can vary by 100% or more depending on technical parameters, site conditions, and project requirements. Moreover, many cost elements are subject to market fluctuations. For instance:

- Loss-related costs depend on energy generation costs.
- Overhead line costs depend on the prices of towers, insulators, and conductor materials.
- Cable costs are influenced by the prices of copper, aluminum, and plastic.

All of these components are subject to variable market prices, further complicating direct comparisons.

Figure 6.35 illustrates the breakdown of total costs for high and extra-high voltage lines, categorized by different cost groups.

The investment costs include not only the expenses for materials and construction, but also those for planning, and often extensive evaluation and approval procedures, as well as numerous environmental impact assessments, including any required

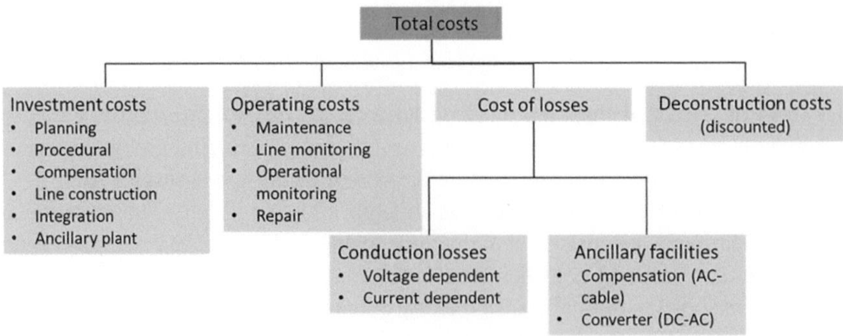

Fig. 6.35 Cost structure of overhead lines and underground cables at the high and extra-high voltage levels

compensation and replacement measures. Also included are the costs of auxiliary infrastructure at the respective terminal points of a line, such as compensation equipment for longer AC cable routes or converter stations for HVDC lines.

In cable construction projects, the dominant cost factor is often the complex civil engineering work, which largely depends on soil conditions and the number and type of crossing structures (e.g., roads, rivers, or other infrastructure) that must be tunneled under or crossed using trenchless methods. Also relevant are the transport costs for the large and heavy cable drums, as well as the costs for restoring surfaces disturbed by the construction activities.

The ongoing operating costs mainly consist of:

- Transmission losses across the entire system, and
- Maintenance and repair costs.

Transmission losses depend on several factors, including:

- The length of the line,
- The power transmitted over time (i.e., load factor, see Chap. 5),
- And the type of transmission technology used.

For cables, apart from occasional inspection of joints and cross-bonding equipment, routine visual inspections and keeping the route clear are typically sufficient.

For overhead lines, additional measures are required, such as:

- Renewal of corrosion protection on towers after approximately 30–45 years (depending on local pollution levels), and
- Replacement of conductors after around 60–80 years of service life.

For HVDC systems, the maintenance and repair costs for the converter stations must also be considered—these can be substantial.

6.10.1 Cost Comparison at the High Voltage Level

An example from Germany illustrates that new lines built on new routes at the high voltage level (up to and including 110 kV) must be implemented as underground cables—provided that their costs do not exceed 2.75 times those of a technically comparable overhead line, in the respective application.

At the medium voltage level, however, both new installations and replacements of existing lines are predominantly carried out as underground cable systems, regardless of the cost factor.

At high voltage levels, line lengths often extend over several tens of kilometers, where the reactive power characteristics of cables only slightly influence transmission capability. From a technical standpoint, there are few restrictions. Therefore, the main challenge lies in the generally higher total costs associated with cable systems.

However, in networks with Petersen coils (Sect. 6.7.5), the integration of cables is limited by the need to stay within the self-extinction threshold for arc-earth faults.

As previously discussed, due to their up to 20 times higher conductor-to-earth capacitance compared to overhead lines, cables reach this threshold much more quickly. In such cases, the additional costs for measures such as subdividing the network into smaller subnetworks or converting to low-resistance neutral grounding must be considered.

To perform a meaningful cost comparison between overhead lines and underground cables, project-specific factors must be considered. This is particularly important because routing, terrain, and line lengths usually differ between the two technologies.

For underground cables, key cost drivers include:

- Soil conditions, and
- The number and type of infrastructure crossings (e.g. roads, railways, waterways) [32].

Although, as mentioned earlier, total costs are subject to significant variation due to numerous influencing factors, reference values are provided as a basis for orientation.

For 110 kV lines, average values are as follows:

- Approximately € 0.5 million per kilometer for a double-circuit overhead line
- Approximately € 1.5 million per kilometer for two parallel-laid underground cable systems of comparable capacity (see Table 6.4).

According to [33], wider cost ranges may apply depending on project conditions:

- For overhead lines, up to around € 0.8 million per kilometer
- For cable systems, up to approximately € 2.5 million per kilometer.

Costs below the indicated lower limits may be achievable in favorable conditions, while significantly higher total costs may occur in particularly complex or constrained projects.

Table 6.4 Reference values for total costs per kilometer of three-phase overhead lines and underground cables at the 110 kV level in Germany (*Sources* Netze BW, [32, 34])

Type of line	Specification	Cross section (per conductor)	Possible power (MVA)	Total costs (million €)		
				From	To	Average
2 Systems OHL (1)	3x1x264-AL1/34-ST1A	264 Al	240	0.4	0.6	0.5
2 Systems UGC (2)	3xN2xS(FL)2Y 1x630RM/50	630 Cu	250	1.0	2.0	1.5

6.10.2 Cost Comparison at the Extra-High Voltage Level

At the extra-high voltage level (typically 380 kV in many countries), the cost structure of different transmission technologies becomes significantly more complex. In this context, not only pure overhead line or underground cable connections must be considered, but also mixed configurations, such as partial undergrounding, and the integration of HVDC systems.

For overhead lines, the focus is not only on new constructions or replacements of existing lines, but also on capacity upgrades, such as:

- Re-conductoring with high-temperature low-sag (HTLS) conductors, or
- Replacing conductors with larger cross-sections to increase current-carrying capacity.

Cost assumptions, such as those in the German Network Development Plan Electricity 2035 (status 2021) [35], are widely used as a reference for network expansion planning. However, these figures represent approximate values and may vary considerably—both upward and downward—depending on project-specific conditions.

For long AC transmission lines, reactive power compensation systems must be included in the cost considerations. In the case of intermediate undergrounding, even short cable segments can become disproportionately expensive due to the need for transition stations at both ends of the cable.

The same principle applies to HVDC cable connections, which require complex and costly converter stations at each terminal to integrate into the existing three-phase AC network.

In both AC and DC applications, the total cost of cables—depending on transmission power and line length—can range from three to nine times the cost of an equivalent overhead line. In the case of AC cables, the additional requirement for reactive power compensation further increases both investment and operating (loss-related) costs, especially over long distances (Fig. 6.36).

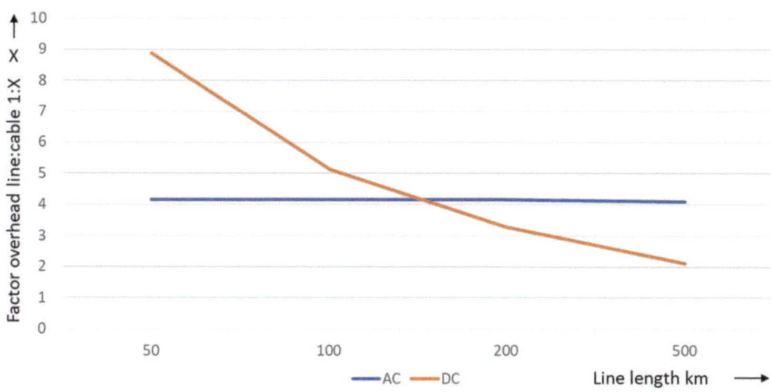

Fig. 6.36 Total cost ratio of AC cables and DC cables compared to an overhead line in the extra-high voltage range for a transmission power of 2000 MW, as a function of line length (based on [18])

As a result, the break-even distance at which an HVDC cable system becomes more economical than an AC cable system is typically below 200 km [34].

Table 6.5 provides reference values for the total costs of various line components at the extra-high voltage level. These figures are based on initial cost estimates and expert assessments prepared for multiple line construction projects in recent years. For each component, a range is presented, including the lowest costs, observed under favorable routing conditions, and the highest costs, incurred under challenging technical or environmental conditions. Additionally, the arithmetic mean is provided as a general point of reference.

It must be emphasized that, particularly at this voltage level, these are only approximate indicative values. They primarily serve as a basis for estimating total network expansion costs and should not be interpreted as fixed prices. The actual costs are heavily dependent on design type, topography, and soil-related boundary conditions, among other factors. Accordingly, a wide cost range must be considered.

Under particularly difficult installation conditions, for example, the total costs of cable systems may reach up to five times the average values listed in Table 6.5 [34].

In cost comparisons, all expenses incurred prior to and during the construction of a transmission line are considered at their full actual value. By contrast, the ongoing costs—such as energy losses, maintenance, and eventual dismantling at the end of the system's service life—are typically discounted using the applicable long-term capital interest rate at the time the investment is made.

For overhead lines, a typical service life of 80 years is assumed. For underground cables, particularly XLPE-insulated cables, where long-term operational experience is still limited, a service life of approximately 40 years is generally used.

Although overhead lines are expected to last twice as long, most comparative cost assessments are still based on a 40-year analysis period. For critical infrastructure projects that are deemed to be of permanent necessity, a longer observation period of 80 years is appropriate. In such cases, the replacement of underground cables after 40 years must be factored in. Similarly, for overhead lines, the renewal of corrosion protection and re-conductoring after roughly the same period should also be considered.

Table 6.5 Reference values for total costs per kilometer of overhead lines and underground cables in the extra-high voltage range for AC and DC systems in Germany (*Sources* 1—Netze BW, 2—[34], 3—[35], 4—ACER)

Type of line		Specification	Cross section per conductor	Possible power	Total costs in million €		
					Min.	Max.	Average
AC	Double overhead line (1), (2), (4)	3x4x264-AL1/ 34-ST1A	1056 Al	1790 MVA	1.0	2.6	**1.8**
	Two cable systems (1), (2), (4)	3xN2XS(FL)2Y 1x2500 RMS/ 250	2500 Cu	1510 MVA	4.0	11.0	**7.5**
	Rewiring per circuit (3)	–	–	–	0.5		
	Rewiring to HTLS per circuit (3)	–	–	–	0.7		
	Double overhead line, new replacement, dismantling the existing line (3)	–	–	–	2.8		
	New double overhead line (3)	–	–	–	2.5		
	Two cable systems as partial wiring with cable transitions without compensation (3)	–	–	–	11.5		
	Compensation coil without switch panel (3)	–	–	–	0.02 per Mvar		
DC	New double overhead line (3)	–	–	4000 MW	20		
	Underground cables, one system (3)	–	–	2000 MW	6.5		
	Converter station(3)	–	–	–	0.3 per MW		

References

1. Flosdorff R, Hilgarth G (2008) Elektrische Energieverteilung, Vieweg-Teubner
2. Fricke K (1992) in Kabel und Freileitungen in überregionalen Versorgungsnetzen, Expert-Verlag
3. Oswald B (2007) 380 kV Salzburgleitung, Energie-Control Wien
4. Grünbaum M, Noroozian M, Thorvaldsson B (1999) FACTS—leistungsfähige Systeme zur flexiblen Energieübertragung, ABB Technik 5/1999
5. Oeding D, Oswald B (2016) Elektrische Kraftwerke und Netzc, Springer
6. Kießling F, Nefzger P, Kaintzyk U (2003) Freileitungen. Springer
7. Anon, Influence of Line Configuration on Environment Impacts of Electrical Origin, CIGRE Paris 2005
8. Mimos E et al (2016) Electric and magnetic fields produced by 400 kV double circuit overhead lines, CIGRE Paris
9. Brakelmann H, Jarass L (2019) Erdkabel für den Netzausbau, BoD Norderstedt
10. Anon, Elektromagnetische Felder. Informationsbroschüre, Bundesamt für Strahlenschutz 2021
11. Planete M et al (2016) 50–60 Hz magnetic fields and cancer, forty Years of Research, CIGRE Paris
12. Planete M et al (2017) Living with electric and magnetic fields, CIGRE Paris
13. Wertheimer E et al (1979) Electrical wiring configurations and childhood cancer. Am J Epidem
14. Thériault M et al (1994) Cancer risks associated with occupational exposure to magnetic fields. Am J Epidem
15. Anon, Grundlagen Grenzwerte—Internationale Grenzwerte für HF/NF, Aaroina AG Strickenscheid 2022
16. Anon, Gesundheitliche Wirkungen elektrischer und magnetischer Feldervon Stromleitungen, femu 2013
17. David E, Palic M (1994) Elektrosmog—Wirkungen elektrotechnischer Felderauf den Menschen, ETZ
18. Novitskiy A (2012) Freileitungen oder Erdkabelleitungen? Eine Metastudie über die Kriterien und Ergebnisse von Netzuntersuchungen zum Netzausbau. Ilmenau
19. Marenbach R, Jäger J, Nelles D (2020) Elektrische Energietechnik: Grundlagen, Energieversorgung, Antriebe und Leistungselektronik, Springer
20. Leprich U et al (2011) Ausbau elektrischer Netze mit Kabel oder Freileitung unter besonderer Berücksichtigung der Einspeisung Erneuerbarer Energien, Saarbrücken
21. Oswald B (2006) Gutachten zur Bewertung einer alternativen Verkabelung der geplanten 110.kV-Hochspannungsfreileitungen Baumstraße-Lüstringen und Pkt. Belm-Powe, Hannover
22. International Workshop on Power Cables in the Transmission Grid, Fredericia, Denmark, (2010)
23. Redina R et al (2006) The new Turbigo-Rho 380 kV transmission line: an example of the use of underground XLPE cables in a meshed transmission grid. CIGRE Paris 2006
24. Schäfer K (2020) Netzberechnung—Verfahren zur Berechnung elektrischer Energieversorgungsnetze, Springer
25. Saadat H (2002) Power system analysis. McGraw Hill
26. Schlabbach J (2009) Elektroenergieversorgung. VDE Verlag
27. Hochspannungs-Gleichstrom-Übertragung, Grundlagen und Integration in die Stromversorgung der Zukunft, ABB Mannheim 2014
28. Sauer D (2016) Stromnetze—Technologiesteckbrief zur Analyse „Flexibilitäts-konzepte für die Stromversorgung 2050“, Nat. Akad. D. Wissenschaften Aachen
29. HVDC PLUS—Der entscheidende Schritt in die Zukunft, Siemens Erlangen 2016
30. Anon, die evolution von HVDC light, ABB Review 2017
31. Anon, High voltage direct current transmission—proven technology for power exchange, Siemens Erlangen 2011
32. Malsch M, Prinz S (2018) Neubau der 110-KV-Hochspannungsfreileitung—Abzweig Oberelsdorf, Gutachten new/Grid Erfurt

33. Hofmann L, Oswald B (2011) Wirtschaftlicher Vergleich von Kabeln und Freileitungen mit Zwischenverkabelung im 110-kV-Hochspannungs- bereich, Hannover
34. Hofmann L et al (2011) Ökologische Auswirkungen von 380-kV-Erdleitungen und HGÜ-Erdleitungen, BMU Studie Berlin
35. Anon, Netzentwicklungsplan Strom 2035, Version 2021, Bundesnetz-agentur Bonn

Chapter 7
Outlook

Summary

In the transmission of electrical energy, several parallel developments can be observed that are closely intertwined and represent a structural change in the entire energy supply, referred to as the *energy transition*, which is unprecedented since the beginning of electricity supply. The main goals of the energy transition are the reduction of CO_2 emissions and the reduction of import dependence on fossil fuels. This affects almost all countries in the world. The move away from fossil energy was decided, among other things, at the UN Climate Conference in Dubai in December 2023.

This chapter discusses anticipated development trends in power supply systems, along with innovative technologies in overhead line and cable technology.

7.1 Development Trends

In addition to the expansion of transmission grids, there is a need for a substantial reinforcement and expansion of distribution grids. Beyond their traditional supply functions, these networks will increasingly need to accommodate significant loads from the heating sector and the growing demand associated with electromobility.

At the same time, the power flow at all voltage levels will increasingly develop bidirectionally, for which extensive digitization and automation of the grids is indispensable. Predictive maintenance-though not yet widely implemented, at least in distribution grids—is expected to become established alongside conventional routine or time-based preventive maintenance. This maintenance strategy is based on the evaluation and use of digitally recorded historical equipment data and enables demand-oriented maintenance, which reduces both maintenance costs and downtime.

The system conversion to a fluctuating, CO_2-free power generation, which is not oriented towards demand as before, but towards fluctuating supply, requires a large number of flexible, primarily natural gas and later hydrogen-powered power

K. O. Papailiou et al., *High-Voltage Overhead Lines and Underground Cables*, Power Systems, https://doi.org/10.1007/978-3-032-07498-0_7

plants as control energy suppliers and at the same time as large storage capacities as possible. Both are currently not available in most power supply systems to the extent necessary in the future and must be expanded and established as quickly as possible. This results in enormous transport tasks for the transmission grids from the generation focal points of renewable energies to the load focal points and for the distribution grids, in addition to the connection of renewable generation plants, an increased orientation towards flexible backup power plants and large-scale storage in the area. In the transformation process towards the use of renewable energies, volatile generation plants, which are not oriented towards demand, but towards supply, will replace thermal power plants, which as base load power plants have very high so-called *full utilization hours*. The period of consideration is usually set at one year; hence the alternative term "annual utilization duration". It is obtained by dividing the energy generated annually by a generation plant by its nominal power. Thus, conventional thermal blocks (nuclear, coal, oil, gas) could theoretically provide over 8000 full load hours, assuming necessary revision times and repairs are disregarded and that fuel and cooling water are continuously available. The same applies to biomass power plants and river power plants with permanent inflow.

Due to the fluctuating supply, PV power plants in Germany only achieve just over 900 full load hours, and offshore wind turbines reach a maximum of 3500 h. Onshore, it is only about 1500 to a maximum of 1800 h at inland locations. It must be distinguished with the full load hours whether they are demand or supply dependent. While conventional power plants were predominantly used in a demand-oriented manner, fluctuating, supply dependent generation plants can only be used when they are available. This makes it clear that the *bridge technology* "power generation with natural gas or hydrogen" with all its geopolitical risks will be used for a very long time, as for successful storage, for example in the form of high capacity pumped storage power plants, a storage volume about a thousand times higher than the existing ones would be required in Germany alone to survive a three-week *dark doldrums* without interruption of supply [1]. The planning of the few new plants has been on hold for years because the interventions in nature and landscape associated with their construction would be so severe that they would most likely be denied permits. Apart from that, even many new pumped storage power plants could not noticeably alleviate the shortage.

As part of sector coupling surplus electricity generated from renewable sources during periods of low grid load will be converted into hydrogen. A portion of this hydrogen will be utilized directly as a primary energy source, while the remainder will be stored and later re-electrified to compensate for periods of insufficient renewable energy generation [2].

All of this is likely to lead to a completely new price structure in power supply in the future. As the cost of electricity is largely determined by the cost of primary energy, there will increasingly be a need for a shift in the construction and operating costs of generation and distribution facilities, with a high proportion of fixed costs. The use of solar and wind energy eliminates the purchase costs for primary energy.

The outlook for the future can be derived from the previous chapters. Some selected techniques will be highlighted in the following sections.

For the Area of Overhead Lines

Overhead lines are the predominant method used for transporting electrical energy worldwide. It is foreseeable that this will continue to play a central role for the foreseeable future. The energy transition is largely based on this transmission technique. The following are particularly noteworthy as modern equipment, operating resources and procedures in overhead line construction:

- Route planning for new lines to minimize environmental impacts
- Corridor design and consideration of biodiversity in new lines and existing ones
- Conductors with higher current carrying capacity for upgrading existing lines
- Ground wires with optical fibers for telecommunications and protection
- Composite insulators for greater flexibility in tower design
- A modern standardization system
- Alternative tower design for better appearance and acceptance
- Corrosion protection and coating to increase service life
- Dynamic thermal rating to increase the capacity of existing lines
- Maintenance with the help of drones, robots, and artificial intelligence.

For the Area of Cables

Cable technology has advanced rapidly in recent decades, particularly in the development and application of cables with synthetic insulation. As seen in Chap. 1, XLPE direct current (DC) cables developed from ± 10 kV to ± 640 kV within just 20 years. Further developments can be expected here.

Progress in semiconductor technology has made it technically feasible and economically viable to construct very long HVDC cable lines with minimal losses and to integrate them flexibly into the existing meshed three-phase AC grid. It seems very likely that an overlay grid with multi-terminal operation will develop based on DC cables in the foreseeable future.

The following main developments can be expected:

- The use of submarine cables will continue to be significant and will continue to develop for the transport of offshore wind energy.
- For line distances of 500 km and more, HVDC cables will increasingly be used. This is particularly expected in densely populated countries and areas where overhead line routes are not possible (e.g. submarine cables).
- Since HVDC technology is still new at the highest voltage level, operational experience will be gathered in the coming years. Particular attention should be paid to the numerous joints; the high number of which results from the maximum cable lengths of about 1000 m currently available.

7.2 Transmission Grids

The development away systems and thermonuclear power generation in power plants near load centers with the present grid structures oriented towards them, towards renewable power generation in gigantic onshore and offshore wind farms as well as extensive photovoltaic systems, requires a massive expansion of the transmission grids [3]. This paradigm shift requires powerful connections between the new generation- and the existing load centers. In addition to conventional and compact new overhead line construction, innovative methods such as monitoring (Chap. 4), the replacement of standard conductors with HTLS conductors (high temperature conductors in Chaps. 4 and 7) and load flow control (Chap. 6) are increasingly being used for grid expansion in accordance with the NOVA principle (Sect. 3.4.7). For power transmission over long distances, the low loss direct current technology was recreated. Using the newest semiconductor technologies at operating voltages above 500 kV, it allows transmitting high power over large distances without reactive power. These currently point-to-point HVDC connections are likely to develop into a mashed so-called *Overlay grid* with integrations into the existing three phase transmission grid soon.

7.3 Distribution Grids

Distribution grids play a central role in ensuring supply security and maintaining the regional balance of generation and demand. In addition to the expansion of the grid, there is a particular need for an extension of the information technology of the energy system transitioning it toward a fully integrated *Smart Grid*. The current situation in distribution grids is characterized by numerous, partially controllable, decentralized generation plants and by the lack of a continuous communication technology. A complete change to a mashed system offers a number of new and important functionalities. For this, the term *Area Power Plant* is used in some literature [4].

The grid expansion in this segment for new lines is expected to be implemented primarily with underground cables up to the 110 kV level and partially by overhead lines where appropriate.

Regardless of the voltage level, the replacement of existing lines and the retention in the existing corridor, possibly also on sections, is also an option. At the same time, the operators of the distribution grids, like those in the transmission grid level, are striving to increase transmission capacities by monitoring existing lines and replacing existing conductors with HTLS conductors. For the inner-city transmission of high power, the use of superconducting cables may also be considered in the future, as is currently being in service and tested in pilot projects for higher power (Sect. 7.5.5).

7.4 Sector Coupling, Decarbonization and Sustainability

7.4.1 *Sector Coupling*

Sector coupling, i.e., the integration of electricity, natural gas/hydrogen, heat, industry, and mobility sectors into a holistic supply structure, is an important brick for the success of the energy transition. It involves the comprehensive connection of all generation and consumption sectors towards an integrated energy system based on renewable energy generation by 2050. Sector coupling is much more than just the replacement of CO_2-emitting conventional power generation plants (Fig. 7.1, gray and crossed out) by renewable ones and the accumulation of all line expansions in the context of future grid expansion. At the same time, it underlines the importance of grid expansion as the backbone of the energy transition. The goal is to expand renewable power generation capacities beyond the existing grid load in order to then generate hydrogen in large electrolysis plants with the expected surplus in power generation.

This option, referred to as *Power-to-Gas*, envisages that the hydrogen generated from renewable sources either directly replaces conventional energy carriers in industrial processes or, after a methanation process, is fed into the natural gas grid and

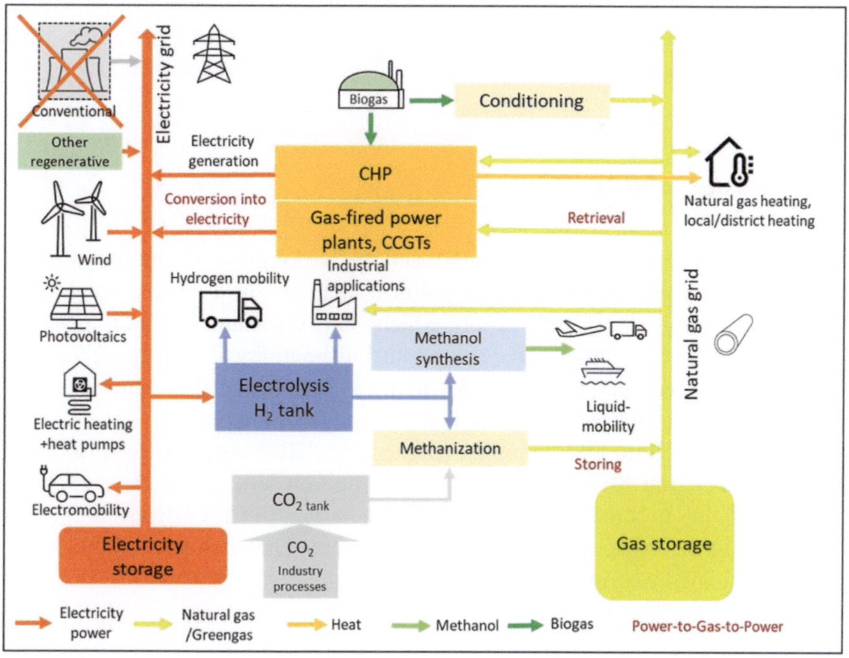

Fig. 7.1 Schematic presentation of sector coupling after the discontinuation of power-feed-in from fossil and nuclear power plants by 2050 in central Europa

thus used as so-called *Greengas* or stored for re-electrification. While the possibilities for electricity storage are very limited, there are sufficient storage capacities in the natural gas grid to withstand a multi-week period of no wind or sun. In this case, we speak of *Power-to-Gas-to-Power*. However, the precondition is sufficiently large gas-based power plant capacities for re-electrification. This path is currently associated with an overall efficiency of less than 50% from renewable primary generation through the double conversion to re-feed into the grid [3].

Parts of the generated hydrogen can be converted into liquid fuel via various processes—such as methanol synthesis—to replace liquid fossil fuels in heavy-duty and aviation. This approach is known as *Power-to-Liquid* [5].

7.4.2 Decarbonization

In this way, the multi-talents electricity and hydrogen are intended to enable a far-reaching decarbonization in all sectors while simultaneously reducing the dependence on primary energy imports.

The needed shift of energy production from fossil to alternative taking the situation in 2018 as base and taking 2050 as goal is presented in Fig. 7.2. The reduction in the use of oil, natural gas and coal can be seen clearly, as well as the increase for electricity, modern biomass and hydrogen. Even though these values are an estimation, the tendency goes clearly in this direction—provided the political and social outlook supports this. To reach this, large developments in the grid become necessary.

7.4.3 Sustainability

One of the explanations of *sustainability* and the background can be found in [6]: "The term sustainability has assumed political contours, in particular through the Brundtland Commission's 1987 report, which defined sustainable development as one that meets current needs without compromising the ability of future generations to meet their own needs and, called for growth that integrates social and environmental aspects into economic considerations, both spatially and temporally."

Heads of state from 193 countries adopted the "2030 Agenda for Sustainable Development" at the 2015 United Nations "Summit on Sustainable Development". Sustainability has become an important aspect in today´s life, for private households, industry, utilities, transportation, nearly everywhere. The situation for the energy sector is explained well in [7].

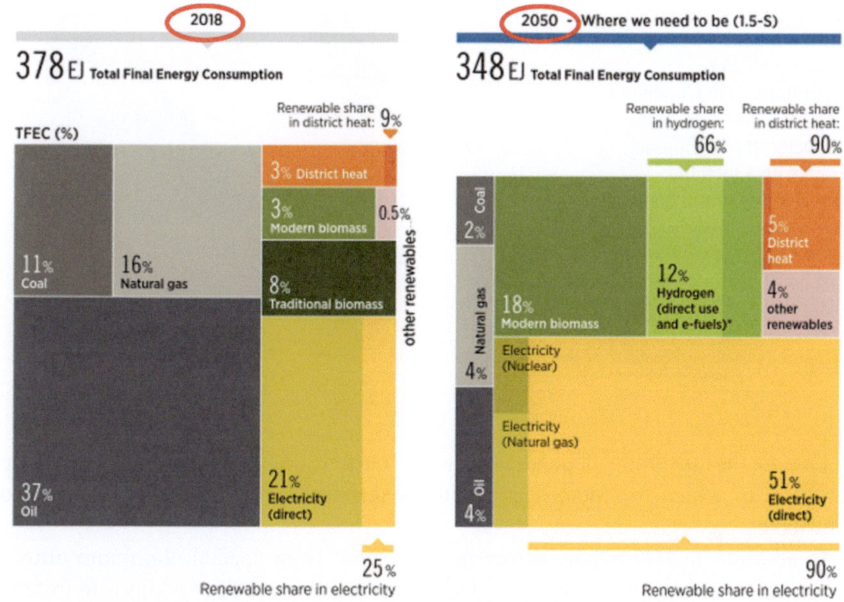

Fig. 7.2 Shift of energy production from fossil to alternative 2018 to 2050. (*Source* CIGRE Session Paris 2022 Presentation: Antonio ILICETO, Terna, IT; *source* World Energy Transition Outlook, March 2021 preliminary)

7.5 Technical Developments in Building the Grid

To achieve efficient electricity transmission within the framework of the desired energy transition, a number of innovations are expected in line technology, automation, and digitization of the power grids. These have largely been presented and described in individual chapters. A summary of some prominent examples follows.

7.5.1 High Temperature Conductors for Overhead Lines

To enable existing lines to transmit higher currents and thus greater power, *High-Temperature Conductors* have been developed in recent years.

With classic conventional aluminum-steel conductors or those with aluminum alloys (Aldrey), the operating temperature is generally limited to 80 °C. In order to not increase the sag compared to conventional conductors despite higher temperatures, materials with a lower thermal expansion are used in high temperature conductors (HTLS, High Temperature Low Sag). This allows operating temperatures of 150–210 °C and even 250 °C in temporary emergency operation.

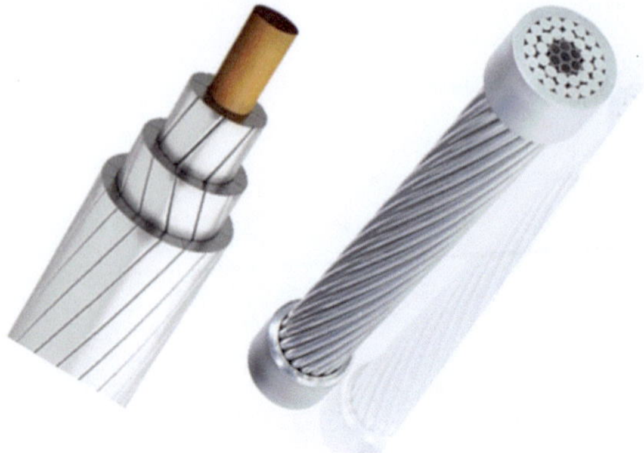

Fig. 7.3 High temperature conductors for up to over 200 °C conductor temperature. *Left*: Conductor with composite core, *right*: Conductor in conventional design with a metallic thermally highly resistant core (*Sources* CTC GLOBAL; Lumpi-Berndorf)

Due to the quadratic dependence on the current, high transmission losses are accepted, to which the increased conductor resistance at these temperatures also contributes.

In addition to HTLS conductors manufactured from special aluminum alloys, conductors that have a core made of hybrid carbon or glass fiber composite instead of a metallic core are used. Such composite cores have a very favorable strength-to-weight ratio and at the same time have a very small coefficient of thermal expansion (Fig. 7.3). Further variants of high temperature conductors and details have already been described in more detail in Sect. 4.3.9.

Since conductors of this type and the associated fittings are comparatively expensive, their use is particularly economical in those cases where the transmission capacity on an existing line can be increased, and the foundations and the tower statics can largely be left unchanged and allow an increase in the cross section of the new conductors.

In countries such as Japan, it is common practice to equip new lines with high temperature conductors, thereby incorporating provisions for future increases in transmission capacity already during construction.

7.5.2 *Compact Overhead Lines and Alternative Tower Design*

Compact lines are lines designed with reduced tower dimensions. Their importance is expected to grow in the future, particularly in scenarios where it is necessary to retain existing corridors of a line and/or to minimize the visual impact on the landscape. Also, alternative tower design with unconventional appearance predominantly uses compact tower design. The structures of compact lines can be steel lattice towers or tubular poles which can be fabricated from steel (Fig. 7.3), concrete or composite material [8].

Their spread goes hand in hand with the positive development of composite insulators. They are indispensable for the compact design with the insulating cross arms

Fig. 7.4 500 kV compact line in Indonesia with insulating cross arms made of composite insulators (*Source* Pfisterer)

increasingly used for this purpose. Today, compact lines, as shown in Fig. 7.4, can be realized with composite insulators up to 500 kV.

Today, public perception is increasingly being considered alongside technology and science in line planning and the selection of tower designs. Examples of this can be found worldwide, although such alternative solutions are often applied only to specific sections of a transmission route rather than the entire line.

Particularly striking towers are often single solutions for only one item. It should also be noted in such considerations that a new tower design, which deviates from a technically and operationally proven tower design, will be more expensive, technically more complicated, and more difficult in maintenance than conventional towers. They also have a different eco-balance and are designed with the aim of being more appealing. Special attention must be paid to the electrical and magnetic field values as well as the audible noise produced due to the different conductor arrangements. Each tower design represents an optimization focused on specific priorities or requirements. Changes to one parameter have effects on other parameters. Figure 7.5 shows this in detail. For example, if the distances between the phases are increased, the magnetic and electric field, the audible noise and radio interference are reduced. If the ground clearances of the conductors are increased, i.e. the towers are raised, the electric and magnetic field at ground level, as well as audible noise, decrease—the visibility of the line is of course increased (the latter is not entered in the table).

The consideration of aesthetics must also be considered in line planning; this is e.g. prescribed for new lines for European CENELEC countries by a standard. In the European standard EN 50341-1:2012 "Overhead electrical lines exceeding AC 1 kV-Part 1: General requirements-Common specifications" it is stated in 3.2 "Requirements of overhead lines"—3.2.1 Basic requirements: "...... *An overhead line shall also be designed, constructed, and maintained in such a way that*

PARAMETER		EF	MF	RI	AN
Phase to phase distance	↑	↑	↑	↘	↓
Conductor height above ground	↑	↓	↓	↘	↘
Number of sub-conductors (for a given total cross-section)	↑	↑	=	↓	↓
Sub-conductor spacing	↑	↗	=	↗	↗
Total conductor cross-section	↑	↗	=	↘	↘

↑ Strong increase ↓ Strong decrease = no significant effect

↗ Slight increase ↘ Slight decrease

Fig. 7.5 Relationship of the dimensioning parameters electric field (EF), magnetic field (MF), audible noise (AN), radio interference (RI) [9]

due regard is given to safety of the public, durability, robustness, maintainability, environmental considerations and appearance." Designers observe this as one of the highest premises when developing alternative towers. Examples of this will be shown below.

The tower design "Wintrack", which has been positively received by the population and is much noticed in professional circles, is used by the Dutch grid operator TenneT as a new standard. It is a double system 380 kV line and can also carry two additional 110 kV systems. The final version followed years of trials and intensive research work, also with a variant as composite towers. The built lines consist of two closely adjacent individual tubular steel poles. Each pole carries one 380 kV system on the inner side and can take additional 110 kV systems on both the outer side if necessary. The poles consist of several approx. 2.5 m high rolled metal sheet rings, which are welded. The towers can be climbed from the inside and outside and have a screwed connection with the foundation. The magnetic field was minimized due to the narrow phase arrangement. Special tools and facilities had to be developed for maintenance. Figure 7.6 shows the transition from the previous standard steel lattice tower to the new Wintrack design and the arrangement of the two electrical systems.

In 1994, alternative towers have been developed in France through a design competition by RTE. The condition was that the submitted designs must be technically feasible. 121 proposals were submitted. Figure 7.7 shows a selection of the 380 kV towers from the design competition by RTE 1994 [10].

Industrial architects develop very appealing overhead lines in many countries, here are examples of 380 kV solutions from Italy, Denmark and Great Britain (Fig. 7.10).

Fig. 7.6 *Left image*: 380 kV double circuit lines of TenneT in the Netherlands: in the foreground design "Wintrack" with tubular steel poles, in the background standard lattice steel tower; *right image*: Wintrack suspension tower with insulating crossarms

Fig. 7.7 380 kV towers from the design competition by RTE (France) 1994 [10]

It is important that the erected towers exactly follow the design specifications in order to achieve the desired visual impression.

For the planners and the executing technicians, this can be a significant challenge, as for example, Fig. 7.9 shows.

Fig. 7.8 Some of the built 380 kV towers from the design competition of RTE [10]

South of Florence in Italy runs an elegant lattice steel construction with the aim of optimally integrating it into the surrounding landscape (Design by Normen Foster). The design "eagle" has been using in Denmark over tens of kilometers with hundreds of towers. It consists predominantly of tubular metallic poles. As a variant, this tower design is used on some kilometers with bodies made of lattice steel, as these appear more "transparent" which was more advantageous in the considered area.

In Great Britain, the tower "T-Pylon" has been tested on a test line. The tower body and the crossarms consist of tubes, the arrangement of the insulators is designed in the shape of a diamond. The line is very low due to this arrangement and because

Fig. 7.9 Complicated bracings, nodes and panel points on an alternative tower, 380 kV line near Florence, Italy (*Source* A. Ernst, Vienna)

Fig. 7.10 Alternative tower design for 380 kV lines: *left* Design Norman Foster in Italy (Rendering), *middle* Tower "eagle" in Denmark, *right* Tower "T-Pylon" in Great Britain, both by Bystrup [11])

Fig. 7.11 Alternative 380 kV tower shapes, left image: assembly with mobile crane (*Source* Bystrup Design)

of the short spans. Maintenance can, as with many alternative towers, only be carried out from a mobile crane, which requires appropriate terrain and access roads. This limits the field of application [8] (Figs. 7.10 and 7.11).

There are also unconventional and artistic approaches to exploring what a transmission tower could look like. The example in Fig. 7.12 comes from the Netherlands and shows a tower that is literally "slanted". Even if only a few towers were built, it proves the willingness to be extraordinary.

Fig. 7.12 "Slanted" 380 kV double circuit tower in the Netherlands

Fig. 7.13 Study on modern tower design for lattice steel double circuit towers 380 kV [12]

A research project at the University of Arts in Linz/Danube, Austria, demonstrated that appealing tower designs are also possible using steel lattice structures. In this project, students-supported by a structural engineer and experts from the grid operator APG-were tasked with developing buildable tower concepts. Out of the many proposals submitted, four were selected as fundamentally executable and technically feasible designs (Fig. 7.13).

Towers can also work without an earth wire peak, as proven by a 380 kV tower design in Sweden. One of the three phases is mounted at the tower top via a post insulator. In the event of a lightning strike, a surge arrester reacts and discharges the overvoltage in that phase (Fig. 7.14). This allows a very compact tower design, with attention to the usual design parameters electrical and magnetic field as well as audible noise.

Composite towers made of fiber reinforced polymers (FRP) have been in use at medium voltage in many countries for years, especially because they are light and durable, and the fact that they require no wood protection treatments, unlike wooden poles. They are also considered for higher voltage levels-for 220 kV they have been used in the USA in Ormond Beach since 1997 and constructions from China are also known. Synthetic towers have the advantage that they are electrically non-conductive and light, they do not corrode, and they promise a high life expectancy. Project ideas also exist for 380 kV and would open up new possibilities for tower design (Fig. 7.15). Until then, there are still mechanical and electrical challenges to be clarified and those for maintenance. Science and industry are dealing with these aspects [12, 13].

In 2019 the European Union supported five international partners by the Innovation Program “European Composite Pylon” aiming to design and produce three test pylons made of composite material for a 400 kV double circuit line. The goal was to reduce the visual and environmental impacts. This was achieved end of 2021 when the towers have been erected in Denmark.

The towers have one level of single crossarms. They are made of composite material and have the function of insulators. This allows to reduce the size of the

Fig. 7.14 380 kV tower in Sweden without ground wire, but with surge arrester as lightning protection [29 in Chap. 2]

Fig. 7.15 *Left*: Study on a 380 kV composite tower [11], *right* 220 kV composite tower from the USA [13]

tower, as all three phases of one electric system are fixed on one crossarm. Electrical and mechanical tests in full scale mode have been carried out in laboratories.

The technical data is impressive. The static design followed these specifications [14]:

- European Standard EN50341 „Overhead electrical lines exceeding AC 1 kV" and EN50341-2-4 (NNA for Germany)
- Wind zone 2, Ice load zone E2
- 2 × 400 kV AC, triple bundle, 4400 A
- Site altitude: 100 m
- Basic mean hourly wind speed: 25 m/s (Wind Zone 2)
- Terrain category: II
- Max single span: 350 m
- Max weight span: 450 m
- Conductor: ACPR Low-Sag 451–AL1/64
- Earth Wire: ACSR 240/40
- Galloping acc. CIGRE TB 322
- Max everyday tension at 10°: 52 N/mm^2
- Max operating temp: 80°
- Lightning acc. IEEE standard 1243:1997

The shaft is constructed from a single piece of tubular steel. Multiple layers of protective coating are applied, giving it an expected lifespan of up to 80 years. During this time, no parts of the tower are anticipated to require replacement. Also, the crossarm was designed according to EN 50341-2-4 as explained above. The coordination of insulation followed IEC 60071-1 and IEC TS 60815-3, additional mechanical requirements from IEC 61462 for hollow core insulators were considered.

The foundation consists of a steel monopile which can be driven within one day, depending on the soil conditions. For the installation of the towers the complete pylons were assembled on the ground and then lifted by cranes. Depending on the environment and weather conditions, one tower can be erected in one day. The 400 kV towers are smaller (app. 30 m) than standard towers made e.g. of lattice steel (Figs. 7.16 and 7.17).

Fig. 7.16 2 × 400 kV composite towers, 30 m, span: 350 m span (*Source* "European Composite Pylon, Fast Track to Innovation 2019–2021")

Fig. 7.17 Erection of a double circuit 400 kV composite tower [14]

The technology of composite towers is discussed in Chap. 4.

Steel towers must be protected against corrosion. It is environmentally friendly and effective to coat the tower's components directly after galvanizing in the factory with water-soluble coating materials (Fig. 7.18). Coating increases the lifespan of corrosion protection. The quality is enhanced by the precisely controlled coating process carried out in the factory, which is unaffected by weather conditions. This memberwise coating process has become known as the "in-factory-DUPLEX-system". It can be carried out all year round regardless of the season and weather. The lifespan of the corrosion protection can be extended by this coating process from previously 20–25 years to 45 years and beyond [15]. The DUPLEX system is widely used by transmission line operators around the world. This coating allows transmission lines to be color-matched to specific requirements, for example, to achieve a "camouflage" effect in sensitive landscapes. Figure 7.19 shows two 380 kV towers. Clearly visible on the left in the picture is a galvanized tower in gray color. To the right stands a dark, green-coated tower that blends into the background, effectively camouflaged so that only its concrete foundations remain visible.

This camouflage line is equipped with very thick conductors (twin bundle Al/St 1288/183) coated in dark green. This has been applied in the factory of the conductor manufacturer immediately after stranding. Conductor coatings were originally developed to improve heat dissipation and thus for an increase in transmission performance. Also, with appropriate coatings, corona noises can be reduced in unfavorable weather conditions.

The coloring of transmission towers is not only used for camouflage but for another way, namely, to deliberately make towers visible in the environment. The idea behind this approach is to present transmission lines as visible symbols of modern infrastructure—much like other technical facilities—and, in some cases, even highlight them through illumination. Such solutions exist as few single solutions with the function as "eyecatchers" and can invite discussion. Especially in Scandinavian countries, this opinion is followed. Examples can be found in Finland and France (Fig. 7.20). In the case of the French tower, not only were the angles painted in color, but colorful ropes were also added later to enhance its visual impact. Interestingly, in Finland,

Fig. 7.18 *Left*: Tower with coated components, the panel points are uncoated. *Right*: Delivery of the members which were DUPLEX coated in the factory

Fig. 7.19 380 kV “camouflage line” in Tyrol/Austria, on the *left* an uncoated galvanized gray tower, to the *right* of it is a dark green coated camouflage tower only recognizable by the concrete foundation bases [15]

blue is commonly used for painted towers and other structures—likely a nod to the national flag, where blue holds symbolic significance.

Overhead line construction occasionally features extraordinary solutions, such as the so-called “invisible tower” (Fig. 7.21). The conductors are suspended from a support rope attached at two fixed points, resembling shirts hanging on a clothesline. These are custom solutions when special circumstances demand them. Two examples are mentioned here. One was suitable in South Africa for crossing a mountain ridge, as shown in Fig. 7.21. The insulators for the extra high voltage phases are integrated into

Fig. 7.20 *Left* in Finland illuminated 380 kV tower, *middle*: in France illuminated colorful 110 kV tower, *right*: in Finland 110 kV tension tower with separated phases (*Sources* Fingrid, RTE EdF)

Fig. 7.21 "Invisible towers" in South Africa: a support rope serves as a supporting element, on which the insulators and conductors are mounted

the support rope and thus provide the required phase clearances, while the conductors of the high voltage phases are tensioned as usual—however, on the support rope.

Another example of invisible towers comes from British Columbia in Canada. Again, a support rope is used, it crosses a valley and is attached to the two mountain flanks (Fig. 7.22). The conductors are mounted on suspension strings. The first crossing was built in 1955 (steel conductors 76 mm, 760 m span, 130 m above ground) and was renewed in 2008 (28 conductors with 19 mm diameter in quadruple bundles, 1200 m span, 150 m above ground). For maintenance purposes, the support ropes can be walked on via a 142 m walkway (catwalk) and are also accessible by helicopter.

Another example of an unusual design is the 110 kV line Ponte a Ema—Varlungo near Florence in Italy. The towers and the insulating crossarms are movable in the longitudinal direction and are stabilized by the conductors and ground wires (Fig. 7.23). A model test at the University of Pisa in the planning phase had shown the feasibility of this idea.

Fig. 7.22 "Catenary Solution" in Canada, suspension of the conductors on a valley spanning support cable (*Source* Cigré SC B2 Reykjavik Colloquium 2011)

Fig. 7.23 Movable 110 kV line in Italy, the movable towers are stabilized by the ground wires and the conductors

In Iceland, the spectacular tower design "Human Pylon Sculpture Tower" by the designers Choi + Shine Architects from the USA was the winner of a tower competition in 2008. The international reactions were enormous, newspapers reported on it, museums included it in their exhibitions on (Fig. 7.24). The pylons were ultimately

Fig. 7.24 Design-Project "Land of Giants^{TM}", Human Sculpture Tower in Iceland. (copyright 2008, 2009 Choi+Shine Architects, LCC)

not constructed due to high costs, a decision that many people regretted. Supporters see the design as a big step forward, away from the usual patterns of thought, which dominates in line construction, and it was time that a female aspect has been introduced into the very male dominated line construction business. Skeptics consider this design as an unnecessarily large effort; modern design could be achieved more practically by other means.

7.5.3 Secondary Uses of Overhead Lines

Wireless communication with mobile phones and data traffic requires a dense grid of antennas. From the early days of this technology, it was a logical step to consider installing such antennas on existing overhead line towers (Fig. 7.25). Initially, some line operators resisted this due to the additional effort for planning, testing and for legal reasons, and particularly because third parties would be granted access to the towers and climbing facilities. Meanwhile, telecommunication laws in many countries oblige line operators to allow the installation of antennas on their towers. This approach reduces the need for new "greenfield towers" for telecommunications

Fig. 7.25 Antennas for telecommunication and data transmission. *Left*: Directional antenna on a tower, *center*: Transmitting antennas on a 380 kV tension tower, *right*: Base station at the foundation area of the tower

in open areas by making use of existing structures. This requires intensive, time-consuming and costly coordination between line operators and telecommunications companies and affects access and ascent to the antennas—this should also be possible when the line is in operation—worker protection and the requirements and special training of the telecommunication company's staff. Also, the installation of the base stations within the tower area and their power supply with the associated grounding considerations are important topics. It is especially important that such new installations do not interfere with the maintenance of the transmission line. This has led to the fact that often two separate climbing facilities were mounted on one tower. Contractual agreements are necessary. A precondition for this is the suitability of the tower and its environment from a static and electrical engineering point of view (e.g. distances to live parts, lightning protection, grounding systems). This also affects the dimensioning of the antenna support structure for attaching the antennas, as well as climbing aids, standing grates, platforms and the cables which run from the base stations to the antennas. Another key aspect is the clear definition of responsibilities when a tower is jointly used by multiple parties. The loads emanating from the antennas, acting on the antenna support structure and the tower, can often lead to reinforcements on the structures. In some cases, adaptions or annexes to standards become necessary on the design of overhead lines to define these loads and the design methods.

In the 1980s, line operators began using fiber optic technology for internal data transmission, i.e. for line protection, for internal telephony and data transmission. To this end, plastic cables with internal optical fibers were initially wound onto existing ground wires ("Sky-wrap") or attached to them with clips We speak of optical fiber cables, or optical ground wires OPGW. With technical progress, ground wires with internal optical fibers soon came onto the market and their use became standard (Fig. 7.26). Also, conductors with optical fibers were developed and used on lines of the lower high voltage levels (Optical Phase Conductors (OPPC). Such conductors or earth wires can accommodate well over a hundred fibers. Some line operators rent parts of their fibers to third parties, and fiber optic cables are today one of the basic elements of communication technology.

Fig. 7.26 Ground wires with optical fibers in the core, *left* fibers in a common tube, *right* fibers in separate tubes (*Source* Lumpi-Berndorf, Linz, Austria)

Due to the sensitivity of optical fibers, extreme care must be taken during the transport, stringing, and installation of OPGW and OPPC. Manufacturers provide specific guidelines to ensure proper handling throughout these processes. The fibers are connected in splice boxes, which are mounted high above ground level in the towers for protection against vandalism.

7.5.4 Gas Insulated Lines (GIL) AC and DC

With gas insulated three phase lines, nominal currents up to 5000 A can be transmitted. Thus, a single GIL system with up to 3700 MVA can transmit the same electrical power as several parallel underground cable systems. The impact on the environment can be significantly minimized in such applications due to the narrower corridor width. The electrical capacity is lower compared to underground cables at high and extra high voltage level, for GIL no compensation of reactive power is necessary for long lines, significantly longer than 70 km. Gas insulated lines represent a well-developed alternative to transmission via overhead lines, however, GIL are usually significantly more expensive than overhead lines and underground cables in the same power class. Therefore, GIL are mainly considered where much power needs to be transmitted when space is limited or under extreme environmental conditions or when special environmental requirements exist. As an example for this, reference is made to the 220 kV GIL in Lausanne, Switzerland, which was built in 2001 in the area of the airport with a length of 430 m in a concrete tunnel as partial undergrounding of a 220 kV overhead line, as shown in Fig. 7.27. The tunnel runs under an exhibition hall of the Palexpo fair.

One of the world's longest 380 kV AC GIL is approximately 1 km in length at Frankfurt Airport in Kelsterbach, Germany, and was commissioned in 2010. Its pipes are laid directly in the ground. GIL with route lengths of 500–600 m exist, for example, in switchgears in Saudi Arabia, Wales, and Bangkok, which partly run elevated above the ground, partly in trenches.

Other examples of AC GIL include since 1978 a 145 kV GIL has been in place in Joshua Falls USA. A 550 kV GIL exists in Baxter Wilson USA 2001, 550 kV in

Fig. 7.27 220 kV GIL in the area of Lausanne airport, Switzerland

Bowmanville Canada 1985–87, 800 kV GIL in Huanghe Laxiwa China 2009, and numerous power plant connections. In Germany, a two system 420 kV power plant connection has been in operation as GIL over 700 m in length since 1975.

For the future underground transmission of large amounts of power, GIL for DC are currently being developed. This is based on the technology of AC-GIL, which has been tested and developed for decades, and the latest findings from the development of the ±320 kV DC technology. Due to the high current carrying capacity of up to 5000 A at ±500 kV, the DC-GIL is expected to enable low-loss and cost-effective transmission of very large amounts of energy over long distances. Modern AC and DC GIL use a gas mixture of sulfur hexafluoride SF_6 and nitrogen N_2 in a mixing ratio of usually 1:7, with a pressure of 7 bar. SF_6 is classified as dangerous and is only permitted for use in electrical engineering under strict conditions. At the European level, a new regulation on the future use of fluorinated greenhouse gases came into force in spring 2024, which regulates the phasing out of the use of the strong greenhouse gas SF_6 in electrical installations [16]. The regulation states that in the European Union, emissions of fluorinated greenhouse gases account for 2.5% of the total greenhouse gas emissions and have doubled between 1990 and 2014, unlike other greenhouse gas emissions, which have decreased. Although its use has been banned in recent years in many areas such as refrigerators, air conditioning systems, shoe manufacturing, and soundproofing windows. The use of new SF_6 in electrical switchgears should be avoided if it is technically feasible, and if reprocessed or recycled SF_6 is available, and the safe operation of the power grids and power plants is not endangered. In addition, it is determined that from January 1st, 2035, the use of SF_6 for the maintenance or servicing of electrical switchgear is prohibited, unless it has been reprocessed or recycled, unless it is proven that SF_6 (a) cannot be used for technical reasons or (b) is not available in the event of an emergency repair. The regulation therefore allows exceptions in the event that replacement gases are not available in time.

When electrical switchgear is commissioned for the first time, the use of fluorinated greenhouse gases is prohibited in the following cases:

(a) medium voltage electrical switchgear for primary and secondary distribution up to and including 24 kV: from January 1, 2026,
(b) medium voltage electrical switchgear for primary and secondary distribution of over 24 kV and up to and including 52 kV: from January 1, 2030,
(c) high voltage electrical switchgear with a voltage from 52 kV and up to and including 145 kV and a short circuit current up to and including 50 kA with a greenhouse warming potential (GWP) of 1 or more: from January 1, 2028,
(d) high voltage electrical switchgear with a voltage of more than 145 kV or a short circuit current of more than 50 kA with a GWP of 1 or more: from January 1, 2032.

GIL are not explicitly included in the list. Nevertheless, experts in the industry understand that these specifications for extensions and new installations of GIL must also be taken into account.

Intensive research is being conducted worldwide on the development of an unproblematic gas to replace SF_6. Switchgears and busbars with up to 145 kV are already on the market with non-fluorinated gases [17]. These are gas mixtures of nitrogen and oxygen. If they are available for higher voltages, an increased application of GIL in the high voltage range is to be expected in the future, both for HVAC and for HVDC.

7.5.5 *Superconducting Cables*

Metallic materials lose their electrical resistance near absolute zero at 0–4 Kelvin and become superconducting. These low temperatures can be achieved with liquid helium, whose cooling is very complex and expensive. For this reason, applications with metallic superconductors at present are limited to a few and special fields of application.

The discovery of ceramic *high-temperature superconductors* made of Yttrium-Barium-Copper Oxide (YBaCuO) in the mid-80s of the last century enabled the construction of cables, whose so-called *critical temperature*, i.e. the temperature at which they almost instantaneously lose their resistance, is higher than that of liquid nitrogen (Fig. 7.28). Cable constructions cooled with liquid nitrogen also require complex cooling circuits, but they are feasible with today's techniques as constructions, which are cooled with helium.

Superconducting cables can be a suitable solution in densely populated areas with acute lack of space, as they allow high currents at lower transmission voltages. Thus, transmission capacities can be realized at the medium voltage level, for which otherwise conventional high voltage cables would be necessary. In 2014, the first superconducting cable over one kilometer length was put into operation in the city of Essen, Germany [19]. This pilot project known as *AmpaCity* used a 10 kV three phase cable with a concentric structure, which can transport about five times as much current as a comparable conventional copper cable (Fig. 7.29).

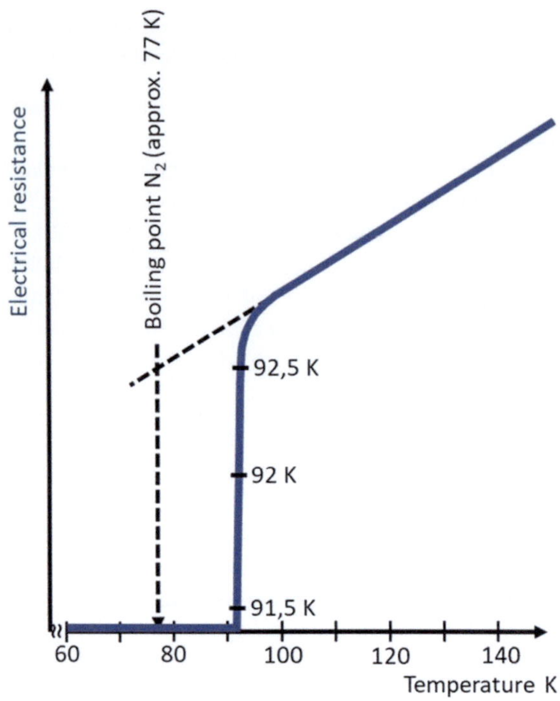

Fig. 7.28 Electrical resistance of YBaCuO during cooling [18]

A nitrogen (N_2) filled insulating jacket keeps the core of the cable at approx. –180 °C. This line with a transmission capacity of 40 MVA replaces a high voltage cable between two substations in the city center of Essen. A study previously conducted came to the conclusion: Modern high temperature superconductors (HTS) are currently technically and economically a useful option to avoid further expansion of inner-city power grids with high voltage cables. In the project, the operators combined the cable with a resistive superconducting current limiter as overload protection. The future spread of HTS cable technology, however, depends on the extent to which it succeeds in improving the ratio of price and performance of HTS conductor materials, the cable production as well as the reliability of the cooling technology and how to reduce the costs [18]. With the currently available technology a cooling station is needed every 5–10 km. DC-HTS cables are currently in development stage and cooling stations would be required at intervals of about 20 km.

HTS cables have been installed in grids since the beginning of the 2000s, among them in Denmark, USA, Korea, Japan with voltages between 13 and 138 kV and with transmission capacities from 48 to 475 MVA.

In Long Island (USA), the world's first HTS superconducting cable was put into operation in the transmission grid in 2008 (Holbrook Superconductor Project). According to the manufacturer, this 600 m long 138 kV cable laying in a tunnel can

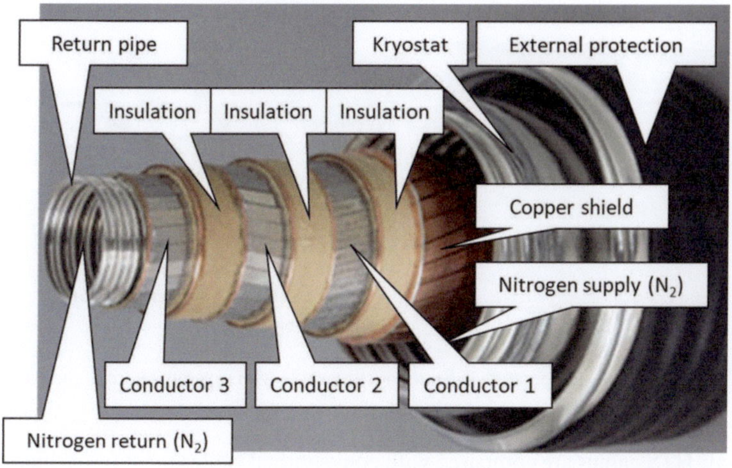

Fig. 7.29 Structure of a concentrically constructed superconducting 10 kV three phase cable [20]

transmit up to 575 MVA (2400 A), which is about five times as much current as for usual cables [21].

ENTSO-E (European Network of Transmission System Operators for Electricity) cite examples as "best practice" for superconducting cables, including the following facilities: In Shingal, South Korea, a 1 km long 23 kV AC HTS cable was laid in 2019 as a connection between two substations. In 2015, an internet data center in Ishikari, Japan, was connected by two 500 and 1000 m long HTS-DC cables.

Research on HTS cables continues intensively. In Munich, Germany, the project of a fifteen kilometer long 110 kV route "Super-Link" between the main substation Menzing and the load center Munich-South was started in 2020. Manufacturers, scientists and grid operators want to investigate possibilities of "retrofitting", i.e. the replacement of existing standard cables with new, more powerful ones. The HTS cable would be the longest in the world and is supposed to transmit 500 MVA. Such a cable is not affected by "hotspots" in urban areas (e.g. bottlenecks, crossings and proximities to heat sources) and the transmission capacity is not reduced. It is a three-phase cable with a separate nitrogen return cable, which can be led on a separate route (Fig. 7.30).

This construction can achieve large lengths of cable and cooling system. One joint with an intermediate cooling station is sufficient for the 15 km long line. In the event of a cooling system failure, the line could remain in operation for several hours. After passing the qualification test, a test operation with 150 m cable length is running in a substation to test the newly developed elements, as terminations, joints, cables and cooling system (2023). After completion of the tests, a decision will be made about the construction [23, 22].

The use of HTS cables is being considered in several major cities. From the USA, a 230 kV project is reported, 4000 A are to be transmitted with two systems. The approx. 2 km long route runs through an industrial area, crosses a railway line and a

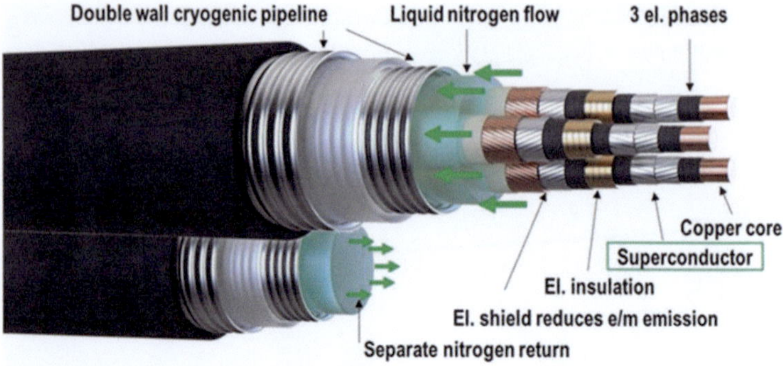

Fig. 7.30 110 kV high-temperature superconducting cable with separate return of the coolant [22]

highway. HTS cables are compared with conventional cables in terms of route width, route selection, jointing systems, laying methods, costs. With standard cables, five-fold cables per system would be required (30 phases), with the HTS cable on the other hand, single cables (6 phases). Pros and cons of the new technology are highlighted, including the additional effort for the cooling systems [24].

In addition to technical considerations, it is equally important to take financial aspects into account. The costs of implementing such a new cable route can be weighed against the long-term expenditures required for substation modifications and line reinforcements when using conventional cables. In some cases, this comparison may clearly favor the adoption of new technology. More definitive statements about the lifespan of HTS cables will become possible once several such systems have been in operation over extended periods [25].

7.5.6 Overlay Grid

In the further development of the grids, two trends are emerging, in which direct current technology will also become very important. On the one hand, large line connections will increase in order to connect different regions with each other and to transmit energy from generation hotspots (e.g. offshore wind farms) and to interconnect the energy markets more strongly. On the other hand, areas are forming that can largely supply themselves with renewable energy including storage and load management. These two models will most likely achieve the politically high-set goals regarding the relief of environmental burden, supply security and economic efficiency. For reasons of acceptance, the necessary cable technology cannot be limited only to the highest voltage levels in the future. Thanks to modern power electronics, it is already possible today, for example, to run extensive data centers exclusively in direct current technology.

The further development of powerful DC switches indicates that instead of the previously realized point-to-point connections, those with three and more converter stations can be realized. This technology is referred to as *Multiterminal system.* As in the conventional three phase grid, the faulty part can be disconnected in the event of a disturbance, and the intact part can continue to operate. Via additional connections, the multiterminal system can be further expanded into a meshed DC grid in the next step. Since this grid will expand into a structure that lies above the existing three phase high voltage grid and relieves it, it will also be referred to as an *Overlay grid* in the future.

A few decades ago, project *DESERTEC* was suggested for the future, possibilities for the joint use of resources, especially the wind power from the north and the hydropower from the south of Europe as well as the solar energy from North Africa and the Middle East. The envisaged connections of the power grids between Europe and North Africa are shown in Fig. 7.31.

To create such extended grids, however, the political and regulatory framework conditions would have to be adapted, but also in the field of technology, the components and interfaces would have to be standardized. This was not necessary in the previous applications of direct current technology. At the same time, geopolitical risks must also be taken into account. Regional groups worldwide continue to deal with considerations for a global overlay grid, as Fig. 7.32 shows.

A concrete example of the approach of a global grid is shown in Fig. 7.33. The line corridors illustrated show where transmission lines could be routed. This concept,

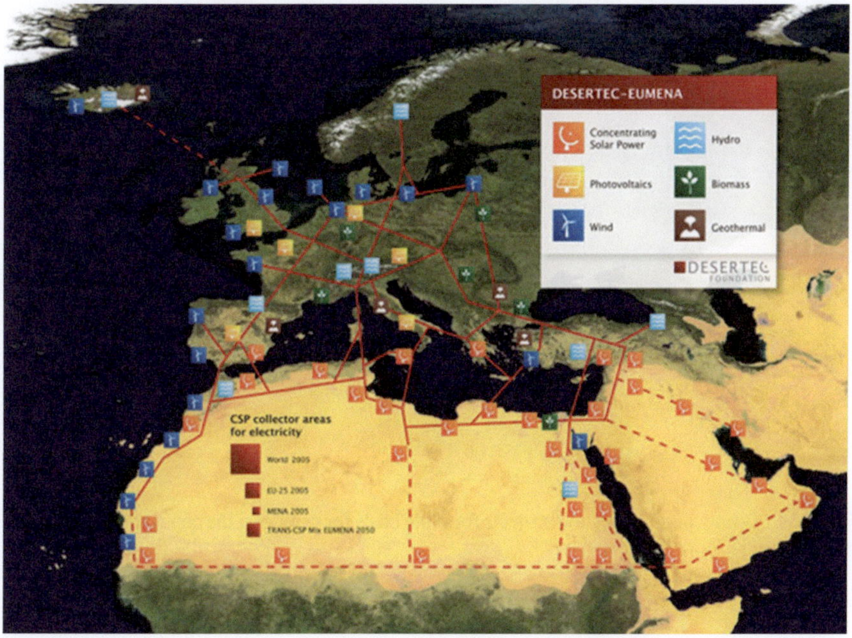

Fig. 7.31 DESERTEC project for energy exchange between Europe and North Africa (*Source* M. Schmale FH Cologne 2010, DESERTEC Foundation www.desertec.org)

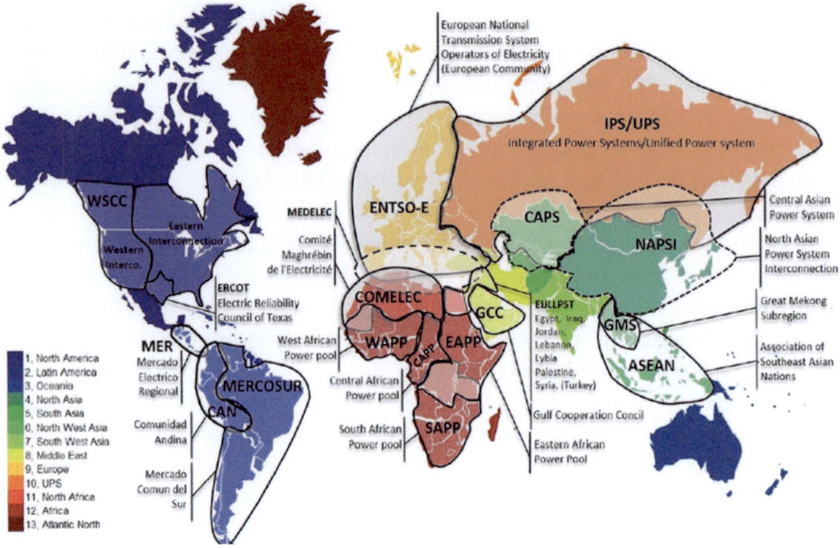

Fig. 7.32 Regional groups for the development of regional grids [26]

Fig. 7.33 Twenty interconnection projects form the basis for exploring the concept of a worldwide overlay grid [27]

as presented by CIGRE, offers a potential framework for large-scale exchange of wind and solar energy across regions, helping to reduce reliance on fossil fuel-based generation. One of the key enablers of this vision is the vast wind energy potential in Central Asia, among other regions.

Overhead lines, submarine cables and cables on land will be required for such projects in large numbers and scopes, with individual line lengths of up to several thousand kilometers. It is assumed that these will predominantly be direct current lines.

In Chap. 4, hybrid overhead lines are discussed, i.e. overhead lines with AC and DC circuits on the same towers, and also pure DC lines for large transmission capacities and large route lengths. Worldwide, there are considerations for converting existing AC overhead lines to DC operation. This could save a lot of time compared to the building of new lines. DC cables in larger lengths exist as submarine cables and are discussed in Chap. 5. The first long DC cable routes on land over several hundreds of kilometers are currently being planned in Germany (2024). The whole world is eagerly watching these projects.

7.5.7 The "Intelligent" Line

Overhead Lines

The current transmission capacity of overhead lines is monitored by monitoring systems, as explained in Sect. 4.3.10. This serves the optimal utilization of the system. This is particularly relevant for long overhead line routes, over hundreds of kilometers in length, which are on large routes through uninhabited areas, from where no reports of anomalies or disturbances to the system are received from residents along the route. Conventional inspection methods reach their limits here—modern automated systems can contribute to obtaining data about the condition of the line and possible operational management. Long lines cross different climatic and topographic, political and social regions. Such *condition monitoring* and *remote sensing* is a great help for maintenance. The establishment of such systems must consider the following aspects, among others:

- Selection of sensor installation locations
- Data collection, storage, and analysis
- Power supply, data transmission
- Security of the devices against vandalism, terrorism, security of the route of transmission
- Test procedures and evaluation
- And finally: available devices.

The optical fibers carried in modern lines, mostly in the ground wires as OPGW, can be used for data transmission. An installation in Slovenia can be seen as an

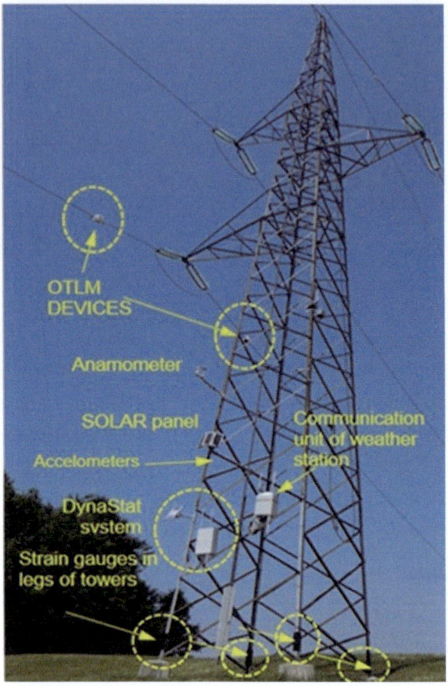

Fig. 7.34 Tower equipped with sensors for detection and transmission of various condition data (CIGRE Symposium Ljubljana, Slovenia 2021)

example of such an application for monitoring icing and thus endangering the line. Triggered by ice disasters, towers were equipped with sensors that enable permanent monitoring of the meteorological data such as speed and direction of wind, icing, ambient temperature, conductor temperature, also changes of mechanical loads in tower bases and the analysis of static and dynamic loads on the structures (Fig. 7.34).

From other countries, sensors are known that can detect when the tower is being sawed or bracing angles are being unscrewed, which can occur in cases of sabotage. Sabotage, terrorism, theft, and vandalism must also be expected in the future, especially with long lines in uninhabited or sparsely populated areas.

Underground Cables

Modern underground cable lines carry optical fibers that provide information about the condition of the line, especially the temperature distribution (Sect. 5.11). The operating condition, further operational management, and load of the system can thus be assessed. Possible fault locations can also be located with such systems. This also applies to joints, buried in the ground or in bunkers. For the monitoring and maintenance of transition stations from cables to overhead lines, basically the same applies as was explained above for overhead lines.

7.6 Artificial Intelligence in High Voltage Grids

In recent years, Artificial Intelligence (AI) has been increasingly used worldwide in the maintenance and operation of transmission lines. AI is a technology that enables computers and machines to simulate human learning, understanding, problem-solving, decision-making, creativity, and autonomy [28]. AI algorithms are being developed that mimic the decision-making processes of the human brain. These algorithms can "learn" from available data and make increasingly precise classifications or predictions over time [29].

Figure 7.35 illustrates the nature and development of AI.

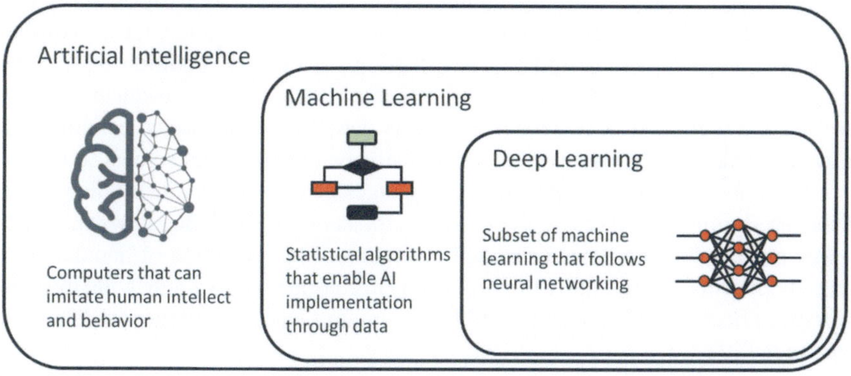

Fig. 7.35 Artificial Intelligence (AI) [29]

7.6.1 State of Development and Classification of AI

Currently, AI is divided into different stages of development, which are described differently in the general literature depending on the perspective and their performance. In simplified terms, three stages of AI can be characterized [29].

1. Narrow AI: Solves narrow set of problems at human or superhuman levels.
2. General AI: Possesses generalized human cognitive abilities; can perform any intellectual task that a human can do.
3. Superintelligent AI: or Artificial Superintelligence (ASI): surpasses human intelligence across all domains.

Current level of AI is somewhere between Narrow AI and General AI [29]. Structurally, this means that all previous AI applications can only be as good and effective as the data they have previously been supplied with. As development is progressing rapidly, advances are expected in all areas of current AI use, as well as its penetration into areas where it has previously played no role.

In maintenance, AI enables systematic early detection of damage to overhead line components such as towers, conductors, insulators, and fittings. In grid operation, it can help reduce or avoid shutdowns of renewable generation plants to avoid costly redispatch measures. Another important area of application is improving the resilience of grids by enabling early, curative measures to be taken based on previous line defects.

7.6.2 AI in Maintenance and Repair

To ensure high supply reliability, regular visual inspections are performed on high voltage and extra high voltage overhead lines. For many years, helicopters and drones have been used for this purpose. They capture the current condition of the line components in photographs and videos, which are then manually evaluated during or after the flight and documented in reports. Due to the enormous technological advances in drones, they have replaced the costly use of helicopters in most countries.

By using AI in drone inspections, the images can be automatically diagnosed. This can significantly increase operational efficiency. Figure 7.36 shows the diagnostic process for line accessories and insulators. Early detection of anomalies in line components is particularly important, e.g. in coastal regions where wind induced vibrations occur more frequently, and high humidity leads to a higher tendency for corrosion [30].

Some companies also detect corona behaviour, which can provide evidence of damage to conductors and fittings. Thermal imaging cameras are also used to detect thermal anomalies and hotspots. Anomalous findings are identified and serve as the basis for decisions about whether and how repairs need to be carried out. The evaluation effort can vary greatly from country to country and from network operator to network operator.

Manual evaluation, especially of photographs, is a very monotonous, boring, and tiring process, with the result that the consistency of the results varies from individual to individual. Even with several experienced people reviewing the material independently, the efficiency in detecting anomalies is 50%. Expanding the image evaluation process to include image-based AI, which is now capable of processing

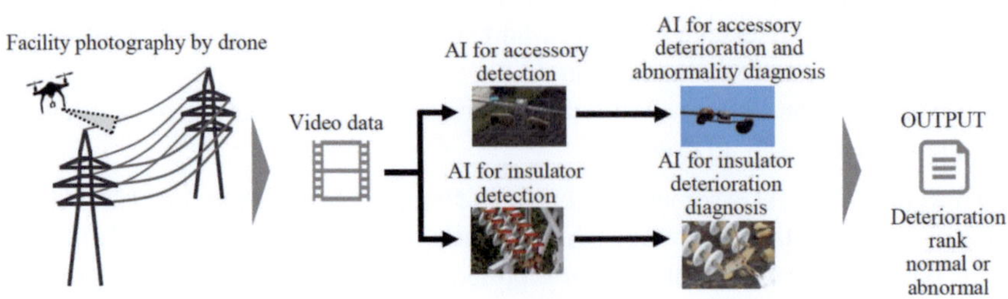

Fig. 7.36 Automatic diagnosis flow for conductor accessories and insulators [30]

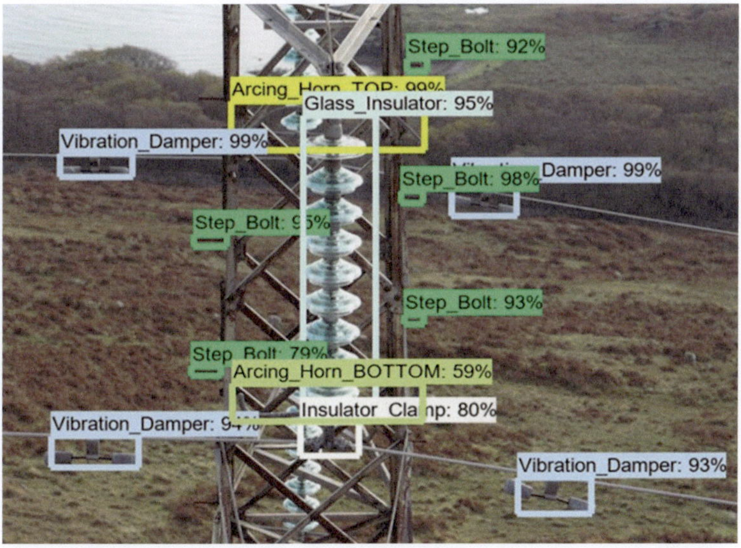

Fig. 7.37 An image with AI recognition of components on a transmission line. Courtesy of Recognise Technologies Inc. (www.recognaise.com)

several thousand images per minute, has a confidence level of at least 85% [29]. The confidence level indicates how certain the AI is that the result it found corresponds to what it learned from the data provided. A confidence level of 100% therefore means that the AI is completely certain of its identification. Many images of intact line components serve as the learning basis. Figure 7.37 shows the confidence level with which the AI recognizes components of a high-voltage overhead line.

Figure 7.38 shows two examples of AI-assisted detection of potential problems on porcelain cap and pin insulators.

AI can also detect anomalies in conductors. Figure (a) and (b) in Fig. 7.39 show examples of such detected damages [30].

a **b**

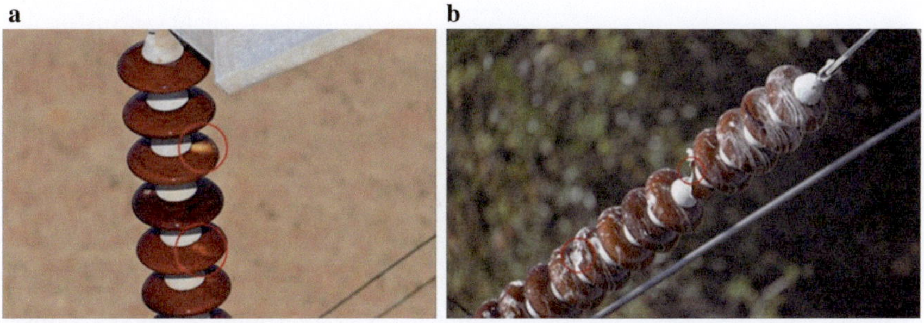

Fig. 7.38 Identified anomalies on porcelain insulators: **a** detected discoloration, **b** chipped porcelain caps [29]

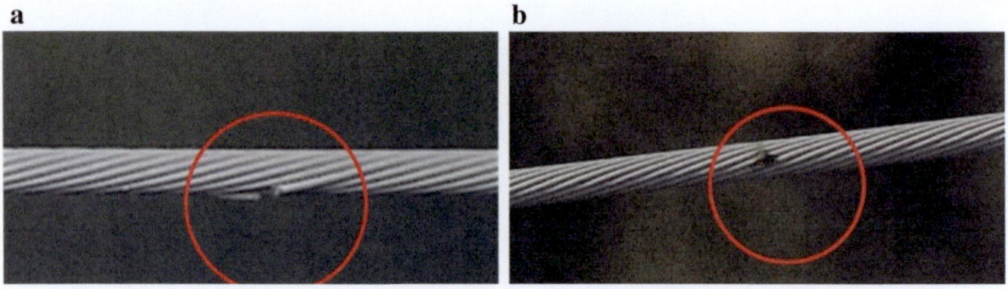

Fig. 7.39 Wire damage to a conductor **a** broken single wire, **b** melted single wires [29]

Fig. 7.40 Corrosion assessment of a steel lattice tower using AI, green: no corrosion, red: heavily corroded [29]

AI is also applied in the corrosion assessment of steel lattice towers to determine the degree of corrosion. An example is shown on Fig. 7.40.

7.6.3 AI in Defect Diagnosis in Line Operation

One of the biggest challenges in ensuring good transmission system availability is the time it takes to diagnose the causes of line defects. Depending on the cause, these defects can lead to permanent transmission line outages, for example, in cases that require the deployment of maintenance personnel. AI is now being used worldwide

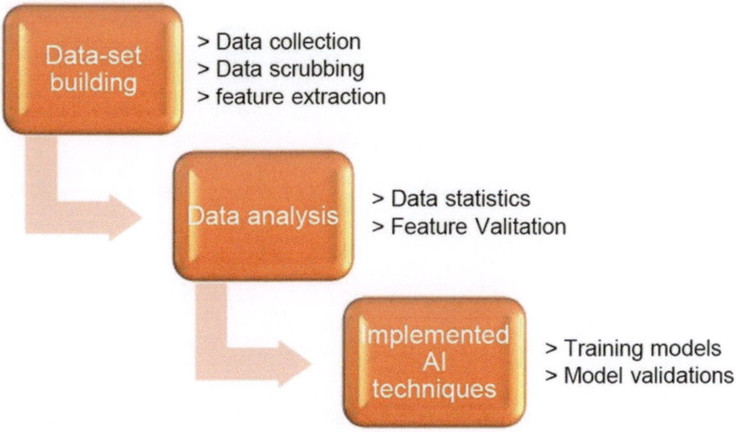

Fig. 7.41 Methodology of AI implementation [32]

to perform automated diagnostics that enable rapid decision making in real-time operation. In many cases, it is based on oscilloscopic recordings of the original defects, which serve as a training basis for AI [32].

The diagnosis required to determine the cause of the defect involves a detailed and complex analysis process. The most important source of information is the oscilloscopic data recordings, as they contain the electrical variables involved in the phenomena under investigation. The methodology proposed as an example in [32] consists of three basic steps (Fig. 7.41) to implement the diagnosis of transmission line faults using AI.

The information available for analysis in [32] includes 70 defect cases in a network operator's electrical circuits. Single phase faults, which in this case were caused by a fire under the line and the approach of vegetation, were particularly difficult to be detected. The similarity of the defect resistances made classification between fault types difficult.

Although the model does not provide a prognosis by quickly diagnosing the causes of the defects, it helps operators make rapid decisions for safe line operation. It is expected that the previously established diagnostic systems will soon evolve into prognostic systems.

7.6.4 AI in the Optimization of Grid Operation

By measuring the temperature close to or in the conductors, it is possible to adapt the operating mode of the lines to the meteorological conditions (see also Sect. 4.3.10). The recent development towards a "smart conductor", in which the conductor temperature can be measured in the conductor itself via integrated fibre optic cables, will enable an adapted mode of operation to be achieved in the

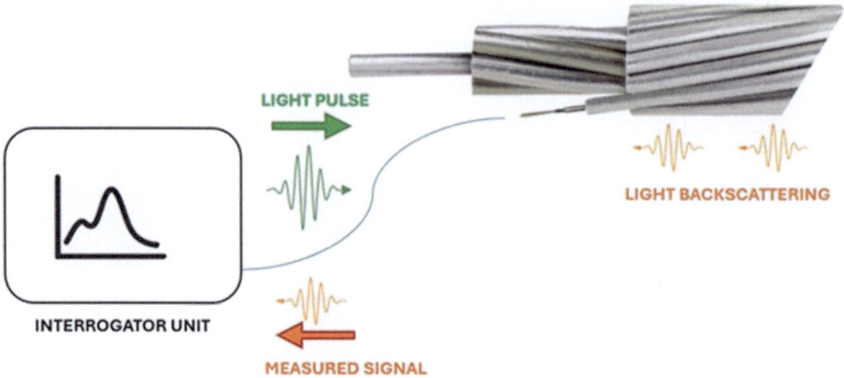

Fig. 7.42 Fiber optic-supported temperature measurement in the conductor (*Source* De Angeli Prodotti)

future, allowing optimum utilization of the networks. This technology allows the temperature to be detected according to amount and location. The system of integrated temperature measurement in the conductor is shown in Fig. 7.42. The principle is similar to temperature measurement for underground cables (Sect. 5.10).

This development enables a shift from purely preventative grid operation based on the (n-1) criterion to curative grid operation, thereby increasing the utilization of the grids in line with demand. Here, AI algorithms can accelerate load flow calculations and evaluate and suggest grid-friendly switching actions for the grid topology.

A further step could be automated grid operation using deep reinforcement learning (DRL), which shows great potential in learning to run a power network (L2RPN) [33]. So far, only individual actions at the substation level have been subjected to topology optimization by most existing DRL algorithms. In [33], a holistic approach is proposed that includes specific Target Topologies (TTs) as actions. These topologies are selected according to their robustness. In this work, a search algorithm is presented to find the TTs and to further develop previously developed DRL agents Curriculum Agent (CAgent) into a novel topology agent.

7.6.5 Resilience of the Transmission Grid

In technical and engineering contexts, resilience refers to the ability of systems to not fail completely in the event of partial failures or disturbances, but to maintain essential system functions. Resilience is often understood as the ability to create a new state that can even have an improved behaviour of the system compared to the initial state. Examples for this approach can be found in many practices. In our case it is the un-interruptible supply of electric energy in case of a failure of the primary power supply.

A universally accepted definition seems not to exist, as it is a dynamic and multi-dimensional understanding. Nevertheless, the resilience of a transmission network is an important factor of the quality and availability of this network. Transmission networks are exposed to several operating risks like climatic events (wind, storms, typhoons, lightnings, high floods, landslides, avalanches, etc.), fires, vehicle collision, component failures, ageing, cyber-attacks, vandalism and others. Such incidents bring aspects of reliability into the discussion. While reliability is a well-understood concept in investment decisions, design and operation of power systems, the concept of resilience is either not fully understood or commonly mistaken for reliability. Power system resilience is the ability to limit the extent, severity and duration of system degradation following an extreme event [34]. This objective is achieved through a set of key measures to be taken before, during and after extreme events, such as anticipation, preparation, absorption, adaptation, rapid recovery and sustainment of critical system operation. This includes the application of lessons learnt after an event occurred.

An organization seeking to be highly resilient also needs to continuously focus on aspects related to the potential for multiple failures at all levels of the organization, to find opportunities to improve its emergency preparedness and operational practices prior to, during, and following major disturbances, and service interruptions, as well as improvements based on lessons learnt from past events [35].

This leads to the following resilience concept:

- it requires a comprehensive evaluation of system response to disturbances, including not only the system degradation but also the system behaviour during the restoration phase, as well as all the measures taken to preventively improve system performance;
- it supports the characterization and design of actionable measures aimed at improving the performances of the power system response following extreme events triggered by adverse weather conditions, malicious acts, cyber-attacks, etc. with due consideration to past extreme events.

The quantification of resilience began in civil engineering. The process is shown in a very simplified form in Fig. 7.43 from [36]. It shows the changes of function or quality R(t) in the concerned system. R(t) is normally 1.0 and accomplishes the requirements of the system. If a malfunction occurs in time t_0 this means that R(t) decreases to a lower level depending on the damage and gradually comes back to normal situation at time t_1 and the system is in normal operation again. The form of the changes of the operation reminds of a triangle-the resilience triangle. Improved resilience should reduce the area of the triangle-the smaller the better.

Resilience should be the overall strategy for ensuring business continuity. When focusing on resilience in general, organizations must consider safety, security, and reliability. Resilience includes security measures that can mitigate impacts, not only before incidents (identify and prevent), but also during such incidents (detect and respond) and after incidents have been resolved (recover). For resilience of cyber assets, organizations must similarly consider safety, security, and reliability for cyber

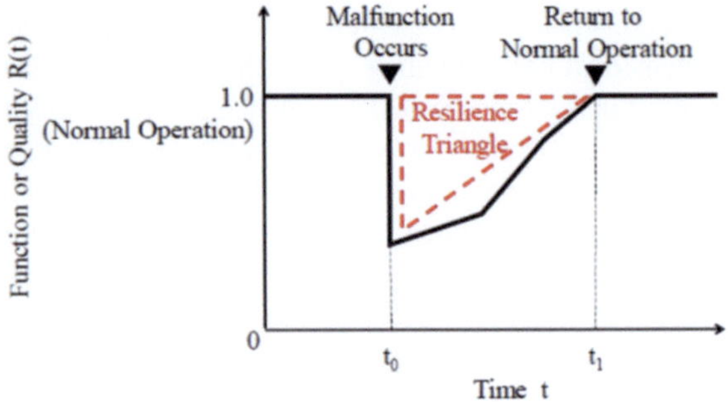

Fig. 7.43 Concept of resilience triangle [36]

assets. Resilience thus involves a continuous improvement process to support business continuity. Resilience is not just a technical issue but must involve an overall business approach that combines cyber security techniques with system engineering and operations to prepare for and adapt to changing conditions, and to withstand and recover rapidly from disruptions. Information sharing and interoperability within and across organizations is also becoming a crucial part of resilience [37].

Eurelectric (Union of the Electricity Industry) issued the report "The Coming Storm-Building electricity resilience to extreme weather". In the associated presentation it is stated "All assets in the electricity value chain are exposed to the effects of this growing number of extreme weather events, from electricity generation and transmission to distribution and the final customer". Resilience is considered as a growing component of the grid investment. Table 7.1 shows the power system assets transmission and distribution lines as exposed to danger from the climate change. Tables 7.1 and 7.2 indicate adaptation measures of transmission and distribution assets in Europe necessary to obtain the desired resilience [38].

Table 7.1 Exposed power system assets: transmission and distribution [38]

Fire weather	Extreme heat	Drought and aridity	Severe storms	Mean precipitation and floods	Cold spell, snow, ice and frost
Power lines disconnected by emergency responders	Loss of load due to reduced grid capacity and overheating transformers	Postponed maintenance or repair due to working conditions	Loss of remote system control, mountain lines become difficult to reach	Loss of interconnection lines	Avalanche risks for transmission towers
Burning trees fall on grid infrastructure	Hot spots in cable insulation and isolation failure	Increase in underground grid fault rate	Lines damaged by falling trees, Distribution towers collapse	Landslides weaken infrastructure foundations	Conductors broken by icing or snow sleeves

Table 7.2 Adaptation measures of transmission and distribution assets in Europe [38]

Cold spell, snow, ice and frost	Extreme heat and drought	Severe storms	Pluvial and coastal floodings
Adoption of mechanical fuses to reduce conductor breakages	Remote control, grid automation and digitalisation to promptly recon-figure and restore the network	Planning of network according to the "n-1" standard with alternative back feeds available in the event of damage	Flood risk assessment must be carried out to ensure assets are well placed and sufficiently protected
Provision of alternative power network paths (network meshing, back feeding for laterals with many customers)	Remote control to isolate only the faulted network	"Meshing" of network should be considered where feasible	Coordination with local authorities to help ensure a coherent approach to flood risk assessment
		Design of overhead lines to account for increased wind speeds and icing, also considering vegetation management	

7.6.6 *Future Development in Grid Construction and Grid Operation*

Artificial Intelligence holds immense potential in the field of transmission and distribution line engineering. As AI continues to expand its capabilities, an increasing number of tasks and solutions can initially be augmented and, eventually, fully managed by AI systems. Key areas such as transmission line design, construction, operation, and maintenance management are prime candidates for AI integration.

The advancement of AI in this domain will heavily rely on progress in hardware (e.g. chips), software, and system innovations. Some applications may prove to be disruptive, not only to current utility and grid operations but also to society as a whole, reshaping how energy systems are managed and maintained.

It is also worth to know that current AI systems require large amount of electricity and without a robust electrical grid there are no well operated large data centres containing AI [29].

References

1. Palic M (2015) Grundlagen Erneuerbare Energieen, bdew akademie, Seminarunterlage. Erfurt
2. Anon., Power-to-Gas, EnBW Informationsbroschüre (2015)

3. Sauer DU et al (2016) Stromnetze Technologiesteckbrief zur Analyse Schriftenreihe Energiesysteme der Zukunft. Aachen
4. Scheffler J (2016) Verteilnetze auf dem Weg zum Flächenkraftwerk. Springer
5. Erbslöh FD (2021) Der Weg zur Energiewende. Expert-Verlag
6. Papailiou KO (ed) (2021) Springer handbook of power systems. Springer, Singapore
7. Staschus K, Vazquez M, Sanders H (2018) Cigré Reference Paper "Sustainability—At the Heart of CIGRE's Work" in ELECTRA nr. 300, Oktober 2018
8. Anon., Electricity supply systems of the future. CIGRE Green Books, Springer (2020)
9. CIGRE (2020) Technischer Bericht 792 "Compact AC overhead lines", Paris
10. Anon., RTE EdF, CIGRE 2005—B2—Colloquium Brasilien
11. Skouboe H (2019) Bystrup, Going offshore—Challenges of the future powergrid. CIGRE Symposium Aalborg
12. Anon., Industrial Design Education Austria Linz "DER TOWER—Gegenwart & Zukunft", APG—Austrian Power Grid in Kooperation mit scionic I.D.E.A.L.
13. Anon., Transmission line structures with Fiber Reinforced Polymer (FRP) composite, CIGRE Technischer Bericht TB 818
14. Skouboe H et al (2022) 2 × 400 kV composite pylon ready for use, innovative and compact—reducing the impact of OHTL considerably, Cigre Session Paris, paper 10889. https://www.powerpylons.com/composite-pylon
15. Lugschitz H, Ernst A, Gros T (2004) Corrosion protection of steel towers and camouflage of lines using the DUPLEX-system. B2-213, CIGRE Session Paris
16. Verordnung (EU) 2024/573 Des Europäischen Parlaments und des Rates vom 7. Februar 2024 über fluorierte Treibhausgase, zur Änderung der Richtlinie (EU) 2019/1937 und zur Aufhebung der Verordnung (EU) Nr. 517/2014
17. Petz I (2022) Auf Wiedersehen SF6! Clean Air-Technologie aus Berlin, SIEMENS energy
18. Steckel RD (1992) in Kabeltechnik in Kabel und Freileitungen in Überregionalen Netzen. Expert-Verlag
19. Merschel F (2017) Supraleiterkabel—Lösung für dicht besiedelte Gebiete, Vortrag anlässlich der 10. Fachtagung Freileitungen, Wiesbaden
20. Anon., Gerapid—DC-Gleichstromschnellschalter. ABB Connect, Zürich (2022)
21. JICABLE (2007) Development and demonstration of a long length transmission voltage cold dielectric superconducting cable to operate in the long island power authority grid
22. Bach R et al (2024) Superconducting power cable for 500 MVA at 110 kV in Munich—first insights in the test run. CIGRE B1-11426, Paris
23. SWM-Magazin „Das können Hochspannungs-Supraleiter" März 2023
24. Edwards C et al (2024) High-temperature superconducting cable systems as a solution to underground transmission line routing in congested project areas. B1-10328 Paris
25. Kashcheev A (2024) Feasibility assessment of solutions for the introduction of high-temperature superconducting AC cable lines in megacities. CIGRE B1-10786, Paris
26. CIGRE (2019) Technischer Bericht 775 "Global electricity network Feasibility study", Paris
27. CIGRE ELECTRA 306-4 (2019) Global electricity network—Feasibility study. WG C1.35, Paris
28. What Is Artificial Intelligence (AI)? IBM company
29. Toth J (2024) Emerging technologies for transmission lines. RecognAIse Technologies Inc., Budapest
30. Takahashi S et al (2023) Development of automatic diagnostic AI for overhead wire accessories and insulators. CIGRE 2023 Sendai Colloquium, Japan, 3–7 October 2023
31. Iwamoto M et al (2023) Development of power line abnormality detection system utilizing AI and image analysis. CIGRE 2023 Sendai Colloquium, Japan, 3–7 October 2023
32. Arenas O, Gomez M (2022) Artificial intelligence applied to explore the causes of transmission line faults. CIGRE 2022, Paper ID: 10577
33. Lehna M et al (2024) HUGO—highlighting unseen grid options: combining deep reinforcement learning with a heuristic target topology approach

34. CIGRE (2021) Technical Brochure 833: "Operating strategies and preparedness for system operational resilience", Paris
35. CIGRE (2019) Reference Paper: "Defining power system resilience" in ELECTRA 306, Paris
36. Sakai N et al (2023) Development of evaluation process for demand-side resilience against power outages, paper 5302885, CIGRE Colloquium Sendai, Japan
37. International Electrotechnical Commission IEC (2019) Cyber security and resilience guidelines for the smart energy operational environment. Geneva, Switzerland
38. Eurelectric (Union of the Electricity Industry) report (2022) The coming storm—building electricity resilience to extreme weather. Bruxelles, Belgium. https://resilience.eurelectric.org/

Further References

39. Anon., Genehmigung des Szenariorahmens 2021–2035, Bundesnetzagentur, Bonn (2020)
40. Anon., Gerapid—DC-GleichstromschalterABB Connect, Zürich (2022)
41. Hatziargyriou N, Hrsg (2020) Electricity supply systems of the future. CIGRE green books. Springer
42. Jensen C, Argaut P (2006) 400 kV underground cables in rural areas. CIGRE
43. CIGRE (1996) Technical Brochure 110 "Comparison of high voltage overhead lines and underground cables, report and guidelines", Paris
44. Anon., Technical Report on the Future Expansion and Undergrounding of the Electricity Transmission Grid, Elinfrastrukturudvalget (2008)
45. J. Verseilles, T. Neesen: "Feasibility and technical aspects of partial undergrounding of extra high voltage power transmission lines" ENTSO-E - Europacable, Bruxelles 2011.